Paul Ackman

Concepts and Methods of
'HEORETICAL PHYSICS

by

ROBERT BRUCE LINDSAY
Hazard Professor of Physics
in Brown University

DOVER PUBLICATIONS, INC.
NEW YORK

Published in Canada by General Publishing Company, Ltd., 30 Lesmill Road, Don Mills, Toronto, Ontario.
Published in the United Kingdom by Constable and Company, Ltd., 10 Orange Street, London WC 2.

This Dover edition, first published in 1969, is an unabridged and slightly corrected republication of the work originally published in 1951. It is published by special arrangement with D. Van Nostrand Company, Inc., the original publisher.

Standard Book Number: 486-62354-8
Library of Congress Catalog Card Number: 70-93198

Manufactured in the United States of America
Dover Publications, Inc.
180 Varick Street
New York, N.Y. 10014

PREFACE

This book is intended as an introduction to the methods of theoretical physics for students who have had intermediate college courses in physical mechanics and electricity as well as advanced calculus and elementary partial differential equations. As the title implies the emphasis is on the development and clear definition of physical concepts and the analysis of methods rather than the detailed presentation of all the factual content of theoretical physics which the student will naturally need in pursuing graduate physics to the doctorate. In this respect it does not pretend to cover the whole ground as elaborately as many of the present-day standard texts. In its approach the book follows the spirit of *Foundations of Physics* by Lindsay and Margenau (Wiley, 1936)* and tries to maintain the philosophical flavor of this work. It is written, however, in a form more suitable for classroom and private study use and with more attention to pedagogical devices such as concrete examples and numerous problems. Attention is focused on the postulational character of physical theories, and the endeavor is made to stress those points of view which have proved to have the greatest application in the description of physical phenomena. The mathematical analysis, although not rigorous in the sense of pure mathematics, is believed to be sufficiently sound for the purposes of most physicists and is presented in general in sufficient detail to render excessive reference to mathematical texts unnecessary. In certain developments, to be sure, some specific steps are deliberately omitted to enable the student to improve his power of mathematical manipulation. Throughout the book the emphasis is primarily on the physical meaning of the results of analysis rather than on the analysis itself, i.e., the book deals primarily with theoretical physics and not with the branch of mathematics commonly called mathematical physics.

After a brief introduction on the logical structure of theoretical physics, the book is divided into two main parts, viz., *particle* physics and *field* physics. These two points of view have competed for primacy throughout the historical development of the science. The logical development proceeds from the single particle to collections of particles and thence to the statistical method. Thermodynamics is treated as a deduction from physical statistics. Part I closes with a brief

*Reprinted by Dover Publications.

survey of the principles of quantum mechanics. Part II discusses the application of the field concept to both material media (deformable solids and fluids) and nonmaterial media (electromagnetic fields). There is more stress on wave propagation than is usual in theoretical physics texts. The book closes with a treatment of classical electron theory and special relativity.

The nature and method of presentation reflect the author's views of the relative importance of the various principles. The treatment is sufficiently flexible to permit the teacher in a course in which the book is used to introduce alternative and supplementary developments and illustrations without destroying the continuity. At the same time the author hopes and believes that the conscientious reader who masters the book will be amply prepared to study with profit the advanced treatises in both classical and modern physics.

As is so often the case, the writing of the book grew out of a graduate course given for several years. The author owes a debt of gratitude to the many students who by their questions and criticisms have helped to clarify his ideas and improve the presentation. It is of course impossible to acknowledge adequately the writer's debt to the fundamental sources of the ideas and analysis presented in the book. He desires to express his particular thanks to the physicists who have read the manuscript in advance of publication and made very valuable suggestions. Numerous footnote references have been inserted to direct the reader's attention to other treatments and in particular to original articles.

The author wishes to acknowledge his deep indebtedness to his wife, Rachel T. Lindsay, for her unfailing encouragement and in particular for her assistance with the proof and index.

R. B. L.

Providence, R. I.
July, 1951

CONTENTS

INTRODUCTION

Chapter I—The Nature of Theoretical Physics

PART I—PARTICLE PHYSICS

Chapter II—Kinematics. The Geometry of Motion

Chapter III—Concepts of the Theory of Mechanics. The Principles of Dynamics

Chapter IV—Motion of a Particle in a Straight Line. Oscillations

Chapter VIII—Minimal Principles. Lagrange's Equations

Chapter IX—The Transformation Theory of Mechanics. Atomic Structure

Chapter X—Statistical Mechanics and Thermodynamics

Chapter XI—Quantum Theory

PART II—FIELD PHYSICS

Chapter XII—Scalar and Vector Fields

Chapter XIII—Deformable Material Media and Tensor Fields

Concepts and Methods of

THEORETICAL PHYSICS

CHAPTER I

THE NATURE OF THEORETICAL PHYSICS

1.1. What is Physics? It is important at the outset to dispel a prevalent misconception concerning the nature of the subject matter of this book. In spite of the symbol-starred pages, theoretical physics is *not* mathematics and is not primarily concerned with mathematical methods. It is just as much physics as elementary physics and deals with the same fundamental material, with the same ultimate goal: the description of a certain portion of human experience. There is, of course, a difference, else no justification would exist for studying theoretical physics. The difference is twofold: (a) in *language* and (b) in *methodology*.

Elementary physics is usually taught on the descriptive level with the principal emphasis on the actual physical phenomena encountered both in common experience and in the laboratory. In this process it usually happens that the student misses the real significance of what the physicist is about. It is true that there is much talk about concepts like mass, force, and energy, and the student learns how to manipulate in formal fashion certain simple relations involving them, without gaining much insight into the reasons why the physicist considers them useful in describing nature. Since in theoretical physics he is asked to use more precise and powerful mathematical tools in dealing with the same concepts, it is only fair that he should first find out what physics really is.

Let us therefore begin by asking ourselves the question: "What is Physics?" It is not a question here of those "stunt" definitions which propose to depict the subject in one sentence. The problem is to try to understand just what a physicist does. One of the characteristic features of his activity and an indication of his modesty is that he contents himself with a decidedly limited portion of the totality of his sense impressions which we call experience. Philosophers have devoted much attention to the analysis of the sensations and nowadays it is a fertile field of psychological inquiry in which interestingly enough physical apparatus is being used more and more extensively. It will not be practicable to enter on a discussion of this here.[1] We

[1] A classic work on this subject from the point of view of a physicist is Ernst Mach's *The Analysis of Sensations* (English translation, Dover reprint).

shall however stress the fact that the physicist is not satisfied with a purely passive attitude toward those sense impressions which excite his interest. He insists on obtaining new sensations by a process called *experiment,* which involves conscious activity and which also definitely limits and conditions the fundamental data of physics. It is customary in popular books and articles to emphasize that in science we want to find out the *truth* about *nature,* as if nature were something quite apart from us and we were on the outside looking in at it. Actually, nature is simply what happens when we perform our experiments, and a different method of carrying out experiments would presumably lead to knowledge of a different kind of nature. This is already evident in the studies of so-called extra-sensory perception which have caused so much controversy in scientific circles. It may be objected that there are certain limitations to the way we may perform experiments and still call them scientific. Actually no one is really in a position to lay down dogmatically these limitations. The conventional but certainly arbitrary method of experimentation in physics is to bring things that we call physical objects in juxtaposition, and then to observe what happens that did not happen while they were apart. The number of possible operations of this kind is enormous and it seems unlikely that we shall complete the trial of all of them for a very long time.

For at least the past 350 years physical science has been committed to the conviction that experiment must be quantitative and involve *measurement.* Now all measurement ultimately reduces to attaching a number to the coincidence of a pointer and a scale and is therefore founded directly on arithmetical and geometrical concepts. These concepts at one time were considered imposed on the mind as necessary technics of human thinking. We now realize that the necessity of any particular geometry or number system is illusory and that the choice of our common cardinal numbers for the representation of the results of our experiments is purely a matter of convenience.

Physics might very well consist of the description of experiments in terms of everyday language accompanied by the numbers resulting from the associated measurements. As such, to be at all precise it would have to be verbose, for ambiguity is fatal to scientific knowledge. To be sure, economy and precision in the description of the operations performed in carrying out experiments can be obtained by the liberal use of diagrams to provide concrete pictorial representation of the terms employed. The numbers representing pointer coincidences pose a rather more difficult problem, however. We desire to attach *meaning* to the results of measurement. Why is this? Strictly this is a philosophical or psychological question, but since it is fundamental for physics we are justified in seeking a common sense reason which satisfies physicists, if no one else. The reason we do not consider we have done enough in simply tabulating the numbers resulting from measurement is that: (1) we find, on repetition of certain operations under conditions maintained as nearly the same as is humanly possible, that we obtain results which are nearly the same, thus suggesting that this portion of our experience is not chaotic but displays a certain order; (2) further examination persuades us to the belief that many kinds of operations are not so isolated or intrinsically unrelated as the way we carry them out or the way we describe them in ordinary language would appear to indicate. It is therefore inevitable that we shall seek *relations* among apparently diverse sets of operations. This search for meaning puts us definitely on the way to physical theorizing: our description takes on a more sophisticated character. Let us give an illustration.

1.2. The Meaning of a Physical Law. Imagine we are performing the operation of stretching a vertically suspended wire by attaching a weight at the free end. Stimulated by the qualitative observation that an increase in the attached load leads to an increase in the stretch, we decide to measure the stretch with appropriate experimental devices for various values of the load. The result is two sets of associated numbers in which examination reveals that when the stretch doubles the corresponding weight increase does approximately the same: This is interesting but might be of no general significance since it might be descriptive of the single experience only. So we try other wires of different lengths and with different attached loads. The numbers we get are now different, but the striking thing is that as long as the loads are not too large the same approximate numerical association prevails. We conclude that we have hit on a *routine* of experience. How should we describe it? We might leave the description as it has just been given, i.e., in terms of sets of corresponding numbers. But this seems unhandy and uneconomical. As long as we are using numbers we may as well employ the other apparatus which the mathematicians have presented to us and describe the measurements in terms of *algebraic* symbolism and the concept of *functional relation.* It is clear that as soon as we do this we are transcending the original raw experience involved in the experimental operations. Already we are introducing hypotheses fundamental for the whole course of physical theory. There is no *a priori* guarantee that the introduction of the symbol x to denote the result of the operation which we call measuring the stretch of the wire and the symbol W to denote the result of the operation which we call increasing the load on the wire, and finally the treatment of x and W as variables subject to all the conventional rules of the game of mathematics, will provide a successful description of physical experience. The decision to take this step was historically a bold experiment, however natural it may seem to us today because of the familiarity engendered by an educational process which practically forces us to look at things in this way whether we wish to or not.

The simple linear functional relation between x and W with which we thus decide to represent the results of many such series of operations on stretched wires and rods then becomes a *physical law* (in this case the familiar one known by the name of Hooke), a symbolic description of a routine of experience. Hooke's law is a shorthand expression of our confidence that within certain limits we can by a single measurement predict the stretch of any given wire under any applied load. The single measurement provides a numerical value for the parameter k in the equation

$$x = kW.$$

At once we have subsumed in a single formula the results of a host of measurements. Of course we pay a penalty for the apparent generality: Hooke's law and indeed all physical laws fail somewhere to be descriptive of experience. We are all familiar with the fact that if W becomes too great the law no longer predicts the observed facts. Thus for large W, there may be no steady value of x, but a continual increase of x may be observed, the so-called *creep*. Ultimately, of course, the wire will break. The best we can say is that Hooke's law is a good description of the stretching of a wire under certain more or less narrowly limited circumstances. Our interest in the law is predicated on the fact that just these circumstances are often encountered in practice.

We might continue to study from the viewpoint just illustrated various experimental observations and describe them in terms of laws. The simple examples of Boyle's law and Ohm's law will occur to the reader among many others already encountered in elementary physics. Would this constitute physics? It might, but it actually does not satisfy. A mere collection of laws in symbolic form is doubtless more economical than a collection of verbal statements about physical operations and laboratory experiments. But it ignores the possibility of the existence of *relations* among apparently diverse operations. In the case of Hooke's law, for example, it overlooks the possibility of the existence of other operations by which a solid may be deformed in addition to stretching by a suspended weight. In other words, it disregards the possibility that other stresses may exist besides those due to weight, as well as the possibility that weights may produce other effects besides stretching wires. In the case of Boyle's law, to consider another case, it overlooks the possible existence of *other* operations to the results of which the symbol pressure (p) can be just as appropriately applied as to the common operation of noting relative levels of mercury in a U-tube. Treating physics as a mere collection of laws, in short, overlooks the possibility of subsuming large realms of phenomena in terms of a relatively small number of fundamental concepts, transcending particular operations and subject to certain *postulated* mathematical relations. When we have reached the stage of conceiving physical description in this more elaborate way we are ready for a *physical theory.*

1.3. The Meaning of a Physical Theory. The Methodology of Physics. The construction of a theory marks the third and culminating stage in the development of the method of physics. In spite of more than half a century of discussion of the foundations of physics it remains the least understood of all the activities of the physicist. Yet the process of physical theorizing is perfectly straightforward from a methodological point of view, as can be seen by even a cursory inspection of the earliest and one of the most famous of all physical theories, that of *geometry.*

It may be objected that geometry is mathematics, not physics. Whatever the modern mathematical attitude may be, the fact remains that historically geometry is a physical theory, as indeed its very name suggests. It is based fundamentally on certain primitive notions of space, like the concepts of point, line, etc., which are strictly indefinable but must be accepted as intuitively given. From these more elaborate and precise concepts are constructed such as angles, lengths, circles, triangles. This is followed by the introduction of certain postulated relations which such entities must satisfy. These are known historically as the axioms and postulates of Euclidean geometry. Actually they are the hypothetical principles on which the theory rests. They cannot be verified directly and completely by experiment, although they may be suggested strongly by such experimental measurements as have been carried out on geometrical figures. But from them it is possible to deduce by logical reasoning the so-called theorems of geometry which are actually the experimental laws of the science. These are the results (e.g., the theorem that the sum of the angles of any plane triangle equals two right angles) which are experimentally verifiable and were undoubtedly originally discovered by experiment.

If the reader will examine any of the famous physical theories with which he is acquainted (e.g., the theory of classical mechanics, the kinetic theory of gases, the theory of electromagnetism, the quantum theory of atomic structure, the theory of relativity, etc.), he will find that they all have the same logical structure as that of geometry. They all begin with certain primitive, intuitive, undefined notions out of which more precise concepts are constructed and given symbolic representation. Relations are then assumed to connect these concepts: these are the hypothetical principles of the theory (e.g., the so-called "laws" of motion in mechanics, the relation between kinetic energy and temperature in kinetic theory, the field equations in electromagnetic theory, etc.). In every case these relations take the form of statements: "Let us assume . . . ." The postulates may be and usually are suggested by experimental observations, but are not directly descriptive of experiments. When expressed in symbolic form they predict many specialized results. If the theory is a successful one the latter must be directly descriptive of experience, i.e., be physical laws. The greater the number of correct predictions, the more successful is the theory, particularly if the theory predicts laws not hitherto discovered experimentally. A thorough methodological study of the nature of a physical theory would involve such problems as the historical evolution of physical concepts, the necessity for precise definitions, the relation between concepts and postulates on the one hand and experience on the other, the special mathematical devices used in the development of theories, and the criteria for the success of theories. We shall not find it profitable to discuss these things further at this point, although they will enter the subsequent discussion in connection with specific applications.[1]

It is the aim of this book to provide a thorough survey of those theories of physics which furnish the best introduction to the method of physical theorizing. We shall discuss these theories from the methodological standpoint but with enough wealth of detail so that the reader will understand not only how they are constructed but the extent to which they successfully describe a wide range of natural phenomena.

[1] For a more detailed discussion cf. Lindsay and Margenau, *Foundations of Physics* (reprinted by Dover Publications), Chapter I. Reference may also be made to the article by Lindsay in *Philosophy of Science* 5,452, 1938. Cf. also H. Margenau, *The Nature of Physical Reality* (New York, McGraw-Hill, 1950), for an exhaustive treatment of physical methodology.

The reader will observe that the material which follows is divided into two main parts, referred to respectively as *particle* physics and *field* physics. The justification for this subdivision will become more apparent with extended study. Yet it is worth while pointing out here that it reflects an important element in the historical evolution of physical theories. From earliest times there has been lively competition between the point of view which looks upon all physical phenomena as various manifestations of the behavior of discrete particles and that which insists that the fundamental substratum of physical experience is a wholly continuous medium or a set of such media. It is well to realize at the outset that neither view has prevailed entirely. Each possesses obvious advantages in describing certain portions of experience, and it seems likely that they will continue to form complementary aspects of physical theorizing for some time to come. The particle point of view seems the natural starting point for the development of atomic theory, whereas the continuum or field concept seems the natural method for describing large-scale properties of material fluids as well as many of the older phenomena of electricity and light. So-called classical mechanics has leaned so heavily on the particle concept that we shall naturally begin our detailed study with particle physics and develop the theory of mechanics from this angle. We shall discuss the theory of electricity mainly from the field point of view. Actually it will be made clear as we go along that both points of view have made valuable contributions to our understanding of all physical experience.

QUESTIONS

1. Give several illustrations of "passive" observation of physical phenomena and suggest corresponding *experiments* that have been or might be carried out.

2. Analyze two theories of gravitation from the methodological standpoint of this chapter.

3. Analyze two different theories of light, both of which lead to Snell's law of refraction.

4. Make a list of ten physical laws with a careful description of the meaning of each from an elementary point of view. Name the theory in which each one appears as a logical deduction.

5. Compare the *inductive* and *deductive* aspects of physics from the standpoint of this chapter.

Part I

PARTICLE PHYSICS

CHAPTER II

KINEMATICS

2.1. Introduction. Concept of Motion. The Material Particle. Motion has long been regarded as the most fundamental of all physical phenomena. The contemplation of terrestrial objects like projectiles and vehicles of all kinds, as well as the changing aspect of the heavens, all bear witness to the significance of motion in daily life. So thoroughly a part of our existence are these phenomena and so completely taken for granted that we scarcely realize at first the difficulty involved in a careful description of them. In fact it is not at all an easy thing to give a precise definition of motion, and philosophers from time immemorial have given much attention to this problem. We shall content ourselves here with the observation that motion is a variety of sensation characterized by the relative ease with which the human organism can compensate it by simple bodily activity. This can perhaps best be grasped by contrasting the change in the flux of sensation which we call a chemical reaction with that inherent in what we call a car going down the street.

More commonly in physics motion is termed simply change of position and the branch of physics concerned with it is called mechanics. In embarking on the study of mechanics, it is therefore important to understand what position is and what it is that changes its position. It is simpler to take the second question first. Rather than base the study of motion on the behavior of large-scale bodies which may differ much in size and shape and other characteristics the physicist prefers to introduce as the fundamental moving entity the so-called *material particle.* The best we can do with its definition in the present place is the following:

1. A material particle has position but no extension. In this respect it is analogous to a geometrical point.

2. A material particle has inertia, measured by its mass. This concept will be precisely defined in the next chapter.

3. A particle may have certain relations with respect to other particles, e.g., gravitation, electrostatic attraction, etc.

4. The large-scale bodies of ordinary experience are aggregates of particles.

We shall be concerned in the first part of this book with the motion of material particles. It will develop that a large part of physics finds its adequate description in this way. This will include not only the obvious motion of large-scale objects which are of interest because of more or less practical applications, but also physical phenomena attributed to the motion of unseen particles of theoretical postulation. Among the latter the kinetic theory of gases is an excellent example. We shall then refer to the material in Part I as particle physics.[1]

It is obviously simplest to begin our study of particle physics with the motion of a single particle. When considered with reference to change of geometrical position without consideration of any other property of the particle, this is the subject of kinematics—the geometry of motion.

Since motion is change of position it is essential at first to develop some ideas of the specification of position.

2.2. Specification of Position of a Particle. Coordinate Systems. The concept of position is a primitive notion undoubtedly connected with the specification of a method of reaching the particle from wherever one happens to be. It is refined by the introduction of geometry in which the position of a point is symbolized relative to a collection of other points constituting a so-called *frame of reference.* Thus the earth might constitute a reference frame, although more generally in classical mechanics the so-called "fixed" stars furnish a more useful one. These have been said to form the *primary inertial* system. There are infinitely many ways of specifying the position of a point or particle in a reference frame. But in every case some single reference point must be arbitrarily chosen. If one chooses this as O in Fig. 2.1 and passes through it three mutually perpendicular planes, the position of any point P can be specified by the three perpendicular distances of P from the planes respectively. These distances are called the rectangular or cartesian coordinates of P and are usually denoted by x, y, z. The rectangular reference frame in Fig. 2.1 is termed *right handed* since as one looks along the z axis a right-hand screw rotation through 90° takes the x axis into the y axis, etc. It is fair to say that rectangular coordinates form the most generally useful set for classical mechanics. For certain purposes

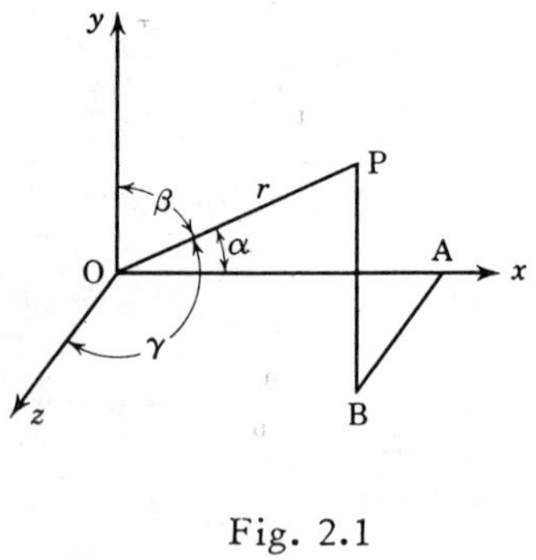

Fig. 2.1

[1]Cf. end of Sec. 1.3.

others are more valuable, however, and we mention them here for future reference.

In cylindrical coordinates (Fig. 2.2) the position of the point P is specified on the surface of a right circular cylinder of radius ρ and

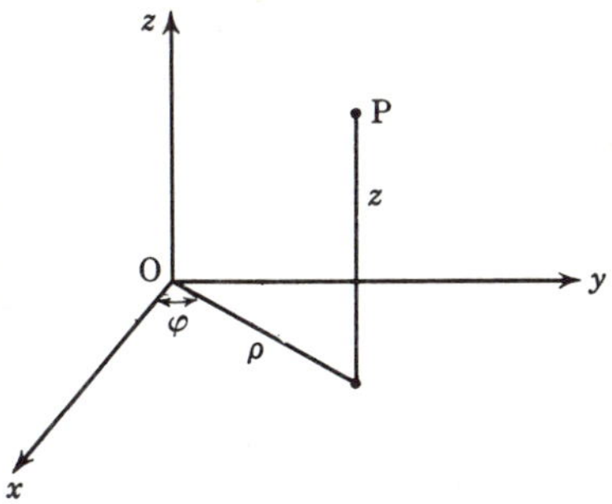

Fig. 2.2

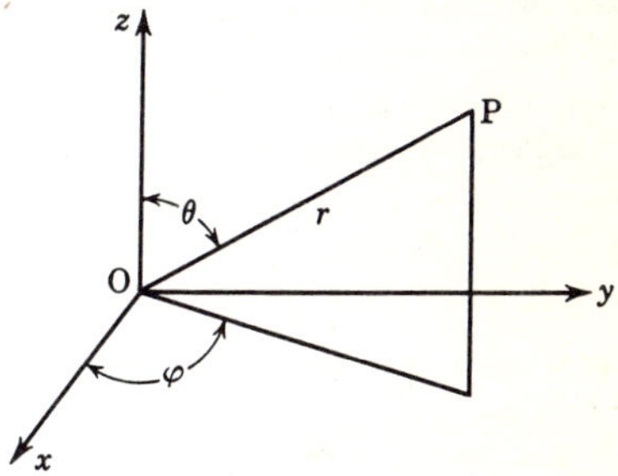

Fig. 2.3

having the z axis as its axis. The coordinates are then ρ, z, and ϕ, where z is the perpendicular distance of P above the xy plane and ϕ is the angle between the plane containing P and the z axis and the xz plane. The transformation equations between cylindrical and rectangular coordinates are given as follows:

$$
\begin{aligned}
x &= \rho \cos \phi , \\
y &= \rho \sin \phi , \\
z &= z .
\end{aligned}
\tag{2.1}
$$

In cylindricals $\rho =$ constant is the equation of a cylinder, while $\phi =$ constant is the equation of a plane containing the z axis, and $z =$ constant is the equation of a plane parallel to the xy plane. The cylinder and plane are thus the two types of surface characteristic of cylindrical coordinates.

In spherical coordinates (Fig. 2.3) the position of the point P is specified on the surface of a sphere of radius r with center at O. The three coordinates are r, θ, ϕ, where θ is the so-called co-latitude and ϕ the longitude. (Recall the coordinates for a point on the surface of the earth.) The z axis is here taken as the polar axis. The transformation equations between spherical and rectangular coordinates are as follows:

$$
\begin{aligned}
x &= r \sin \theta \cos \phi , \\
y &= r \sin \theta \sin \phi , \\
z &= r \cos \theta ,
\end{aligned}
\tag{2.2}
$$

In sphericals ϕ = constant is the equation of a plane containing the z axis. The position of P in such a fixed plane will then be specified by r and θ, which are sometimes called *polar* coordinates. The equation θ = constant is the equation of a right circular cone, and r = constant is of course the equation of a sphere. Hence the plane, cone, and sphere are the three surfaces characteristic of spherical coordinates.

Other intersecting families of surfaces can be used to define coordinate systems. A rather useful set is obtained by considering the two families of confocal parabolas in the xz plane

$$x^2 = 2\xi^2(z + \xi^2/2), \tag{2.3}$$

$$x^2 = -2\eta^2(z - \eta^2/2), \tag{2.4}$$

where ξ and η are the defining parameters. In Fig. 2.4, the curves denoted by ξ_1 and ξ_2 are members of the family represented by (2.3), with η_1 and η_2 giving two members of the family represented by (2.4). All these parabolas have a common focus at the origin O. If they are rotated about the z axis, confocal paraboloids of revolution result. The intersection of a ξ paraboloid with an η paraboloid is a circle in a plane perpendicular to the z axis, and a plane through the z axis specified by a constant value of ϕ, the longitude as in cylindricals and sphericals, will cut this circle in two points, only one of which need be associated with the given ϕ. Hence the three parameters ξ, η, ϕ specify a point P and can serve as its coordinates. These are usually referred to as parabolic coordinates. The transformation equations connecting them with rectangulars are

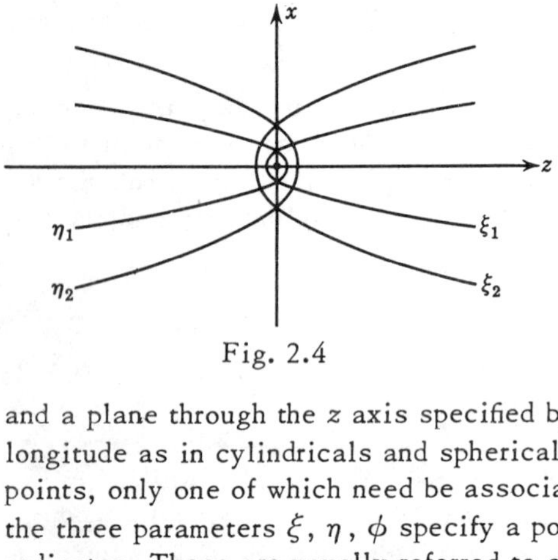

Fig. 2.4

$$\begin{aligned} x &= \xi\eta \cos\phi, \\ y &= \xi\eta \sin\phi, \\ z &= (\eta^2 - \xi^2)/2. \end{aligned} \tag{2.5}$$

Details regarding other types of coordinate systems will be found in Margenau and Murphy: *The Mathematics of Physics and Chemistry* (New York, Van Nostrand), Chapter 5.

2.3. The Vector Method of Specifying Position. Vector Notation. If we refer again to Fig. 2.1, we see that it is possible to specify the position of P by means of the straight line distance r from O to P and the three angles α, β, γ which OP makes with the x, y, and z coordinate axes respectively. These are sometimes called the generalized polar coordinates in three dimensional space. Since

$$x = r\cos\alpha\,, \quad y = r\cos\beta\,, \quad z = r\cos\gamma\,, \tag{2.6}$$

and from the Pythagorean theorem

$$x^2 + y^2 + z^2 = r^2\,, \tag{2.7}$$

it follows that

$$\cos^2\alpha + \cos^2\beta + \cos^2\gamma = 1\,, \tag{2.8}$$

and α, β, γ are not all independent. Any two of them can serve with r as the three coordinates of P.

But now we wish to introduce a more important notation involving the use of r, α, β, γ. We have already stressed that fundamentally specification of position means specifying a way of reaching a point from wherever one happens to be. Now P (in Fig. 2.1) can be reached by going from O a distance r in a certain direction, namely that specified by any two angles α, β, γ. We may therefore find it convenient to think of the position of P as represented by a quantity having a magnitude (namely the distance r) and a *direction* in space relative to a definite reference frame. Such a quantity in elementary physics is termed a *vector*. Since we shall have much to do with vectors in the sequel it is desirable at this point to provide a more precise definition.

Referring to Fig. 2.5, let us suppose the rectangular axes O-xyz are rotated so as to assume the position O-$x'y'z'$ (still remaining orthogonal). The position of P relative to the new set of axes is given by the transformation equations

$$\left.\begin{aligned} x' &= l_1x + l_2y + l_3z\,, \\ y' &= m_1x + m_2y + m_3z\,, \\ z' &= n_1x + n_2y + n_3z\,, \end{aligned}\right\} \tag{2.9}$$

where l_1, l_2, l_3 denote the direction cosines of the x' axis in the original system; m_1, m_2, m_3 are the direction cosines of the y' axis in the original system; and n_1, n_2, n_3 are the direction cosines of the z' axis. From analytic geometry l_1, l_2, etc., must satisfy the following relations

$$\left.\begin{aligned} l_1^2 + m_1^2 + n_1^2 &= 1\,, \\ l_2^2 + m_2^2 + n_2^2 &= 1\,, \\ l_3^2 + m_3^2 + n_3^2 &= 1\,. \end{aligned}\right\} \tag{2.10 a}$$

$$\sum_{i=1}^{3} l_i m_i = \sum_{i=1}^{3} m_i n_i = \sum_{i=1}^{3} n_i l_i = 0 \,,$$

$$\begin{aligned} 0 &= l_1 l_2 + m_1 m_2 + n_1 n_2 \,, \\ 0 &= l_2 l_3 + m_2 m_3 + n_2 n_3 \,, \\ 0 &= l_3 l_1 + m_3 m_1 + n_3 n_1 \,. \end{aligned} \tag{2.10 b}$$

Any set of three quantities A_x, A_y, A_z which under a rotation of rectangular axes transform as x,y,z do in (2.9) and (2.10) constitute a *3-vector*, more commonly called simply a *vector*. The coordinates x,y,z of a point evidently form the simplest illustration of a vector, but it is clear that the concept would not have achieved prominence in physics were it not for the fact that many physical quantities have precisely this property. That this is indeed the case will be brought out in what follows. The important thing here is to introduce a convenient notation for a vector. The collection of the three rectangular coordinates x,y,z of P will be called the *position vector* of P and denoted by $\mathbf{r}$ if r is the distance OP, also called the *magnitude* of the vector. The coordinates x,y,z are called the rectangular *components* of the vector, or more accurately the magnitudes of the rectangular components.

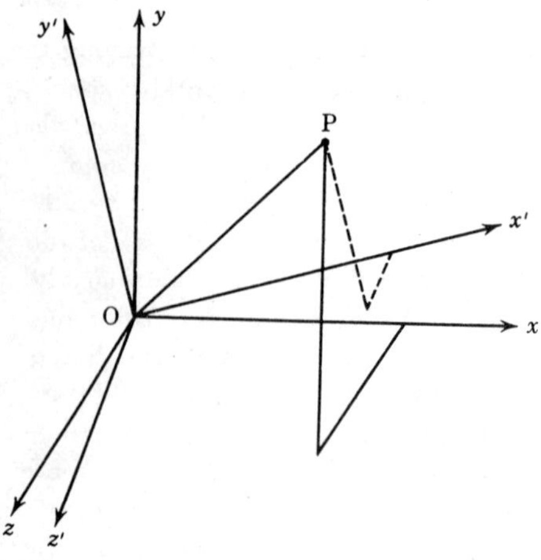

Fig. 2.5

There is a certain advantage in visualizing a vector geometrically. Thus in Fig. 2.1, the position vector $\mathbf{r}$ will be represented by the directed line segment from O to P. The magnitude of the vector is the distance r in appropriate length units, and its direction is given, as stated before, by the angles α, β, γ. The magnitudes of the components of $\mathbf{r}$ are given by eq. (2.6).

For the sake of defining certain properties of vectors, we shall also introduce a schematic algebraic notation for a vector. The vector $\mathbf{r}$ will be said to be equivalent to (x,y,z), and in general the vector $\mathbf{A}$, which is the set of three quantities A_x, A_y, A_z transforming according to eqs. (2.9) and (2.10), will be denoted by (A_x, A_y, A_z), so that the relations

$$\mathbf{r} = (x,y,z), \quad \mathbf{A} = (A_x, A_y, A_z) \tag{2.11}$$

represent identities. The *magnitudes* of **r** and **A** respectively are defined to be

$$|\mathbf{r}| = r = \sqrt{x^2 + y^2 + z^2}; \quad |\mathbf{A}| = A = \sqrt{A_x{}^2 + A_y{}^2 + A_z{}^2}. \tag{2.12}$$

A vector may be multiplied by an ordinary number, i.e., one that we shall call a scalar, in accordance with the following definition

$$a\mathbf{A} = (aA_x, aA_y, aA_z). \tag{2.13}$$

Examination of (2.9) and (2.10) discloses that the quantity on the right is a vector if **A** is a vector. The magnitude of $a\mathbf{A}$ is simply aA.

The *sum* of two vectors **A** and **B** is defined as

$$\mathbf{A} + \mathbf{B} = (A_x + B_x, A_y + B_y, A_z + B_z). \tag{2.14}$$

Once again, (2.9) and (2.10) assure that the quantity on the right is a vector. Since the commutative law holds for ordinary numbers, it follows that it must also hold for vectors, i.e.,

$$\mathbf{A} + \mathbf{B} = \mathbf{B} + \mathbf{A}. \tag{2.15}$$

The reader should have no trouble in showing that the associative law also holds, i.e.,

$$(\mathbf{A} + \mathbf{B}) + \mathbf{C} = \mathbf{A} + (\mathbf{B} + \mathbf{C}). \tag{2.16}$$

The difference between two vectors **A** and **B** is defined analogously to (2.14), viz.,

$$\mathbf{A} - \mathbf{B} = (A_x - B_x, A_y - B_y, A_z - B_z). \tag{2.17}$$

The reader should show that (2.14) and (2.17) are equivalent to the usual parallelogram law for the addition and subtraction of vectors considered as directed line segments.

In general, the sum of any number of vectors **A**, **B**, **C**, **D** · · · is defined by the generalization of (2.14), viz.,

$$\begin{aligned}\mathbf{A} + \mathbf{B} + \mathbf{C} + \mathbf{D} + \cdots = (&A_x + B_x + C_x + D_x + \cdots, \\ &A_y + B_y + C_y + D_y + \cdots, \\ &A_z + B_z + C_z + D_z + \cdots,)\end{aligned} \tag{2.18}$$

A special case of (2.18) is of considerable importance. Consider the *unit* vectors defined as follows

$$\mathbf{i} = (1,\ 0,\ 0):\quad \mathbf{j} = (0,\ 1,\ 0);\quad \mathbf{k} = (0,\ 0,\ 1). \tag{2.19}$$

Geometrically $\mathbf{i}$ is a unit line segment directed from the origin along the x axis, etc. (cf. Fig. 2.6). It follows from(2.13) and (2.18) that

$$\mathbf{r} = x\mathbf{i} + y\mathbf{j} + z\mathbf{k}, \tag{2.20}$$

so that $\mathbf{r}$ can be thought of as the *sum* of three vectors of magnitudes x, y, z, respectively. In general we may express any vector $\mathbf{A}$ in the form

$$\mathbf{A} = A_x\mathbf{i} + A_y\mathbf{j} + A_z\mathbf{k}. \tag{2.21}$$

In this notation it follows that

$$\mathbf{A} + \mathbf{B} = (A_x + B_x)\mathbf{i} + (A_y + B_y)\mathbf{j} + (A_z + B_z)\mathbf{k}. \tag{2.22}$$

Other properties of vectors will be developed as occasion requires.

It is perhaps appropriate at this point to stress the fundamental physical significance of vector quantities. Suppose we have three vectors $\mathbf{A}$, $\mathbf{B}$, $\mathbf{C}$, where

$$\mathbf{A} = (A_x,\ A_y,\ A_z),\ \text{etc.},$$

and A_x, A_y, A_z are functions of x, y, z. Further suppose that there exists a relation connecting $\mathbf{A}$, $\mathbf{B}$, $\mathbf{C}$, e.g.,

$$\mathbf{A} = \mathbf{B} + \mathbf{C}.$$

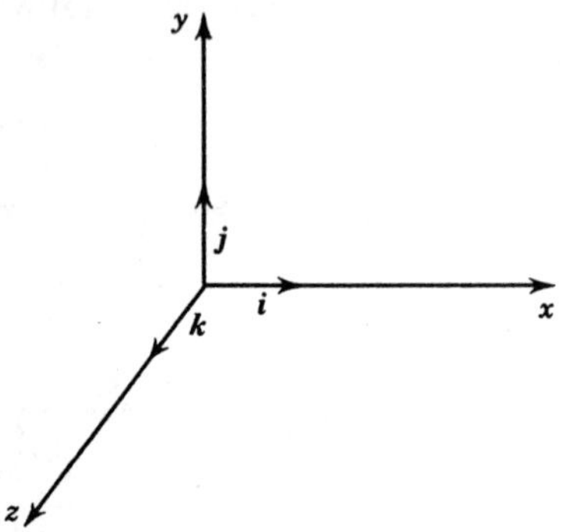

Fig. 2.6

Now transform the axes by rotation. From the fact that $\mathbf{A}$, $\mathbf{B}$, $\mathbf{C}$ are all vectors it follows that all three transform in the same way under the rotation. Hence if we denote by $\mathbf{A}'$ the form taken by $\mathbf{A}$ after the transformation, and similarly for $\mathbf{B}$ and $\mathbf{C}$, it is clear that after the transformation the above relation becomes

$$\mathbf{A}' = \mathbf{B}' + \mathbf{C}'.$$

In words the relation connecting the vectors remains *invariant* under the transformation. A brief examination shows that the same is true if the transformation is a simple translation of the axes. This is an important result. It says that if the equations of physics are vector equations they will preserve the same form independently of the orientation of the axes of the coordinate system involved. By the use of vector quantities and vector equations we insure that physical statements are independent of rotation and translation of the coordinate system. The physicist cannot help but think that this is a very desirable result: it would seem strange indeed if the form of physical equations describing physical phenomena were to depend on the position and orientation of axes in space. The full value of this invariance consideration will not become apparent until we have considered the generalization of the 3-vector here under discussion to the tensor concept (cf. Chapter 13).

2.4. The Concepts of Displacement, Velocity, and Acceleration. When a particle changes its position in a certain reference frame it is said to suffer a *displacement*. Consulting Fig. 2.7, suppose that the particle is at position P_1 with position vector $\mathbf{r}_1 = (x_1, y_1, z_1)$ at time t, and at the slightly later time $t + \Delta t$ is at P_2 with position vector $\mathbf{r}_2 = (x_2, y_2, z_2)$. The question now is: what quantity shall we use as a representation of the displacement from P_1 to P_2? This is not at all an obvious matter, since in getting from P_1 to P_2 the particle may have wandered around a great deal. We are going to make the fundamental assumption that, if Δt is a *sufficiently short* time interval, the displacement may be represented by the directed straight line segment P_1P_2, which from the definition (2.17) is simply the *difference* between the vectors $\mathbf{r}_1$ and $\mathbf{r}_2$. We shall denote this difference by $\Delta\mathbf{r}$ and write for the displacement

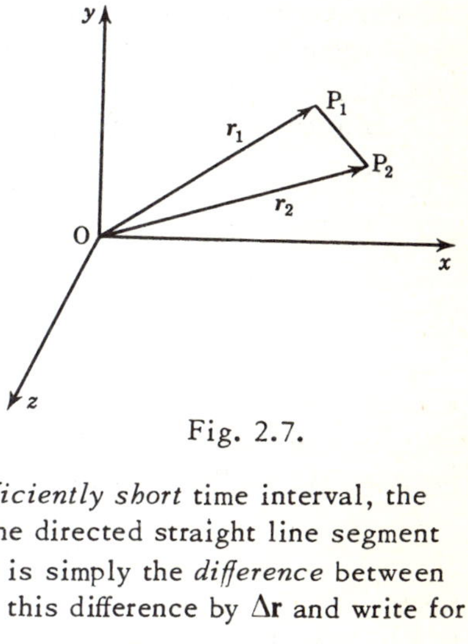

Fig. 2.7.

$$\Delta\mathbf{r} = \mathbf{r}_2 - \mathbf{r}_1, \tag{2.23}$$

where $\Delta\mathbf{r}$ denotes an increment in the position vector $\mathbf{r}$. It must be admitted that the words "sufficiently short" in describing the time interval Δt are not precise. As a matter of practical fact the qualification depends on the particle. For example, $\Delta t = 10^{-3}$ second may be sufficiently short for an ordinary large-scale particle like a bullet in flight, whereas $\Delta t = 10^{-8}$ second would be none too short for an atomic projectile like an electron, proton, or neutron. In our treatment here there is the tacit assumption that the displacement of a particle through relatively long time intervals is a continuous curve bounded in space; and that we can represent it as the limit of the sum of elementary vector displacements of the form of $\Delta\mathbf{r}$ in (2.23) as $\Delta t \to 0$.

From eq. (2.13) it is possible to construct a new vector of the form $\Delta\mathbf{r}/\Delta t$. This will be defined as the *average velocity* of the particle during the displacement $\Delta\mathbf{r}$. It is now assumed that as $\Delta t \to 0$ a unique limit exists for $\Delta\mathbf{r}/\Delta t$, which will be called the instantaneous velocity of the particle at the beginning of the time interval Δt. We shall write for this quantity

$$\frac{d\mathbf{r}}{dt} = \lim_{\Delta t \to 0} \frac{\Delta\mathbf{r}}{\Delta t} = \mathbf{v} = \dot{\mathbf{r}}. \tag{2.24}$$

Note that the dot notation to indicate differentiation with respect to time will be used extensively, although we shall also employ $\mathbf{v}$ to denote instantaneous velocity. It must be emphasized that $\dot{\mathbf{r}}$ is the instantaneous velocity of the particle *relative to* the origin 0 of the frame of reference being used.

It is often convenient to represent $\dot{\mathbf{r}}$ in terms of its components. Since (eq. 2.20)

$$\mathbf{r} = x\mathbf{i} + y\mathbf{j} + z\mathbf{k},$$

it follows from the rules of calculus that

$$\dot{\mathbf{r}} = \dot{x}\mathbf{i} + \dot{y}\mathbf{j} + \dot{z}\mathbf{k}, \tag{2.25}$$

where, to be sure, we have assumed that the unit vectors $\mathbf{i}$, $\mathbf{j}$, $\mathbf{k}$ are constant in time. Later we shall remove this restriction and introduce *moving axes*. We shall sometimes use the notation

$$v_x = \dot{x}, \quad v_y = \dot{y}, \quad v_z = \dot{z} \tag{2.26}$$

for the components of the velocity along the axes. From eq. (2.12) it follows that the magnitude of the instantaneous velocity, often called the *speed* is given by

$$v = |\mathbf{v}| = \sqrt{v_x^2 + v_y^2 + v_z^2}. \tag{2.27}$$

We encounter particles with *constant* velocity (i.e., $\dot{\mathbf{r}}$ independent of time) only rarely in mechanics. Hence it is of importance to introduce the rate of change of velocity with time. For this purpose it is convenient to represent instantaneous velocity vectors in a velocity space, i.e. one in which a point represents a particular instantaneous velocity and its three rectangular coordinates are v_x, v_y, v_z (cf. eq. 2.26). The position vector of any point in this space is the instantaneous velocity **v**. (cf. Fig. 2.8). Hence, if at successive instants t and $t + \Delta t$ the instantaneous velocities of a particle are $\mathbf{v}_1$ and $\mathbf{v}_2$, respectively, the change in velocity during Δt is $\mathbf{v}_2 - \mathbf{v}_1$, which we shall denote by $\Delta\mathbf{v}$. We shall define $\Delta\mathbf{v}/\Delta t$ as the *average acceleration* of the particle during the interval in question. It is also a vector and we assume we have a right to proceed to a limit as in the case of the average velocity. Thus we shall write for the *instantaneous acceleration* of the particle at the beginning of the interval Δt

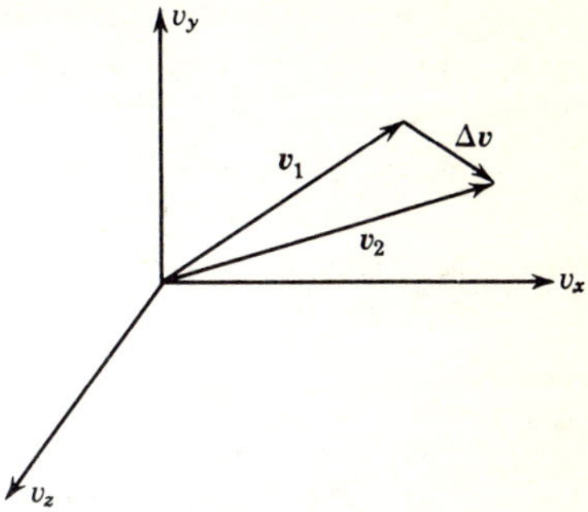

Fig. 2.8

$$\frac{d\mathbf{v}}{dt} = \lim_{\Delta t \to 0} \frac{\Delta\mathbf{v}}{\Delta t} = \mathbf{a} = \ddot{\mathbf{r}} = \dot{\mathbf{v}}. \tag{2.28}$$

Once again it is convenient to set

$$\mathbf{a} = \ddot{x}\mathbf{i} + \ddot{y}\mathbf{j} + \ddot{z}\mathbf{k} = \dot{v}_x\mathbf{i} + \dot{v}_y\mathbf{j} + \dot{v}_z\mathbf{k}, \tag{2.29}$$

where we often write for the acceleration components

$$a_x = \ddot{x} = \dot{v}_x, \quad a_y = \ddot{y} = \dot{v}_y, \quad a_z = \ddot{z} = \dot{v}_z. \tag{2.30}$$

The magnitude of the instantaneous acceleration is

$$|\mathbf{a}| = a = \sqrt{a_x^2 + a_y^2 + a_z^2}. \tag{2.31}$$

The thoughtful reader will ask, why do we stop our study of kinematics with the acceleration? Why do we not proceed to set up the time rate of change of the acceleration, etc? For an answer we fall back on the criterion of simplicity or the principle of parsimony: it is folly to introduce more where less will serve. Since it has proved possible to describe motion adequately with the concepts already introduced we forego the luxury of adding new ones. Note that we should be perfectly ready and willing to introduce new ones if the old were to prove incompetent to handle all the problems we encounter.

2.5. Motion in a Plane. Since we shall have occasion to deal with this type of motion extensively it will pay to consider its kinematics here. We shall still use the vector notation but specialize it for the x and y axes only. Thus the position vector **r** now is given by (cf. Fig. 2.9)

$$\mathbf{r} = r\cos\theta\mathbf{i} + r\sin\theta\mathbf{j}, \tag{2.32}$$

whence the velocity of the particle is

$$\mathbf{v} = \dot{\mathbf{r}} = (\dot{r}\cos\theta - r\sin\theta.\dot{\theta})\mathbf{i} + (\dot{r}\sin\theta + r\cos\theta.\dot{\theta})\mathbf{j}. \tag{2.33}$$

From this we deduce that

$$v^2 = \dot{r}^2 + r^2\dot{\theta}^2, \tag{2.34}$$

whence we may conclude that **v** can be considered as the sum of two vectors at right angles with magnitudes $\dot{r}$ and $r\dot{\theta}$. The direction of these vectors can be seen by rewriting (2.33) in the form

$$\dot{\mathbf{r}} = \dot{r}(\mathbf{i}\cos\theta + \mathbf{j}\sin\theta) + r\dot{\theta}(-\mathbf{i}\sin\theta + \mathbf{j}\cos\theta). \tag{2.35}$$

Now the vector $\mathbf{i}\cos\theta + \mathbf{j}\sin\theta$ is just the unit vector directed along **r** (cf. Fig. 2.9). We shall denote it by $\mathbf{r}_1$. Similarly $-\mathbf{i}\sin\theta + \mathbf{j}\cos\theta$ is the unit vector directed at right angles to **r** in the direction of increasing θ. We shall call this unit vector $\boldsymbol{\theta}_1$, and hence can write

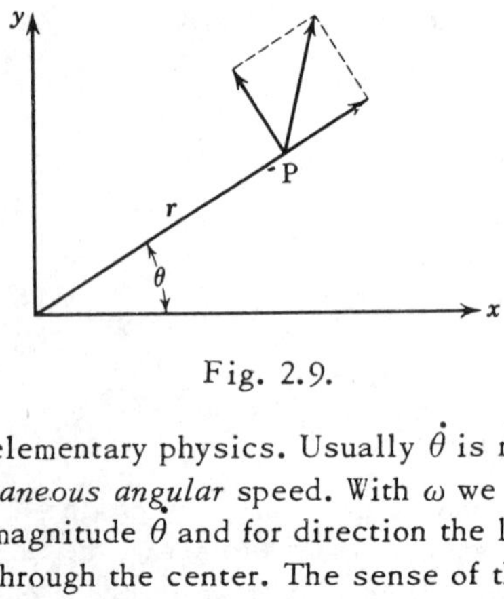

Fig. 2.9.

$$\dot{\mathbf{r}} = \dot{r}\mathbf{r}_1 + r\dot{\theta}\boldsymbol{\theta}_1. \tag{2.36}$$

The special case of circular motion with r = constant and therefore $\dot{r} = 0$, while $\dot{\mathbf{r}} = r\dot{\theta}\boldsymbol{\theta}_1$, will occur to the reader from his recollection of elementary physics. Usually $\dot{\theta}$ is represented by ω and called the *instantaneous angular* speed. With ω we can associate a vector $\boldsymbol{\omega}$ having the magnitude $\dot{\theta}$ and for direction the line normal to the plane of the circle through the center. The sense of the vector is such that as one looks along it the angular motion is clockwise. Strictly speaking we must justify the association of a vector with angular velocity by showing that it is a quantity possessing the properties laid down in Sec. 2.3. We shall postpone this consideration to Chapter 7 on rigid bodies.

We can now proceed a step further and obtain the acceleration, viz., $\ddot{\mathbf{r}}$. We have

$$\mathbf{a} = \ddot{\mathbf{r}} = (\ddot{r}\cos\theta - 2\dot{r}\dot{\theta}\sin\theta - r\dot{\theta}^2\cos\theta - r\ddot{\theta}\sin\theta)\mathbf{i}$$
$$+ (\ddot{r}\sin\theta + 2\dot{r}\dot{\theta}\cos\theta - r\dot{\theta}^2\sin\theta + r\ddot{\theta}\cos\theta)\mathbf{j}. \tag{2.37}$$

From this it follows that

$$a^2 = (\ddot{r} - r\dot{\theta}^2)^2 + (2\dot{r}\dot{\theta} + r\ddot{\theta})^2. \tag{2.38}$$

We conclude that the acceleration **a** can be considered as the sum of two perpendicular vectors with magnitudes

$$\ddot{r} - r\dot{\theta}^2 \quad \text{and} \quad 2\dot{r}\dot{\theta} + r\ddot{\theta}, \tag{2.39}$$

respectively. The direction of these vectors can be seen by the method employed above for $\dot{\mathbf{r}}$. We get

$$\ddot{\mathbf{r}} = (\ddot{r} - r\dot{\theta}^2)\mathbf{r}_1 + (2\dot{r}\dot{\theta} + r\ddot{\theta})\boldsymbol{\theta}_1. \tag{2.40}$$

It is interesting to observe that the component of the acceleration along r is not $\ddot{r}$ simply but is modified by the term $-r\dot{\theta}^2$. The reader will recognize this as his old friend the so-called *centripetal acceleration* in elementary physics. Note the significance of the *minus* sign. In the component of the acceleration at right angles to r the term $r\ddot{\theta}$ is the familiar tangential acceleration in circular motion. (The angular acceleration $\ddot{\theta}$ is usually denoted by α in elementary treatments.) The additional term $2\dot{r}\dot{\theta}$ is not generally discussed in elementary physics, for indeed it disappears for circular motion in which $\dot{r} = 0$. However it has importance and is called the *Coriolis acceleration*. Imagine an object dropped freely from rest at a point on the earth's equator. Then if the center of the earth is the origin of a system of axes fixed in space (i.e., not rotating with the earth), $\ddot{r} = -g$, the acceleration of gravity, while $\ddot{\theta} = 0$, since the angular velocity of the earth, $\dot{\theta}$, is practically constant. In addition to the downward acceleration there is thus a Coriolis acceleration $2\dot{r}\dot{\theta}$ at right angles to the meridian. A further discussion based on the study of moving axes (cf. Chapter 7, Sec. 7.7) indicates that the Coriolis acceleration for the freely falling body on the earth's surface is directed east, and hence the object should fall somewhat east of the plumb line. This is actually observed.

2.6. Kinematics in Spherical and Cylindrical Coordinates. Spherical coordinates in three dimensions often offer the same convenience as polar coordinates in the plane. We therefore develop here expressions for the various kinematical quantities in terms of them. The position vector now becomes

$$\mathbf{r} = \mathbf{i}r \sin\theta \cos\phi + \mathbf{j}r \sin\theta \sin\phi + \mathbf{k}r \cos\theta, \tag{2.41}$$

whence

$$\begin{aligned} \dot{\mathbf{r}} = {} & \mathbf{i}(\dot{r} \sin\theta \cos\phi + r\dot{\theta} \cos\theta \cos\phi - r\dot{\phi} \sin\theta \sin\phi) \\ & + \mathbf{j}(\dot{r} \sin\theta \sin\phi + r\dot{\theta} \cos\theta \sin\phi + r\dot{\phi} \sin\theta \cos\phi) \\ & + \mathbf{k}(\dot{r} \cos\theta - r\dot{\theta} \sin\theta). \end{aligned} \tag{2.42}$$

Now if we denote the unit vector along the radius vector by $\mathbf{r}_1$, as usual, that perpendicular to $\mathbf{r}$ in the direction of increasing θ by $\boldsymbol{\theta}_1$ and that perpendicular to the plane of $\mathbf{r}_1$ and $\boldsymbol{\theta}_1$ in the direction of increasing ϕ by $\boldsymbol{\phi}_1$ we have

$$\begin{aligned} \mathbf{r}_1 &= \mathbf{i} \sin\theta\cos\phi + \mathbf{j}\sin\theta\sin\phi + \mathbf{k}\cos\theta, \\ \boldsymbol{\theta}_1 &= \mathbf{i}\cos\theta\cos\phi + \mathbf{j}\cos\theta\sin\phi - \mathbf{k}\sin\theta, \\ \boldsymbol{\phi}_1 &= -\mathbf{i}\sin\phi + \mathbf{j}\cos\phi. \end{aligned} \tag{2.43}$$

Consequently in terms of the new unit vectors we have

$$\dot{\mathbf{r}} = \mathbf{v} = \dot{r}\mathbf{r}_1 + r\dot{\theta}\boldsymbol{\theta}_1 + r\sin\theta\cdot\dot{\phi}\boldsymbol{\phi}_1. \tag{2.44}$$

Differentiating once more we have

$$\begin{aligned} \ddot{\mathbf{r}} = \ddot{r}\mathbf{r}_1 &+ \dot{r}\dot{\mathbf{r}}_1 + \dot{r}\dot{\theta}\boldsymbol{\theta}_1 + r\ddot{\theta}\boldsymbol{\theta}_1 + r\dot{\theta}\dot{\boldsymbol{\theta}}_1 \\ &+ \dot{r}\sin\theta\,\dot{\phi}\boldsymbol{\phi}_1 + r\cos\theta\cdot\dot{\theta}\dot{\phi}\boldsymbol{\phi}_1 \\ &+ r\sin\theta\cdot\ddot{\phi}\boldsymbol{\phi}_1 + r\sin\theta\cdot\dot{\phi}\dot{\boldsymbol{\phi}}_1. \end{aligned} \tag{2.45}$$

If now we determine $\dot{\mathbf{r}}_1$, $\dot{\boldsymbol{\theta}}_1$, and $\dot{\boldsymbol{\phi}}_1$ from (2.43) and substitute into (2.45) we can show that the vector acceleration is

$$\begin{aligned} \ddot{\mathbf{r}} = \mathbf{a} = \mathbf{r}_1&[\ddot{r} - r\dot{\theta}^2 - r\sin^2\theta\cdot\dot{\phi}^2] \\ &+ \boldsymbol{\theta}_1[2\dot{r}\dot{\theta} + r\ddot{\theta} - r\sin\theta\cos\theta\cdot\dot{\phi}^2] \\ &+ \boldsymbol{\phi}_1[2\sin\theta\cdot\dot{r}\dot{\phi} + 2r\cos\theta\cdot\dot{\theta}\dot{\phi} + r\sin\theta\cdot\ddot{\phi}] \end{aligned} \tag{2.46}$$

The reader will note the various types of accelerations entering into this expression. Thus $r\dot{\theta}^2$, $r\sin^2\theta\cdot\dot{\phi}^2$ and $r\sin\theta\cos\theta\cdot\dot{\phi}^2$ are centripetal acceleration components, whereas $2\dot{r}\dot{\theta}$, $2\sin\theta\cdot\dot{r}\dot{\phi}$, $2r\cos\theta\cdot\dot{\theta}\dot{\phi}$ are Coriolis acceleration components.

The reader should show by the application of the above method that in cylindrical coordinates if the unit vectors are $\boldsymbol{\rho}_1$ in the direction of increasing ρ, $\boldsymbol{\phi}_1$ perpendicular to the plane of ρ and z in the direction of increasing ϕ, and $\mathbf{k}$ in the direction of increasing z we have

$$\mathbf{v} = \dot{\rho}\boldsymbol{\rho}_1 + \rho\dot{\phi}\boldsymbol{\phi}_1 + \dot{z}\mathbf{k}, \tag{2.47}$$

$$\mathbf{a} = \boldsymbol{\rho}_1(\ddot{\rho} - \rho\dot{\phi}^2) + \boldsymbol{\phi}_1(\rho\ddot{\phi} + 2\dot{\rho}\dot{\phi}) + \mathbf{k}\ddot{z}. \tag{2.48}$$

2.7. Kinematics in General Curvilinear Coordinates. For many purposes it is desirable to generalize the kinematical formulas of the previous section to apply to any set of three coordinates associated with intersecting mutually perpendicular surfaces in three dimensions. Such coordinates are usually called curvilinear orthogonal coordinates. Thus we imagine three sets of surfaces given by the equations

$$q_1(x,y,z) = C_1; \quad q_2(x,y,z) = C_2; \quad q_3(x,y,z) = C_3. \tag{2.49}$$

A member of each family passes through every point of space and at every point the three surfaces through that point are orthogonal. Consequently each point in space is characterized by the values of the q's for the surfaces passing through it.

It is clear that the rectangular coordinates of any point in space can be expressed in terms of any particular set of curvilinear coordinates by the transformation equations

$$x = x(q_1,q_2,q_3); \quad y = y(q_1,q_2,q_3); \quad z = z(q_1,q_2,q_3). \tag{2.50}$$

Associated with changes dq_1, dq_2, dq_3 in the coordinates are the actual displacements in the directions of increasing q_1, q_2, q_3, respectively. These we shall denote by ds_1, ds_2, ds_3, respectively, and write

$$ds_1 = h_1 dq_1; \quad ds_2 = h_2 dq_2; \quad ds_3 = h_3 dq_3,$$

where h_1, h_2, h_3 are functions of q_1, q_2, q_3 in general.

We can now write for the position vector of any point

$$\mathbf{r} = \mathbf{i}x(q_1,q_2,q_3) + \mathbf{j}y(q_1,q_2,q_3) + \mathbf{k}z(q_1,q_2,q_3), \tag{2.51}$$

whence

$$\dot{\mathbf{r}} = \mathbf{i}\sum_{n=1}^{3} \frac{\partial x}{\partial q_n}\dot{q}_n + \mathbf{j}\sum_{n=1}^{3} \frac{\partial y}{\partial q_n}\dot{q}_n + \mathbf{k}\sum_{n=1}^{3} \frac{\partial z}{\partial q_n}\dot{q}_n, \tag{2.52}$$

Now the unit vectors in the directions of increasing q_1, q_2, q_3, respectively, are

$$\begin{aligned}
\mathbf{q}_{11} &= \left(\mathbf{i}\frac{\partial x}{\partial q_1} + \mathbf{j}\frac{\partial y}{\partial q_1} + \mathbf{k}\frac{\partial z}{\partial q_1}\right)\frac{\partial q_1}{\partial s_1} = \frac{1}{h_1}\left(\mathbf{i}\frac{\partial x}{\partial q_1} + \mathbf{j}\frac{\partial y}{\partial q_1} + \mathbf{k}\frac{\partial z}{\partial q_1}\right), \\
\mathbf{q}_{12} &= \frac{1}{h_2}\left(\mathbf{i}\frac{\partial x}{\partial q_2} + \mathbf{j}\frac{\partial y}{\partial q_2} + \mathbf{k}\frac{\partial z}{\partial q_2}\right), \\
\mathbf{q}_{13} &= \frac{1}{h_3}\left(\mathbf{i}\frac{\partial x}{\partial q_3} + \mathbf{j}\frac{\partial y}{\partial q_3} + \mathbf{k}\frac{\partial z}{\partial q_3}\right).
\end{aligned} \tag{2.53}$$

Consequently we can write for the velocity $\mathbf{v}$

$$\dot{\mathbf{r}} = \mathbf{v} = h_1\dot{q}_1\mathbf{q}_{11} + h_2\dot{q}_2\mathbf{q}_{12} + h_3\dot{q}_3\mathbf{q}_{13}. \tag{2.54}$$

For the acceleration we have

$$\ddot{\mathbf{r}} = \mathbf{a} = \frac{d}{dt}(h_1\dot{q}_1)\mathbf{q}_{11} + \frac{d}{dt}(h_2\dot{q}_2)\mathbf{q}_{12} + \frac{d}{dt}(h_3\dot{q}_3)\mathbf{q}_{13}$$
$$+ h_1\dot{q}_1\dot{\mathbf{q}}_{11} + h_2\dot{q}_2\dot{\mathbf{q}}_{12} + h_3\dot{q}_3\dot{\mathbf{q}}_{13}. \tag{2.55}$$

Now

$$\left.\begin{aligned} \dot{\mathbf{q}}_{11} &= \mathbf{i}\frac{d}{dt}\left(\frac{1}{h_1}\frac{\partial x}{\partial q_1}\right) + \mathbf{j}\frac{d}{dt}\left(\frac{1}{h_1}\frac{\partial y}{\partial q_1}\right) + \mathbf{k}\frac{d}{dt}\left(\frac{1}{h_1}\frac{\partial z}{\partial q_1}\right), \\ \dot{\mathbf{q}}_{12} &= \mathbf{i}\frac{d}{dt}\left(\frac{1}{h_2}\frac{\partial x}{\partial q_2}\right) + \mathbf{j}\frac{d}{dt}\left(\frac{1}{h_2}\frac{\partial y}{\partial q_2}\right) + \mathbf{k}\frac{d}{dt}\left(\frac{1}{h_2}\frac{\partial z}{\partial q_2}\right), \\ \dot{\mathbf{q}}_{13} &= \mathbf{i}\frac{d}{dt}\left(\frac{1}{h_3}\frac{\partial x}{\partial q_3}\right) + \mathbf{j}\frac{d}{dt}\left(\frac{1}{h_3}\frac{\partial y}{\partial q_3}\right) + \mathbf{k}\frac{d}{dt}\left(\frac{1}{h_3}\frac{\partial z}{\partial q_3}\right). \end{aligned}\right\} \tag{2.56}$$

But we can also get

$$\left.\begin{aligned} \dot{\mathbf{q}}_{11} &= A_1\mathbf{q}_{11} + B_1\mathbf{q}_{12} + C_1\mathbf{q}_{13}, \\ \dot{\mathbf{q}}_{12} &= A_2\mathbf{q}_{11} + B_2\mathbf{q}_{12} + C_2\mathbf{q}_{13}, \\ \dot{\mathbf{q}}_{13} &= A_3\mathbf{q}_{11} + B_3\mathbf{q}_{12} + C_3\mathbf{q}_{13}. \end{aligned}\right\} \tag{2.57}$$

Using (2.53), (2.56), and (2.57) we can solve for the A, B, C, and get

$$A_1 = \frac{\begin{vmatrix} \dfrac{d}{dt}\left(\dfrac{1}{h_1}\dfrac{\partial x}{\partial q_1}\right), & \dfrac{1}{h_2}\dfrac{\partial x}{\partial q_2}, & \dfrac{1}{h_3}\dfrac{\partial x}{\partial q_3} \\ \dfrac{d}{dt}\left(\dfrac{1}{h_1}\dfrac{\partial y}{\partial q_1}\right), & \dfrac{1}{h_2}\dfrac{\partial y}{\partial q_2}, & \dfrac{1}{h_3}\dfrac{\partial y}{\partial q_3} \\ \dfrac{d}{dt}\left(\dfrac{1}{h_1}\dfrac{\partial z}{\partial q_1}\right), & \dfrac{1}{h_2}\dfrac{\partial z}{\partial q_2}, & \dfrac{1}{h_3}\dfrac{\partial z}{\partial q_3} \end{vmatrix}}{\Delta}, \tag{2.58}$$

$$B_1 = \frac{\begin{vmatrix} \dfrac{1}{h_1}\dfrac{\partial x}{\partial q_1}, & \dfrac{d}{dt}\left(\dfrac{1}{h_1}\dfrac{\partial x}{\partial q_1}\right), & \dfrac{1}{h_3}\dfrac{\partial x}{\partial q_3} \\ \dfrac{1}{h_1}\dfrac{\partial y}{\partial q_1}, & \dfrac{d}{dt}\left(\dfrac{1}{h_1}\dfrac{\partial y}{\partial y_1}\right), & \dfrac{1}{h_3}\dfrac{\partial y}{\partial y_3} \\ \dfrac{1}{h_1}\dfrac{\partial z}{\partial q_1}, & \dfrac{d}{dt}\left(\dfrac{1}{h_1}\dfrac{\partial z}{\partial q_1}\right), & \dfrac{1}{h_3}\dfrac{\partial z}{\partial q_3} \end{vmatrix}}{\Delta}, \tag{2.59}$$

$$C_l = \frac{\begin{vmatrix} \frac{1}{h_1}\frac{\partial x}{\partial q_1}, & \frac{1}{h_2}\frac{\partial x}{\partial q_2}, & \frac{d}{dt}\left(\frac{1}{h_l}\frac{\partial x}{\partial q_l}\right) \\ \frac{1}{h_1}\frac{\partial y}{\partial q_1}, & \frac{1}{h_2}\frac{\partial y}{\partial q_2}, & \frac{d}{dt}\left(\frac{1}{h_l}\frac{\partial y}{\partial q_l}\right) \\ \frac{1}{h_1}\frac{\partial z}{\partial q_1}, & \frac{1}{h_2}\frac{\partial z}{\partial q_2}, & \frac{d}{dt}\left(\frac{1}{h_l}\frac{\partial z}{\partial q_l}\right) \end{vmatrix}}{\Delta}, \tag{2.60}$$

where

$$\Delta = \begin{vmatrix} \frac{1}{h_1}\frac{\partial x}{\partial q_1}, & \frac{1}{h_2}\frac{\partial x}{\partial q_2}, & \frac{1}{h_3}\frac{\partial x}{\partial q_3} \\ \frac{1}{h_1}\frac{\partial y}{\partial q_1}, & \frac{1}{h_2}\frac{\partial y}{\partial q_2}, & \frac{1}{h_3}\frac{\partial y}{\partial q_3} \\ \frac{1}{h_1}\frac{\partial z}{\partial q_1}, & \frac{1}{h_2}\frac{\partial z}{\partial q_2}, & \frac{1}{h_3}\frac{\partial z}{\partial q_3} \end{vmatrix}, \tag{2.61}$$

and l takes on the values 1, 2, 3. We can then write (2.55) in the general form

$$\begin{aligned} \ddot{\mathbf{r}} = \mathbf{a} = \mathbf{q}_{11}&\left[\frac{d}{dt}(h_1\dot{q}_1) + h_1\dot{q}_1A_1 + h_2\dot{q}_2A_2 + h_3\dot{q}_3A_3\right] \\ + \mathbf{q}_{12}&\left[\frac{d}{dt}(h_2\dot{q}_2) + h_1\dot{q}_1B_1 + h_2\dot{q}_2B_2 + h_3\dot{q}_3B_3\right] \\ + \mathbf{q}_{13}&\left[\frac{d}{dt}(h_3\dot{q}_3) + h_1\dot{q}_1C_1 + h_2\dot{q}_2C_2 + h_3\dot{q}_3C_3\right]. \end{aligned} \tag{2.62}$$

The reader should show that for spherical and cylindrical coordinates (2.62) reduces to (2.46) and (2.48), respectively.

PROBLEMS

1. A car is traveling with a velocity $\mathbf{v}$ with respect to the ground. A man jumps off the car in such a way that his velocity relative to the car is $\mathbf{V}$. Determine the vector representing the velocity of the man with respect to the ground. (Assume the general case in which the angle between $\mathbf{v}$ and $\mathbf{V}$ is different from zero.)

2. A person traveling due east at 3 miles/hr. finds that the wind appears to blow directly from the north. After doubling his speed he finds that the wind appears to come from the northeast. Find the vector representing the wind velocity.

3. Find the magnitude and direction of the resultant of the four vectors

$$\mathbf{r}_1 = 6\mathbf{i} - 4\mathbf{j} + 10\mathbf{k}$$

$$\mathbf{r}_2 = -6\mathbf{i} + 4\mathbf{j} - 10\mathbf{k}$$

$$\mathbf{r}_3 = 4\mathbf{i} - 6\mathbf{j} - 10\mathbf{k}$$

$$\mathbf{r}_4 = 10\mathbf{j} + 4\mathbf{k}.$$

4. Use the vector equation $\mathbf{a} + \mathbf{b} = \mathbf{c}$ to derive by vector methods the cosine theorem of trigonometry.

5. Prove by a study of its transformation properties under rotation of axes that the aggregate

$$(y\dot{z} - z\dot{y},\ z\dot{x} - x\dot{z},\ x\dot{y} - y\dot{x})$$

is a vector. What is its physical significance?

6. Prove that the magnitude of the vector in Prob. 5 is an invariant under rotation of axes.

7. Derive the expression for the square of the magnitude of the velocity of a particle in terms of $\dot{r}$, $\dot{\theta}$ and $\dot{\phi}$ (spherical coordinates).

8. In motion in a central field the general law of areas can be written in the form $r^2\dot{\theta} = k$, where k is the area constant. In the special case of an inverse square field the path is the ellipse

$$r = \frac{l}{1 - \epsilon \cos \theta},$$

where ϵ = eccentricity and l = semi latus rectum. Utilizing the fact that in this case

$$k = \frac{2\pi a^2}{T}\sqrt{1 - \epsilon^2},$$

where a = semi major axis, T = period of revolution, and $l = (1 - \epsilon^2)a$, find the Coriolis acceleration as a function of θ, ϵ, a and T. Find its maximum and minimum values.

9. Apply the results of Prob. 8 to determining the maximum Coriolis acceleration for an electron in the Bohr model of hydrogen with principal

quantum number 2 and azimuthal quantum number 1. Use $\epsilon^2 = 1 - n_1{}^2/n^2$, where n = principal quantum number and n_1 = azimuthal quantum number. Also $a = n^2h^2/4\pi^2me^2$, and $T = n^3h^3/4\pi me^4$. Data for electron:

$$m = 9 \times 10^{-28} \text{ gram}$$
$$e = 4.8 \times 10^{-10} \text{ e.s.u.}$$
$$h = 6.6 \times 10^{-27} \text{ erg sec}$$

10. Two particles move with equal speeds v in opposite directions on the same circle. In what position will their relative velocity have a maximum magnitude and what will it be? Answer the same question for the minimum magnitude. What is the magnitude of the relative velocity when the radius vectors of the two particles are at right angles to each other? Draw a clear diagram showing the actual relative velocities.

11. Comparing with Prob. 5 show by a consideration of its transformation properties that the aggregate $(y\dot{z} + z\dot{y},\ z\dot{x} + x\dot{z},\ x\dot{y} + y\dot{x})$ is *not* a vector.

12. Derive expressions (2.47) and (2.48) for **v** and **a** in cylindrical coordinates.

13. A particle moves in a plane with respect to fixed rectangular axes. Express its velocity and acceleration components relative to a set of rectangular axes rotating about the same origin with constant angular velocity ω in terms of the corresponding components relative to the fixed axes. Specialize to the case where the particle is at rest in the fixed system.

14. Show that in spherical coordinates eq. (2.62) reduces to eq. (2.46).

15. A particle moves in a plane so that the angular speed $\dot{\theta}$ of its radius vector r about the origin is constant and $\ddot{r} = Kr$ where K is a positive constant. Find the equation of the path of the particle. Evaluate the arbitrary constants by assuming that the particle starts at rest at $r = 0$, $\theta = -\infty$.

CHAPTER III

CONCEPTS OF THE THEORY OF MECHANICS

THE PRINCIPLES OF DYNAMICS

3.1. The Concepts of Mass. We have introduced in the preceding chapter the concepts of displacement, velocity, and acceleration. The question arises at once: Are these sufficient for a description of observed motions? Thus by setting the acceleration (the second time derivative of the displacement) equal to various functions of the position vector, the velocity and the time, we should get differential equations, which when solved for the position vector as a function of the time might be found to correspond to the actually observed motions of a single particle under different conditions. This is a plausible procedure but it suffers from at least two drawbacks: (1) it provides no physical meaning for the functions to which the acceleration is equated, and (2) it would encounter particular difficulty as soon as the motion of more than one particle was in question. It therefore proves desirable to concoct some new concepts to take care of these situations. These are of course the concepts of *force* and *mass*.

Since mechanics has had a long history and a slow growth it is not surprising that its concepts have been looked at in different lights by different generations of investigators. In elementary physics the Newtonian formulation of mechanics is usually employed. However it is generally admitted that this is decidedly wanting in logical clarity.[1] Hence we shall proceed at once to define mass and force essentially in accordance with the proposal of Ernst Mach.[2]

It is necessary to assume that particles always exert mutual influence on each other (cf. Sec. 2.1). We shall further assume that this influence takes the form of accelerated motion. Consider any two particles A and B very far removed from all other particles in the universe. By the action of B, let A have the acceleration $\mathbf{a}_{AB}$ relative to the primary inertial system (the frame of reference attached to the fixed stars). By the action of A let B have the acceleration $\mathbf{a}_{BA}$ in the same system. It is now postulated that the ratio

$$\mathbf{a}_{AB}/-\mathbf{a}_{BA}$$

---

[1] R. B. Lindsay and H. Margenau, *Foundations of Physics* (1936, Dover reprint) Sec. 3.4.

[2] Ernst Mach, *The Science of Mechanics* (English translation published by Open Court Publishing Co., La Salle, Ill., 1907), p. 216.

is a positive scalar constant characteristic of the two particles but independent of the nature of their mutual interaction (e.g., whether gravitation, electrical attraction or repulsion, connection with a spring, etc.) and also independent of their relative positions, their relative velocity (as long as it is small compared with the velocity of light) as well as of the time and temperature. It will be observed that we are excluding any reference to relativistic mechanics: if the velocities are very large our definition of mass will fail to be significant and must be replaced by the appropriate relativistic one (Sec. 17.5).

To appreciate the meaning of the above assumptions it will be wise to have a concrete situation in mind, e.g., two small spheres of the same size but of different material, say of lead and copper, attached to each other by a rubber band. When the two are pulled apart, and let go the two accelerations are readily observed not to be equal in magnitude.

We shall express the fundamental assumption in symbolic form as

$$\frac{\mathbf{a}_{AB}}{-\mathbf{a}_{BA}} = m_{BA}. \tag{3.1}$$

We can now imagine the experiment leading to (3.1) repeated with many isolated pairs of particles A, B, C, D, . . . and ascertain the pair-ratios m_{CA}, m_{CB}, m_{DA}, m_{DB}, . . . etc., each defined analogously to (3.1). Now in "any old kind" of world one might not expect to find any systematic relation connecting these ratios: they might indeed be wholly independent. However, experience with such particles as we have been able to test (admittedly under less ideal conditions than our theory demands) *suggests* that the ratios *are* related and indeed in the following fashion

$$m_{CB} = m_{CA}/m_{BA} \tag{3.2}$$

a relation which will be postulated for any three particles A, B, C. Since $m_{CB} = \mathbf{a}_{BC}/-\mathbf{a}_{CB}$, by definition, we get on substitution from (3.2)

$$m_{BA}\mathbf{a}_{BC} = -m_{CA}\mathbf{a}_{CB}. \tag{3.3}$$

Similarly if we considered the particles A, B, D we should be led to

$$m_{BA}\mathbf{a}_{BD} = -m_{DA}\mathbf{a}_{DB} \tag{3.4}$$

Inspection of (3.3) and (3.4) reveals the possibility of choosing a standard particle (A in our case) and, by letting the particles B, C, D . . . interact with it in turn, assigning to the latter the constants m_{BA}, m_{CA}, m_{DA} . . . , *each of which will remain associated with its respective particle no matter with what other particle interaction takes place.* This important consequence of our fundamental postulate suggests that the constant denoted by m should be treated as a significant physical quantity characteristic of each particle. We shall indeed name m_{BA} the *mass* of B relative to A. If now we carefully preserve the particle A in a museum and agree to call it the standard *unit* of mass, m_{BA} becomes simply the mass of B in these units and we may leave off the subscript A and call it simply m_B.

The reader will recall that the most accurate method of measuring mass is by means of a balance, but a little reflection will probably convince him that since the use of the balance depends on weight it is not a satisfactory means of *defining* mass (cf. Sec. 3.3).

The definition of mass presented here should serve to stress the essentially hypothetical character of the concepts of mechanics. This is important, since from long familiarity mechanical ideas have achieved an obviousness not shared by other physical concepts, like those in electricity, for example. To be sure, the concept of mass in mechanics is directly suggested by the everyday experience of inertia, but before this notion can be used it must be precisely defined and this inevitably introduces the element of hypothesis.

3.2. The Concept of Force and Equation of Motion. If the reader will revert to eq. (3.3) he will see that the fundamental postulates are there expressed in the following form: In the interaction of any two particles, e.g., B and C, otherwise isolated, the product of mass times acceleration has the same magnitude for the two particles and differs only in sign. We introduce now the further assumption that this mass-acceleration product depends only on the position of the one particle relative to the other, the relative velocity and the time: This dependence we symbolize by the vector function

$$\mathbf{F}(\mathbf{r}_{BC}, \dot{\mathbf{r}}_{BC}, t) \equiv \mathbf{F}_{BC}, \tag{3.5}$$

and write

$$m_B \mathbf{a}_{BC} = \mathbf{F}_{BC}, \tag{3.6}$$

with the analogous equation

$$m_C \mathbf{a}_{CB} = \mathbf{F}_{CB}. \tag{3.7}$$

We shall define $\mathbf{F}_{BC}$ as the *force* on particle B due to particle C, and $\mathbf{F}_{CB}$ as the force on particle C due to particle B. Hence eq. (3.3) leads to

$$\mathbf{F}_{BC} = -\mathbf{F}_{CB}, \tag{3.8}$$

which in words says that in the interaction of the two particles B and C the force on B due to C is equal and opposite to the force on C due to B.

This is equivalent to Newton's third law of motion.

There has been much discussion of the logical status of (3.6). Some prefer to treat the mass-acceleration product itself as the force. Then (3.6) becomes essentially an identity. We have preferred to treat (3.6), however, as a genuine equation, which we shall call the *equation of motion* of B due to C. If we knew what to insert for the force function $\mathbf{F}_{BC}$ we should be able to solve (3.6) as a differential equation of the second order and ultimately determine the position vector of B relative to C. Of course we really do not *know* what to insert for the force function. Our only recourse is to postulate various functions and see what happens. If we arrive at solutions which correspond to actually observed relative motions we shall conclude that the corresponding force functions are of importance in mechanics. We can then make a catalogue of useful force functions for further reference and study in the hope that those which have been found to work in certain well-known situations will also work in other less well-known ones. The special case of a force function depending only on the magnitude of the relative position vector and directed along this vector (central field) comes to mind at once, and the still more specialized case in which the dependence is on the inverse square of the distance between the particles has had enormous significance in the history of physics.

The question at once arises: What shall we do if the particle B is in the presence of not merely particle C, but also other particles D, E, etc? We shall have to introduce an additional postulate to cover this situation: Corresponding to the force functions $\mathbf{F}_{BC}$, $\mathbf{F}_{BD}$, $\mathbf{F}_{BE}$ etc., we shall assume the existence of a *resultant force* on B which is the sum of these vectors, ie.,

$$\mathbf{F} = \mathbf{F}_{BC} + \mathbf{F}_{BD} + \mathbf{F}_{BE}. \tag{3.9}$$

This is sometimes called the principle of *superposition of forces.* We shall not expect it to work in every case, but it does so often enough to justify its introduction. We now generalize (3.6) to the form

$$m\ddot{\mathbf{r}} = \mathbf{F}, \tag{3.10}$$

where m is now written for the mass of the particle B (i.e., *any* particle) and $\ddot{\mathbf{r}}$ is the *resultant* acceleration of the particle in the primary inertial system. Eq. (3.10) is the fundamental equation of motion for any particle. Its resemblance to the so-called second law of Newton is obvious. It will be assumed that the reader is familiar with the standard cgs system of units employed in the use of eq. (3.10).

An interesting question arises as to the generality of eq. (3.10). We have assumed it to hold in the primary inertial system. Does it keep its form invariant in any reference system? Since it is a vector equation, its form is certainly independent of rotation and translation of axes. But one can imagine transformations to systems of axes which are not merely at rest with respect to the original ones but move with respect to them. Such motion may be either with constant or varying velocity. Consider the special case in which the transformed axes move relative to the original ones with constant velocity $\mathbf{v}$. The position vector $\mathbf{r}'$ in the moving system then becomes in terms of the position vector $\mathbf{r}$ in the original system,

$$\mathbf{r}' = \mathbf{r} - \mathbf{v}t - \mathbf{b},$$

where $\mathbf{b}$ is a constant vector giving the initial separation of the origins of the two sets of axes. From this it follows from the fact that $\mathbf{v}$ is constant that

$$\ddot{\mathbf{r}}' = \ddot{\mathbf{r}}.$$

At the same time the force $\mathbf{F}$ on the particle, since it depends only on the relative position vector of the particle with respect to other particles, is unaltered by the transformation above, i.e., $\mathbf{F}' = \mathbf{F}$. Hence eq. (3.10) remains invariant in form under the transformation in question.

The physical consequence of the foregoing result is that no experiment based on the purely mechanical behavior of particles is competent to distinguish between any reference systems moving with respect to each other with constant velocity. Such systems are called inertial systems: in them the equation of motion has the same form and therefore they are indistinguishable so far as any results from this equation are concerned. This is the Newtonian principle of relativity.

3.3. Mass and Force as Illustrations of the Nature of Physical Definition. The two previous sections have presented definitions of mass and force which, with the purely kinematical concepts, make up the working basis of mechanics. As has been stressed in Chapter I, all physical theories describe or explain physical phenomena in terms of concepts which must be clearly defined. In particular their meaning must be made clear in two ways. In the first place the definition must tell us how the concept is to be used in the analytical development of the theory, and in the second place the definition must make clear how the concept is to be identified with or related to actual experimental operations. The definitions of mass and force with their importance for the methodology of mechanics and consequently of all physics provide good illustrations of these properties and make this an appropriate place for some observations on the nature of physical definition.

We can perhaps do this best by examining how the definition of mass presented in Sec. 3.1 meets the requirements just mentioned. Since the definition leads ultimately to the equation of motion (eq. 3.10) or what in the Newtonian formulation of mechanics is referred to as the second law of motion, it makes clear the role of mass in the analytical development of the whole theory. Specifically it provides theoretically for the existence of a property of physical particles descriptive of their interaction and tells us how to employ the associated symbolic parameter in solving mechanical problems. At the same time the definition provides a means of attaching numbers to the parameter by measurement. In other words, it identifies the concept with a precise operation. The fact that this is not the operation commonly used for the actual laboratory measurement is not important. Actually all practical mass determining devices (balances, etc.) operate in ways deducible from the Mach definition.

What shall we say about other definitions of mass sometimes found, particularly in elementary texts? Thus for a long time it was common to define mass as the "quantity of matter in a body." One still encounters occasionally "mass is the measure of inertia." When we test these "definitions" by the criteria before mentioned, it is clear that they are failures. It is true that they may convey some vague notion, they may make a lay person think he has grasped the idea of mass, but they are without value as part of the physicist's stock in trade. But the question may be raised: Why does one have to go to all the trouble Mach went to in framing a definition? Is it not sufficient to say that mass is a property of a material particle represented by a measurable parameter m which enters into the equation $m\mathbf{a} = \mathbf{F}$, which itself is completely able to describe the motion of all particles? Reflection leads to the conclusion that whereas this satisfies the first of the criteria for a satisfactory definition, it does not satisfy the second. There is no way to produce a connection with experience sufficient for assignment of a number to m unless we have an independent definition of force. This of course can be achieved, but it is a considerable merit of the Mach definition that it effectively defines mass without the specific introduction of force and only introduces the latter concept as a means of simplifying the theory. It of course provides a definition of force at the same time.

Another possible definition of mass treats it directly as something measured on a balance, i.e., by weight. This certainly satisfies the operational criterion, but it just as clearly does not provide a theoretical significance that is general enough for the whole of mechanics, which after all must deal with particles which are not merely terrestrial. This definition is again inadequate.

Other methodological questions are pertinent in this connection. We shall consider only one. Is it proper to say that there must be but *one* definition of any particular physical concept? This view is widely held by teachers of physics, who are at any rate impressed by its pedagogical virtue. However, it must be admitted that working physicists have not felt themselves bound by it. We are all aware, for example, that the definition of temperature in kinetic theory is not the same as the definition in thermodynamics. The definition of impedance in mechanical systems is not the same as in electrical systems, although to be sure there is a close mathematical analogy. Actually it is possible to take the stand that physics could not function as a going concern without a plurality of definitions of the same concept. This point has been argued with some force by Margenau,[3] who develops the interesting idea that every physical concept will be represented by definitions falling into two large classes. The first he calls *epistemic* definitions: these relate the concept to actual experience; they might equally well be called operational, for they also tell how to assign numbers to the concept if it is a quantity. The second he calls *constitutive:* these relate the concept to other concepts in the same theory; they might be called theoretical. This type of critique opens up vast opportunities for methodological analysis into which we shall not go here. But it will be observed that in our discussion we have effectively demanded that every satisfactory physical definition shall possess both epistemic and constitutive elements. For some this might seem more logically economical than to treat the two criteria as describing entirely different types of definition. The interested reader should be on the lookout for further opportunities of studying the nature of definition in the various theories discussed in the subsequent sections of this book.

3.4. The Concept of Momentum. Before proceeding to the study of special cases of the equation of motion (3.10) we can profitably investigate some general consequences. A possible question at this stage is: What is the cumulative effect of a resultant force $\mathbf{F}$ on a particle? The answer will depend on whether the effect is thought of as extending over time or space. Let us first concentrate on the time-cumulative effect, which will be naturally defined as the time integral of the force over the time interval in question, e.g., from t_1 to t_2. Eq. (3.10) then yields

$$\int_{t_1}^{t_2} \mathbf{F}\, dt = \int_{t_1}^{t_2} m\ddot{\mathbf{r}}\, dt = m\mathbf{v}_2 - m\mathbf{v}_1, \qquad (3.11)$$

where $\mathbf{v}_1$ is the instantaneous velocity of the particle at t_1 and $\mathbf{v}_2$ is the instantaneous velocity at t_2. The quantity $m\mathbf{v}$ is called the *momentum*

[1]H. Margenau, *The Nature of Physical Reality* (New York, McGraw Hill, 1950), pp. 220ff.

of the particle. According to (3.11) the time-cumulative effect of a force on a single particle is equal to the *change in momentum*. Since the time integral of the force is usually termed the *impulse* of the force, the theorem (3.11) is often known as the *impulse-momentum* theorem. The equation of motion (3.10) can be written in a form more closely associated with this theorem as follows

$$\frac{d(m\mathbf{v})}{dt} = \mathbf{F}. \qquad (3.10')$$

The utility of the impulse-momentum theorem is best appreciated in considering the motion of an aggregate of particles. The general case is best postponed to a later chapter, but it will be instructive to consider a simple special case involving the interaction by collision of two particles of masses m_1 and m_2. Let the force exerted on m_1 by m_2 during the time τ that the collision lasts be $\mathbf{F}_{12}$. The reaction force of m_2 on m_1 is $\mathbf{F}_{21}$. Let the velocity of m_1 immediately prior to the impact be $\mathbf{u}_1$ and immediately after the impact $\mathbf{v}_1$, with similar quantities with appropriate subscripts for m_2. From the theorem (3.11) we therefore have (counting time from the instant of impact)

$$\int_0^\tau \mathbf{F}_{12}\, dt = m_1(\mathbf{v}_1 - \mathbf{u}_1) \qquad (3.12)$$

and

$$\int_0^\tau \mathbf{F}_{21}\, dt = m_2(\mathbf{v}_2 - \mathbf{u}_2). \qquad (3.13)$$

But from the third law of Newton (eq. 3.8)

$$\mathbf{F}_{12} = -\mathbf{F}_{21}. \qquad (3.14)$$

Hence (3.12) and (3.13) yield

$$m_1\mathbf{u}_1 + m_2\mathbf{u}_2 = m_1\mathbf{v}_1 + m_2\mathbf{v}_2. \qquad (3.15)$$

In words the total momentum of the system of two particles remains unaltered by the collision. This is a simple illustration of the principle of conservation of momentum which holds in the mutual interaction of particles when no external forces act. For the general theorem, see Chapter 6.

3.5. The Concept of Energy. Multiplication of Vectors. The space-cumulative effect of a force on a particle is naturally measured by the space integral of the force over the path of the particle. Since a space interval or displacement is a vector, we are at once faced with the necessity of defining what we shall mean by the product of two vectors, i.e., the resultant force **F** and the displacement $d\mathbf{r}$. Mathematically speaking, this is an arbitrary procedure and we are therefore entitled to set up any definition which is not inconsistent with the properties of vectors and which is physically useful. We shall do it as follows. Reverting to the definition of a vector given in eq. (2.11) and (2.12), we *define* the *scalar product* of two vectors **A** and **B** as follows

$$\begin{aligned}\mathbf{A} \cdot \mathbf{B} &\equiv (A_x, A_y, A_z) \cdot (B_x, B_y, B_z) \\ &\equiv A_x B_x + A_y B_y + A_z B_z .\end{aligned} \tag{3.16}$$

If we denote by θ the angle between **A** and **B**, analytic geometry gives

$$\cos \theta = \frac{A_x}{A} \cdot \frac{B_x}{B} + \frac{A_y}{A} \cdot \frac{B_y}{B} + \frac{A_z}{A} \cdot \frac{B_z}{B} , \tag{3.17}$$

and hence

$$\mathbf{A} \cdot \mathbf{B} = AB \cos \theta . \tag{3.18}$$

In particular for the unit vectors (3.18) leads to

$$\mathbf{i} \cdot \mathbf{i} = \mathbf{j} \cdot \mathbf{j} = \mathbf{k} \cdot \mathbf{k} = 1, \tag{3.19}$$

and

$$\mathbf{i} \cdot \mathbf{j} = \mathbf{j} \cdot \mathbf{k} = \mathbf{k} \cdot \mathbf{i} = 0. \tag{3.20}$$

The distributive law holds, i.e., we can write

$$\begin{aligned}(\mathbf{i}A_x + \mathbf{j}A_y + \mathbf{k}A_z) \cdot (\mathbf{i}B_x + \mathbf{j}B_y + \mathbf{k}B_z) = \\ A_x B_x + A_y B_y + A_z B_z\end{aligned} \tag{3.21}$$

in agreement with the definition (3.12). Moreover the commutative law holds, for clearly

$$\mathbf{B} \cdot \mathbf{A} = \mathbf{A} \cdot \mathbf{B}. \tag{3.22}$$

We shall now agree to give physical significance to the scalar product $\mathbf{F} \cdot d\mathbf{r}$, which will denote the cumulative effect of the resultant force **F** acting on a particle as it makes the displacement $d\mathbf{r}$. We shall call this quantity the *work* done by the force **F** while the particle moves under its influence through $d\mathbf{r}$ and write

$$\mathbf{F} \cdot d\mathbf{r} = dW. \tag{3.23}$$

The work dW is a *scalar* quantity. The space-cumulative effect of **F** while the particle is displaced from position $\mathbf{r}_1$ to position $\mathbf{r}_2$ is then

$$\int_{\mathbf{r}_1}^{\mathbf{r}_2} \mathbf{F} \cdot d\mathbf{r} = W. \tag{3.24}$$

It should be noted that we can form this quantity and attach meaning to it even if **F** is *not* the resultant force acting on the particle. If **F** is perpendicular to $d\mathbf{r}$, no work is done by **F**. The maximum work done by **F** for given $d\mathbf{r}$ occurs when $d\mathbf{r}$ is parallel to **F**.

If **F** is the *resultant* force acting on the particle, it is possible to express the work done in a way that depends on the mass and velocity of the particle by substituting for **F** in (3.24) its equivalent $m\ddot{\mathbf{r}}$. Thus, utilizing (3.22) we get

$$\int_{\mathbf{r}_1}^{\mathbf{r}_2} m\ddot{\mathbf{r}} \cdot d\mathbf{r} = \frac{m}{2} \int_{t_1}^{t_2} \frac{d}{dt}(\dot{\mathbf{r}} \cdot \dot{\mathbf{r}})dt. \tag{3.25}$$

Now $\dot{\mathbf{r}} \cdot \dot{\mathbf{r}} = v^2$, where v is the magnitude of the *velocity* of the particle. Consequently (3.24) becomes

$$W = \int_{r_1}^{r_2} \mathbf{F} \cdot d\mathbf{r} = \frac{m{v_2}^2}{2} - \frac{m{v_1}^2}{2}. \tag{3.26}$$

The formal analogy between this and (3.11) is obvious and instructive. The quantity $mv^2/2$ is evidently going to be of some importance in mechanics. We shall call it the *kinetic energy* of the particle. Eq. (3.26) then is a theorem relating the work done by the *resultant* force during a displacement of the particle with the change in kinetic energy experienced by the particle during the displacement. It is therefore usually called the work-kinetic energy theorem. The integrals (3.11) and (3.26) are historically of great interest since they were effectively the basis of the celebrated controversy between Descartes and Leibniz during the late seventeenth and eighteenth centuries. Descartes considered the *impulse* fundamental whereas Leibniz preferred the *work*. D'Alembert later pointed out that both are perfectly legitimate modes of description of the same problem.

3.6. Potential Energy and Total Energy. In the integral in eq. (3.26) let us write the force and displacement in terms of their rectangular components, thus

$$\mathbf{F} = \mathbf{i}F_x + \mathbf{j}F_y + \mathbf{k}F_z, \tag{3.27}$$

$$d\mathbf{r} = \mathbf{i}\,dx + \mathbf{j}\,dy + \mathbf{k}\,dz. \tag{3.28}$$

Eq. (3.26) then becomes

$$\int_{P_1}^{P_2} (F_x\,dx + F_y\,dy + F_z\,dz) = \frac{1}{2}mv_2^{\,2} - \frac{1}{2}mv_1^{\,2}, \tag{3.29}$$

where P_1 and P_2 are the initial and final positions of the particle. In general the force components F_x, F_y, F_z are functions of x, y, z, and the integrand on the left is a linear differential form. Let us suppose it is a perfect differential, i.e., let there exist a function $V(x, y, z)$ such that

$$F_x\,dx + F_y\,dy + F_z\,dz = -dV, \tag{3.30}$$

when, of course,

$$F_x = -\frac{\partial V}{\partial x},\quad F_y = -\frac{\partial V}{\partial y},\quad F_z = -\frac{\partial V}{\partial z}. \tag{3.31}$$

The integrations may then be immediately carried out and (3.29) becomes ultimately

$$V_1 + \frac{1}{2}mv_1^{\,2} = V_2 + \frac{1}{2}mv_2^{\,2}, \tag{3.32}$$

where

$$V_1 = V(x_1, y_1, z_1); \quad V_2 = V(x_2, y_2, z_2) \tag{3.33}$$

are the values of V at the two positions P_1 and P_2, respectively, between which the displacement has taken place. Since these are *any* two arbitrary positions, eq. (3.32) means that during the motion of the particle

$$\frac{1}{2}mv^2 + V(x, y, z) = \text{constant}. \tag{3.34}$$

The constant we shall call the *total energy* of the particle and denote it by U. The function V is named the *potential energy* of the particle. The existence of V guarantees the existence of U, and the result is known as the theorem of the *conservation of mechanical energy*. Note that if the force **F** does not allow the existence of V, eq. (3.34) fails to hold and the concept of energy is strictly meaningless. We shall have occasion in the sequel to discuss this at greater length.

As an illustration of a case in which eq. (3.30) holds, consider

$$F_x = -k_1 x, \quad F_y = -k_2 y, \quad F_z = -k_3 z. \tag{3.35}$$

Clearly

$$V = \frac{1}{2}(k_1 x^2 + k_2 y^2 + k_3 z^2) + \text{constant}, \tag{3.36}$$

where the constant is arbitrary and may conveniently be set equal to zero. This type of force has considerable practical value as will be made clear in the next chapter. Consider, on the other hand, the innocent enough looking variant of eq. (3.35)

$$F_x = -k_1 y, \quad F_y = -k_2 z, \quad F_z = -k_3 x. \tag{3.37}$$

Here it is impossible to find a function V satisfying (3.30) and hence for this type of force we, strictly speaking, are unable to define the total energy. It seems intuitively clear that if the components of a force are taken at random, the chances of the existence of a potential energy are slim. It turns out indeed (cf. Chapter 14) that the components must satisfy the following conditions in order that eq. (3.30) may hold

$$\frac{\partial F_z}{\partial y} - \frac{\partial F_y}{\partial z} = \frac{\partial F_x}{\partial z} - \frac{\partial F_z}{\partial x} = \frac{\partial F_y}{\partial x} - \frac{\partial F_x}{\partial y} = 0. \tag{3.38}$$

The full physical significance of (3.38) will appear later (Sec. 14.7).

Another physical consequence of the considerations of this section results from a closer inspection of (3.29). The left-hand side is of course the work done as the particle moves from P_1 to P_2. The evaluation of the integral depends in general on the path taken. Thus in Fig. 3.1 we shall in most cases find that the integral along the path P_1AP_2 is different from that along the path P_1BP_2. This means that in general the work done by a force during a displacement of a particle depends on the path between the initial and final positions. However, when (3.38) and (3.30) hold

$$\begin{aligned}\int_{P_1}^{P_2} (F_x\,dx + F_y\,dy + F_z\,dz) &= -\int_{P_1}^{P_2} dV \\ &= V(x_1, y_1, z_1) - V(x_2, y_2, z_2).\end{aligned} \tag{3.39}$$

In this case the value of the integral depends only on the end points of the path and not on the path itself. The existence of V is thus a guarantee that the work done by the associated force is independent of the path.

One important further consequence of (3.34) may now be noted. We can solve it for v and have

$$v = \sqrt{\frac{2}{m}(U - V)}. \qquad (3.40)$$

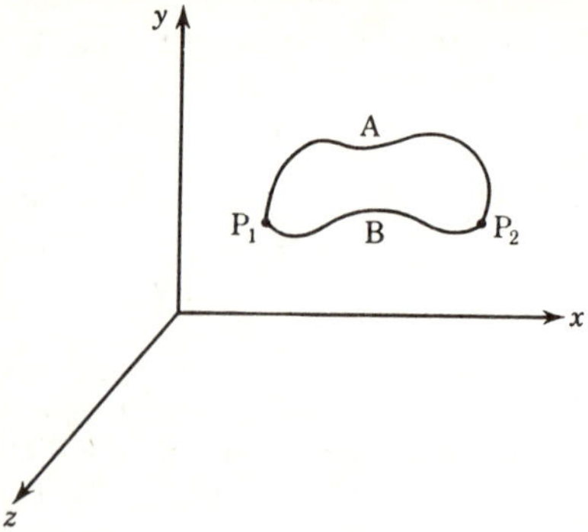

Fig. 3.1.

This provides a connection between the instantaneous speed of the particle and its instantaneous position once U has been assigned by means of the knowledge of v and V for one position. In particular it tells us that actual motion of the particle can take place only for $U - V > 0$. This provides a limitation on the region in which the particle can move if the potential energy function exists. Thus assume that V is a function of x alone and when plotted has the appearance shown in Fig. 3.2. Suppose the total energy U has the value indicated by the dotted line. Then (3.40) shows that the intervals Ox_1 and x_2x_3 are intervals in which motion is possible, whereas motion in x_1x_2 and x_3x_4 is not permitted. At the ends of the allowed motion intervals the speed becomes zero and reaches a maximum at that point of the interval where V is a minimum. It is clear that the potential energy curve can thus shed a good deal of light on the motion of the particle. The question might indeed be raised why it does not provide the whole story, i.e., why energy considerations alone are not adequate to describe completely the motion of a particle. The reader is advised to think this over for himself. The answer will be given in Chapter 6.

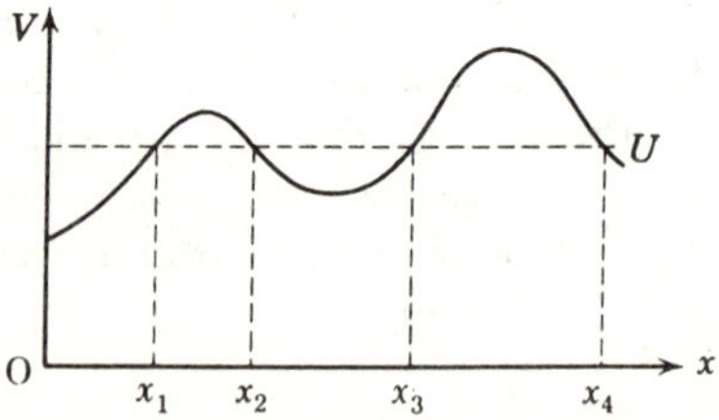

Fig. 3.2.

PROBLEMS

1. Two balls A and B are tied together by an elastic band, then pulled apart and released in a nonresisting medium. With respect to a fixed scale, A is found to have an acceleration of 10 cm/sec^2, and B, -5 cm/sec^2. What is the mass of B relative to A? What is the acceleration of B *relative* to A? B is now replaced by another

ball C and the accelerations become: for C, -10 cm/sec^2, and for A, 8 cm/sec^2. Find the mass of C relative to A and the mass of C relative to B.

2. Show that the two vectors

$$\mathbf{r}_1 = 9\mathbf{i} + \mathbf{j} - 6\mathbf{k}$$
$$\mathbf{r}_2 = 4\mathbf{i} - 6\mathbf{j} + 5\mathbf{k}$$

are perpendicular to each other. Find the area of the rectangle determined by them.

3. A force $\mathbf{F} = kte^{-at}(\mathbf{i} + \mathbf{j})$ acts on a particle. Find the change in momentum produced in 1/10 sec if $k = 100$ dyne/sec and $a = 20$ sec^{-1}. Plot the magnitude of $\mathbf{F}$ as a function of t.

4. The constant force $\mathbf{F} = 5\mathbf{i} + 4\mathbf{j} + 3\mathbf{k}$ (units in dynes) acts on a particle as it moves from

$$\mathbf{r}_1 = 3\mathbf{i} + 4\mathbf{j} + 5\mathbf{k}$$

to

$$\mathbf{r}_2 = 6\mathbf{i} + 7\mathbf{j} + 8\mathbf{k}$$

(where units are in cm). How much work is done by the force?

5. The force $\mathbf{F} = k\mathbf{r}$ acts on a particle, where k is a constant. How much work is done by this force as the particle moves between the two positions given in Prob. 4? Does this work depend on the path followed by the particle? Explain!

6. Does a potential energy function exist for the force in Prob. 5? If so, what is it? Could this be used in *every* case to calculate the change in kinetic energy of the particle in going from one position to another? Explain!

7. Two particles A and B with masses m_A and m_B, respectively, move subject only to their mutual interaction. What is true of their instantaneous velocities? Suppose the particles move along the same straight line. Find the ratio of their instantaneous velocities. Specialize to the case where they are initially at rest.

If A moves in a circle of radius r_A with constant angular velocity, show that B must likewise move in a circle with the same angular velocity. What is the radius of this second circle? Specialize to the lowest energy state in the Bohr model of the hydrogen atom and give numerical values.

8. Two particles B and C with masses m_B and m_C respectively move subject only to their mutual interaction. Find the expression for the acceleration of C *relative* to B in terms of the acceleration of C in the reference frame. Find the relative fractional acceleration of C. What are its maximum and minimum possible values?

What will be the effective relative mass, i.e., the mass of a particle which, moving with the *relative* acceleration, has the same unbalanced force acting on it as acts on B and C themselves?

9. A force $\mathbf{F} = \mathbf{i}F_0 + \mathbf{j}kte^{-at}$ acts on a particle, where F_0 is constant and equals $10/e$ dynes, $a = 1/20\ \text{sec}^{-1}$ and $k = 200$ dynes/sec. Find the momentum produced (direction and magnitude) in 1/10 sec. Plot the magnitude of $\mathbf{F}$ as a function of t.

10. A particle of mass m is acted on by a force $\mathbf{F} = kr^{-3}\mathbf{r}$ where k is a constant. How much work is done by the force when the particle moves in the xy plane in the arc of a circle from the position

$$\mathbf{r}_1 = a\mathbf{i}$$

to the position

$$\mathbf{r}_2 = a\mathbf{j}?$$

Obtain the answer in two different ways.

11. Two particles of masses m_1 and m_2 move along the same straight line subject only to their mutual interaction. Find the coordinate of a point on the line such that if a particle with mass equal to the sum of the two masses were placed there, it would have zero acceleration.

12. A force $F = kte^{-at}$, where k and a are positive constants, acts on a particle of mass m constrained to move in the x direction. If the particle starts from rest, how far will it go in the time interval from zero to t_0 when the force reaches its maximum value? Find the kinetic energy attained in time t_0 and the maximum kinetic energy together with the time at which it is attained. Power being the rate of doing work, find the instantaneous power at time t_0 and the *average* power over the time interval from zero to t_0.

13. Show that the unit vectors

$$\mathbf{r}_1 = \mathbf{i}\sin\theta\cos\phi + \mathbf{j}\sin\theta\sin\phi + \mathbf{k}\cos\theta,$$
$$\theta_1 = \mathbf{i}\cos\theta\cos\phi + \mathbf{j}\cos\theta\sin\phi - \mathbf{k}\sin\theta,$$
$$\phi_1 = -\mathbf{i}\sin\phi + \mathbf{j}\cos\phi$$

are mutually perpendicular. Prove that the time rate of change of a unit vector must always be perpendicular to the unit vector.

14. Let mass be defined in such a way that

$$\mathbf{a}_{AB}/-\mathbf{a}_{BA} = m_{AB},$$

in place of 3.1. This amounts effectively to defining the ratio of the masses of two particles as the negative of the *direct* ratio of their accelerations on interaction. Investigate the consequences of this assumption. What kind of equation of motion will result? If two particles were joined together, how would the masses compound to give the resultant mass?

15. A particle of mass m moves along the x axis in such a way that its potential energy for any value of x is $V_0 \sin kx$, where V_0 and k are positive constants. Find the portion of the x axis to which the motion of the particle is restricted if its total energy is V_0. Find the corresponding result if the total energy is $V_0/2$ or $2V_0$ respectively. Find maximum and minimum velocity values for each case. Could the motion ever be simple harmonic? Discuss the circumstances.

CHAPTER IV

MOTION OF A PARTICLE IN A STRAIGHT LINE. OSCILLATIONS

4.1. Equation of Motion. There are many cases of interest in which a particle is restricted to move along a straight line. The fundamental equation of motion (3.10) then becomes the purely scalar equation

$$m\ddot{x} = F(x, \dot{x}, t). \tag{4.1}$$

A general treatment presents some difficulty, so we specialize at first by assuming that the force is independent of x and $\dot{x}$, and hence a function of t alone. One integration with respect to the time yields

$$m\dot{x} = G_1(t) + C_1, \tag{4.2}$$

where

$$G_1(t) = \int F(t)\, dt, \tag{4.3}$$

and C_1 is the constant of integration. A second integration yields

$$mx = G_2(t) + C_1 t + C_2, \tag{4.4}$$

where

$$G_2(t) = \int G_1(t)\, dt. \tag{4.5}$$

This completes the solution of the equation of motion, in that it provides an expression for the x coordinate of the particle at any time t. It will at once be noted that *two* arbitrary constants are involved. They can be given physical significance in terms of the values of x and $\dot{x}$ for $t = 0$, i.e., the so-called *initial* conditions. Thus from (4.2) we have

$$C_1 = m\dot{x}_0 - G_1(0), \tag{4.6}$$

and from (4.4)

$$C_2 = mx_0 - G_2(0). \tag{4.7}$$

Hence the solution in terms of the initial position and initial velocity takes the form

$$m(x - x_0) = G_2(t) - G_2(0) + t(m\dot{x}_0 - G_1(0)). \tag{4.8}$$

The initial values x_0 and $\dot{x}_0$ provide what may be called the temporal boundary conditions of the problem. Without them it is impossible to state the complete solution of any problem in mechanics.[1]

The special case of $F(t) = \text{constant} = -mg$, for a free particle near the surface of the earth at once gives

$$G_1(t) = -mgt; \qquad G_2(t) = -\frac{mgt^2}{2}, \tag{4.9}$$

and (4.8) takes the familiar form

$$x - x_0 = -\frac{gt^2}{2} + \dot{x}_0 t. \tag{4.10}$$

(Note that x is measured positively upward, whereas g is a downward acceleration.) Any uniformly accelerated motion in a straight line can be handled similarly. For an *impulsive* force, cf. Prob. 3 in Chapter 3.

In the next case of importance the force F is a function of x alone; thus

$$m\ddot{x} = F(x). \tag{4.11}$$

Here it is always possible to carry out one integration by multiplying both sides of the equation by $\dot{x}dt = dx$. The result is

$$\frac{1}{2}m\dot{x}^2 = \int F(x)\, dx + c_1, \tag{4.12}$$

where the constant of integration is now represented by c_1. The instantaneous velocity is

$$\dot{x} = \pm\sqrt{\frac{2}{m}\{\int F(x)\, dx + c_1\}}. \tag{4.13}$$

Eq. 4.12 is called a first integral of the equation of motion, or more commonly the energy integral. Note that (4.12) is a special case of (3.26). In particular if we carry out the integration between definite limits x_1 and x_2, the constant c_1 disappears and we have

$$\int_{x_1}^{x_2} F(x)\, dx = \frac{1}{2}m\dot{x}_2^{\,2} - \frac{1}{2}m\dot{x}_1^{\,2}. \tag{4.14}$$

---

[1]For a detailed methodological discussion of the significance of boundary conditions in physics see Lindsay and Margenau, *Foundations of Physics*, pp. 48ff.

The fundamental theorem of integral calculus assures us that for all decently behaved functions such as we should be likely to use for forces in mechanics, the indefinite integral in (4.12) exists and we may write

$$\int F(x)\,dx = -V(x), \tag{4.15}$$

so that (4.14) becomes

$$\frac{1}{2}m\dot{x}_1^{\,2} + V(x_1) = \frac{1}{2}m\dot{x}_2^{\,2} + V(x_2). \tag{4.16}$$

$V(x)$ is the potential energy function for this special case.

The evaluation of x as a function of t demands a second integration of (4.13): Thus

$$\int \frac{dx}{\sqrt{\frac{2}{m}\left(c_1 - V(x)\right)}} = \pm \int dt + c_2. \tag{4.17}$$

In principle therefore all straight-line motions in which the force is of the form $F(x)$ can be determined from the equation of motion. Actually the integration in (4.17) may be impossible in terms of simple functions in closed form; series or numerical integration may have to be employed in special cases.

Much information about the motion can often be gathered from the energy eq. (4.16) in the form

$$\dot{x} = \pm\sqrt{\frac{2}{m}\left(U - V(x)\right)}\,. \tag{4.18}$$

By plotting $V(x)$, one can readily see to what part of the x axis the motion is confined, where the velocity has a maximum and where a minimum. This has already been discussed at the end of Sec. 3.6.

4.2. The Simple Harmonic Oscillator. Probably the most important straight-line motion of a particle after the freely falling body is *simple harmonic motion.* Here in (4.11) $F(x) = -kx$, where k is a positive constant. This is exemplified by the motion of a weight on the end of a perfectly elastic spring, the bob of a simple pendulum swinging through a very small arc, atoms in a crystal lattice, the nuclei of atoms in molecules, and many other particles of significance in pure and applied physics.

The energy equation for this type of motion is

$$\frac{1}{2}m\dot{x}^2 + \frac{1}{2}kx^2 = U. \tag{4.19}$$

The constant k is usually known as the *stiffness* or equivalent stiffness of the system under study, and U, as usual, is the total energy. If we follow the suggestion at the end of Sec. 4.1 and plot the potential energy $V(x) = \frac{1}{2}kx^2$ as a function of x we obtain the parabola indicated in Fig. 4.1. If now we further draw the horizontal line denoting the total energy U, we see that the motion is confined to the region on the x axis from $x = -\sqrt{\frac{2U}{k}}$ to $x = +\sqrt{\frac{2U}{k}}$. We call $\sqrt{\frac{2U}{k}}$, which is the largest excursion the particle makes on either side of the origin, the amplitude A of the motion. At $x = \pm A$, the velocity is zero. At $x = 0$, the velocity is a maximum and equal to $\pm\sqrt{\frac{2U}{m}}$.

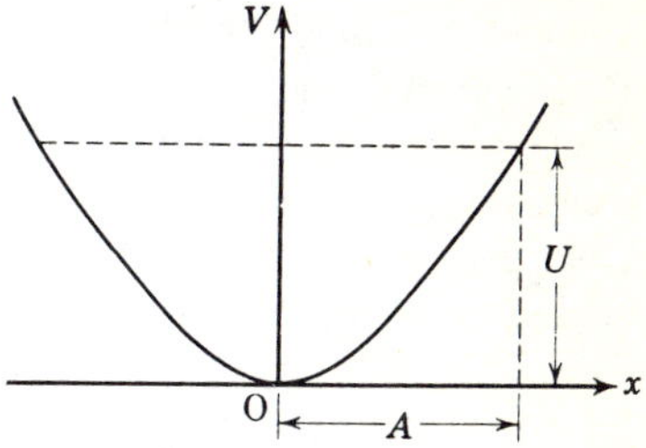

Fig. 4.1.

To study the motion in greater detail it is necessary to integrate the first-order equation (4.19). This finally yields

$$x = \sqrt{\frac{2U}{k}} \cos\left(\sqrt{\frac{k}{m}}t + \alpha\right), \tag{4.20}$$

where α is the second constant of integration, the total energy U appearing as the first. The motion is oscillatory with frequency

$$\nu = \frac{1}{2\pi}\sqrt{\frac{k}{m}}, \tag{4.21}$$

and with period

$$P = 2\pi\sqrt{\frac{m}{k}} = \frac{1}{\nu}. \tag{4.22}$$

The argument of the cosine in (4.20) is the so-called phase and α is the initial phase or *epoch*. Both amplitude and epoch can be expressed in terms of the initial displacement x_0 and initial velocity $\dot{x}_0$, and it may be shown that

$$A = \sqrt{\frac{2U}{k}} = \sqrt{x_0^2 + \frac{\dot{x}_0^2}{\omega^2}}, \tag{4.23}$$

$$\alpha = \arc\tan\,(-\dot{x}_0/\omega x_0), \tag{4.24}$$

where $\omega = 2\pi\nu$ is often used for simplicity and is called the *angular frequency*. Of course the sine might be used in (4.20) in place of the cosine, although in that case α would be different.

4.3. Damped Oscillations. In most practical cases of simple harmonic oscillations the motion, once started, does not persist forever, as eq. (4.20) would indicate. Rather the amplitude is observed to die down in the course of time. This has been ascribed in general to resistance which the medium offers to the motion, although in certain cases the explanation may have to be more complicated. Fortunately it proves possible to describe the effect, called generally *dissipation*, by relatively simple mathematical means through the introduction of a force which varies directly as the instantaneous velocity and always opposes the motion. In terms of the general treatment of Sec. 4.1, the force function to which $m\ddot{x}$ must be equated now becomes

$$F(x, \dot{x}) = -kx - R\dot{x}, \tag{4.25}$$

where R is termed the *damping* or *dissipation* coefficient (sometimes the *resistance factor*). The equation of motion then becomes

$$m\ddot{x} + R\dot{x} + kx = 0. \tag{4.26}$$

Inspection shows that it is now impossible to secure a first integral in the form of the energy equation (4.19). The closest we can come to it is

$$R\dot{x}^2 = -\dot{U}, \tag{4.27}$$

where U is given by (4.19). Strictly speaking there is no longer an energy because there no longer exists a potential energy function. However, in many practical problems the rate of decrease in the amplitude of motion is small and it becomes convenient to introduce the fiction of an energy which decreases with the time at the rate indicated in (4.27).

The simplest method of solution of (4.26) (a linear differential equation with constant coefficients) is to employ the substitution

$$x = Ae^{\theta t}. \tag{4.28}$$

If this is to serve as a solution for nonvanishing A we must have

$$\theta = -\frac{R}{2m} \pm \sqrt{\frac{R^2}{4m^2} - \frac{k}{m}}. \tag{4.29}$$

It is clear that if $R^2/4m^2 > k/m$, θ is real and no oscillations result. However, if $k/m > R^2/4m^2$, θ is complex and oscillations take place. We shall be concerned here entirely with this case (cf. the problems at the end of the chapter for the former case). Let us put

$$\sqrt{\frac{k}{m} - \frac{R^2}{4m^2}} = \omega. \tag{4.30}$$

Then the solution becomes

$$x = e^{-Rt/2m} \cdot [A' e^{i\omega t} + B' e^{-i\omega t}], \tag{4.31}$$

where A' and B' are the arbitrary constants. By employing the usual expansion for the imaginary exponential, the real part of (4.31) can be expressed in the form

$$x = e^{-Rt/2m} \cdot [A'' \cos \omega t + B'' \sin \omega t], \tag{4.32}$$

where A'' and B'' are new arbitrary constants. That (4.32) is a solution can readily be verified by direct substitution into (4.26). It can also be shown that the imaginary part of (4.31) is likewise a solution and that its physical meaning is not essentially different from (4.32). In general we shall employ the real part. For many purposes the complex form (4.31) is very convenient, particularly in differentiating and integrating. We must be careful, however, to revert to the real form before carrying out multiplications. A somewhat more useful form than (4.32) is

$$x = Ae^{-Rt/2m} \cdot \cos (\omega t + \alpha), \tag{4.33}$$

which reminds us of the undamped case (4.20). The epoch is again α, but A is now the *initial* amplitude.

In time $2m/R$, the amplitude is reduced in the ratio $1/e$. This quantity is known as the *decay modulus*. More commonly used is the *logarithmic decrement*, which is the logarithm of the ratio of two successive amplitudes. The reader should show that the decrement δ is given by

$$\delta = \frac{RP}{2m}. \tag{4.34}$$

The following relations are readily established for A and α in terms of the initial displacement x_0 and initial velocity $\dot{x}_0$,

$$A = \sqrt{x_0^2 + (\dot{x}_0 + Rx_0/2m)^2/\omega^2}, \tag{4.35}$$

$$\alpha = \text{arc tan} \left[-\frac{\dot{x}_0 + Rx_0/2m}{\omega x_0} \right]. \tag{4.36}$$

The reader will find it illuminating to plot x in eq. (4.33) as a function of t for some appropriately chosen values of x_0, $\dot{x}_0$, m, R, and ω.

Let us now revert to eq. (4.27) and introduce some energy considerations. Using x in eq. (4.33) we have for the velocity

$$\dot{x} = -\frac{AR}{2m} e^{-\frac{Rt}{2m}} \cdot \cos(\omega t + \alpha) - A\omega e^{-\frac{Rt}{2m}} \cdot \sin(\omega t + \alpha) \tag{4.37}$$

whence

$$\dot{U} = -R\dot{x}^2 = -RA^2 e^{-\frac{Rt}{m}} \cdot \left[\frac{R^2}{4m^2}\cos^2(\omega t + \alpha) + \omega^2 \sin^2(\omega t + \alpha) + \frac{R}{m}\omega \cdot \cos(\omega t + \alpha) \cdot \sin(\omega t + \alpha)\right]. \tag{4.38}$$

It is thus clear that the rate of energy dissipation varies periodically with the time in a rather complicated fashion. To secure a practical measure of $\dot{U}$ it is advisable to average it over the time. Let τ be a time considerably greater than a single period P of the motion. The average of $\dot{U}$ over the interval from 0 to τ (denoted by the bar over $\dot{U}$) is

$$\bar{\dot{U}} = -\frac{1}{\tau}\int_0^\tau R\dot{x}^2\, dt. \tag{4.39}$$

Carrying out the integration with the use of (4.38) discloses that as τ becomes indefinitely great, $\dot{U} \to 0$, which is hardly helpful although not surprising in view of the presence of the factor $e^{-Rt/m}$ in the integrand. However, if we assume that the damping is small so that $R/2m \ll \omega$, we can arrange the terms in the integral in decreasing orders of magnitude. If we retain only the terms of highest magnitude, we get

$$\bar{\dot{U}} = -\frac{RA^2\omega^2}{2} \cdot \frac{m}{R\tau}(1 - e^{-R\tau/m}). \tag{4.40}$$

Of course this still vanishes as τ grows indefinitely great. However, let us form the average of the square of the amplitude of the displacement x_{max} over the interval from 0 to τ. The result is

$$\overline{x^2_{max}} = \frac{1}{\tau}\int_0^\tau A^2 e^{-\frac{Rt}{m}}\, dt = \frac{mA^2}{R\tau}(1 - e^{-\frac{R\tau}{m}}). \tag{4.41}$$

Consequently we may write (4.40) in the form

$$\bar{\dot{U}} = -\frac{R\omega^2}{2} \cdot \overline{x_{max}^2}, \qquad (4.42)$$

i.e., the average rate of dissipation of energy over the interval from 0 to τ is proportional to the mean square amplitude over this same interval. The dependence on the square of the frequency is also an important characteristic.

4.4. Forced Oscillations of a Dissipative System. If, in spite of the dissipation, an oscillator is to be kept oscillating continuously with constant amplitude, it is essential that the energy loss be made good from an external source. This means the application of an external force to do work on the system. The resulting oscillations are therefore called *forced oscillations*. The force function to which $m\ddot{x}$ must be equated now is the sum of three terms, viz., the restoring force $-kx$, the damping force $-R\dot{x}$, and the maintaining force, which we shall assume to be oscillatory and of the form $F_0 \cos \omega_0 t$. It is mathematically more convenient to write the latter in the complex form $F_0 e^{i\omega_0 t}$. The equation of motion of the system then becomes

$$m\ddot{x} + R\dot{x} + kx = F_0 e^{i\omega_0 t}. \qquad (4.43)$$

The complete solution consists of the sum of the complementary function (4.33) and the particular integral which, as will be seen, represents the *steady state*. We have already discussed the former sufficiently and can now concentrate on the latter. From an inspection of (4.43), a solution of the form

$$x = Be^{i\omega_0 t} \qquad (4.44)$$

is indicated. Substitution checks this, provided B has the form

$$B = \frac{F_0}{i\omega_0 R + k - m\omega_0{}^2}. \qquad (4.45)$$

For many purposes the velocity $\dot{x} = i\omega_0 Be^{i\omega_0 t}$ is more valuable than the displacement. The real parts of x and $\dot{x}$ turn out to be

$$x_{real} = \frac{F_0 \cos (\omega_0 t - \beta)}{\sqrt{\omega_0{}^2 R^2 + (k - m\omega_0{}^2)^2}}, \qquad (4.46)$$

and

$$\dot{x}_{real} = \frac{F_0 \cos(\omega_0 t - \gamma)}{\sqrt{R^2 + (m\omega_0 - k/\omega_0)^2}}, \tag{4.47}$$

where the angles β and γ are phase angles between force and displacement and force and velocity, respectively, given as follows

$$\beta = \text{arc tan}\ \frac{\omega_0 R}{k - m\omega_0{}^2}, \tag{4.48}$$

$$\gamma = \text{arc tan}\ \frac{m\omega_0 - k/\omega_0}{R}. \tag{4.49}$$

Clearly x_{real} and $\dot{x}_{real}$ differ in phase by $\pi/2$. Examination discloses that the amplitude of x_{real} becomes a maximum for

$$\omega_0 = \sqrt{k/m - R^2/2m^2}. \tag{4.50}$$

On the other hand, the amplitude of $\dot{x}_{real}$ is a maximum for

$$\omega_0 = \sqrt{\frac{k}{m}} = \omega_{res}. \tag{4.51}$$

We shall agree to call

$$\nu_{res} = \frac{1}{2\pi}\sqrt{k/m} \tag{4.52}$$

the *resonance frequency* of the oscillating system. Note that the resonance frequency is not strictly the frequency for maximum amplitude, although if the damping is small so that $R^2/2m^2 \ll k/m$, the difference may be negligible. The resonance frequency is considered the more fundamental. For resonance $\gamma = 0$ and $\beta = \pi/2$. The physical significance of this should be noted. Anyone who has ever pushed a swing realizes that he provides maximum velocity amplitude when the maximum force coincides with maximum velocity.

It is of value at this point to remark on the analogy between mechanical oscillations and those of an electrical circuit containing capacity C, inductance L, and resistance R in series, for which the fundamental equation (if the circuit has an impressed emf $E_0 \cos \omega_0 t$) is

$$L\ddot{q} + R\dot{q} + q/C = E_0 \cos \omega_0 t, \tag{4.53}$$

where q is the instantaneous charge on the condenser. The solution yielding the current, usually the quantity of greater significance, is

$$I = \dot{q} = \frac{E_0 \cos(\omega_0 t - \gamma)}{\sqrt{R^2 + \left(L\omega_0 - \dfrac{1}{C\omega_0}\right)^2}} \tag{4.54}$$

in analogy with (4.47), where γ is defined analogously to the mechanical phase angle in (4.49). Since it is customary to refer to $L\omega_0 - \dfrac{1}{C\omega_0}$ as the *electrical reactance,* we may call

$$m\omega_0 - k/\omega_0$$

the *mechanical reactance*. The complex quantity

$$R + i\left(m\omega_0 - \frac{k}{\omega_0}\right) = Z = Z_1 + iZ_2 \tag{4.55}$$

is called the *mechanical impedance*, with a real part which will be named the *mechanical resistance* and the imaginary part, the mechanical reactance as just defined. Evidently the velocity amplitude (cf. eq. 4.47) depends on the absolute value of Z. This so-called impedance notation has proved rather useful in the study of oscillations of all kinds. Electrical engineers and others familiar with electric oscillating circuits have developed the tendency to describe oscillations of all kinds in terms of the impedances of analogous electric circuits. This has certain obvious advantages but we must never forget that the analogy in question is purely mathematical.

Since the force maintains the oscillations at constant amplitude, it is clear that on the average the force must do enough work on the system to compensate for the loss of energy through dissipation. Now the average rate of dissipation computed by averaging $R\dot{x}^2$ over the time turns out to be

$$\overline{R\dot{x}^2} = \frac{F_0^{\,2}}{2R} \cos^2 \gamma. \tag{4.56}$$

On the other hand, the average rate at which energy is supplied to the system is

$$\overline{\dot{x}F \cos \omega_0 t} = \frac{F_0^{\,2}}{|Z|} \overline{\cos(\omega_0 t - \gamma) \cos \omega_0 t}$$

$$= \frac{F_0^{\,2}}{2R} \cos^2 \gamma, \tag{4.57}$$

the two quantities turning out to be the same, in accordance with the physical requirements of the situation. It is of interest to note that the maximum average rate of energy supply and dissipation occurs for resonance. It is instructive to plot the average rate of energy dissipation as a function of

$$\frac{\omega_{res}}{\omega_0} - \frac{\omega_0}{\omega_{res}} = z,$$

and to observe that, other things being equal, the resonance peak is high and sharp for small R, whereas it is low and broad for large R. Important illustrations of the forced oscillations of a dissipative system are found in sound sources and resonators as well as in the theory of the dispersion of light in material media (cf. Probs. 12, 13, 14, 15 at end of chapter. Also see Sec. 17.7).

4.5. The Anharmonic Oscillator. The simple restoring force proportional to the displacement and leading to the oscillatory motion discussed in the immediately preceding paragraphs is not always realized in experience. Often we encounter a restoring force which is not symmetrical about the position of equilibrium: the eardrum is a good illustration of this. The force in question can be written in this case to a good approximation as $-k_1x - k_2x^2$ where k_1 and k_2 are positive constants, and the equation of undamped motion appears as

$$m\ddot{x} + k_1x + k_2x^2 = 0. \tag{4.58}$$

This contains the square of the dependent variable and is therefore a *nonlinear* equation which demands a different method of solution from that employed in Sec. 4.2. A scheme which is often valuable in such cases involves the use of successive approximations. We suppose x is expanded in a series of terms involving an arbitrary parameter ϵ as follows:

$$x = \epsilon f_1(t) + \epsilon^2 f_2(t) + \epsilon^3 f_3(t) + \cdots. \tag{4.59}$$

Direct substitution into (4.58) yields

$$\begin{aligned} m[\epsilon\ddot{f}_1 + \epsilon^2\ddot{f}_2 + \epsilon^3\ddot{f}_3 + \cdots] = &-k_1[\epsilon f_1 + \epsilon^2 f_2 + \epsilon^3 f_3 + \cdots] \\ &-k_2[\epsilon^2 f_1^2 + \epsilon^4 f_2^2 + 2\epsilon^3 f_1 f_2 + \cdots]. \end{aligned} \tag{4.60}$$

Since ϵ is arbitrary, if the above is an identity the coefficients of each power of ϵ must be identically equal to zero. Hence we have

$$m\ddot{f}_1 = -k_1 f_1, \tag{4.61}$$

$$m\ddot{f}_2 = -k_1 f_2 - k_2 f_1^2, \tag{4.62}$$
$$m\ddot{f}_3 = -k_1 f_3 - 2k_2 f_1 f_2, \tag{4.63}$$
$$\cdots\cdots\cdots\cdots\cdots$$

for the evaluation of $f_1, f_2, f_3, \ldots$ successively. Clearly the solution of (4.61) is in the form

$$f_1 = A \cos(\omega t - \alpha) \tag{4.64}$$

with $\omega = \sqrt{k_1/m}$. This is then the first approximation to the solution of (4.58). Let us now make the assumption that at $t = 0$, $x = x_0$, and $\dot{x} = \dot{x}_0$: The problem then is to solve the successive equations (4.61), (4.62), ... subject to this condition. We shall carry the approximation only as far as the term in f_2, i.e., the second approximation. Direct substitution into (4.62) yields

$$m\ddot{f}_2 = -k_1 f_2 - k_2 A^2 \cos^2(\omega t - \alpha). \tag{4.65}$$

Now by the use of a standard trigonometric identity this becomes

$$m\ddot{f}_2 = -k_1 f_2 - \frac{k_2 A^2}{2}[1 + \cos 2(\omega t - \alpha)]. \tag{4.66}$$

We try the solution

$$f_2 = B_1 \cos 2(\omega t - \alpha) + C_1. \tag{4.67}$$

Substitution shows that this satisfies eq. (4.66) with

$$B_1 = k_2 A^2/6k_1,$$
$$C_1 = -k_2 A^2/2k_1. \tag{4.68}$$

To the approximation being considered we have

$$x = \epsilon A \cos(\omega t - \alpha) + \frac{k_2 A^2 \epsilon^2}{6k_1} \cos 2(\omega t - \alpha) - \frac{k_2 A^2 \epsilon^2}{2k_1}. \tag{4.69}$$

There are two constants, viz., ϵA and α at our disposal. These can be evaluated in terms of the initial conditions, a task left for the reader to perform. The important thing to note about (4.69) is that the solution to this approximation involves the first *harmonic* of the natural frequency of the original oscillator, for which $k_2 = 0$. Obviously the validity of the approximation increases as the ratio k_2/k_1 decreases. If the oscillator is badly anharmonic, further successive approximations are necessary or possibly an entirely new method of solution.

4.6. Relaxation Time Phenomena. An important concept in theoretical physics is illustrated by a damped oscillator in which the mass is so small that the inertia term $m\ddot{x}$ is negligible compared with the elastic restoring force and the dissipative force. Imagine a very light weight on the end of a stiff spring pulled down and let go in a viscous medium like molasses. The equation of motion becomes effectively

$$R\dot{x} + kx = 0, \tag{4.70}$$

with the solution

$$x = x_0 e^{-k/R \cdot t}. \tag{4.71}$$

The initial condition that $x = x_0$ for $t = 0$ has already been imposed. It is seen that the displacement in this case approaches zero exponentially. When $t = R/k$, $x = x_0/e$ and it is said that the system has "relaxed" to $1/e$ of its original displacement. Consequently R/k is called the "relaxation time." Large damping and small stiffness correspond to a long relaxation time and conversely. We can of course still speak of a relaxation time when the mass is not negligible but still $R^2/4m^2 \geqq k/m$ so that the system is over damped or critically damped. The problem and result are more complicated and the work is left as an exercise.

The notion of a relaxation time is important in many physical phenomena. As another illustration consider the stretching of a wire by a weight suspended at its free end. In elementary physics this problem is handled by the application of Hooke's law, as has already been mentioned in Sec. 1.2, i.e., with each stretching force a definite strain is associated. However this takes no account of the *process* of stretching in which the length starts with its initial value and *in time* attains the final value corresponding to the attached weight. In other words, the process of stretching is not instantaneous and a more careful description of it must take this into account. The equation

$$F = kx, \tag{4.72}$$

where k is the stiffness, x the stretch, and F the *constant* applied force, must now be replaced by

$$F = kx + R\dot{x}, \tag{4.73}$$

where $R\dot{x}$ represents the simplest corrective term (R is a positive constant) to introduce dynamics into the problem. This amounts to saying that the constant applied force must give some velocity to the end of the wire at which it is applied. It might be argued that the force must also provide acceleration and hence a term $m\ddot{x}$ should also be included. We shall neglect this, however, by assuming k and R large enough to outweigh the influence of the mass. This can certainly be realized readily enough in the process of stretching a wire.

The solution of (4.73) is seen to be:

$$x = F/k \cdot (1 - e^{-kt/R}), \tag{4.74}$$

if the initial stretch is zero. Physically (4.74) says that x does not immediately attain the equilibrium value (given by the simple Hooke's law relation (4.72)) but approaches it exponentially and indeed reaches it only after infinite time. However after time

$$\tau = R/k, \tag{4.75}$$

x has reached to within $1/e$ of F/k. We therefore again term R/k the relaxation time of the stretching process.

The reader who recalls his electricity will at once observe that the process of charging a condenser of capacity C by means of a constant electromotive force E through a wire of resistance R and negligible inductance is mathematically describable by an equation of the form (4.73) and hence should yield a relaxation time. The relevant equation is

$$E = q/C + R\dot{q}, \tag{4.76}$$

where we have denoted the charge on the condenser at the instant t by q and the instantaneous current by $\dot{q}$, of course. The solution with $q = 0$ at $t = 0$ is, analogously to (4.74),

$$q = EC(1 - e^{-t/RC}). \tag{4.77}$$

Here the relaxation time is

$$\tau = RC, \tag{4.78}$$

sometimes referred to also as the "time constant" of the circuit. If the resistance and capacity are large it takes a long time for the charge to reach within $1/e$ of its equilibrium steady state value EC.

Obviously every physical phenomenon in which there is a time lag in the attainment of the effect of a given cause will show a relaxation time. Since this is true for most physical phenomena when observed with care, the notion of a relaxation time is of widespread importance. It turns up in such widely diverse fields as magnetic hysteresis, absorption of acoustic radiation in material media, and the collision of molecules in the kinetic theory of gases.

4.7. Relaxation Oscillations. Another illustration of the relaxation time concept which involves oscillations as well as the idea of non-linearity already mentioned in Sec. 4.5 is provided by the following considerations.

In Sec. 4.4 in discussing the forced oscillations of a dissipative system we supposed the force to be periodic and quite independent of the system – an external source of energy, in fact, necessary to maintain the vibrations and uncontrolled by the system itself. There exist, however, important cases of vibrations in which the motion of the system produces or controls the force which sustains the oscillations. These are usually termed *self-excited* oscillations. A mechanical illustration is found in the oscillatory motion of the piston in the cylinder of a reciprocating steam or internal combustion engine: the force sustaining the motion is of course supplied by the expanding steam or burning vapor, but the oscillations themselves are controlled by the working of the valves which are a part of the moving system. A similar situation is exemplified electrically in a vacuum tube circuit in which the energy is supplied by a nonperiodic source, viz., the B battery or d.c. power supply, while the oscillations are excited by the vacuum tube itself through the agency of the grid. The latter acts as a kind of *negative* damping which serves to feed energy into the circuit instead of dissipating it.

The concept of negative damping is useful in the study of self-excited oscillations. It would appear to possess the obvious limitation of leading to infinite amplitude of vibration (cf. eq. 4.33 with R negative). That this actually never happens in practical cases means that the damping must be considered negative only for small amplitudes and then becomes positive again for sufficiently large amplitudes. This will lead to a kind of stabilization of the amplitude and a settling down of the vibration into a steady state.

Lord Rayleigh[1] visualized a simple example of such a "negative-positive damping" oscillation as represented by an equation of the form

$$m\ddot{x} - R_1\dot{x} + R_2\dot{x}^3 + kx = 0, \tag{4.79}$$

where R_1 and R_2 are both positive constants. Without the presence of R_2, the system would have negative dissipation. Even if R_2 is small, $\dot{x}$ will eventually reach such a figure that the resulting damping becomes positive. This happens indeed for

[1]*Theory of Sound* (New York, Dover, 1945), Sec. 68a.

$$\dot{x} \geqq \sqrt{R_1/R_2}. \tag{4.80}$$

For $\dot{x}$ less than this figure, the damping is negative. Calling $k/m = \omega_0{}^2$, an approximate solution of (4.79) for $R_1/m \ll 1$ and $R_2/m \ll 1$ may be obtained in the form

$$x = A \sin \omega_0 t + R_2/m \cdot \omega_0 A^3/32 \cdot \cos 3\omega_0 t, \tag{4.81}$$

where

$$A = 2/\sqrt{3}\,\omega_0 \cdot \sqrt{R_1/R_2}. \tag{4.82}$$

In spite of the fact that the damping is originally negative, the motion builds up to a steady state characterized by A. Obviously if R_1 and R_2 were to differ in sign, no steady motion would be possible. In particular we see at once from (4.79) that if R_2 were negative and R_1 positive, the value of x would increase without limit, since the greater the velocity the greater would be the effective force driving the system with mass m and stiffness k. On the other hand, if R_1 were negative and R_2 positive, any motion initially given to the system would be damped to zero ultimately.

If we do not wish to restrict the discussion of (4.79) to the case of very small values of R_1 and R_2 we find it advisable to transform the coordinates. Thus let

$$\begin{aligned} t' &= \omega_0 t, \\ x' &= \dot{x}\sqrt{3R_2/R_1}, \\ \epsilon &= R_1/\sqrt{km}. \end{aligned} \tag{4.83}$$

We then have from the first two equations (note that $\dot{x} = dx/dt$)

$$\begin{aligned} \frac{dx}{dt'} &= \frac{\dot{x}}{\omega_0} = \frac{1}{\omega_0}\sqrt{R_1/3R_2} \cdot x', \\ x &= \frac{1}{\omega_0}\sqrt{R_1/3R_2} \cdot \int x'\,dt', \\ \dot{x} &= \sqrt{R_1/3R_2} \cdot x', \\ \ddot{x} &= \omega_0\sqrt{R_1/3R_2} \cdot \frac{dx'}{dt'}. \end{aligned} \tag{4.84}$$

Hence substitution into (4.79) gives

$$m\omega_0 \frac{dx'}{dt'} - R_1 x' + R_1/3 \cdot x'^3 + k/\omega_0 \cdot \int x' dt' = 0. \tag{4.85}$$

Differentiating with respect to t' now yields

$$m\omega_0 \frac{d^2x'}{dt'^2} - R_1 \frac{dx'}{dt'} + R_1 x'^2 \frac{dx'}{dt'} + \frac{kx'}{\omega_0} = 0. \tag{4.86}$$

Dividing through by $m\omega_0$ we finally secure

$$\frac{d^2x'}{dt'^2} - \epsilon(1 - x'^2)\frac{dx'}{dt'} + x' = 0. \tag{4.87}$$

This is known as Van der Pol's equation.[1] It is of course nonlinear. Note that ϵ is a dimensionless quantity. Let us first suppose it is small compared with unity. We may then hope to find an approximation to the solution by setting

$$x' = f(t') \cos t', \tag{4.88}$$

where $f(t')$ is a slowly varying function of t'. In the limiting case of $\epsilon = 0$, it reduces to a constant. If we carry out the indicated operations and substitute into (4.79), neglecting d^2f/dt'^2 compared with the other terms, $\cos 3t'$ compared with $3 \cos t'$, etc., the final result is that f must satisfy the equation

$$-2\frac{df}{dt'} + \epsilon f - \epsilon/4 \cdot f^3 = 0. \tag{4.89}$$

This can be put into the simpler form

$$\frac{d(f^2)}{dt'} - \epsilon(f^2 - f^4/4) = 0, \tag{4.90}$$

which is readily solved with result

$$f^2 = \frac{4}{1 - \dfrac{1}{K} \cdot e^{-\epsilon t'}}, \tag{4.91}$$

where $K = f_0^{\,2}/(f_0^{\,2} - 4)$, f_0 being the value of f for $t' = 0$. An approximate solution of (4.87) for $\epsilon \ll 1$ then becomes

[1] B. Van der Pol, *Phil. Mag.* 2, 978, 1926.

$$x' = \frac{2 \cos t'}{\sqrt{1 - \frac{1}{K} \cdot e^{-\epsilon t'}}}. \tag{4.92}$$

Reverting to x and t, this gives

$$\dot{x} = \frac{2\sqrt{R_1/3R_2} \cdot \cos \omega_0 t}{\sqrt{1 - \frac{1}{K} \cdot e^{-\epsilon\omega_0 t}}}. \tag{4.93}$$

We shall not try to find x, but merely note that $\dot{x}$ is a harmonic function of the time with an amplitude changing with the time. If the initial velocity

$$\dot{x}_0 < 2\sqrt{R_1/3R_2},$$

the amplitude of $\dot{x}$ increases with the time to a maximum steady state value of $2\sqrt{R_1/3R_2}$. If the initial value is greater than $2\sqrt{R_1/3R_2}$, the amplitude of $\dot{x}$ decreases with time to this as its steady state. The behavior of the system again shows relaxation.

Let us now go to the other extreme and suppose that $\epsilon \gg 1$. It is quite clear that the assumption of a harmonic solution even as a first approximation will not do. Although the motion corresponding to eq. (4.87) in this case will still be presumably periodic (due to the negative damping at small amplitudes and the positive damping at larger amplitudes), it will clearly possess many harmonics and the frequency will be different from $\omega_0/2\pi$.

The solution can best be studied graphically. Let us put $\frac{dx'}{dt'} = v'$, whence (4.87) becomes

$$\frac{dv'}{dx'} = \epsilon(1 - x'^2) - x'/v'. \tag{4.94}$$

Axes of x' (abscissa) and v' (ordinate) are now set up, and a value of ϵ chosen (say $\epsilon = 10$, as in Fig. 4.2). Isoclines, or curves of constant dv'/dx', are then plotted for a variety of constant slopes. An integral curve can be found by starting at any arbitrary point in the x', v' plane and passing from point to point by going along segments having the correct slope for that region. It is found that no matter where one starts the integral curve finally settles down to a repeating one. Fig. 4.2 shows the resulting x', v' curve for $\epsilon = 10$. It will be noted from (4.94) that for $\epsilon = 0$ (simple harmonic solution of 4.87) the integral curve in the $x'v'$ plane is a circle. The deviation between this and the closed curve for

$\epsilon = 10$ is an indication of the departure from harmonicity. From Fig. 4.2 it is not difficult to construct x' as a function of t'. In the present case the result is shown in Fig. 4.3.

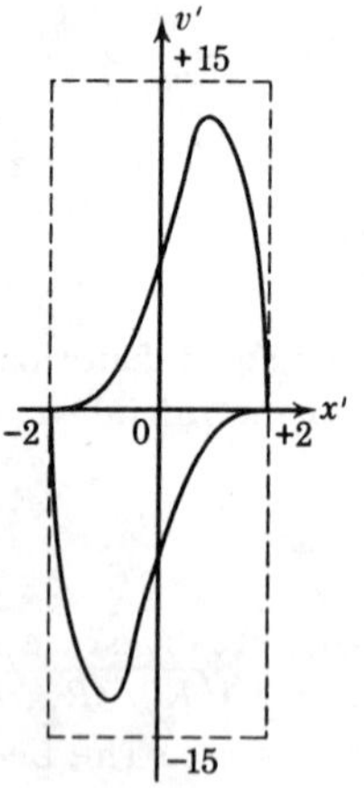

Fig. 4.2.

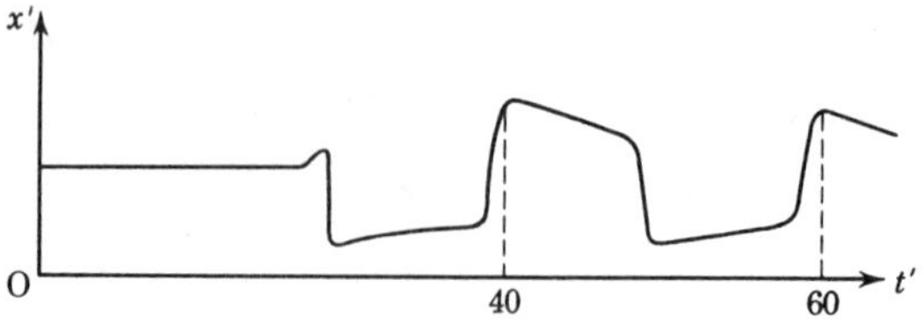

Fig. 4.3.

The interesting fact, which more careful investigation establishes in general, is that the period of x' as a function of t' is very approximately 2ϵ. Hence the period in terms of t, from the first equation of (4.83), is

$$2\epsilon/\omega_0 = 2R_1/k. \tag{4.95}$$

But R_1/k (from 4.71) is just the relaxation time of a system with positive damping corresponding to R_1 and with stiffness K. For this reason, Van der Pol named these oscillations "relaxation" oscillations. In his paper to which reference has been made above it is shown that the standard multivibrator vacuum tube circuit produces oscillations of this kind because its behavior is describable in terms of an equation of the form (4.87). A bibliography of references to relaxation oscillations and other nonlinear phenomena is found in L. A. Pipes, *Applied Mathematics for Engineers and Physicists* (New York, McGraw-Hill, 1946), p. 611.

PROBLEMS

1. A ball of mass m is thrown straight up into the air with initial speed v_0. Neglecting the air resistance show that the time spent in going up equals the time spent in coming down and that the speed of return to the ground equals v_0. What is the potential energy of the ball at height h above the ground? What is the energy equation for the ball?

2. A ball of mass m is thrown straight up into the air with initial speed v_0. Assuming air resistance of the form kv, where k is a constant and v the instantaneous speed, solve the equation of motion to find the time taken by the particle to reach its highest point and the height attained.

Show that, as $k \to 0$, the results reduce to those of the nonresisting medium case (cf. Prob. 1). How long does the ball take to reach the ground again and with what velocity will it return? (Hint: Use approximate series expansions if necessary to handle transcendental equations.)

3. A particle of mass m is attracted to a force center with a force following the law k/x^2 where k is a constant. Find the energy integral. How much work is done as the particle moves from distance a_1 from the force center to distance $a_2 (a_2 < a_1)$? If the particle starts from rest at $x = a_1$, how long will it take it to reach $x = a_2$?

4. An alpha particle is projected directly at a stationary gold nucleus (atomic number = 79) with a velocity 2×10^9 cm/sec at a distance of 10^{-3} cm. Assuming a force of repulsion following Coulomb's law, to what distance from the nucleus will the alpha particle penetrate? Take the mass of the alpha particle as 6.6×10^{-24} grams. Its charge is $2e$, where $e = 4.8 \times 10^{-10}$ e.s.u. The charge on the gold nucleus = $79e$.

5. Calculate and compare the time averages of the kinetic and potential energies of a simple harmonic oscillator.

6. A particle of mass m falls under gravity through a medium which resists its motion with a force varying directly as the square of the speed (with a damping coefficient R). Show that the downward velocity will approach a limiting value and find this value. Find the velocity at time t if the initial velocity is zero.

7. In Prob. 6 find the work done against the resisting force during t sec from rest.

8. A particle of mass m moves along a straight line subject to the force

$$F = -k_1/x^2 + k_2/x^3$$

where k_1 and k_2 are positive $(k_1 > k_2)$, and x is the distance from the fixed force center.

Find the energy integral. How much work is done as the particle moves from distance a_1, from the force center to distance $a_2 (a_2 < a_1)$? If the particle starts from rest at $x = a_1$, how long will it take it to reach $x = a_2$?

9. In a Helmholtz acoustic resonator the dissipative force which makes the oscillations decay is due to the radiation of sound from the opening. The damping coefficient is

$$R = \rho\omega^2 S^2/2\pi c,$$

where ρ = density of the air, S = area of cross section of the opening, ω = angular frequency, and c = velocity of sound in air. Find the average rate of dissipation of energy by a resonator in air at 20°C at the *resonance* frequency of 300 cycles/sec if the maximum excess pressure at the opening is 100 dynes/cm^2.

10. Write the relaxation equation corresponding to the imposition of a constant excess pressure p_e on a fluid of mean density ρ_0. This is the differential equation for the excess density ρ_e. Find the expression for the relaxation time. From elementary physics what physical significance can be associated with the ratio p_e/ρ_e?

11. In a Rayleigh oscillator governed by eq. (4.79) find the expression for the rate of energy dissipation by the system. If m = 1000 grams, ω_0 = 200 sec^{-1}, and R_1 = 10 dynes sec/cm while R_2 = 10 dynes sec^3/cm^3, plot the above rate as a function of x.

Express the rate of energy dissipation as a function of time by the use of the approximate solution (4.81). Find the time average.

12. A mass of 50 grams is attached to the center of a horizontal string 20 cm long which is stretched with a tension of 2000 g dynes. If the mass of string is negligible, what is the resonance frequency of the system as a vibrator? What is the equivalent stiffness of the system? If in the vibrating system here considered the amplitude of vibration is reduced in the ratio 1 to e in 5 sec, find the damping coefficient and logarithmic decrement. What is the *natural* frequency of vibration of the system and what is its percentage difference from the resonance frequency?

13. If the mass in Prob. 12 is given an initial amplitude of 1/2 cm, what is the approximate average rate of energy dissipation?

14. If the system in Prob. 12 if forced to vibrate by the application of an external force with a frequency of 50 cycles/sec, what must be the amplitude of the force in order that an amplitude of 1/2 cm shall be maintained? What will be the phase difference between the force and the velocity.

15. The number of periods it takes a certain tuning fork of natural frequency 256 cycles/sec to have its amplitude diminished in the ratio 1 to e when vibrating freely is 5900. Calculate the logarithmic decrement. What is the resonance frequency of the fork? If the fork is forced

to vibrate at angular frequency ω find the values of the ratio ω/ω_{res} for which the value of the velocity amplitude falls to one half its maximum value.

16. Find the energy equation for the anharmonic oscillator discussed in Sec. 4.5. Choosing reasonable values of k_1 and k_2 (with $k_2 < k_1$), plot the potential energy as a function of x and see how much information about the motion can be obtained from the plot.

17. The string of a simple pendulum is gradually shortened at a rate very slow compared with the period of the pendulum. Calculate the *average* work done in the shortening over a single period. From the expressions for the kinetic and potential energies of the pendulum show that during the shortening process the value of E/ν remains unaltered, where E is the total energy of the bob and ν is the frequency of the oscillation. (N.B. E/ν is an *adiabatic* invariant of the motion.)

Apply similar reasoning to the gradual increase in the tension in the string in Prob. 12 and determine whether E/ν is also an adiabatic invariant in this case.

18. Discuss the solution of eq. (4.26) for the special case in which $k/m = R^2/4m^2$ and show that it can be written in the form

$$x = e^{-R/2m \cdot t}(A + Bt).$$

Find the physical significance of the constants A and B and plot x as a function of t for various cases.

19. Discuss the solution of eq. (4.26) for the special case in which $k/m < R^2/4m^2$. Plot x as a function of t for some simple initial conditions.

CHAPTER V

MOTION OF A PARTICLE IN A PLANE

5.1. Motion in a Uniform Field. If the resultant force on a particle lies in the plane of its initial motion, it is clear from the equation of motion (3.10) of Chapter 3 that its path will continue to lie in this plane. An important special case is that of a particle projected in any way in a uniform field, e.g., the constant gravitational field near the surface of the earth. If we treat the plane of the motion as the xy plane (Fig. 5.1) the force $\mathbf{F}$ on a particle of mass m becomes, neglecting resistance forces due to air and wind,

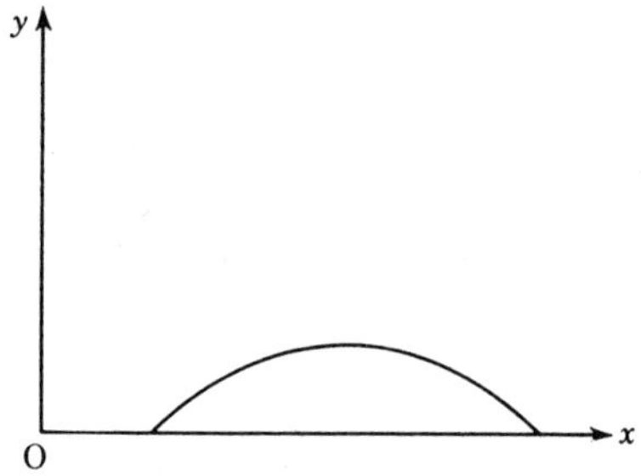

Fig. 5.1.

$$\mathbf{F} = -\mathbf{j}mg. \tag{5.1}$$

The equation of motion is here most conveniently treated by breaking it up into its cartesian components, viz.,

$$\begin{aligned} \ddot{x} &= 0, \\ \ddot{y} &= -g, \end{aligned} \tag{5.2}$$

leading to the solutions

$$\begin{aligned} x &= \dot{x}_0 t + x_0, \\ y &= -gt^2/2 + \dot{y}_0 t + y_0, \end{aligned} \tag{5.3}$$

where the initial position of the particle is (x_0, y_0) and the initial velocity $(\dot{x}_0, \dot{y}_0)$. The latter components can be expressed in terms of the initial speed v_0 and θ, the initial projection angle with the x axis. Thus

$$\begin{aligned} \dot{x}_0 &= v_0 \cos\theta, \\ \dot{y}_0 &= v_0 \sin\theta. \end{aligned} \tag{5.4}$$

The path of the particle is the parabola (resulting from the elimination of t between the two equations in (5.3) assuming $x_0 = y_0 = 0$, for simplicity)

$$y = x \tan \theta - gx^2/2v_0{}^2 \cos^2 \theta. \tag{5.5}$$

The time of flight, range, etc., are left to the reader to determine.

In the actual motion of a projectile the resistance of the air cannot be neglected. This force is usually dependent on the velocity in the orbit as well as the mass. Let us denote its magnitude per unit mass as R. Its introduction is facilitated by using a different method of resolving the motion into components, viz., along the horizontal direction and perpendicular to the tangent to the orbit, respectively. In Fig. 5.2 let APB denote a portion of the path in the neighborhood of the point P. The resistance force at P has the magnitude mR and is directed along the tangent which makes angle ϕ with the horizontal. The horizontal equation of motion then becomes

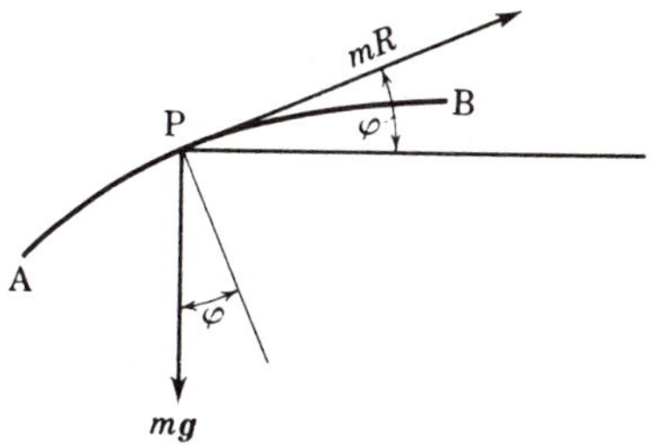

Fig. 5.2.

$$m\dot{v}_\phi = -mR \cos \phi, \tag{5.6}$$

where

$$v_\phi = v \cos \phi. \tag{5.7}$$

Perpendicular to the tangent the force is the centripetal force mv^2/ρ, where ρ is the radius of curvature at P. From the calculus study of plane curves we recall that

$$\rho = ds/d\phi, \tag{5.8}$$

where ds is the element of arc associated with the change $d\phi$ in ϕ. The centripetal force then becomes

$$mv^2/\rho = mv^2 d\phi/ds = mv\dot{\phi}, \tag{5.9}$$

since

$$v = \dot{s}. \tag{5.10}$$

Hence the component equation perpendicular to the tangent becomes

$$mv\dot{\phi} = -mg \cos \phi. \tag{5.11}$$

The combination of (5.6) and (5.11) gives

$$d(v \cos \phi) = Rv/g \cdot d\phi. \tag{5.12}$$

If R is known as a function of v, this is a differential equation connecting v with ϕ. After it has been solved, the path can in principle be found in parametric form by integrating the equations

$$\begin{aligned} dx &= -v^2/g \cdot d\phi, \\ dy &= -v^2/g \cdot \tan\phi \, d\phi. \end{aligned} \tag{5.13}$$

The reader is advised to apply (5.13) to the special case $R = 0$, when $v \cos\phi =$ constant. Actually the analysis is here more complicated than that leading to eq. (5.5). When $R \neq 0$, however, the eqs. (5.13) have the advantage that one can plot the trajectory step by step by use of the angle parameter ϕ. In fact, it turns out that when R is of the form Cv^n, eq. (5.12) is exactly integrable and the problem becomes greatly simplified in the present formulation.

5.2. Motion in a Central Field. The Vector Product. Let us suppose that the force **F** is directed along the position vector **r** and is a function of r alone. This is called a *central* force. The equation of motion becomes

$$m\ddot{\mathbf{r}} = f(r)\mathbf{r}_1, \tag{5.14}$$

where $\mathbf{r}_1$ is the unit vector along **r**, i.e., $\mathbf{r} = r\mathbf{r}_1$. It is clear that the force always lies in the plane of projection of the particle and hence the motion lies in this plane. For simplicity suppose this to be the xy plane. Let us try to establish some general results without further specifying the nature of $f(r)$.

We shall now find it convenient to introduce a second kind of vector multiplication. We shall define the *vector product* of two vectors **A** and **B** in terms of rectangular components as follows

$$\begin{aligned} \mathbf{A} \times \mathbf{B} &= (A_x, A_y, A_z) \times (B_x, B_y, B_z) \\ &\equiv \begin{vmatrix} \mathbf{i} & \mathbf{j} & \mathbf{k} \\ A_x & A_y & A_z \\ B_x & B_y & B_z \end{vmatrix} \end{aligned} \tag{5.15}$$

Expanding the determinant we have

$$\mathbf{A} \times \mathbf{B} = \mathbf{i}(A_yB_z - A_zB_y) + \mathbf{j}(A_zB_x - A_xB_z) + \mathbf{k}(A_xB_y - A_yB_x). \tag{5.16}$$

The vector product (sometimes called the *cross* product to distinguish it from the scalar or *dot* product (3.16)) is thus itself a vector. Since by inspection we have

$$\mathbf{A} \cdot (\mathbf{A} \times \mathbf{B}) = \mathbf{B} \cdot (\mathbf{A} \times \mathbf{B}) = 0, \tag{5.17}$$

the vector product is perpendicular to the plane formed by A and B. From analytical geometry it develops that

$$|\mathbf{A} \times \mathbf{B}| = AB \sin \phi, \tag{5.18}$$

where ϕ is the angle between **A** and **B** (cf. eq. (3.17)). We may therefore express the vector product in the form

$$\mathbf{A} \times \mathbf{B} = \boldsymbol{\epsilon} AB \sin \phi, \tag{5.19}$$

where $\boldsymbol{\epsilon}$ is the unit vector perpendicular to the plane of **A** and **B** and pointed so that **A** goes into **B** by a right-handed screw motion directed along $\boldsymbol{\epsilon}$. Clearly $|\mathbf{A} \times \mathbf{B}|$ is equal to the area of the parallelogram of which **A** and **B** are the adjacent sides.

The cross product is not commutative. In fact we have

$$\mathbf{A} \times \mathbf{B} = -\mathbf{B} \times \mathbf{A}. \tag{5.20}$$

The distributive law however holds for the vector as for the scalar product, i.e.,

$$\mathbf{A} \times (\mathbf{B} + \mathbf{C}) = \mathbf{A} \times \mathbf{B} + \mathbf{A} \times \mathbf{C}. \tag{5.21}$$

The proof is left as an exercise for the reader.

We are now ready to make application of these results to central field motion. Taking the cross product of both sides of (5.14) with **r** we have

$$m\ddot{\mathbf{r}} \times \mathbf{r} = f(r)\mathbf{r}_1 \times \mathbf{r}. \tag{5.22}$$

But clearly $\mathbf{r}_1 \times \mathbf{r} = 0$, and hence

$$\ddot{\mathbf{r}} \times \mathbf{r} = 0 \tag{5.23}$$

for motion in any central field. Making use of the fact that the usual rule holds for the derivative of the cross product of two vectors, as long as we keep the order unchanged, i.e.,

$$\frac{d}{dt}(\mathbf{A} \times \mathbf{B}) = \dot{\mathbf{A}} \times \mathbf{B} + \mathbf{A} \times \dot{\mathbf{B}}, \tag{5.24}$$

we have

$$\frac{d}{dt}(\dot{\mathbf{r}} \times \mathbf{r}) = \ddot{\mathbf{r}} \times \mathbf{r} + \dot{\mathbf{r}} \times \dot{\mathbf{r}}. \tag{5.25}$$

But $\dot{\mathbf{r}} \times \dot{\mathbf{r}} = 0$ and hence (5.25) leads to

$$\dot{\mathbf{r}} \times \mathbf{r} = \mathbf{C}, \tag{5.26}$$

where $\mathbf{C}$ is a vector, which is independent of time and normal to the plane of $\mathbf{r}$ and $\dot{\mathbf{r}}$. Now from the definition (5.15)

$$\dot{\mathbf{r}} \times \mathbf{r} = \begin{vmatrix} \mathbf{i} & \mathbf{j} & \mathbf{k} \\ \dot{x} & \dot{y} & 0 \\ x & y & 0 \end{vmatrix} = \mathbf{k}(\dot{x}y - \dot{y}x), \tag{5.27}$$

since $z = \dot{z} = 0$. Using polar coordinates r, θ we readily show that

$$|\dot{\mathbf{r}} \times \mathbf{r}| = r^2\dot{\theta}, \tag{5.28}$$

which is twice the rate at which the position vector sweeps out area as it moves in the plane. Hence

$$r^2\dot{\theta} = C \tag{5.29}$$

is true for any central motion. It is the so-called *law* of *areas*. Differentiation of (5.29) with respect to the time reveals an interesting fact about the Coriolis acceleration for a particle in central motion (cf. eq. 2.40).

5.3. Inverse Square Central Field. Historically the most famous case of eq. (5.14) is that for which

$$f(r) = -K/r^2, \tag{5.30}$$

where K is a positive constant. Here the central force is one of attraction to the fixed origin and varying inversely with the square of the distance thereto. Its connection with Newton's principle of gravitation is obvious. Let us see what motion corresponds to it. Letting $K/m = A$, eq. (5.14) now becomes

$$\ddot{\mathbf{r}} = -A\mathbf{r}_1/r^2. \tag{5.31}$$

We form the cross product with $\mathbf{C}$ from (5.26). Thus

$$\ddot{\mathbf{r}} \times \mathbf{C} = -\frac{A}{r^2}\mathbf{r}_1 \times \mathbf{C} = -\frac{A}{r^2}\mathbf{r}_1 \times (\dot{\mathbf{r}} \times \mathbf{r}). \tag{5.32}$$

Utilizing the readily verifiable result that for any three vectors $\mathbf{a}$, $\mathbf{b}$, $\mathbf{c}$

$$\mathbf{a} \times (\mathbf{b} \times \mathbf{c}) = \mathbf{b}(\mathbf{a} \cdot \mathbf{c}) - \mathbf{c}(\mathbf{a} \cdot \mathbf{b}), \tag{5.33}$$

we have

$$\ddot{\mathbf{r}} \times \mathbf{C} = -\frac{A}{r^2}[\dot{\mathbf{r}}(\mathbf{r}_1 \cdot \mathbf{r}) - \mathbf{r}(\mathbf{r}_1 \cdot \dot{\mathbf{r}})]. \tag{5.34}$$

Hence

$$\ddot{\mathbf{r}} \times \mathbf{C} = -A\dot{\mathbf{r}}_1. \tag{5.35}$$

A single integration with respect to the time yields

$$\dot{\mathbf{r}} \times \mathbf{C} = -A\mathbf{r}_1 + \mathbf{a}. \tag{5.36}$$

where $\mathbf{a}$ is a vector independent of time and in the plane of the motion. If we cross both sides of (5.36) with $\mathbf{C}_1/C$, where $\mathbf{C}_1$ is the unit vector along $\mathbf{C}$,

$$\frac{\mathbf{C}_1}{C} \times (\dot{\mathbf{r}} \times \mathbf{C}) = \frac{\mathbf{C}_1}{C} \times (-A\mathbf{r}_1) + \frac{\mathbf{C}_1}{C} \times \mathbf{a}. \tag{5.37}$$

The result is

$$\dot{\mathbf{r}} = -\frac{A}{C}\mathbf{C}_1 \times \mathbf{r}_1 + \frac{1}{C}\mathbf{C}_1 \times \mathbf{a}. \tag{5.38}$$

Introducing the cross product with $\mathbf{r}$ again we have

$$\begin{aligned} \mathbf{C} = \dot{\mathbf{r}} \times \mathbf{r} &= -\frac{A}{C}(\mathbf{C}_1 \times \mathbf{r}_1) \times \mathbf{r} + \frac{1}{C}(\mathbf{C}_1 \times \mathbf{a}) \times \mathbf{r} \\ &= \frac{Ar}{C}\mathbf{C}_1 - \frac{(\mathbf{r} \cdot \mathbf{a})}{C}\mathbf{C}_1. \end{aligned} \tag{5.39}$$

Dot multiplication by $\mathbf{C}$ gives finally

$$C^2 = Ar - ra\cos\theta, \tag{5.40}$$

where θ is the angle between $\mathbf{r}$ and $\mathbf{a}$. Eq. (5.40) may be written

$$r = \frac{C^2/A}{1 - \dfrac{a}{A}\cos\theta}. \tag{5.41}$$

This is the equation of a conic section with the focus at the origin, with semi-latus rectum C^2/A and eccentricity $\epsilon = a/A$. If $\epsilon = 1$, the conic is a parabola; if $\epsilon < 1$, an ellipse, and if $\epsilon > 1$, a hyperbola. The precise character of the orbit in central field motion thus depends on the constant vector $\mathbf{a}$, which in turn depends on the initial conditions.

To see this let us write the magnitude of the resultant velocity in the form (cf. eq. 2.34)

$$v^2 = \dot{r}^2 + r^2\dot{\theta}^2,$$

which from (5.41) and (5.29) becomes

$$v^2 = \frac{A^2\epsilon^2 \sin^2\theta}{C^2} + \frac{C^2}{r^2}. \tag{5.42}$$

Further use of (5.41) leads to

$$\epsilon^2 - 1 = \frac{C^2}{A^2}\left(v^2 - \frac{2A}{r}\right). \tag{5.43}$$

Now suppose that the particle starts at $r = r_0$ with initial speed v_0. Clearly according as

$$v_0{}^2 = \frac{2A}{r_0},\; v_0{}^2 < \frac{2A}{r_0},\; v_0{}^2 > \frac{2A}{r_0}, \tag{5.44}$$

the eccentricity

$$\epsilon = 1, \quad \epsilon < 1, \quad \epsilon > 1, \tag{5.45}$$

respectively, and the nature of the orbit is determined accordingly. If $v_0 = 0$ for $r = \infty$, i.e., if the particle starts from rest at an infinitely great distance, $\epsilon^2 = 1$ always and the path will be a degenerate parabola, i.e., the straight line joining the initial position to the force center. For such a particle the velocity at $r = r_0$ will be

$$v = \sqrt{2A/r_0} = \sqrt{2K/mr_0}. \tag{5.46}$$

Hence we can say that, according as the initial velocity of projection at r_0 is equal to, less than, or greater than the velocity attained at r_0 by a particle starting from rest at infinity in this field, the orbit is a parabola, ellipse, or hyperbola.

The above method of studying motion in an inverse square central field has the disadvantage that it does not stress the energy equation. The latter is, of course, obtained from (5.31) by taking the dot product with $\dot{\mathbf{r}}\,dt = d\mathbf{r}$ and integrating. Then we get

$$\frac{1}{2}mv^2 - K/r = U, \tag{5.47}$$

where the potential energy to an arbitrary additive constant appears as

$$V = -K/r, \tag{5.48}$$

and the total energy is U, if we agree to let the potential energy be zero at $r = \infty$.

The reader should show by the use of (5.43) that for an elliptical orbit the total energy is negative, whereas for a hyperbolic orbit it is positive. The physical significance of this should be noted. The association of negative total energy with an elliptical orbit, i.e., a closed path in which the particle forever remains within a finite distance from the force center, prompts us to consider this energy as a form of *binding* energy, insuring the localized motion of the particle. If we form the time average of each term in eq. (5.47) we can interpret the total energy U, if it is positive, as the excess of the average kinetic energy over the magnitude of the average potential energy; if U is negative we can interpret it as the excess of the magnitude of the average potential energy over the average kinetic energy. In the latter case this is precisely the energy which it is necessary to contribute to the particle in order to bring its average kinetic energy up to its average potential energy in magnitude. This will make its total energy zero and from the way the boundary conditions have been assigned will assure that the particle will be found at rest infinitely far from the force center.

If the particle is an electron and the force center a heavy positively charged nucleus, as in the Bohr-Rutherford atomic model, the process of removing the electron from the vicinity of the nucleus is physically realizable and is called ionization. It is brought about by contributing to the atom an amount of energy (by collision or otherwise) equal to the binding energy. This is therefore sometimes referred to as the ionization energy.

The analysis leading to the conic section as the orbit corresponding to an inverse square central field has been carried out above in terms of vector notation. The reader may find it instructive to carry through the analysis in the scalar form, employing polar coordinates. A feature of the latter method is the elimination of the time while the equations are still in differential form, thus producing a differential equation for r in terms of θ. The vector notation method lends itself nicely to the inverse square case but is not so tractable in other cases. The scalar method is illustrated in Sec. 5.4, and applied to some other types of central field motion.

5.4. Other Types of Central Field Motion. Perturbations. Following the lead of the last paragraph of the preceding section we may write the general differential equation of a central orbit in terms of $u = 1/r$ and θ. Here it is convenient to resolve the motion into components along the radius vector and perpendicular to the radius vector. The component force along the radius vector is of course the central force, whereas the component force at right angles thereto vanishes. Using eq. (2.40) we therefore may replace (5.14) by the two equations

$$m(\ddot{r} - r\dot{\theta}^2) = f(r), \tag{5.49}$$

$$m(r\ddot{\theta} + 2\dot{r}\dot{\theta}) = 0. \tag{5.50}$$

Utilizing eq. (5.29) we have

$$r\dot{\theta}^2 = C^2u^3. \tag{5.51}$$

Moreover by a bit of manipulation involving the use of $\dot{u} = du/d\theta \cdot \dot{\theta}$, we get

$$\ddot{r} = -C^2u^2\frac{d^2u}{d\theta^2}. \tag{5.52}$$

Consequently (5.49) turns into the differential equation connecting u and θ

$$\frac{d^2u}{d\theta^2} + u + \frac{F(u)}{mC^2u^2} = 0, \tag{5.53}$$

where $F(u) = f(r)$. This equation takes on a particularly simple form for the inverse square force field and leads rather directly by a simple integration to the usual conic section solution.

Even when $F(u)$ does not have the form Ku^2 corresponding to the inverse square field, eq. (5.53) can give useful information about the nature of the motion in a central force field. For many purposes we can expand $f(r)$ in a series of the form

$$f(r) = A_1/r^2 + A_2/r^3 + A_3/r^4 + \cdots \tag{5.54}$$

corresponding to

$$F(u) = A_1u^2 + A_2u^3 + A_3u^4 + \cdots. \tag{5.55}$$

Thus we might in certain circumstances regard A_2 and A_3, etc., as small enough so that the corresponding terms act merely as *perturbations* on the effect of the first term. Eq. (5.53) now becomes

$$\frac{d^2u}{d\theta^2} = -A_1/mC^2 - (1 + A_2/mC^2)u - A_3/mC^2 \cdot u^2 \cdots. \tag{5.56}$$

We introduce the substitution

$$w = (1 + A_2/mC^2)u + A_1/mC^2. \tag{5.57}$$

Hence (5.56) takes the form, if for the moment we neglect the A_3 and higher terms,

$$\frac{d^2w}{d\theta^2} = -(1 + A_2/mC^2)w. \tag{5.58}$$

Calling

$$1 + A_2/mC^2 = \gamma^2,$$

we can write the solution of (5.58) in the form

$$w = G \cos (\gamma\theta - \alpha), \tag{5.60}$$

where G and α are arbitrary constants. In terms of r this becomes

$$r = \frac{-mC^2\gamma^2/A_1}{1 - mC^2G/A_1 \cdot \cos (\gamma\theta - \alpha)}. \tag{5.61}$$

If we assume A_1 to be negative and use $\phi = \gamma\theta$ as the polar angle, this becomes the equation of an ellipse with eccentricity mC^2G/A_1. In terms of θ the equation no longer represents an ellipse but rather a rosette as indicated in Fig. 5.3. We may if we wish look upon this as an ellipse with rotating perihelion, i.e., with an advance in the perihelion of $2\pi(1-1/\gamma)$ for each complete cycle of the radius vector r. Note that the motion is still plane motion entirely confined between two circles of radii r_{max} and r_{min} respectively.

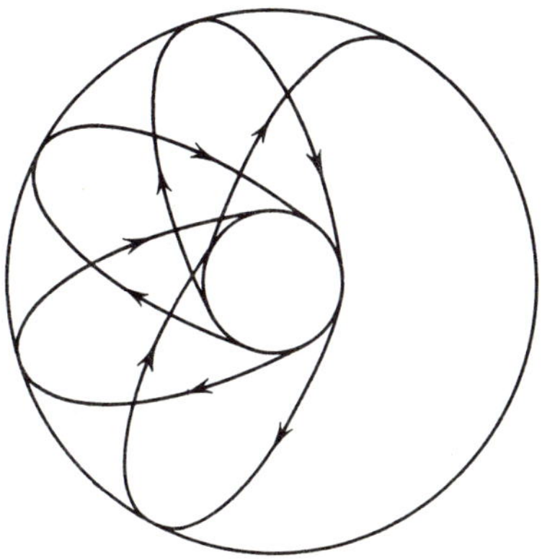

Fig. 5.3.

If $A_3 \neq 0$ but still small enough so that the contribution of the term A_3/r^4 in (5.54) is still small compared with that of the earlier terms, we see that eq. (5.56) can be handled by the successive approximation technique already used for the anharmonic oscillator (Sec. 4.5). The result will be that the solution (5.60) will have to be corrected by terms of the form $\cos 2(\gamma\theta - \alpha)$, etc. The motion becomes more complicated but it is still confined to the region between two circles as in Fig. 5.3.

PROBLEMS

1. A projectile travels in a vertical plane through a medium which resists its motion with a force varying as the square of the resultant velocity. Show that the equations of motion can be written in the form

$$\ddot{x} = -k\dot{x}\dot{s},$$
$$\ddot{y} = -g - k\dot{y}\dot{s},$$

where k is a constant and $\dot{s}^2 = \dot{x}^2 + \dot{y}^2$. Integrate to find the differential equation of the path in the form

$$\frac{d^2y}{dx^2} = \frac{-ge^{2ks}}{v_0^2 \cdot \cos^2 \alpha},$$

where v_0 is the initial speed, α the angle of projection, and s is the distance along the path from the point of projection.

Assume a flat trajectory, i.e., $\left|\frac{dy}{dx}\right| \ll 1$ and determine the equation of the trajectory in the form of a correction to the standard parabola obtained by neglecting air resistance.

2. A particle moves in a central field under an attractive force varying directly as the distance from the force center. Discuss the motion. (N.B. It will be most simple to split the motion into components along rectangular axes.)

3. A particle moves in a plane and satisfies the following component equations of motion

$$m\ddot{x} = -c_1x,$$
$$m\ddot{y} = -c_2y,$$

where c_1 and c_2 are different positive constants. Discuss the motion.

4. Prove that in any central field motion the speed of the particle at any point in its path is inversely proportional to the length of the normal drawn from the force center to the tangent to the path at that point.

5. An alpha particle of mass m and with initial velocity v_0 approaches a nucleus with charge Ze in such a way that the perpendicular distance between the line of approach and the nucleus is p. Show that the angle of deflection θ is given by

$$\cot \frac{\theta_0}{2} = \frac{mv_0^2p}{2Ze^2}.$$

6. Referring to Prob. 5, show that the ratio of the number of alpha particles scattered through unit area at distance r from the nucleus and in the neighborhood of angle θ_0 to the number incident per unit area is

$$\left(\frac{Ze^2}{mv_0{}^2 r}\right)^2 \csc^4 \frac{\theta_0}{2}.$$

What important quantity can be determined by measurements based on this formula?

7. Use the law of areas to deduce Kepler's third law of planetary motion. Compute the mass of the sun.

8. Solve eq. (5.12) for the motion of a projectile in the case in which $R = 0$. Also show that if $R = Cu^n$, where C and n are constants, the equation is integrable exactly. Specialize to $n = 1$ and 2.

9. Solve the problem of central field motion by resolving the motion into components along and perpendicular to the radius vector and obtaining a second order equation connecting $u = 1/r$ and θ. Discuss the orbit for the force function

$$f(r) = A_1/r^2 + A_2/r^3,$$

where A_1 and A_2 are constants.

10. Obtain the modification of the u, θ equation of Prob. 9 if there is a perturbing force T at right angles to the radius vector. Let the particle be an electron with charge e and

$$T = e\dot{r}H/c,$$

or the force produced by a magnetic field H oersteds (c = velocity of light in free space). Show that under the influence of H, the electron performs the same kind of motion relative to axes rotating about an axis parallel to the field with angular velocity

$$\omega = eH/2mc,$$

as it does relative to stationary axes in the absence of the field (Larmor precession).

11. Two points A and B lie neither in the same horizontal nor vertical line. Find the curve joining them down which a particle sliding under gravity and starting from rest at the higher will reach the lower point in the minimum time.

12. Prove the distributive law for the vector product.

13. Establish the vector identity

$$\mathbf{a} \times (\mathbf{b} \times \mathbf{c}) = \mathbf{b}(\mathbf{a} \cdot \mathbf{c}) - \mathbf{c}(\mathbf{a} \cdot \mathbf{b}).$$

14. Prove that for any three vectors **a**, **b**, **c**,

$$\mathbf{a} \cdot (\mathbf{b} \times \mathbf{c}) = (\mathbf{a} \times \mathbf{b}) \cdot \mathbf{c}.$$

Give an interpretation in terms of the geometrical meaning of either side.

15. Show that $\mathbf{i} = \mathbf{j} \times \mathbf{k}$ and derive the other analogous relations for the unit vectors. What is the geometrical significance of $\mathbf{i} \cdot (\mathbf{j} \times \mathbf{k})$?

16. Use the analysis in Sec. 5.3 to show that $\dot{\mathbf{r}}$ in an inverse square central field can never be zero save for unit eccentricity. Comment on the geometrical significance.

17. Find the energy equation for central field motion in which the force function $f(r)$ has the form indicated in Prob. 9. Connect the information about the motion provided by the energy equation with that inherent in the perihelion precession formula (5.61).

CHAPTER VI

MOTION OF A COLLECTION OF PARTICLES

6.1. Some General Conservation Theorems. Momentum and Moment of Momentum. In the discussion so far we have been concerned with the motion of a single particle. Since large-scale bodies are thought of as composed of a multitude of particles, it becomes necessary to consider the motion of a collection of particles. We shall first establish some general results which are independent of the precise forces of interaction between the particles. The use of vector notation will simplify the analysis.

Suppose we have a system of n particles of masses $m_1, m_2, \ldots m_n$, respectively, and acted on by resultant forces $\mathbf{F}_1, \mathbf{F}_2, \ldots \mathbf{F}_n$, respectively. For convenience we consider each force to be the sum of an external force and the resultant of the interaction forces between the particle in question and all the other particles. Thus for the ith particle

$$\mathbf{F}_i = \mathbf{F}_{ie} + \sum_{k \neq i = 1}^{n} \mathbf{F}_{ik}, \tag{6.1}$$

where $\mathbf{F}_{ik}$ is the force on the ith particle due to the kth particle. Note that the term for $k = i$ in the sum is excluded. The equation of motion of the ith particle is then

$$m_i \ddot{\mathbf{r}}_i = \mathbf{F}_{ie} + \sum_{k \neq i = 1}^{n} \mathbf{F}_{ik}, \tag{6.2}$$

Let us sum the equation over all particles and obtain

$$\sum_{i=1}^{n} m_i \ddot{\mathbf{r}}_i = \sum_{i=1}^{n} \mathbf{F}_{ie} + \sum_{i=1}^{n} \sum_{k \neq i = 1}^{n} \mathbf{F}_{ik}. \tag{6.3}$$

The double sum in (6.3) contains an $\mathbf{F}_{ki}$ for every $\mathbf{F}_{ik}$. But from the principle of action and reaction

$$\mathbf{F}_{ik} = -\mathbf{F}_{ki}. \tag{6.4}$$

Hence the double sum vanishes and (6.3) reduces to

$$\frac{d^2}{dt^2}\left(\sum_{i=1}^{n} m_i \mathbf{r}_i\right) = \sum_{i=1}^{n} \mathbf{F}_{ie}. \tag{6.5}$$

This suggests the introduction of a point whose position vector $\bar{\mathbf{r}}$ is given by

$$\bar{\mathbf{r}} = \sum_{i=1}^{n} m_i \mathbf{r}_i / M, \tag{6.6}$$

where $M = \sum_{i=1}^{n} m_i$ = the total mass of the collection of particles. The motion of this point is then given by the equation

$$M\ddot{\bar{\mathbf{r}}} = \sum_{i=1}^{n} \mathbf{F}_{ie}. \tag{6.7}$$

The point with position vector $\bar{\mathbf{r}}$ is called the *center of mass* of the system of particles. It moves as if all the mass of the system were concentrated at it and as if all the forces were acting on it. Clearly (6.7) says that if the resultant of the external forces on the collection of particles vanishes, the center of mass remains at rest or moves with constant velocity in a straight line. Evidently eq. (6.7) contains the important implication that even if we are unable to specify directly the motions of the individual particles of the aggregate we can yet have some idea *where* the aggregate is at any instant since, if we know where the center of mass is, we can be sure that the aggregate cannot be far away. It may be argued indeed that the various particles may still be dispersed widely from the center of mass, but in this case there is some question whether we ought to consider the particles as forming an aggregate at all.

Eq. (6.5) can be rewritten in the form

$$\frac{d}{dt}\left(\sum_{i=1}^{n} m_i \dot{\mathbf{r}}_i\right) = \sum_{i=1}^{n} \mathbf{F}_{ie}. \tag{6.8}$$

Now $\sum_{i=1}^{n} m_i \dot{\mathbf{r}}_i$ is by definition the *total momentum* of the collection of particles. Hence eq. (6.8) says that the time rate of change of momentum of the system is equal to the *resultant* force acting. If the resultant force vanishes, the total momentum remains constant. This is the principle of conservation of momentum for a collection of particles subject only to their mutual interaction forces. It is a fundamental principle of the mechanics of aggregates. Its importance might seem to be somewhat diminished from the fact that we never find in nature a collection of particles subject to no external forces. Fortunately there exist many examples where the external forces are so small compared with the internal forces that they can be neglected. Thus, although the planets of the solar system are certainly acted on by forces due to bodies outside the system, the internal gravitational forces between them and the sun and each other are far more important, and we are justified in applying the principle of conservation of momentum to this system. Somewhat similar considerations apply to the movement of billiard balls on a flat horizontal table. They are all acted on by the force of gravity in the vertical direction. This produces only frictional forces on the balls in the horizontal plane and, if these are small compared with the impact forces involved in collisions of the balls, we are justified again in applying the principle. One of the best illustrations is the case of an atomic model considered to be made up of many charged particles. If the atom is considered as far away from neighboring atoms, as in a gas, for example, we can satisfactorily apply the principle to each atom separately. Of course the approximation involved must always be kept in mind.

In a later chapter we shall apply the principle of momentum conservation to a rigid body, i.e., a collection of particles so tightly bound together by their mutual interactions that the interparticle distance never changes or at most changes by a negligible amount.

Going back to (6.2) let us form the cross product of both sides with $\ddot{\mathbf{r}}_i$ and get

$$m_i \mathbf{r}_i \times \ddot{\mathbf{r}}_i = \mathbf{r}_i \times \mathbf{F}_{ie} + \sum_{k \neq i=1}^{n} \mathbf{r}_i \times \mathbf{F}_{ik}. \tag{6.9}$$

It is customary to call $\mathbf{r}_i \times \mathbf{F}_{ie}$ the *moment* of the force $\mathbf{F}_{ie}$ about the axis through the origin normal to the plane of $\mathbf{r}_i$ and $\mathbf{F}_{ie}$. This is also referred to as the *torque* due to the force $\mathbf{F}_{ie}$ about the axis in question.

As such it is usually denoted by the symbol $\mathbf{L}_{ie}$. We therefore rewrite (6.9)

$$m_i\mathbf{r}_i \times \ddot{\mathbf{r}}_i = \mathbf{L}_{ie} + \sum_{k\neq i=1}^{n} \mathbf{L}_{ik}. \tag{6.10}$$

Once more we sum over all the particles in the collection

$$\sum_{i=1}^{n} m_i\mathbf{r}_i \times \ddot{\mathbf{r}}_i = \sum_{i=1}^{n} \mathbf{L}_{ie} + \sum_{i=1}^{n}\sum_{k\neq i=1}^{n} \mathbf{L}_{ik}. \tag{6.11}$$

Inspection shows that if the interaction forces are in each case directed along the straight line connecting the particles in question the double sum vanishes. Moreover from eq. (5.24) we can write

$$\frac{d}{dt}\left[\mathbf{r}_i \times (m_i\dot{\mathbf{r}}_i)\right] = \mathbf{r}_i \times (m_i\ddot{\mathbf{r}}_i), \tag{6.12}$$

so that (6.11) finally takes the form

$$\frac{d}{dt}\sum_{i=1}^{n}[\mathbf{r}_i \times (m_i\dot{\mathbf{r}}_i)] = \sum_{i=1}^{n} \mathbf{L}_{ie}. \tag{6.13}$$

From the formation of $\mathbf{r}_i \times (m_i\dot{\mathbf{r}}_i)$ we naturally term it the *moment of momentum* of the ith particle about the origin. The physical significance of (6.13) then is: the time rate of change of the total moment of momentum of the collection of particles about the origin is equal to the total torque of all the external forces about the same point. If the resultant torque vanishes the total moment of momentum remains constant: a second conservation theorem analogous to that of conservation of momentum previously mentioned. The sun and planets of the solar system provide an illustration of this principle, insofar as external forces can be neglected. Other illustrations occur in the atomic domain.

The significance of the requirement that the interaction forces lie along the line connecting the interacting particles in the establishment of the conservation of the moment of momentum of an aggregate deserves more than casual comment. In Sec. 3.2 we noted the fact that, in the logical development of physics, forces are a free invention of the physicist. Nevertheless one is naturally impelled to introduce force functions which lead to observed motions. Conservation theorems have such a powerful attraction for the mind and seem so closely related to experience (even when they are not precisely verified) that the force functions for which they are satisfied assume a peculiar importance. Central forces which not only lie along the line joining the interacting particles but are also functions only of the distance between them belong in this category and it is no accident that physicists have tended to employ them in general for the interaction of particles of all kinds with particular emphasis on the inverse square or Coulombic force. However, other forms of dependence on distance have also been used (recall Sec. 5.4). The success of this program has been conspicuous in the particle theory of the constitution of large-scale bodies. It meets considerable difficulty, however, in the treatment of the interaction of nuclear particles.[1] (cf. Sec. 11.14).

The importance of the center of mass in the consideration of moment of momentum is well brought out by writing the position vector $\mathbf{r}_i$ of the ith particle in the form

$$\mathbf{r}_i = \overline{\mathbf{r}} + \mathbf{r}_i', \tag{6.14}$$

where $\mathbf{r}_i'$ is the position vector of the ith particle relative to the center of mass. If we substitute into (6.13) we obtain

$$\frac{d}{dt}\sum_{i=1}^{n} \mathbf{r}_i' \times (m_i \dot{\mathbf{r}}_i') = \sum_{i=1}^{n} \mathbf{r}_i' \times \mathbf{F}_{ie}. \tag{6.15}$$

This says that the time rate of change of the total moment of momentum about the center of mass is equal to the total torque about the center of mass. Consequently we can describe the motion of the collection of particles by the motion of the center of mass (eq. 6.7) and the motion relative to the center of mass (eq. 6.15). This scheme turns out to be particularly appropriate when the particles form a rigid body, as we shall see a little later.

6.2. Energy of a Collection of Particles. Go back to eq. (6.2) and form the dot product of both sides with $\dot{\mathbf{r}}_i dt$. Integrating from the initial position of the system $\mathbf{r}_{10}, \mathbf{r}_{20}, \ldots \mathbf{r}_{n0}$ to the final position $\mathbf{r}_{11}, \mathbf{r}_{21}, \ldots \mathbf{r}_{n1}$, we finally have

$$K_1 - K_0 = \int_{\mathbf{r}_{i0}}^{\mathbf{r}_{i1}} \left[\sum \mathbf{F}_{ie} \cdot d\mathbf{r}_i + \sum\sum \mathbf{F}_{ik} \cdot d\mathbf{r}_i \right], \tag{6.16}$$

[1]Cf. H. A. Bethe, *Elementary Nuclear Theory* (New York, Wiley, 1947), pp. 23, 73 ff.

where

$$K_0 = \sum \frac{1}{2} m_i v_{i0}^2, \quad K_1 = \sum \frac{1}{2} m_i v_{i1}^2 \tag{6.17}$$

are the total initial and final kinetic energies, respectively. Eq. (6.16) is the work-kinetic energy theorem for a collection of particles (i.e., the generalization of eq. 3.26).

Let us suppose that the interaction forces $\mathbf{F}_{ik}$ are central forces, i.e. they act along the line joining the ith and kth particles, and depend only on $\mathbf{r}_{ik}$. Thus let us write

$$\mathbf{F}_{ik} = f(r_{ik})\mathbf{r}_{ik}. \tag{6.18}$$

It then follows readily that the double sum in (6.16) becomes

$$\sum \sum \mathbf{F}_{ik} \cdot d\mathbf{r}_i = \sum \sum f(r_{ik})\mathbf{r}_{ik} \cdot d\mathbf{r}_{ik}, \tag{6.19}$$

where the double sum now means the sum taken over all *different* pairs of particles. Note we have set

$$\mathbf{r}_{ik} = \mathbf{r}_i - \mathbf{r}_k. \tag{6.20}$$

Since the interaction force is central it is conservative and an interaction potential always exists. Indeed if we integrate

$$\int r_{ik} f(r_{ik})\, dr_{ik}$$

and call the result $-V(r_{ik})$, we shall have

$$f(r_{ik})\mathbf{r}_{ik} = -\left[\mathbf{i}\frac{\partial V(r_{ik})}{\partial x_{ik}} + \mathbf{j}\frac{\partial V(r_{ik})}{\partial y_{ik}} + \mathbf{k}\frac{\partial V(r_{ik})}{\partial z_{ik}}\right]. \tag{6.21}$$

It will be appropriate at this point to introduce a new bit of vector terminology and call the right-hand side of (6.21) $-\nabla V(r_{ik})$ or $-\text{grad}\, V(r_{ik})$ (called the negative *gradient* of $V(r_{ik})$), where ∇ then appears as the *differential vector operator*

$$\nabla \equiv \mathbf{i}\frac{\partial}{\partial x} + \mathbf{j}\frac{\partial}{\partial y} + \mathbf{k}\frac{\partial}{\partial z}. \tag{6.22}$$

It is clear that ∇ always operates on a scalar function of the coordinates. Also

$$\begin{aligned} \nabla V \cdot d\mathbf{r} &= (\mathbf{i}\frac{\partial V}{\partial x} + \mathbf{j}\frac{\partial V}{\partial y} + \mathbf{k}\frac{\partial V}{\partial z}) \cdot (\mathbf{i}\,dx + \mathbf{j}\,dy + \mathbf{k}\,dz) \\ &= dV(x, y, z), \end{aligned} \tag{6.23}$$

or the total differential of V.

We may therefore write, returning to (6.19)

$$\sum\sum \mathbf{F}_{ik} \cdot d\mathbf{r}_i = -\sum\sum dV(r_{ik}), \tag{6.24}$$

where the expression is summed, of course, over all different pairs of i and k. Eq. (6.16) then becomes

$$\begin{aligned} K_1 - K_0 = \int_{\mathbf{r}_{i0}}^{\mathbf{r}_{i1}} \sum \mathbf{F}_{ie} \cdot d\mathbf{r}_i - \sum\sum V(r_{ik})_1 \\ + \sum\sum V(r_{ik})_0, \end{aligned} \tag{6.25}$$

where $V(r_{ik})_0$ denotes the value of the interaction potential of the ith and kth particles in the initial position, etc. Let us further assume that the external forces are also conservative so that

$$\mathbf{F}_{ie} = -\nabla V_{ie}(r_i). \tag{6.26}$$

Then (6.25) becomes

$$\begin{aligned} K_0 + \sum\sum V(r_{ik})_0 + \sum (V_{ie})_0 \\ = K_1 + \sum\sum V(r_{ik})_1 + \sum (V_{ie})_1. \end{aligned} \tag{6.27}$$

We may call $\sum V_{ie}$ the total potential energy of the collection of particles due to the external forces and $\sum\sum V(r_{ik})$ the total potential energy of the collection due to the internal forces. Eq. (6.27) then constitutes the theorem of the *conservation of energy* of a system of conservatively interacting particles also subject to conservative external forces. It is a first integral of the equation of motion of the system. But since it is a *single* equation, it is inadequate by itself to describe completely even the velocity of each particle of the system. However, with the momentum and moment of momentum equations it is able to shed considerable light on the behavior of simple systems. The *general* solution of the equations of motion of a system containing more than two particles has hitherto defied analysis, although many special results in the problem of three or more particles are known and are of extreme value in connection with the motion of the planets of the solar system and electron motions in atoms.

In Sec. 6.1 we found it convenient to introduce the center of mass. Let us see how this affects the consideration of energy. Using eq. (6.14) in the expression for the kinetic energy of the system of particles we have

$$K = \sum \frac{1}{2} m_i \dot{\mathbf{r}}_i \cdot \dot{\mathbf{r}}_i = \frac{1}{2} \sum m_i (\dot{\bar{\mathbf{r}}} + \dot{\mathbf{r}}_i') \cdot (\dot{\bar{\mathbf{r}}} + \dot{\mathbf{r}}_i'). \tag{6.28}$$

Carrying out the indicated operations yields

$$K = \frac{1}{2} M \bar{v}^2 + \frac{1}{2} \sum m_i \dot{\mathbf{r}}_i' \cdot \dot{\mathbf{r}}_i', \tag{6.29}$$

since further use of (6.14) shows that $\sum m_i \dot{\mathbf{r}}_i' = 0$. The physical interpretation of (6.29) is simply that the total kinetic energy of the system is equal to the kinetic energy of the center of mass plus the kinetic energy of all the particles relative to the center of mass. This has an important application to molecules consisting of several atoms.

6.3. The Principle of Virtual Work. Systems Under Constraint. D'Alembert's Principle. If in a collection of particles the resultant force on every particle vanishes the system is said to be in equilibrium and every particle will move with constant velocity in a straight line or remain at rest. In particular, if the ith particle undergoes a displacement $\delta\mathbf{r}_i$, the work done is zero since the resultant force is zero. Thus, summing up over all the particles.

$$\sum \mathbf{F}_i \cdot \delta\mathbf{r}_i = 0. \tag{6.30}$$

Now it may be that the particles are not entirely free to move but are subjected to certain *constraints*. Thus the particles may be tied to each other by inextensible strings (e.g., Atwood's machine or variants thereof) or forced to move on certain surfaces (e.g., inclined planes) or forced by certain internal cohesive forces to maintain invariant their mutual distances (rigid bodies). In all such cases the resultant force on each particle may be resolved into a resultant *external impressed* force and a resultant *internal constraint* force. Thus

$$\mathbf{F}_i = \mathbf{F}_{ie} + \mathbf{F}_{ic}. \tag{6.31}$$

Now it is a characteristic of the constraint forces that when the particles are given displacements $\delta\mathbf{r}_i$ *compatible with the constraints*, the total work done by the constraint forces vanishes, i.e.,

$$\sum \mathbf{F}_{ic} \cdot \delta\mathbf{r}_i = 0.$$

We may indeed use this as a definition of the constraint forces. It follows that for a system in equilibrium, when the particles undergo displacements $\delta\mathbf{r}_i$ compatible with the constraints

$$\sum \mathbf{F}_{ie} \cdot \delta\mathbf{r}_i = 0, \tag{6.32}$$

i.e., the work done by the impressed forces also vanishes, even if the individual impressed forces do not vanish. This is the principle of *virtual work*. Since it is an important principle, it will be wise to give a simple illustration.

We may first note that the displacements in question need not be actual displacements performed by the system but only displacements which *might* be performed without violating the constraints. For that reason we have denoted the displacements by $\delta\mathbf{r}_i$ rather than by $d\mathbf{r}_i$, which we shall always reserve for *actual* displacements performed in time. The name *virtual* displacement arises in this way.

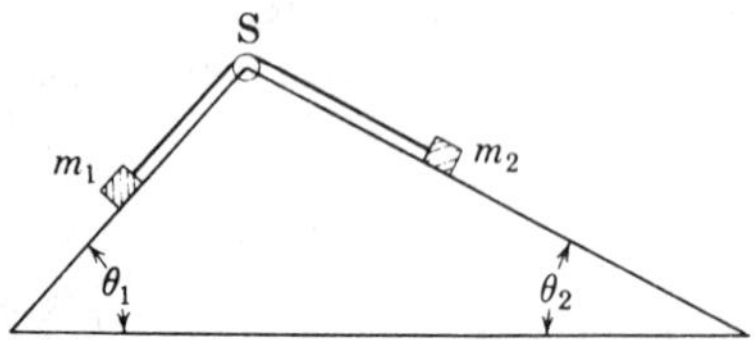

Fig. 6.1.

Consider now the two masses m_1 and m_2, respectively, on the two slopes of a smooth double inclined plane, as indicated in Fig. 6.1. The masses are tied together by an inextensible string, which passes over the smooth peg S at the vertex of the planes. If the system of two particles is in equilibrium, we may at once apply (6.32), with the impressed forces

$$\mathbf{F}_{1e} = m_1 \mathbf{g}; \quad \mathbf{F}_{2e} = m_2 \mathbf{g}. \tag{6.33}$$

Now a displacement compatible with the constraint is one directed *along* the plane. We therefore write

$$m_1 \mathbf{g} \cdot \delta \mathbf{r}_1 + m_2 \mathbf{g} \cdot \delta \mathbf{r}_2 = 0, \tag{6.34}$$

which becomes

$$m_1 g \sin \theta_1 \delta r_1 + m_2 g \sin \theta_2 \delta r_2 = 0. \tag{6.35}$$

Now since the string is inextensible

$$\delta r_1 + \delta r_2 = 0, \tag{6.36}$$

and therefore

$$m_1 g \sin \theta_1 = m_2 g \sin \theta_2 \tag{6.37}$$

is the condition of equilibrium. This could of course have been obtained by equating the resultant force on each particle to zero separately, but this would have entailed the introduction of the reaction force of the plane and the tension in the string. Neither of these enter into the expression of the principle of virtual work since they are constraint forces only.

The principle of virtual work is of value not merely in cases of static equilibrium but also in dynamic problems where equilibrium does not exist. This takes place through the agency of *D'Alembert's principle.*

Let us imagine a system of n particles with masses $m_1, m_2, \ldots m_n$ and acted on by external impressed forces $\mathbf{F}_{1e} \ldots \mathbf{F}_{ne}$, respectively. Concentrating attention on the jth particle (cf. Fig. 6.2), we note that if it were not for the constraints imposed on the system this particle would move with an acceleration given by $\mathbf{F}_{je}/m$. However, because of the constraints the actual acceleration is $\ddot{\mathbf{r}}_j$ and the corresponding force

$m_j\ddot{\mathbf{r}}_j$ may be called the *effective* force on the jth particle. The difference between the impressed force and effective force is denoted by $\mathbf{V}_j$, so that

$$\mathbf{F}_{je} - m_j\ddot{\mathbf{r}}_j = \mathbf{V}_j. \qquad (6.38)$$

D'Alembert's principle is the hypothesis that the system of particles is in equilibrium under the forces $\mathbf{V}_j$. We can at once apply the principle of virtual work and get

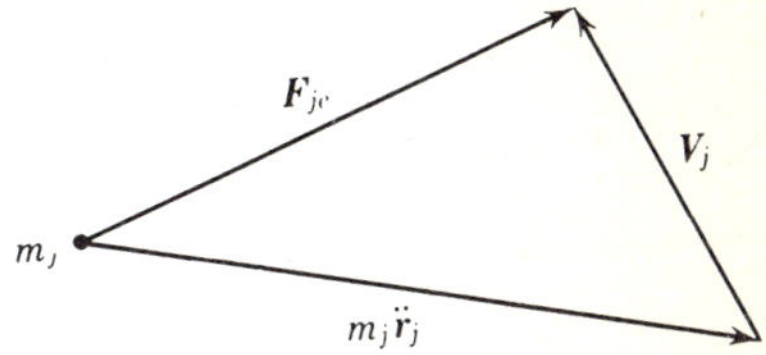

Fig. 6.2.

$$\sum_{j=1}^{n}(\mathbf{F}_{je} - m_j\ddot{\mathbf{r}}_j)\cdot\delta\mathbf{r}_j = 0 \qquad (6.39)$$

as the analytical expression of the principle. It effectively reduces every dynamical problem to a corresponding one in statics. As a simple illustration we may apply it to the case shown in Fig. 6.1, for which (6.39) becomes

$$(m_1\mathbf{g} - m_1\ddot{\mathbf{r}}_1)\cdot\delta\mathbf{r}_1 + (m_2\mathbf{g} - m_2\ddot{\mathbf{r}}_2)\cdot\delta\mathbf{r}_2 = 0. \qquad (6.40)$$

Realizing that $\ddot{\mathbf{r}}_1$ and $\ddot{\mathbf{r}}_2$ as well as $\delta\mathbf{r}_1$ and $\delta\mathbf{r}_2$ are directed along the two planes, respectively, we obtain, setting $|\ddot{\mathbf{r}}_1| = a_1$ and $|\ddot{\mathbf{r}}_2| = a_2$,

$$(m_1 g \sin\theta_1 - m_1 a_1)\delta r_1 + (m_2 g \sin\theta_2 - m_2 a_2)\delta r_2 = 0. \qquad (6.41)$$

But since $\delta r_1 = -\delta r_2$, it also follows that $a_1 = -a_2 = a$, and therefore

$$a = \frac{m_1 g \sin\theta_1 - m_2 g \sin\theta_2}{m_1 + m_2}, \qquad (6.42)$$

which, of course, agrees with the result of applying Newton's law, taktaking into account the constraint forces. The reader is referred to *Foundations of Physics* (pp. 102 ff.) for further methodological discussion of D'Alembert's principle and the associated Gaussian principle of least constraint.

6.4. The Double Pendulum as an Illustration of D'Alembert's Principle. As another illustration of D'Alembert's principle, which will lead us into an important class of problems involving the motion of a collection of particles, consider the double pendulum (Fig. 6.3) in which a flexible, inextensible weightless string has masses m_1 and m_2 attached to it at distances l_1 and $l_1 + l_2$, respectively, from the end O, which is then rigidly fixed. The system, which in equilibrium would hang along the vertical ON, is drawn aside so that the top string makes an angle ϕ_1 with the vertical and the bottom section an angle ϕ_2. It is assumed that ϕ_1 and ϕ_2 are always less than 10°. The particles are then released. Since gravity is the only impressed external force the motion takes place in a vertical plane. The instantaneous displacements of m_1 and m_2 from the vertical along circular arcs will be denoted by s_1 and s_2, respectively. If we use position vectors $\mathbf{r}_1$ and $\mathbf{r}_2$ based on O as origin to fix the instantaneous positions of the particles, the *effective* forces on m_1 and m_2 become $m_1\ddot{\mathbf{r}}_1$ and $m_2\ddot{\mathbf{r}}_2$, respectively. At first glance one would assign the weights $m_1\mathbf{g}$ and $m_2\mathbf{g}$ as the *impressed* forces. But this assignment is incomplete since it overlooks the force with which m_2 acts on m_1 by the tension it produces in the string joining it to m_1. Call this force $\mathbf{T}$. This is a good illustration of a possible source of difficulty in applying D'Alembert's principle: we must be sure to take account of *all* the impressed forces. Only experience in the solution of problems can guarantee accuracy in this regard. There appears to be no foolproof method. In general, if we are in doubt whether a particular force is a constraint force (many if not most tensions are) or an impressed force, it is the wiser course to include it in the set up of the principle. If it is a genuine constraint force it will in any case disappear.

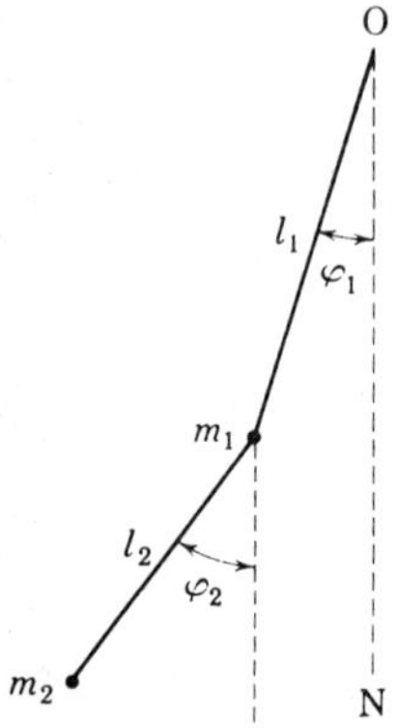

Fig. 6.3.

We are now in a position to express the principle in the form

$$(m_1\ddot{\mathbf{r}}_1 - m_1\mathbf{g} - \mathbf{T}) \cdot \delta\mathbf{r}_1 + (m_2\ddot{\mathbf{r}}_2 - m_2\mathbf{g}) \cdot \delta\mathbf{r}_2 = 0. \tag{6.43}$$

Now in $\ddot{\mathbf{r}}_1 \cdot \delta\mathbf{r}_1$ only the component of $\ddot{\mathbf{r}}_1$ along $\delta\mathbf{r}_1$, i.e., *along* the arc, will give a nonvanishing result. Hence we write

$$m_1\ddot{\mathbf{r}}_1 \cdot \delta\mathbf{r}_1 = m_1\ddot{s}_1\delta s_1, \quad m_2\ddot{\mathbf{r}}_2 \cdot \delta\mathbf{r}_2 = m_2\ddot{s}_2\delta s_2, \tag{6.44}$$

in which the positive direction of s_1 and s_2 is measured *away* from the vertical. We next have

$$\begin{aligned} m_1\mathbf{g} \cdot \delta\mathbf{r}_1 &\equiv -\frac{m_1 g s_1}{l_1}\delta s_1, \\ m_2\mathbf{g} \cdot \delta\mathbf{r}_2 &= -\frac{m_2 g(s_2 - s_1)}{l_2}\delta s_2. \end{aligned} \tag{6.45}$$

Finally

$$\mathbf{T} \cdot \delta \mathbf{r}_1 = m_2 g \sin(\phi_2 - \phi_1)\, ds_1, \tag{6.46}$$

in which we are indeed making the approximation that **T** is directed along the vertical, whereas it is really directed along l_2. However this approximation is valid since we are dealing with small angles and actually in (6.46) shall write

$$\sin(\phi_2 - \phi_1) \doteq \phi_2 - \phi_1 = \frac{s_2 - s_1}{l_2} - \frac{s_1}{l_1}. \tag{6.47}$$

Therefore

$$\mathbf{T} \cdot \delta \mathbf{r}_1 = m_2 g \left(\frac{s_2 - s_1}{l_2} - \frac{s_1}{l_1} \right) \delta s_1. \tag{6.48}$$

Eq. (6.43) then becomes

$$\left[m_1 \ddot{s}_1 + \frac{m_1 g s_1}{l_1} - m_2 g \left(\frac{s_2 - s_1}{l_2} - \frac{s_1}{l_1} \right) \right] \delta s_1 + \left[m_2 \ddot{s}_2 + m_2 g \frac{(s_2 - s_1)}{l_2} \right] \delta s_2 = 0. \tag{6.49}$$

Within the limit of small displacements to which we restrict our attention δs_1 and δs_2 are independently arbitrary and therefore we must have

$$m_1 \ddot{s}_1 = -\frac{(m_1 + m_2)}{l_1} g s_1 + m_2 g \frac{(s_2 - s_1)}{l_2}, \tag{6.50}$$

$$m_2 \ddot{s}_2 = -m_2 g \frac{(s_2 - s_1)}{l_2}. \tag{6.51}$$

These are the equations of motion of the system under the assumed approximation. The presence of both coordinates s_1 and s_2 in each equation is the mathematical expression that the two pendulums are here considered to influence each other, or are, we may say, *coupled*. In fact we shall say that this is an illustration of the motion of coupled systems. Instead of working out the details of this specific motion, let us examine the general properties of vibrations of coupled systems of which the double pendulum is a special case.

6.5. The Vibrations of Coupled Systems. Normal Coordinates. Let us suppose we have two particles of mass m_1 and m_2 subject to restoring forces due to stiffness coefficients k_1 and k_2. But let the systems be coupled together so that their equations of motion take the form

$$\begin{aligned} m_1\ddot{q}_1 + k_1 q_1 + c(q_1 - q_2) &= 0, \\ m_2\ddot{q}_2 + k_2 q_2 + c(q_2 - q_1) &= 0, \end{aligned} \tag{6.52}$$

where c is called the coupling coefficient. It is clear that (6.50) and (6.51) are in this form. If c becomes vanishingly small the two particles (or systems) vibrate independently with frequencies $\frac{1}{2\pi}\sqrt{k_1/m_1}$ and $\frac{1}{2\pi}\sqrt{k_2/m_2}$, respectively. However, as c gets larger the two oscillators influence each other more and more. In the case of the double pendulum the coupling coefficient is

$$c = m_2 g/l_2, \tag{6.53}$$

while the stiffness coefficients are $k_1 = (m_1 + m_2)g/l_1$ and $k_2 = 0$. Let us now investigate the motion. The form of the equations suggests the solution (Sec. 4.3),

$$q_1 = Ae^{i\omega t}; \qquad q_2 = Be^{i\omega t} \tag{6.54}$$

It might seem as if this were too highly specialized and that we should assume $q_1 = Ae^{i\omega_1 t}$; $q_2 = Be^{i\omega_2 t}$ where ω_1 and ω_2 are not necessarily the same. But substitution shows this will not work unless both A and B vanish identically. We might indeed satisfy our intuitive feeling that *two* independent frequencies are involved by assuming a solution in the form

$$\begin{aligned} q_1 &= A_1 e^{i\omega_1 t} + A_2 e^{i\omega_2 t}, \\ q_2 &= B_1 e^{i\omega_1 t} + B_2 e^{i\omega_2 t}. \end{aligned} \tag{6.55}$$

However, inspection shows that the result is ultimately the same as with the choice of (6.54). Hence we substitute (6.54) into (6.52) and obtain the equations for A and B

$$\begin{aligned} A[-m_1\omega^2 + k_1 + c] - Bc &= 0, \\ A[-c] + B[-m_2\omega^2 + k_2 + c] &= 0. \end{aligned} \tag{6.56}$$

The use of (6.55) will give *two* sets of equations of precisely this form for A_1 and B_1, and A_2 and B_2, respectively. In order that (6.56) shall have a nonvanishing solution for A and B it is necessary that the determinant of the coefficients shall vanish, i.e.,

$$\begin{vmatrix} -m_1\omega^2 + k_1 + c, & -c \\ -c, & -m_2\omega^2 + k_2 + c \end{vmatrix} = 0, \tag{6.57}$$

whence ω^2 must satisfy a quadratic equation with the roots

$$\omega_1{}^2 = k_1/2m_1 + k_2/2m_2 + c(1/2m_1 + 1/2m_2) - 1/2\sqrt{(k_1/m_1 - k_2/m_2 + c/m_1 - c/m_2)^2 + 4c^2/m_1m_2}, \tag{6.58}$$

$$\omega_2{}^2 = k_1/2m_1 + k_2/2m_2 + c(1/2m_1 + 1/2m_2) + 1/2\sqrt{(k_1/m_1 - k_2/m_2 + c/m_1 - c/m_2)^2 + 4c^2/m_1m_2}. \tag{6.59}$$

Since the radicand in (6.58) and (6.59) is always positive the roots are always real.

The equations (6.52) are linear and the general solution therefore is

$$\begin{aligned} q_1 &= A_1e^{i\omega_1t} + A_1'e^{-i\omega_1t} + A_2e^{i\omega_2t} + A_2'e^{-i\omega_2t}, \\ q_2 &= B_1e^{i\omega_1t} + B_1'e^{-i\omega_1t} + B_2e^{i\omega_2t} + B_2'e^{-i\omega_2t}. \end{aligned} \tag{6.60}$$

The resultant motion of each mass is thus essentially the composition of two simple harmonic motions with frequencies $\omega_1/2\pi$ and $\omega_2/2\pi$, respectively. These are termed the fundamental frequencies of the system. The reader may work out the frequencies for the special case of the double pendulum and show that here

$$\begin{aligned} \omega_1{}^2 &= \frac{g\mu}{2}\left[1/l_1 + 1/l_2 - \sqrt{(1/l_1 + 1/l_2)^2 - 4/\mu l_1 l_2}\right], \\ \omega_2{}^2 &= \frac{g\mu}{2}\left[1/l_1 + 1/l_2 + \sqrt{(1/l_1 + 1/l_2)^2 - 4/\mu l_1 l_2}\right], \end{aligned} \tag{6.61}$$

where $\mu = (m_1 + m_2)/m_1$.

In (6.60) the A's and B's are not all independent. In fact there can be only four arbitrary constants, since there are two equations of the second order. If we utilize (6.56) we get

$$B_1/A_1 = \frac{-m_1\omega_1{}^2 + k_1 + c}{c} = B_1'/A_1', \tag{6.62}$$

$$B_2/A_2 = \frac{c}{-m_2\omega_2{}^2 + k_2 + c} = B_2'/A_2', \tag{6.63}$$

which enable us to fix B_1, B_2, B_1' and B_2' in terms of arbitrarily chosen A_1, A_2, A_1', and A_2'. The special case in which $m_1 = m_2 = m$ and $k_1 = k_2 = k$, i.e., the two coupled systems are identical, is of interest. Here

$$\omega_1 = \sqrt{k/m}; \quad \omega_2 = \sqrt{\frac{k+2c}{m}}. \tag{6.64}$$

Moreover for this case $B_1 = A_1$ and $B_1' = A_1'$, whereas $B_2 = -A_2$ and $B_2' = -A_2'$. Hence (6.60) can now be boiled down to the more convenient form

$$\begin{aligned} q_1 &= A\cos(\omega_1 t - \alpha) + B\cos(\omega_2 t - \beta), \\ q_2 &= A\cos(\omega_1 t - \alpha) - B\cos(\omega_2 t - \beta), \end{aligned} \tag{6.65}$$

where the arbitrary constants are A, B, α, and β. This special case suggests a tranformation of coordinates which is often useful in the motion of coupled systems. Let

$$\frac{q_1 + q_2}{2} = X_1 = A\cos(\omega_1 t - \alpha), \tag{6.67}$$

$$\frac{q_1 - q_2}{2} = X_2 = B\cos(\omega_2 t - \beta). \tag{6.67}$$

The advantage of these coordinates is that each one is a harmonic function of the time with a frequency equal to one of the two fundamental frequencies of the motion. They are known as *normal coordinates*. In terms of them the equations of motion take the simple form

$$\begin{aligned} m\ddot{X}_1 + kX_1 &= 0, \\ m\ddot{X}_2 + (k+2c)X_2 &= 0. \end{aligned} \tag{6.68}$$

In terms of the normal coordinates the kinetic energy and potential energy become, respectively,

$$K = 1/2 \cdot (2m)(\dot{X}_1^2 + \dot{X}_2^2), \tag{6.69}$$

$$V = kX_1^2 + (k+2c)X_2^2. \tag{6.70}$$

The important thing to note here is the fact that V contains only squared terms. Written in terms of the q_1 and q_2, on the other hand, V will contain product terms as well. Further reference to normal coordinates is made in the next section (cf. also Sec. 8.5).

6.6. Vibrations of a Loaded String. An interesting generalization of the two coupled systems discussed in the preceding section is provided by a set of n particles all of mass m distributed uniformly along and attached firmly to a weightless horizontal string stretched with tension

T. This situation is depicted schematically in the accompanying diagram (Fig. 6.4). The string is fastened at the two ends O and O'. The separation between each pair of successive particles is a and the total length of the string is $(n+1)a = l$. The problem is to determine the motion of the system if one or more particles are displaced in the vertical plane and released. The displacement of the jth particle is denoted by q_j.

Fig. 6.4.

If we examine Fig. 6.5 which indicates the displacements of the $(j-1)$st, jth, and $(j+1)$ particles, we see that the resultant force on the jth particle due to the string and neglecting gravity is

$$-\frac{T}{a}(q_j - q_{j-1}) + \frac{T}{a}(q_{j+1} - q_j). \tag{6.71}$$

The further assumptions are made here that the displacements are so small that the sine of the angle made by each section of the string between consecutive particles with the horizontal may always be replaced by its tangent and that the tension is not altered by the displacement. The equation of motion of the jth particle then becomes

$$m\ddot{q}_j + k(-q_{j-1} + 2q_j - q_{j+1}) = 0, \tag{6.72}$$

where k has been written for T/a. We have n equations of the form (6.72), where j runs from 1 to n. The form of the equations again suggests the solution

$$q_j = A_j e^{i\omega t}, \tag{6.73}$$

and substitution gives

$$-A_{j-1} + (2 - m\omega^2/k)A_j - A_{j+1} = 0, \tag{6.74}$$

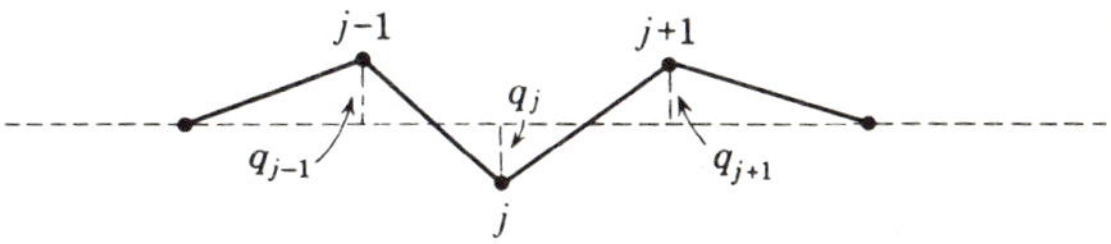

Fig. 6.5.

as the recursion formula for the coefficients. By the same argument used in Sec. 6.5 we are led to the necessity for the vanishing of the following nth order determinant, viz.,

$$\begin{vmatrix} C & -1 & 0 & 0 & 0 & \cdots \\ -1 & C & -1 & 0 & 0 & \cdots \\ 0 & -1 & C & -1 & 0 & \cdots \\ \cdots & \cdots & \cdots & \cdots & \cdots & \cdots \\ \cdots & \cdots & \cdots & \cdots & \cdots & \cdots \\ 0 & 0 & \cdot & \cdot & -1 & C \end{vmatrix} = 0, \tag{6.75}$$

where $C = 2 - m\omega^2/k$. The solution of this equation will yield n values of ω which then provide the frequencies of the motion in the same sense as in the simpler problem of the preceding section.

If we denote by Δ_n the value of the determinant in (6.75) with n rows and columns and by Δ_{n-1} the value of the same determinant with $n - 1$ rows, examination discloses the existence of the following functional equation for Δ_n

$$\Delta_n = C\Delta_{n-1} - \Delta_{n-2}. \tag{6.76}$$

If we let

$$C = 2 \cos \theta,$$

this suggests

$$\Delta_n = K \sin (n + 1)\theta. \tag{6.77}$$

In fact substitution shows that (6.77) satisfies (6.76) independently of K. However since $\Delta_1 = C$ by direct inspection, it follows that

$$K = 1/\sin \theta.$$

Hence

$$\Delta_n = \frac{\sin (n + 1)\theta}{\sin \theta}. \tag{6.78}$$

The roots of $\Delta_n = 0$ are

$$\theta = \frac{s\pi}{n + 1} \tag{6.79}$$

where s is any integer save $n + 1$ or multiples thereof. Consequently from the definition of C it results that the allowed frequencies are

$$\nu_s = \frac{1}{\pi}\sqrt{\frac{T}{ma}} \cdot \sin \frac{s\pi}{2(n + 1)}, \tag{6.80}$$

where $s = 1, 2, \cdots n$. There is thus a minimum frequency

$$\nu_1 = \frac{1}{\pi}\sqrt{\frac{T}{ma}} \sin \frac{\pi}{2(n+1)},$$

and a maximum frequency

$$\nu_n = \frac{1}{\pi}\sqrt{\frac{T}{ma}} \cdot \sin \frac{n\pi}{2(n+1)}.$$

Now from an examination of (6.74) we see that

$$A_{js} = \Delta_{j-1} A_{1s}, \tag{6.81}$$

where A_{js} is the value of the amplitude A_j associated with the frequency ν_s. If we let $A_{1s}/\sin\left(\frac{s\pi}{n+1}\right) = a_s$ we can write

$$A_{js} = a_s \sin \frac{js\pi}{n+1}, \tag{6.82}$$

and consequently the complete solution can be written in the form

$$q_j = \sum_{s=1}^{n} a_s \sin\left(\frac{sj\pi}{n+1}\right) \cdot \cos(\omega_s t + b_s), \tag{6.83}$$

which should be compared with the simpler case of (6.65),

It is of interest to see what happens as the number of mass particles grows without limit while the distance $a \to 0$. We still have

$$(n+1)a = l; \qquad \rho l = nm, \tag{6.84}$$

where ρ is the effective line density of the aggregate, the mass of each particle decreasing as n increases, so that the product nm remains finite. If now we let x denote the distance of the jth particle from the left end of the string we have $j = x/a$, and hence (6.83) becomes

$$q(x) = \sum_{s=1}^{n} a_s \sin\left(\frac{s\pi x}{l}\right) \cdot \cos(\omega_s t + b_s), \tag{6.85}$$

where

$$\omega_s = \frac{s\pi}{l}\sqrt{\frac{n}{n+1}} \cdot \sqrt{\frac{T}{\rho}} \cdot \frac{\sin[s\pi/2(n+1)]}{[s\pi/2(n+1)]}. \tag{6.86}$$

As $n \to \infty$ for any fixed s,

$$\lim \omega_s = \frac{s\pi}{l}\sqrt{\frac{T}{\rho}}, \tag{6.87}$$

where $s = 1, 2, 3, \cdots$ indefinitely. The frequencies now become infinite in number and are all harmonics of the fundamental frequency

$$\nu_{fund.} = \frac{1}{2l}\sqrt{\frac{T}{\rho}}. \tag{6.88}$$

The reader will recall this from elementary physics as the fundamental frequency of the continuous vibrating string of length l fastened at the two ends with tension T.

Reverting to (6.83), we see that the values of the a_s and b_s depend on the initial conditions. Denoting by $q_0(ja)$ and $\dot{q}_0(ja)$, respectively, the initial displacement and initial velocity of the jth particle we have

$$q_0(ja) = \sum_{s=1}^{n} a_s \cos b_s \cdot \sin\left(\frac{j\pi sa}{l}\right), \tag{6.89}$$

$$\dot{q}_0(ja) = -\sum_{s=1}^{n} \omega_s a_s \sin b_s \cdot \sin\left(\frac{j\pi sa}{l}\right). \tag{6.90}$$

The evaluation of $a_s \cos b_s$ and $a_s \sin b_s$ is performed by a process of considerable importance in theoretical physics. We multiply (6.89) by $\sin\left(\frac{r\pi ja}{l}\right)$, where r is integral (and $\leq n$) and sum both sides over j. Now from trigonometry

$$\sum_{j=1}^{n} \sin\left(\frac{j\pi r}{n+1}\right) \cdot \sin\left(\frac{j\pi s}{n+1}\right) = 0 \text{ for } s \neq r. \tag{6.91}$$

On the other hand,

$$\sum_{j=1}^{n} \sin^2\left(\frac{r\pi j}{n+1}\right) = \frac{n+1}{2}. \tag{6.92}$$

Consequently

$$a_r \cos b_r = \frac{2}{n+1} \cdot \sum_{j=1}^{n} q_0(ja) \sin\left(\frac{r\pi ja}{l}\right). \tag{6.93}$$

Similarly it can be shown that

$$-a_r \sin b_r = \frac{2}{(n+1)\omega_r} \cdot \sum_{j=1}^{n} \dot{q}_0(ja) \cdot \sin\left(\frac{r\pi ja}{l}\right). \tag{6.94}$$

These results remind us a good deal of the formula for the evalution of the coefficients of a Fourier series. In fact if the number of particles increases indefinitely so that the string becomes effectively continuous, (6.93) is replaced by

$$a_r \cos b_r = \frac{2}{l}\int_0^l q(x) \sin\left(\frac{r\pi x}{l}\right) dx \tag{6.95}$$

in the usual fashion.

Essentially the same method used above to determine the coefficients in the expansion for $q_j = q(ja)$ can be employed to establish the normal coordinates for the motion. We see indeed that from (6.93) we can equally well write

$$X_r = a_r \cos(\omega_r t + b_r) = \left(\frac{2}{n+1}\right) \cdot \sum_{j=1}^{n} q(ja) \sin\left(\frac{r\pi ja}{l}\right), \tag{6.96}$$

and this particular linear combination of the $q(ja)$ will serve as the rth normal coordinate associated with the rth frequency $\omega_r/2\pi$.

Some important general theorems about the vibrations of systems of many degrees of freedom will be found in Rayleigh, *Theory of Sound.*

The possibility of obtaining normal coordinates for the loaded string suggests the question: In any system of particles coupled by strings or springs providing restoring forces of linear type is it possible to find normal coordinates? The answer to this question is in the affirmative. This is discussed in Sec. 8.5.

PROBLEMS

1. Two particles of masses m_1 and m_2, respectively, are subject only to their mutual attraction varying inversely as the square of the

distance between them and directly as the product of their masses. What is the vector equation of motion of m_1 relative to m_2? Express this in cartesian form. Write the equation of motion of each particle relative to the center of mass of the system.

2. Two gravitating particles of masses m_1 and m_2, respectively, move about each other. Calculate the exact expression for the time for one complete period of the motion.

3. Carry through the demonstration of eq. (6.61).

4. A boat with mass 2 tons (including that of the passenger) is moving through the water with a velocity of 20 ft/sec with respect to the shore. The passenger throws a 2 lb ball with a velocity of 30 ft/sec relative to the boat in the direction in which the boat is headed. How much work does the man do in throwing the ball? (A result accurate to two decimal places is requested if the answer is expressed in foot poundals.)

5. Derive the expression for the mechanical advantage of the wheel and axle and the differential pulley by means of the principle of virtual work.

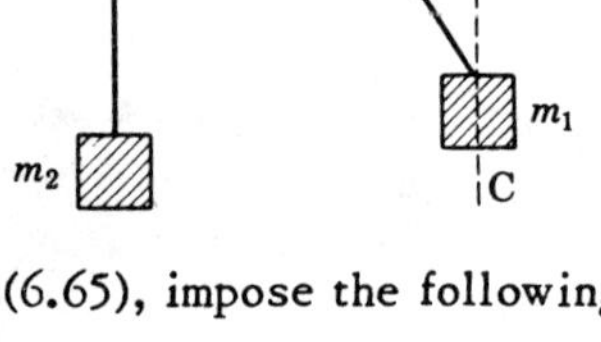

6. A particle of mass m_1 is constrained to move on a vertical straight line BC while attached to a string passing over a smooth peg P, with a particle of mass m_2 attached to the other end of the string. Use D'Alembert's principle to find the expressions for the instantaneous vertical accelerations of m_1 and m_2.

7. In the coupled system described by eqs. (6.65), impose the following boundary conditions:

$$q_1 = q_{10}, \quad \dot{q}_1 = 0 \text{ at } t = 0$$
$$q_2 = 0, \quad \dot{q}_2 = 0 \text{ at } t = 0.$$

Show that in this case both particles oscillate with frequency $(\omega_1 + \omega_2)/4\pi$ and with amplitude modulated by the term $\cos 1/2 . (\omega_1 - \omega_2)t$ and $\pi/2$ out of phase with each other. Could this situation be realized in the case of the double pendulum? What physical behavior would result?

8. Derive the expressions for the potential and kinetic energy of the aggregate of particles in the loaded string in terms of the normal coordinates.

9. Two particles of masses m_1 and m_2, respectively, move in arbitrary fashion. If v_c is the velocity of the center of mass and v the *relative* velocity of the two particles with respect to each other, show that the total kinetic energy of the system is

$$1/2 \cdot (m_1 + m_2)v_c^2 + 1/2 \cdot \left(\frac{m_1 m_2}{m_1 + m_2}\right)v^2.$$

The quantity $\frac{m_1 m_2}{m_1 + m_2}$ is called the *reduced* mass of the two particles. Compare with the results of Prob. 1 and with eq. (6.29).

10. A stationary nucleus of mass m is struck head-on by a neutron of mass m_n moving with velocity v_0. If the collision is perfectly elastic find the expression for the recoil velocity of the nucleus. How could this equation be used to determine the mass of the neutron, even if the velocity v_0 is not known?

11. A ball of mass m and an attached flexible chain of line density ρ and indefinite length originally rest on a smooth horizontal plane. If the ball is tossed vertically into the air with initial velocity v_0, how high will it rise and for how long?

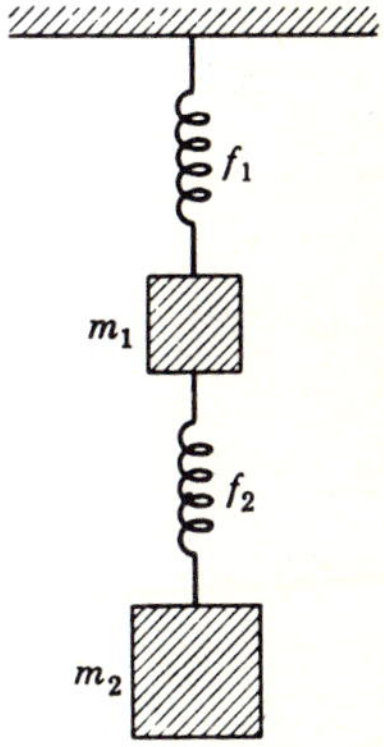

12. The accompanying schematic diagram shows masses m_1 and m_2 suspended by springs with stiffnesses f_1 and f_2, respectively. Assume that a harmonic force of frequency ν acts on m_1. Find the steady state displacement of each mass. What happens when the resonance frequency of m_2 by itself is equal to ν? What practical application might this have?

13. Use D'Alembert's principle to write the equations of motion of the system consisting of three equally spaced particles of masses m_1, m_2, m_3, respectively, attached to a stretched horizontal string of length l. Compare with the treatment of Sec. 6.6. Find the frequencies of the system.

14. Solve Prob. 12 under the assumption that each mass moves subject to a resisting force directly proportional to the velocity, and let the damping coefficients be R_1 and R_2, respectively.

CHAPTER VII

DYNAMICS OF A RIGID BODY

7.1. Definition of a Rigid Body. Types of Motion. In Chapter 6 we discussed in general fashion the motion of a collection of particles subject to external forces and internal forces of interaction. A very important special case of such a collection is that in which the distance between any two particles remains unaltered in time. The aggregate is then termed a *rigid body*. Strictly speaking no such thing exists in our experience, but many bodies approximate sufficiently to it to make worth while the special study of what has come to be called *rigid dynamics*.

Since many large-scale bodies encountered in engineering practice are rigid to a sufficiently good approximation, the mechanics of a rigid body has for a long time had a peculiarly technological flavor. This is true both of rigid bodies in equilibrium, i.e., rigid statics, and rigid bodies in various types of translational and rotational motion. Modern physics with its emphasis on atomic particles and the electromagnetic field has tended gradually to pay less attention to rigid bodies, and there are some physicists who feel the study of them may well be omitted from books on the concepts of theoretical physics. This appears to be ill-advised, however, for several reasons, of which two are worth mentioning here. In the first place the development of the concepts of rigid body motion was an important step in the historical evolution of physical thought, and every physicist should obtain more than a mere bowing acquaintance with it. More important indeed is the fact that many of the concepts invented primarily for the rigid body, e.g., the center of mass, moving axes, etc., have proved to be of considerable value in modern atomic physics. We therefore need not apologize for the introduction of this brief survey of the methods of rigid dynamics.

The motion of an aggregate of particles can be extremely complicated, although we did establish in Chapter 6 certain general theorems concerning the total momentum, moment of momentum, and energy of such a collection. These are of course immediately applicable to rigid bodies with certain simplifications, since the possible types of motion are now considerably restricted. This is clear from an examination of the specification of the *position* of a rigid body. If three non-collinear points are fixed the position of the rigid body is completely determined. Since each point is specified by three coordinates it might therefore seem that nine space coordinates are necessary to fix the position of a rigid body. However, since the three mutual separation distances are fixed and invariant, the number of independent coordinates necessary for the unique determination of the position of the body is reduced to six, which therefore is the number of *degrees of freedom* of the rigid body.

The above considerations enable us to classify the motion of a rigid body into two principal types: (1) *translation*, in which all particles move with the same velocity and acceleration in parallel paths, and (2) *rotation*, in which some point or line of points (an axis) remains fixed in space. It is not difficult to prove that every displacement of a rigid body can be considered as a combination of translations and rotations. Clearly the translation is equivalent in every case to the translation of the center of mass (cf. Sec. 6.1). Hence it is not necessary to consider this further here. We proceed to concentrate on rotation.

7.2. General Equations of Motion. Before proceeding to set up the general equations of motion it will be necessary to consider carefully the meaning of rotation and in particular the definition of angular velocity. In Fig. 7.1, let OQ denote an axis fixed in space. It may be in the rigid body under consideration or it may not. Rotation of the rigid body about this axis then means the following: Every point P of the rigid body moves in a circle with its center on OQ and with radius equal to PN, where N is the perpendicular projection of P on the axis. If $\mathbf{r}$ is the position vector of P relative to the origin O, the circle in question lies in a plane perpendicular to that formed by $\mathbf{r}$ and the axis. The velocity of P, namely $\dot{\mathbf{r}}$, is a vector lying always in the plane of the circle and is always perpendicular to the plane of $\mathbf{r}$ and the axis. In time dt, P will move along the circle through an arc $|\dot{\mathbf{r}}|dt$. From elementary considerations this is just equal to NP multiplied by the angle through which NP moves in time dt. If we call the latter $d\phi$ we then have, since $\overline{NP} = r \sin \theta$,

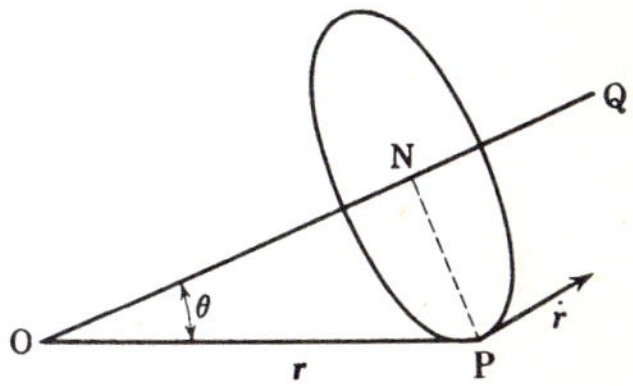

Fig. 7.1.

$$|\dot{\mathbf{r}}|dt = r \sin \theta \cdot d\phi. \tag{7.1}$$

Dividing through by dt and calling $d\phi/dt = \omega$, we have

$$|\dot{\mathbf{r}}| = \omega r \sin \theta. \tag{7.2}$$

Now the form of the right-hand side of (7.2) suggests strongly that it may be treated as the magnitude of the cross product (cf. Sec. 5.2) of the position vector $\mathbf{r}$ with a new vector of magnitude ω making angle θ with $\mathbf{r}$ in such a way that $\dot{\mathbf{r}}$ is perpendicular to the plane of the new vector and $\mathbf{r}$. If the new vector lies along OQ and in the direction indicated it will fill these requirements precisely. We shall call it the *angular velocity* of the rigid body (it is of course the same for every particle by virtue of the rigidity) and denote it by $\boldsymbol{\omega}$. Then the vector formulation corresponding to (7.2) is

$$\dot{\mathbf{r}} = \boldsymbol{\omega} \times \mathbf{r}. \tag{7.3}$$

It may be shown that $\boldsymbol{\omega}$ defined in this way does indeed possess the properties of a three vector laid down in Chapter 2. In particular the resultant of two angular velocities passing through the same point is found by using the parallelogram rule. This is equivalent to saying that we can always resolve any angular velocity $\boldsymbol{\omega}$ into rectangular components ω_x, ω_y, ω_z, so that

$$\boldsymbol{\omega} = \mathbf{i}\omega_x + \mathbf{j}\omega_y + \mathbf{k}\omega_z. \tag{7.4}$$

It is well to emphasize that the vector property of angular velocity is not shared by a finite rotation, even though the latter can be given a direction and a magnitude. The angular velocity $\boldsymbol{\omega}$ defined here is, of course, instantaneous angular velocity and will in general be a function of the time.

Let us now suppose that the particle at P is the jth particle of the rigid body with mass m_j. In Sec. 6.1 (eq. 6.13) we defined the moment of momentum of the jth particle of a collection of particles, about the origin, as

$$\mathbf{r}_j \times m_j\dot{\mathbf{r}}_j.$$

Consequently the total instantaneous moment of momentum of the rigid body about the fixed point is given by

$$\mathbf{M} = \sum_{j=1}^{n} m_j\mathbf{r}_j \times (\boldsymbol{\omega} \times \mathbf{r}_j). \tag{7.5}$$

It is customary to call this the *angular momentum* of the rigid body. Utilizing the expansion formula

$$\mathbf{r} \times (\boldsymbol{\omega} \times \mathbf{r}) = \boldsymbol{\omega}(\mathbf{r} \cdot \mathbf{r}) - \mathbf{r}(\boldsymbol{\omega} \cdot \mathbf{r}),$$

we can expand $\mathbf{M}$ and obtain the following expression in cartesian components

$$\mathbf{M} = \mathbf{i}\left[\omega_x \sum m_j(y_j^2 + z_j^2) - \omega_y \sum m_j x_j y_j - \omega_z \sum m_j x_j z_j\right]$$
$$+ \mathbf{j}\left[-\omega_x \sum m_j x_j y_j + \omega_y \sum m_j(x_j^2 + z_j^2) - \omega_z \sum m_j y_j z_j\right] \tag{7.6}$$
$$+ \mathbf{k}\left[-\omega_x \sum m_j x_j z_j - \omega_y \sum m_j y_j z_j + \omega_z \sum m_j(x_j^2 + y_j^2)\right].$$

This may be written more compactly in the form

$$\begin{aligned} \mathbf{M} = \mathbf{i}(\omega_x I_{xx} - \omega_y I_{xy} - \omega_z I_{xz}) \\ + \mathbf{j}(-\omega_x I_{yx} + \omega_y I_{yy} - \omega_z I_{yz}) \\ + \mathbf{k}(-\omega_x I_{zx} - \omega_y I_{zy} + \omega_z I_{zz}). \end{aligned} \tag{7.7}$$

The reader will recall from elementary physics that I_{xx}, I_{yy}, I_{zz} are the *moments of inertia* of the body about the x, y, and z axes, respectively. The less familiar quantities I_{xy}, I_{yz}, and I_{xz} are termed *products of inertia* associated with the corresponding coordinate planes. Clearly $I_{xy} = I_{yx}$, etc. By proper choice of axes the products of inertia can be made to vanish, and **M** will contain only moments of inertia. In this case the axes are termed *principal* axes of the body.

The fundamental equation (6.13) of Chapter 6 still applies and may now be written in the simple form

$$\dot{\mathbf{M}} = \mathbf{L}, \tag{7.8}$$

where **L** is the total torque or force moment about the instantaneous axis, while **M** is the resultant angular momentum about the same axis. Reference to eq. (6.15) of Chapter 6 reminds us that eq. (7.8) above also applies if the quantities in question refer to the center of mass. This enables us to handle readily the case of a rigid body in which no point is fixed. The translation of the center of mass provides no difficulty, and we can apply (7.8) to get the rotation about the center of mass. For convenience in what follows we shall treat the body as if one point were fixed, realizing that there is really no loss of generality in this assumption.

7.3. Moving Axes. So far we have been using axes fixed in space, for which **i**, **j**, **k** remain unchanged in time. However, as time goes on

ω_x, ω_y, ω_z will in general change and so will the moments and products of inertia. Hence the change of **M** with time will be rather complicated. We can actually simplify matters somewhat by using axes fixed in the rigid body, for with respect to these the I_{xx}, I_{xy}, etc., remain invariant. Of course the axes now move with the body and hence **i**, **j**, **k** will change with the time relative to the primary inertial system. This must be taken into account in calculating **M**. Thus from (7.7) if we denote the components of **M** along the axes by M_x, M_y, and M_z respectively, we get

$$\dot{\mathbf{M}} = \mathbf{i}\dot{M}_x + \mathbf{j}\dot{M}_y + \mathbf{k}\dot{M}_z + \frac{d\mathbf{i}}{dt}M_x + \frac{d\mathbf{j}}{dt}M_y + \frac{d\mathbf{k}}{dt}M_z. \tag{7.9}$$

To evaluate this it is necessary to find $d\mathbf{i}/dt$, etc. The problem is somewhat simplified by the fact that we are still keeping one point fixed and hence the only possible motion of the axes is one of rotation.

Since **i** is a unit vector, $d\mathbf{i}/dt$ must be perpendicular to **i** and hence must lie in the plane of **j** and **k**. Therefore we can write

$$\frac{d\mathbf{i}}{dt} = A_3\mathbf{j} - A_2\mathbf{k}, \tag{7.10}$$

where A_3 and A_2 are at first undetermined coefficients. Similarly,

$$\frac{d\mathbf{j}}{dt} = A_1\mathbf{k} - B_3\mathbf{i}; \quad \frac{d\mathbf{k}}{dt} = B_2\mathbf{i} - B_1\mathbf{j}. \tag{7.11}$$

But we already know from the definition in (5.15) that

$$\mathbf{i} = \mathbf{j} \times \mathbf{k}; \quad \mathbf{j} = \mathbf{k} \times \mathbf{i}; \quad \mathbf{k} = \mathbf{i} \times \mathbf{j}.$$

Hence

$$\frac{d\mathbf{i}}{dt} = \frac{d\mathbf{j}}{dt} \times \mathbf{k} + \mathbf{j} \times \frac{d\mathbf{k}}{dt}, \tag{7.12}$$

and therefore

$$A_3\mathbf{j} - A_2\mathbf{k} = (A_1\mathbf{k} - B_3\mathbf{i}) \times \mathbf{k} + \mathbf{j} \times (B_2\mathbf{i} - B_1\mathbf{j}). \tag{7.13}$$

It follows that

$$A_3 = B_3 \text{ and } A_2 = B_2. \tag{7.14}$$

Similarly we can show that $A_1 = B_1$. Hence finally

$$\frac{d\mathbf{i}}{dt} = A_3\mathbf{j} - A_2\mathbf{k}; \quad \frac{d\mathbf{j}}{dt} = A_1\mathbf{k} - A_3\mathbf{i}; \quad \frac{d\mathbf{k}}{dt} = A_2\mathbf{i} - A_1\mathbf{j}, \tag{7.15}$$

and the 6 coefficients are reduced to 3 independent ones. To see the significance of the coefficients A_1, A_2, A_3, dot multiply $d\mathbf{i}/dt$ with $\mathbf{j}$, etc., and get

$$\mathbf{j} \cdot \frac{d\mathbf{i}}{dt} = A_3; \quad \mathbf{k} \cdot \frac{d\mathbf{j}}{dt} = A_1; \quad \mathbf{i} \cdot \frac{d\mathbf{k}}{dt} = A_2. \tag{7.16}$$

This means, for example, that A_3 is the component of $d\mathbf{i}/dt$ along the y axis. Similarly, $\mathbf{j} \cdot d\mathbf{i}/dt \cdot (dt) = A_3 dt$ is the component of $d\mathbf{i}$ along the y axis. We represent this in Fig. 7.2 in which $\mathbf{j} \cdot d\mathbf{i}$ clearly is the angle $d\theta$ through which the x axis rotates about the z axis due to the change $d\mathbf{i}$ during the time dt. But this $d\theta$ is $\omega_z\, dt$ by definition, and hence we have

$$A_3 = \omega_z, \tag{7.17}$$

with the analogous results obtained similarly

$$A_1 = \omega_x; \quad A_2 = \omega_y. \tag{7.18}$$

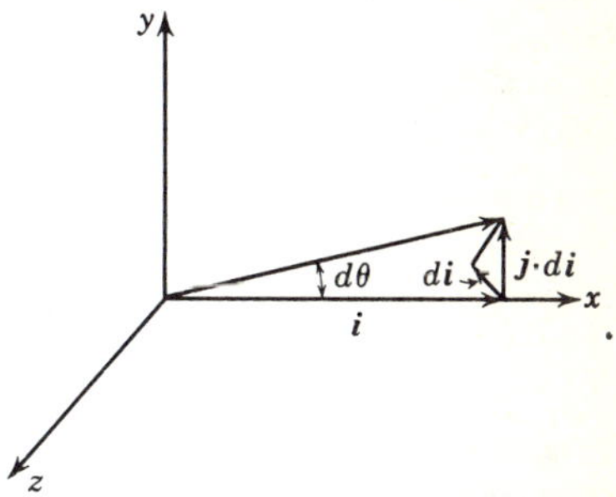

Fig. 7.2.

We can now go back to (7.9) and write

$$\dot{\mathbf{M}} = \mathbf{i}\dot{M}_x + \mathbf{j}\dot{M}_y + \mathbf{k}\dot{M}_z + \mathbf{i}(M_z\omega_y - M_y\omega_z) + \mathbf{j}(M_x\omega_z - M_z\omega_x)$$
$$+ \mathbf{k}(M_y\omega_x - M_x\omega_y) = \mathbf{i}\dot{M}_x + \mathbf{j}\dot{M}_y + \mathbf{k}\dot{M}_z + \boldsymbol{\omega} \times \mathbf{M}. \tag{7.19}$$

This is a general equation whose physical meaning will become clearer if we consider some special cases. First, suppose that instead of a single point being fixed there is a fixed line of points or axis and take this as the z axis. The x and y axes rotate then with the body about the z axis and we have $\omega_x = \omega_y = 0$ and $\omega_z = \omega$. Consequently from (7.7)

$$\mathbf{M} = \omega(-\mathbf{i}I_{xz} - \mathbf{j}I_{yz} + \mathbf{k}I_{zz}), \tag{7.20}$$

and the equation of motion (7.8) becomes

$$\mathbf{L} = \mathbf{i}(-I_{xz}\dot{\omega} + I_{yz}\omega^2) + \mathbf{j}(-I_{yz}\dot{\omega} - I_{xz}\omega^2) + \mathbf{k}I_{zz}\dot{\omega}. \tag{7.21}$$

If we specialize still further and suppose that the body is symmetrical about the xy plane, $I_{yz} = I_{xz} = 0$, and (7.21) reduces to

$$\mathbf{L} = \mathbf{k}I_{zz}\dot{\omega}, \tag{7.22}$$

which is recognized as the standard elementary physics equation for rotational acceleration about a fixed axis.

7.4. Euler's Equations for a Rigid Body. Let us revert to the situation in which one point of the body is fixed. As before we choose this point as origin of a rectangular coordinate system fixed in the body. However, we shall orient these axes so as to make the products of inertia vanish, i.e., choose principal axes. This is always possible. For example, if the body is symmetrical about the fixed point, the three axes of symmetry will suffice. Hence $\mathbf{M}$ will take the form

$$\mathbf{M} = \mathbf{i}\omega_x I_{xx} + \mathbf{j}\omega_y I_{yy} + \mathbf{k}\omega_z I_{zz}, \tag{7.23}$$

and (7.8) will become

$$\begin{aligned} \mathbf{L} = \mathbf{i}[I_{xx}\dot{\omega}_x + (I_{zz} - I_{yy})\omega_y\,\omega_z] \\ + \mathbf{j}[I_{yy}\dot{\omega}_y + (I_{xx} - I_{zz})\omega_z\,\omega_x] \\ +\mathbf{k}[I_{zz}\dot{\omega}_z + (I_{yy} - I_{xx})\omega_x\,\omega_y]. \end{aligned} \tag{7.24}$$

This is called *Euler's equation.* More often the three component equations obtained from (7.24) are known as Euler's equations for a rigid body moving with one point fixed.

The special case of $\mathbf{L} = 0$ is of interest as an illustration. We then get

$$\begin{aligned} I_{xx}\dot{\omega}_x &= (I_{yy} - I_{zz})\omega_y\omega_z, \\ I_{yy}\dot{\omega}_y &= (I_{zz} - I_{xx})\omega_z\omega_x, \\ I_{zz}\dot{\omega}_z &= (I_{xx} - I_{yy})\omega_x\omega_y. \end{aligned} \tag{7.25}$$

Multiplying through these equations by ω_x, ω_y, ω_z, respectively, and adding gives

$$\frac{d}{dt}[I_{xx}\omega_x^2 + I_{yy}\omega_y^2 + I_{zz}\omega_z^2] = 0, \tag{7.26}$$

$$I_{xx}\omega_x^2 + I_{yy}\omega_y^2 + I_{zz}\omega_z^2 = \text{constant}. \tag{7.27}$$

7.5. Further Consideration of Moving Axes. Kinetic Energy. To understand the use of Euler's equations and in particular the significance of (7.27) it is essential to consider further the moving axes we have already used in Sec. 7.3. It will do no harm to make the treatment slightly more general than previously. In Fig. 7.3 we represent a set of rectangular axes fixed with respect to the primary inertial system by x_f, y_f, z_f with origin at O_f. Similarly we introduce a set of moving axes x_m, y_m, z_m with center at O_m, whose position vector in the fixed system is $\mathbf{r}_0$. This of course can change with the time. The position vector of any particular particle P in the moving system is $\mathbf{r}_m$. From the figure we have

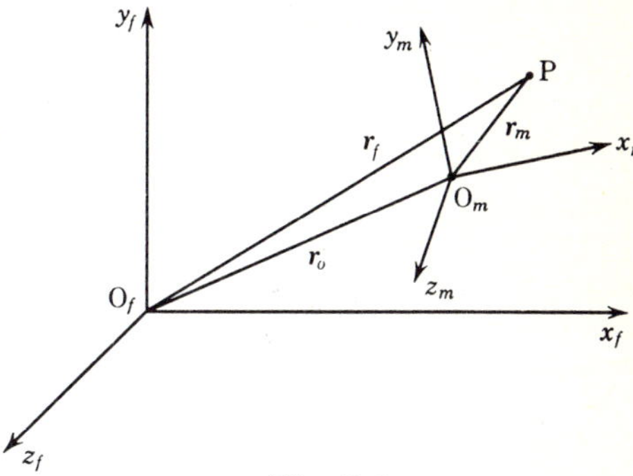

Fig. 7.3

$$\mathbf{r}_m = \mathbf{r}_f - \mathbf{r}_0, \quad (7.28)$$

and likewise for the corresponding velocities

$$\dot{\mathbf{r}}_m = \dot{\mathbf{r}}_f - \dot{\mathbf{r}}_0. \quad (7.29)$$

Let the unit vectors in the fixed system be $\mathbf{i}_f$, $\mathbf{j}_f$, $\mathbf{k}_f$ and those in the moving system be $\mathbf{i}_m$, $\mathbf{j}_m$, $\mathbf{k}_m$. The former remain unchanged in time, whereas the latter change with the time, since the directions of the moving axes vary with time with respect to the fixed axes. Hence we can write in analogy with (7.9)

$$\begin{aligned}\dot{\mathbf{r}}_m &= \mathbf{i}_m\dot{x}_m + \mathbf{j}_m\dot{y}_m + \mathbf{k}_m\dot{z}_m \\ &\quad + x_m\frac{d\mathbf{i}_m}{dt} + y_m\frac{d\mathbf{j}_m}{dt} + z_m\frac{d\mathbf{k}_m}{dt}.\end{aligned} \quad (7.30)$$

The first three years yield the *apparent velocity* of P relative to the moving axes, whereas the last three terms represent the contribution to the velocity of P due to the rotation of the moving axes. We have already worked out the expressions for $\dfrac{d\mathbf{i}_m}{dt}$ etc., in Sec. 7.3, and hence can replace (7.30) with

$$\dot{\mathbf{r}}_m = \dot{\mathbf{r}}_{ma} + \boldsymbol{\omega} \times \mathbf{r}_m, \quad (7.31)$$

where $\dot{\mathbf{r}}_{ma}$ is the apparent velocity of P in the moving system. We need however to make sure of the meaning of $\boldsymbol{\omega}$. Actually it is the angular velocity of the moving axes about the axis of rotation through O_m. We show first that $\boldsymbol{\omega}$ has the direction of the instantaneous axis of rotation through O_m. Since the instantaneous axis is the locus of points which have at that instant *zero* linear velocity as far as rotation is concerned, if P lies on the axis through O_m, we must have from (7.30)

$$x_m \frac{d\mathbf{i}_m}{dt} + y_m \frac{d\mathbf{j}_m}{dt} + z_m \frac{d\mathbf{k}_m}{dt} = 0, \tag{7.32}$$

But this means

$$\boldsymbol{\omega} \times \mathbf{r}_m = 0, \tag{7.33}$$

whence $\boldsymbol{\omega}$ and $\mathbf{r}_m$ are parallel. Thus $\boldsymbol{\omega}$ has the same direction as the axis of rotation through O_m. We can next show that the magnitude of $\boldsymbol{\omega}$ is equal to that of the angular velocity of the axes about the instantaneous axis through O_m. Consider a point P not on the axis of rotation. Its translational velocity due to the rotation of the axes has the magnitude $\omega r_m \sin\theta$, where θ is the angle between r_m and ω. Moreover the direction of the translational velocity is at right angles to the plane of $\boldsymbol{\omega}$ and $\mathbf{r}_m$. This velocity is therefore precisely that of a point moving in a circle of radius $r_m \sin\theta$ with angular speed ω about the center, which in this case lies on the axis of rotation. Finally the sign of $\boldsymbol{\omega}$ agrees with that of the rotation of the axes as is also clear from (7.31).

We can now go back to (7.29) and write

$$\dot{\mathbf{r}}_f = \dot{\mathbf{r}}_0 + \dot{\mathbf{r}}_{ma} + \omega \times \mathbf{r}_m. \tag{7.34}$$

If the rigid body has one point fixed and the moving axes are fixed in the body, clearly $\dot{\mathbf{r}}_0 = 0$ as well as $\dot{\mathbf{r}}_{ma} = 0$, since the various particles of the body have zero velocity with respect to these axes. Hence

$$\dot{\mathbf{r}}_f = \omega \times \mathbf{r}_m, \tag{7.35}$$

where $\mathbf{r}_m$ is now the position vector of P in the axes fixed in the body and $\dot{\mathbf{r}}_f$ is the velocity of the corresponding point with respect to a set of axes fixed in space.

We can now find the kinetic energy of the rigid body relative to the fixed axes. Thus, summing over all particles of the body,

$$K = \sum \frac{1}{2} m_i \dot{\mathbf{r}}_{fi} \cdot \dot{\mathbf{r}}_{fi} = \sum \frac{1}{2} m_i (\omega \times \mathbf{r}_{mi}) \cdot (\omega \times \mathbf{r}_{mi}). \tag{7.36}$$

Utilizing the fact that the cross and dot in the vector product can be interchanged without changing the value of the product, we can write from (7.5)

$$K = \frac{1}{2}\omega \cdot \mathbf{M}. \tag{7.37}$$

Since the rigid body is fixed at one point it is customary to refer to K in (7.37) as the kinetic energy of rotation of the rigid body. We should expect a connection between this and the work done by the resultant *torque* during the rotation of the rigid body, similar to that connecting the kinetic energy of translation and the work done by the resultant *force* during translation (cf. eq. 3.26). From (7.35) it follows that during time dt, the work done is

$$dW = \sum \mathbf{F}_i \cdot \dot{\mathbf{r}}_{fi}\, dt = \sum (\omega \times \mathbf{r}_{mi}) \cdot \mathbf{F}_i\, dt. \tag{7.38}$$

Once more interchanging dot and cross we obtain

$$dW = \sum \mathbf{L}_i \cdot \omega\, dt = \mathbf{L} \cdot \omega\, dt, \tag{7.39}$$

where $\mathbf{L}$ is the resultant torque about the fixed point measured with respect to axes fixed in the body. Now we revert to (7.7) and form $d\mathbf{M}$, remembering that since the axes are fixed in the body the I_{xx}, etc., do not change. We then can show that

$$\omega \cdot d\mathbf{M} = d\left[\frac{1}{2}\omega \cdot \mathbf{M}\right] = dK. \tag{7.40}$$

The combination of (7.8), (7.39), and (7.40) yields on integration between any two configurations of the body

$$W = K_1 - K_0, \tag{7.41}$$

the expected relation.

The special case of zero resultant torque is of particular interest. From (7.8) there follows here

$$\mathbf{M} = \text{Constant}, \tag{7.42}$$

and hence from (7.40) we also have

$$2K = \omega \cdot \mathbf{M} = \text{Constant}. \tag{7.43}$$

Consider an angular velocity space in which every point corresponds to a set of values of ω_x, ω_y, ω_z. Since

$$\boldsymbol{\omega} \cdot \mathbf{M} = I_{xx}\omega_x^2 + I_{yy}\omega_y^2 + I_{zz}\omega_z^2 - 2I_{xy}\omega_x\omega_y - 2I_{yz}\omega_y\omega_z - 2I_{zx}\omega_z\omega_x, \quad (7.44)$$

it follows that (7.43) is the equation of a quadric surface in this space. By proper choice of the axes fixed in the body the products of inertia may be made to vanish and the quadric surface becomes

$$I_{xx}\omega_x^2 + I_{yy}\omega_y^2 + I_{zz}\omega_z^2 = 2K. \quad (7.45)$$

This (which is, of course, identical with eq. 7.27) is the equation of an ellipsoid in the angular velocity space with its center at the origin. The magnitude of the vector from the origin of coordinates in the ω space to any point on the ellipsoid represents a possible value of the angular speed consistent with the given kinetic energy and angular momentum. The ellipsoid has been called the *momental* ellipsoid or ellipsoid of Poinsot. It is clear that in the direction of the minor axis of the ellipsoid (the principal axis of maximum moment of inertia) the angular speed is least, whereas in the direction of the major maxis the angular speed is greatest.

The change in $\boldsymbol{\omega}$ in passing from one point of the momental ellipsoid to another is given by

$$d\boldsymbol{\omega} \cdot \mathbf{M} = 0. \quad (7.46)$$

Now $d\omega$ must lie in the plane tangent to the ellipsoid at the point in question. Hence the tangent plane is perpendicular to **M**. But **M** is a constant vector since the resultant torque is zero. Therefore in this case the tangent plane remains fixed in position relative to a set of fixed axes. If we set

$$\mathbf{p} = \mathbf{M}/2K, \quad (7.47)$$

the vector $\mathbf{p}$ is normal to the tangent plane, and eq. (7.43) yields

$$\boldsymbol{\omega} \cdot \mathbf{p} = 1. \quad (7.48)$$

But $\boldsymbol{\omega} \cdot \mathbf{p}$ is p times the projection of $\boldsymbol{\omega}$ along the normal to the tangent plane. It follows that the tangent plane remains at a constant distance from the origin as the motion goes on. The motion of the rigid body then can be represented by a rolling (without slipping) of the momental ellipsoid on the fixed tangent plane. During the rolling the vector $\boldsymbol{\omega}$ traces out a cone relative to the axes fixed in the body with vertex at the fixed point. This cone intersects the momental ellipsoid in a curve called by Poinsot the *polhode*. It intersects the invariable tangent plane in a curve called the *herpolhode*. For illustrations of these curves, A. G. Webster's *Dynamics* may be consulted.

7.6. The Motion of a Top. In this application of Euler's equations the top is assumed to be a solid of revolution with one end of the axis of symmetry (i.e., the peg) fixed in space. This axis is also taken as the spin axis. The fixed point (cf. Fig. 7.4) is taken as the origin O of a set of axes fixed in space x_f, y_f, z_f. With O as origin we also set up another set of axes x_m, y_m, z_m in which z_m is fixed to the body along its axis of symmetry. Ordinarily we should also expect to fix the axes x_m and y_m to the body, but it now proves more convenient to let x_m and y_m rotate with angular velocity ω which is different from that of spin, which we denote by **s**. This will not affect the application of our previous reasoning since the moments of inertia about all axes perpendicular to the axis of spin are the same (i.e., $I_{xx} = I_{yy}$) and do not change with the time as the axes x_m and y_m rotate. We choose as the x_m axis the line of intersection of the plane through O perpendicular to the z_m axis and the x_f, y_f plane. Let the angle between x_m and x_f be ψ and that between z_m and z_f be θ. These are usually known as the Eulerian angles. There is a third, namely ϕ, which a line OQ rotating with the top makes with the x_m axis. The components ω_x, ω_y, ω_z about the x_m, y_m and z_m axes can be expressed in terms of $\dot{\theta}$ and $\dot{\psi}$ as follows:

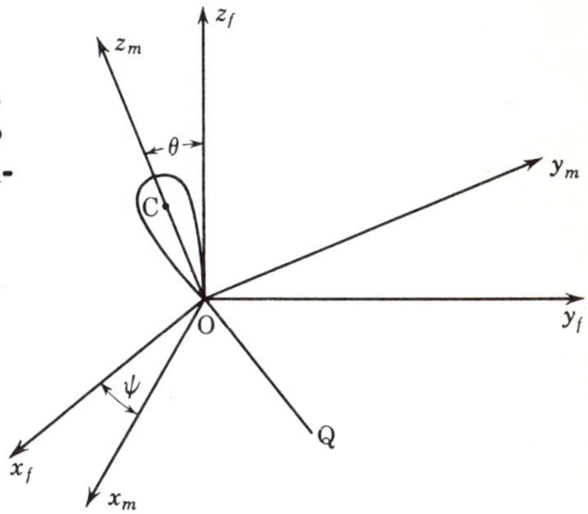

Fig. 7.4

$$\omega_x = \dot{\theta}, \quad \omega_y = \dot{\psi} \sin \theta, \quad \omega_z = \dot{\psi} \cos \theta, \tag{7.49}$$

Note that the x_m axis always lies in the $x_f y_f$ plane. This accounts for the expression for ω_x. Note also that z_f, z_m, and y_m are all in the same plane. As the figure indicates, $\dot{\psi}$ is the angular velocity of the rotating axes about the z_f axis. It is known as the *precessional* velocity of the top.

The resultant torque on the top is due to gravity. If we denote the mass as m and locate it as usual at the center of mass C (with $OC = l$), the component torque about the x_m axis is $mgl \sin 0$. The components about the y_m and z_m axes vanish. Before we can write Euler's equations we must rewrite M to take account of the spin velocity **s**. Thus we now write (recalling that $I_{xx} = I_{yy}$)

$$\mathbf{M} = \mathbf{i}I_{xx}\omega_x + \mathbf{j}I_{xx}\omega_y + \mathbf{k}I_{zz}(\omega_z + s), \tag{7.50}$$

whence

$$\begin{aligned}\dot{\mathbf{M}} &= \mathbf{i}I_{xx}\dot{\omega}_x + \mathbf{j}I_{xx}\dot{\omega}_y + \mathbf{k}I_{zz}(\dot{\omega}_z + \dot{s}) \\ &+ \mathbf{i}\left[\omega_y I_{zz}(\omega_z + s) - \omega_z I_{xx}\omega_y\right] \\ &+ \mathbf{j}\left[\omega_z I_{xx}\omega_x - \omega_x I_{zz}(\omega_z + s)\right] \\ &+ \mathbf{k}\left[\omega_x I_{xx}\omega_y - \omega_y I_{xx}\omega_x\right].\end{aligned} \tag{7.51}$$

Hence Euler's equations now take the form

$$\begin{aligned} I_{xx}\dot{\omega}_x + (I_{zz} - I_{xx})\omega_y\omega_z + I_{zz}\omega_y s &= mgl\sin\theta, \\ I_{xx}\dot{\omega}_y + (I_{xx} - I_{zz})\omega_x\omega_z - I_{zz}\omega_x s &= 0, \\ I_{zz}(\dot{\omega}_z + \dot{s}) &= 0. \end{aligned} \tag{7.52}$$

These can be somewhat simplified by writing $s + \omega_z = S$.

$$\begin{aligned} I_{xx}\dot{\omega}_x - I_{xx}\omega_y\omega_z + SI_{zz}\omega_y &= mgl\sin 0, \\ I_{xx}\dot{\omega}_y + I_{xx}\omega_x\omega_z - SI_{zz}\omega_x &= 0, \\ I_{zz}\dot{S} &= 0. \end{aligned} \tag{7.53}$$

The last equation says that the resultant angular speed about the z_m axis is constant. If we multiply the first equation by ω_x, the second by ω_y, and the third by S and add, we obtain, after integrating

$$1/2 \cdot I_{xx}(\omega_x^2 + \omega_y^2) + 1/2 \cdot I_{zz}S^2 + mgl\cos\theta = E, \tag{7.54}$$

where E is the total energy of the motion, $mgl\cos\theta$ is the potential energy with respect to the horizontal plane and the rest of the left-hand side the resultant kinetic energy.

A special case of the first equation in (7.53) is not without interest. Suppose that the spin speed s is so large that the term $SI_{zz}\omega_y$ dominates the right-hand side. Then $I_{xx}\dot{\omega}_x - I_{xx}\omega_y\omega_z$ may be neglected and the equation becomes

$$SI_{zz}\dot{\psi} = mgl, \tag{7.55}$$

which says that the precessional velocity is inversely proportional to the spin velocity and the moment of inertia about the spin axis. This is the familiar result of the elementary theory of the simple gyroscope.

To treat the problem more generally we note that the torque about the z_f axis is zero, and hence the angular momentum about this axis is constant. But the latter is the sum of the components of M along z_f. We therefore have

$$I_{zz} S \cos\theta + I_{xx}\omega_y \sin\theta = A = \text{constant}. \tag{7.56}$$

Written in terms of the precessional velocity $\dot{\psi}$, this yields

$$\dot{\psi} = \frac{A - I_{zz} S \cos\theta}{I_{xx} \sin^2\theta}. \tag{7.57}$$

Moreover the energy equation (7.54) can be written

$$\dot{\theta}^2 + \dot{\psi}^2 \sin^2\theta = \frac{2E - 2mgl\cos\theta - I_{zz}S^2}{I_{xx}}. \tag{7.58}$$

With the introduction of the new quantities

$$\begin{aligned} \propto &= A/I_{xx}, & \beta &= I_{zz}S/I_{xx}, \\ a &= \frac{2E - I_{zz}S^2}{I_{xx}}, & b &= 2mgl/I_{xx}, \end{aligned} \tag{7.59}$$

eqs. (7.57) and (7.58) take on the simpler form

$$\begin{aligned} \dot{\psi} &= \frac{\propto - \beta\cos\theta}{\sin^2\theta}, \\ \dot{\theta}^2 + \dot{\psi}^2 \sin^2\theta &= a - b\cos\theta. \end{aligned} \tag{7.60}$$

The elimination of $\dot{\psi}$ between these equations yields

$$\dot{\theta}^2 = -\frac{(\propto - \beta\cos\theta)^2}{\sin^2\theta} + a - b\cos\theta. \tag{7.61}$$

Let $u = \cos\theta$ and get

$$\dot{u}^2 = (a - bu)(1 - u^2) - (\propto - \beta u)^2 = f(u). \tag{7.62}$$

Evidently the variation of θ with time depends on the quadrature

$$t = \int \frac{du}{\sqrt{f(u)}} + \text{constant}, \tag{7.63}$$

which in turn can be expressed only in terms of elliptic functions. However, we can learn something about the top's behavior without carrying out the integration in (7.63). Clearly $f(u)$ must be positive to insure the reality of u, and further u must lie between 0 and +1. The function $f(u)$ is a cubic with three roots of which two must lie between 0 and +1. The situation is depicted in Fig. 7.5. We shall call the roots of $f(u)$ lying in

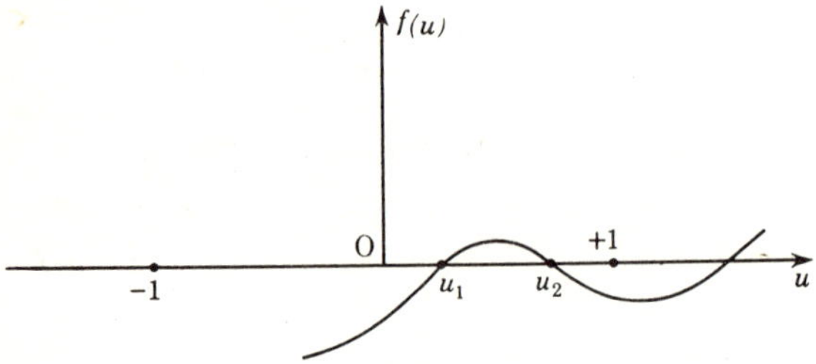

Fig. 7.5.

the allowed interval u_1 and u_2. It follows that the inclination of the top's axis to the vertical is restricted to lie within the limiting angles $\theta_1 = \arccos u_1$ and $\theta_2 = \arccos u_2$. More elaborate consideration of the problem shows that θ varies periodically between these limits. This motion is known as *nutation*. In general, the spinning of a top or gyroscope is accompanied by both precession and nutation. Under certain conditions the nutation vanishes: this demands, of course, that $\theta_1 = \theta_2$ always. Hence $u_1 = u_2$ must be a double root of $f(u)$ and we must be able to write this function in the form

$$f(u) = B(u - u_3)(u - u_1)^2, \tag{7.64}$$

where B is constant. It is clear that $u = u_1$ must satisfy both $f(u) = 0$ and $df/du = 0$. This will suffice to fix u_1, and the precessional velocity $\dot{\psi}$, given by the first equation in (7.60) becomes in terms of u_1

$$\dot{\psi} = \frac{\beta}{2u_1}\left[1 \pm \sqrt{1 - 2bu_1/\beta^2}\right].$$

For each u_1 there are then two possible precessional velocities.

If the spin velocity is so great that $\beta^2 \gg 2bu_1$ the two values of $\dot{\psi}$ are

$$\dot{\psi}_1 = mgl/I_{zz}S, \tag{7.66}$$

$$\dot{\psi}_2 = \frac{I_{zz}S}{u_1 I_{xx}}\left[1 - \frac{mgl\, u_1 I_{xx}}{I_{zz}S^2}\,.\right] \tag{7.67}$$

Evidently $\dot{\psi}_1 \ll \dot{\psi}_2$. The slower precession is usually the one observed.

Suppose the top is started with axis vertical and $S = s$, i.e., with spin velocity only. This leads at once to $\alpha = \beta$ and $a = b$. The roots of $f(u)$ then become

$$u_1 = +1, \quad u_2 = +1, \quad u_3 = \alpha^2/a - 1.$$

As long as $S^2 > 4mgl\, I_{xx}/I_{zz}^2$, $u_3 > 1$, and the motion of the top is confined to simple rotation about the vertical axis with no precession or nutation, i.e., the top sleeps. As a result of friction at the point of support energy may be lost until $S^2 < 4mgl\, I_{xx}/I_{zz}^2$, when u_3 becomes less than unity and precession sets in with nutation between the angles $\theta = 0$ and $\theta = \theta_2$ where

$$\cos\theta_2 = I_{zz}^2 S^2/2mgl\, I_{xx} - 1. \tag{7.68}$$

This means the top begins to wobble. With the continued decrease of S, θ_2 increases until the top falls down.

7.7. Motion of a Particle on the Earth's Surface. Although we have introduced moving axes in this chapter primarily to simplify the treatment of rigid body motion, it is not out of place to emphasize here some applications to the motion of particles which are not rigidly connected. The first refers to the motion of a particle on the surface of the earth. In referring this motion to a set of axes fixed in the earth we are really using rotating axes and must therefore express the motion with respect to a fixed system having the same origin (viz., the center of the earth) by the use of eq. (7.34), noting that $\dot{\mathbf{r}}_0 = 0$ in this case. Since we want to write the equation of motion, we must differentiate (7.34) with respect to the time. Thus the acceleration of the particle in the *fixed* system becomes

$$\ddot{\mathbf{r}}_f = \ddot{\mathbf{r}}_{ma} + \omega \times \dot{\mathbf{r}}_m, \tag{7.69}$$

since $\dot{\omega} = 0$, if we treat the angular velocity of the earth as approximately constant. We now use (7.31) in (7.69) and get

$$\ddot{\mathbf{r}}_f = \ddot{\mathbf{r}}_{ma} + \omega \times \dot{\mathbf{r}}_{ma} + \omega \times (\omega \times \mathbf{r}_m). \tag{7.70}$$

But we must look into $\ddot{\mathbf{r}}_{ma}$ more closely. From (7.30)

$$\dot{\mathbf{r}}_{ma} = \mathbf{i}_m \dot{x}_m + \mathbf{j}\dot{y}_m + \mathbf{k}\dot{z}_m. \tag{7.71}$$

Therefore

$$\ddot{\mathbf{r}}_{ma} = \mathbf{i}_m \ddot{x}_m + \mathbf{j}_m \ddot{y}_m + \mathbf{k}_m \ddot{z}_m + \dot{x}_m \frac{d\mathbf{i}_m}{dt} + \dot{y}_m \frac{d\mathbf{j}_m}{dt} + \dot{z}_m \frac{d\mathbf{k}_m}{dt}. \tag{7.72}$$

But from (7.19) this can be written

$$\ddot{\mathbf{r}}_{ma} = \mathbf{a}_m + \omega \times \dot{\mathbf{r}}_{ma}, \tag{7.73}$$

where $\mathbf{a}_m$ is the *apparent* acceleration of the particle in the moving system, neglecting the rotation of the axes. Then (7.70) becomes

$$\ddot{\mathbf{r}}_f = \mathbf{a}_m + 2\omega \times \dot{\mathbf{r}}_{ma} + \omega \times (\omega \times \mathbf{r}_m), \tag{7.74}$$

and the corresponding resultant force on the particle in the fixed system becomes

$$\mathbf{F} = m\ddot{\mathbf{r}}_f = m\mathbf{a}_m + 2m\omega \times \dot{\mathbf{r}}_{ma} + m\omega \times (\omega \times \mathbf{r}_m). \tag{7.75}$$

We may rewrite this in terms of the apparent acceleration relative to the moving axes (i.e., those fixed in the earth); thus

$$\mathbf{a}_m = \mathbf{F}/m - 2\omega \times \dot{\mathbf{r}}_{ma} - \omega \times (\omega \times \mathbf{r}_m). \tag{7.76}$$

The term $2\omega \times \dot{\mathbf{r}}_{ma}$ will be recognized as the familiar Coriolis acceleration (Sec. 2.5), whereas $\omega \times (\omega \times \mathbf{r}_m)$ is the centripetal acceleration.

Suppose the particle is subject only to the attraction of the earth. Then

$$\mathbf{F} = m\mathbf{g} + m\omega \times (\omega \times \mathbf{r}_m), \tag{7.77}$$

which combined with (7.76) yields

$$\mathbf{a}_m = \mathbf{g} - 2\omega \times \dot{\mathbf{r}}_{ma} \tag{7.78}$$

for the actual apparent acceleration relative to the earth. If we write (7.78) in terms of the rectangular components we get

$$\begin{aligned} \ddot{x}_m &= g_x + 2\omega\dot{y}_{ma}, \\ \ddot{y}_m &= g_y - 2\omega\dot{x}_{ma}, \\ \ddot{z}_m &= g_z, \end{aligned} \tag{7.79}$$

if we recall that $\omega_x = \omega_y = 0$ and $\omega_z = \omega$ in this special case. In Fig. 7.6 let us consider rectangular axes set up at the center of the earth 0 as indicated and let APB be one quarter of the trace of the earth's surface on the yz plane. Suppose a particle is dropped from rest at P in latitude λ. We then have $g_x = 0$, $g_y = -g \cos \lambda$, $g_z = -g \sin \lambda$. Let v be the observed speed of fall at any instant. Eqs. (7.79) then become

$$\begin{aligned} \ddot{x}_m &= -2\omega v \cos \lambda, \\ \ddot{y}_m &= -g \cos \lambda \\ \ddot{z}_m &= -g \sin \lambda. \end{aligned} \tag{7.80}$$

The first equation shows that in addition to the acceleration due to gravity the falling particle also experiences an additional acceleration directed at right angles to the meridian plane in which it starts to fall. This acceleration has the magnitude $2\omega v \cos\lambda$, is due to the Coriolis force, and is always directed toward the east as should be clear from the diagram.

Since $v = gt$ approximately the magnitude of the easterly acceleration becomes

$$|\ddot{x}_m| = 2\omega gt \cos\lambda, \quad (7.81)$$

and hence the total easterly deflection after time t from rest is

$$x_m = \frac{\omega g t^3}{3} \cos\lambda$$

$$= \frac{2}{3}\omega \sqrt{\frac{2h^3}{g}} \cos\lambda, \quad (7.82)$$

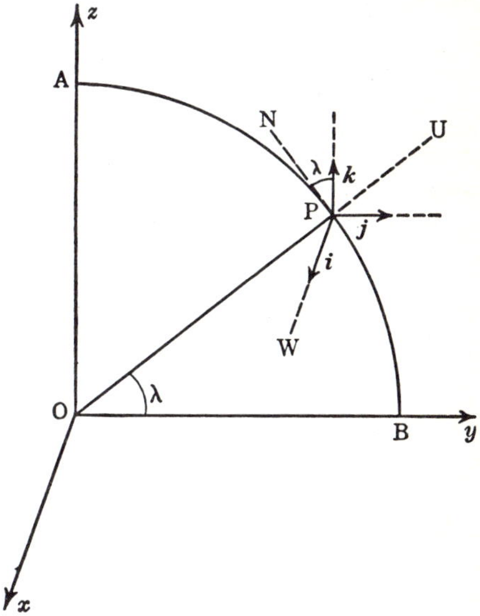

Fig. 7.6.

if we use $h = \frac{1}{2}gt^2$ as the height from which the fall has taken place.

The Coriolis acceleration gives rise to cyclonic wind movements on the earth's surface.[1]

7.8. Application of Moving Axes to Molecular Motions. In atomic theory a molecule is considered to be a collection of atoms which move with respect to each other in various ways. Although by no means rigidly connected, in many important motions of the molecule the separation distances of the atoms change only slightly compared with their equilibrium values. Hence it would seem that the considerations developed in this chapter might find a certain approximate application to molecules. This is indeed the case, particularly with reference to the use of moving axes.

Let us suppose the molecule in question has N atoms of masses m_1, m_2, $\cdots$ m_j, $\cdots$ m_N, respectively. Pick a moving origin O whose

[1] Cf., Page, *Introduction to Theoretical Physics*, 2nd Ed. (New York, Van Nostrand), pp. 107 ff.

distance from the origin of the fixed system is $\mathbf{r}_0$ as in Fig. 7.3. Let the position vector of the jth atom in the moving system be $\mathbf{r}_j$. Further let the equilibrium value of $\mathbf{r}_j$ be denoted by $\overline{\mathbf{r}}_j$. Then the displacement of m_j from its equilibrium position will be

$$\xi_j = \mathbf{r}_j - \overline{\mathbf{r}}_j. \tag{7.83}$$

If we denote by ω as usual the angular velocity of the moving axes and by $\mathbf{v}_j$ the vector whose components are $\dot{x}_j, \dot{y}_j, \dot{z}_j$, i.e., the *apparent velocity* of m_j in the moving system, we can write for $\dot{\mathbf{r}}_{jf}$, the resultant velocity of m_j relative to the axes fixed in space,

$$\dot{\mathbf{r}}_{jf} = \dot{\mathbf{r}}_0 + \mathbf{v}_j + \omega \times \mathbf{r}_j. \tag{7.84}$$

The kinetic energy of the collection of atoms then becomes

$$K = \frac{1}{2}\sum_{j=1}^{N} m_j(\dot{\mathbf{r}}_0 + \mathbf{v}_j + \omega \times \mathbf{r}_j) \cdot (\dot{\mathbf{r}}_0 + \mathbf{v}_j + \omega \times \mathbf{r}_j). \tag{7.85}$$

Expanding the scalar product we get

$$K = \frac{\dot{r}_0^2}{2}\sum m_j + \frac{1}{2}\sum m_j v_j^2 + \frac{1}{2}\sum m_j(\omega \times \mathbf{r}_j) \cdot (\omega \times \mathbf{r}_j)$$

$$+ \dot{\mathbf{r}}_0 \cdot \sum m_j \mathbf{v}_j + \dot{\mathbf{r}}_0 \times \omega \cdot \sum m_j \mathbf{r}_j + \omega \cdot \sum m_j \mathbf{r}_j \times \mathbf{v}_j. \tag{7.86}$$

It is now in order to try to simplify this result. Suppose we place the moving origin O at the center of mass of the system. It follows that

$$\sum m_j \mathbf{r}_j = 0; \quad \sum m_j \mathbf{v}_j = 0. \tag{7.87}$$

Let us suppose that the average angular momentum of the system vanishes. This says that

$$\sum m_j \overline{\mathbf{r}}_j \times \mathbf{v}_j = 0. \tag{7.88}$$

In the last term of the expression (7.86) we replace $\mathbf{r}_j$ by $\overline{\mathbf{r}}_j + \xi_j$ from (7.83) and also avail ourselves of (7.87). The final result is

$$K = \frac{\dot{r}_0^2}{2}\sum m_j + \frac{1}{2}\sum m_j v_j^2 + \frac{1}{2}\sum m_j(\omega \times \mathbf{r}_j) \cdot (\omega \times \mathbf{r}_j)$$

$$+ \omega \cdot \sum m_j \xi_j \times \mathbf{v}_j. \tag{7.89}$$

Of the terms in K, the first is the translational kinetic energy of the collection of atoms while the second is the translational kinetic energy relative to the moving axes or what may appropriately be called the *vibrational* kinetic energy due to the fact that the particles move with respect to each other and are not rigidly connected. The third term is the *rotational* kinetic energy and the last term may be thought of as representing a kind of coupling between rotation and vibration.

Next we expand the rotational kinetic energy term in cartesian form and get

$$\frac{1}{2}\sum m_j(\omega \times \mathbf{r}_j) \cdot (\omega \times \mathbf{r}_j) = \frac{\omega_x^2}{2}\sum m_j(y_j^2 + z_j^2) + \frac{\omega_y^2}{2}\sum m_j(x_j^2 + z_j^2)$$

$$+ \frac{\omega_z^2}{2}\sum m_j(x_j^2 + y_j^2) - \omega_x\omega_y\sum m_j x_j y_j$$

$$- \omega_x\omega_z\sum m_j x_j z_j - \omega_y\omega_z\sum m_j y_j z_j. \tag{7.90}$$

Comparison with (7.6) and (7.7) enables us to write this in the form

$$\frac{1}{2}\sum m_j(\omega \times \mathbf{r}_j) \cdot (\omega \times \mathbf{r}_j) = \frac{1}{2}I_{xx}\omega_x^2 + \frac{1}{2}I_{yy}\omega_y^2 + \frac{1}{2}I_{zz}\omega_z^2$$

$$- I_{xy}\omega_x\omega_y - I_{xz}\omega_x\omega_z - I_{yz}\omega_y\omega_z, \tag{7.91}$$

where we introduce again the moments and products of inertia. We may also perform the scalar product in the coupling term in (7.89) and finally write the kinetic energy of the molecule in the form

$$
\begin{aligned}
K = {} & \frac{1}{2}\Big(\sum m_j\Big)\dot{r}_0^2 + \frac{1}{2}\sum m_j v_j^2 + \frac{1}{2} I_{xx}\omega_x^2 + \frac{1}{2} I_{yy}\omega_y^2 \\
& + \frac{1}{2} I_{zz}\omega_z^2 - I_{xy}\omega_x\omega_y - I_{xz}\omega_x\omega_z - I_{yz}\omega_y\omega_z \\
& + \omega_x \sum m_j(\xi_j \times \mathbf{v}_j)_x + \omega_y \sum m_j(\xi_j \times \mathbf{v}_j)_y + \omega_z \sum m_j(\xi_j \times \mathbf{v}_j)_z.
\end{aligned}
\tag{7.92}
$$

The expression of K in the above form indicates the possibility of employing much of the technique of rigid body dynamics to the motion of the atoms in a molecule.

PROBLEMS

1. Find the period of the small oscillations of a cube of side a swinging about one edge placed horizontal. Find the length of the equivalent simple pendulum.

2. Find the equation of the polhode cone in terms of K and p (cf. eq. 7.47).

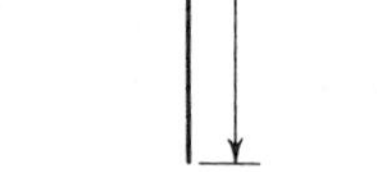

3. Consider the top formed by piercing the center of a uniform disc of radius a and mass m with a thin but stiff wire (as in the accompanying figure). Assume the mass of the wire negligible compared with that of the disc. Compute the minimum spin velocity with which the top will "sleep." At what spin velocity will the top begin to roll on the ground? At what spin velocity will the top experience nutation between the angles $\theta_1 = 0$ and $\theta_2 = 45°$?

4. Find the equation of the momental ellipsoid for a homogeneous cube with its center fixed.

5. Show that the principal axes of the momental ellipsoid are given by values of ω_x, ω_y, ω_z satisfying the linear equations

$$
\begin{aligned}
(I_{xx} - \lambda)\omega_x - I_{xy}\omega_y - I_{xz}\omega_z &= 0, \\
-I_{yx}\omega_x + (I_{yy} - \lambda)\omega_y - I_{yz}\omega_z &= 0, \\
-I_{zx}\omega_x - I_{zy}\omega_y + (I_{zz} - \lambda)\omega_z &= 0,
\end{aligned}
$$

where λ is an arbitrary parameter. Then show that in order that nonvanishing solutions exist, λ must satisfy a certain cubic equation. Obtain formally the expressions for the principal axes.

6. A rotary airplane engine having a mass of 350 lb and radius of gyration of 15 in. makes 1200 r.p.m. when the plane has a velocity of 140 miles/hr. Find: (a) the gyroscopic torque developed when the plane is making a turn of 200 ft radius; (b) the force that must be applied at the tail to balance this torque when the distance from the center of mass of the plane to the center of pressure on the tail is 15 ft.

7. Calculate the magnitude of the easterly deflection from the plumb line experienced by a particle dropped from rest at a height of 1000 ft in latitude 40°N. How will the result be changed if the particle has an upward vertical velocity of 50 miles/hr at the instant it is dropped?

8. A thin homogeneous elliptical sheet of mass m and with semi-major and semi-minor axes a and b, respectively, is held with its plane vertical and with the minor axis vertical. It is then allowed to fall from rest to a horizontal floor. If the point in contact with the floor does not slip, with what velocity will the top end strike the floor? What will be the kinetic energy of the sheet at the instant of striking?

9. Prove that the momental ellipsoid of any regular polyhedron with its center at the origin is a sphere. Specialize to the case of a regular tetrahedron and find the radius of the corresponding sphere.

10. What form do Euler's equations of motion for a rigid body take when the axes are not principal axes?

11. Use eq. (7.37) to show that the kinetic energy of a rigid body rotating about a fixed axis is equal to one-half the product of the moment of inertia about the axis and the square of the angular speed.

12. A simple pendulum has its point of suspension directly over the center of a horizontal turntable rotating with angular velocity ω. If the bob is pulled aside and let go, what sort of curve will it trace out on the table?

13. Show from the general equation of motion of a rigid body rotating about the fixed x axis that in addition to the torque component about the x axis (due to gravity, e.g.) there must be torque components about the y and z axes as well. Comment on the nature of the forces producing the latter torques.

14. A particle is projected with velocity **v** on a smooth horizontal plane. Show that due to the rotation of the earth the particle will move in the arc of a circle of radius $v/2\omega \sin \lambda$, where ω is the angular speed of the earth and λ is the latitude.

15. Prove that every displacement of a rigid body can be considered as a combination of translations and rotations.

CHAPTER VIII

MINIMAL PRINCIPLES. LAGRANGE'S EQUATIONS

8.1. Minimal Principles in Physics. One of the obvious goals of physical theorizing is to build theories with the smallest possible number of postulates, which will therefore possess the widest generality. In particular it is always considered gratifying when a concept or principle which is successful in one branch of physics also proves applicable to another. This attempt at the unification of apparently different physical phenomena has long been an ideal of physicists even when it has failed to produce results of practical value.

A notion which for centuries has had a fascination for human beings is the minimization of effects, i.e., the concept that physical effects take place in such a way that certain important quantities assume their *minimum* values. A very old illustration comes from optics. Hero of Alexandria in the second century B.C. is credited with the idea that light passing in straight lines from a point A to a point B in front of a plane mirror by reflection at the mirror will follow the shortest path. Thus to consult Fig. 8.1, various possible straight line paths by which light can get from A to B after first touching the mirror surface MM' are ADB, ACB (in which the angles ACN and BCN are equal), AEB, etc. Hero's idea was that of all these possible paths Nature prefers the *shortest*. A very simple calculation shows that this is the path ACB, i.e., that for which the angle of incidence is equal to the angle of reflection. In this way Hero established a deduction of the law of the reflection of light (treated as a ray phenomenon with straight-line propagation) by means of a *minimal* principle, i.e., that of the minimum path.

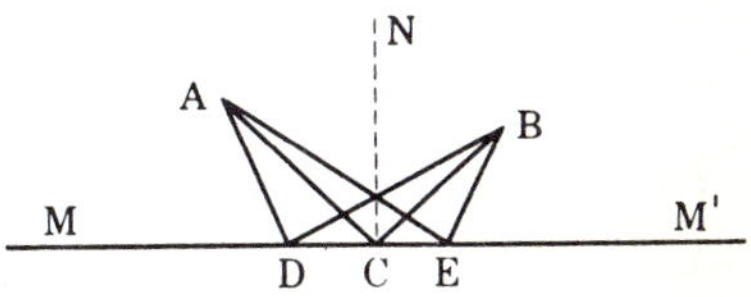

Fig. 8.1.

The attempt to apply the principle of shortest path to the transmission of light from A to B across the interface separating two different media, e.g., air and water, or air and glass, obviously fails since the shortest path in this case is the straight line between A and B and this is negated by the existence of refraction. In the seventeenth century Fermat introduced the postulate that the actual path traveled by light from any point to another in any medium is that traversed in the *least time*. It can readily be shown that this leads to Snell's law of refraction.[1] It also provides the correct law of the reflection of light, since in a medium in which the velocity is constant it is identical with the principle of minimum path.[2]

Concern with the minimization concept was also reflected by the great interest shown by mathematicians of the seventeenth and early eighteenth centuries, e.g., the Bernoullis, Leibniz and Newton, in problems like the *brachistochrone* or curve along which a particle falling under gravity descends from one point to a lower one in the *least possible* time. There was moreover the celebrated problem of the shape assumed in equilibrium by a freely hanging homogeneous flexible chain or cord (the catenary), solved by John Bernoulli under the assumption that for equilibrium the center of mass must be as *low as possible*. The mathematical requirements for the solution of problems of this kind led to the development of the *calculus of variations*.

In 1747 Maupertuis endeavored to establish mechanics on the basis of a principle stating that the wisdom of God is exemplified through having all motions of material particles take place in such way that the total "action" is a minimum. This is the introduction of the principle of *least action*. It was applied by Maupertuis to a number of problems including the collision of particles and the propagation of light. His definition of action was lacking in precision and his deductions were therefore somewhat less than cogent. The principle was put on a better logical basis by Lagrange in 1760. We shall discuss it briefly in Sec. 8.8.

The attractiveness of the minimal idea shows up again in the principle of least constraint developed by Gauss in 1829 to serve as a basis for the solution of all problems in the motion of particles subject to constraint. Here a certain quantity is introduced as a measure of the constraints,[3] and the principle is cast in the form which says that the actual motion of the system of particles under given impressed forces and subject to given constraints takes place in such a way that this quantity, called the total constraint, is a minimum. It is readily shown that D'Alembert's principle is a consequence of the Gaussian principle, or alternatively that the principle of least constraint can be derived from D'Alembert's principle.

[1]Cf. Lindsay and Margenau, *Foundations of Physics*, (New York, John Wiley and Sons, 1936, Dover reprint), p. 135.

[2]The reader may show that Fermat's principle also implies that the incident and refracted rays lie in the same plane, an important part of the law of refraction.

[3]Cf., Lindsay and Margenau, *Foundations of Physics*, p. 113.

Closely associated with Gauss' principle of least constraint is the principle adopted by H. Hertz in his logical treatment of mechanics,[1] usually referred to as the principle of the straightest path. By suitably redefining the Gaussian constraint function as the dynamical curvature of the path of a system, Hertz's principle can be looked upon as the principle of *least curvature* and therefore represents directly another minimal principle. P. Appell[2] has used the principle to write the equations of motion of a dynamical system in a particularly simple differential form.

The foregoing remarks should be sufficient to justify more detailed attention to minimal principles in mechanics. The most important of these is that usually referred to under the name of Hamilton.[3] We discuss this in the next section.

8.2. Hamilton's Principle and Its Deduction from D'Alembert's Principle. We consider again the motion of a collection of particles as described by D'Alembert's principle (eq. 6.39) and examine $\ddot{\mathbf{r}}_j \cdot \delta\mathbf{r}_j$. It will be recalled that $\delta\mathbf{r}_j$ does not represent an actual displacement of the jth particle in time. Nevertheless it represents a *possible* displacement, consistent with the constraints imposed on the system so that it is possible to interpret δ as an operator which produces a *small* change in $\mathbf{r}_j$ not connected with the time. We shall therefore proceed to treat it as a differential operator with all the properties of d. Thus we have

$$\delta(x^2) = 2x\delta x,$$

etc.

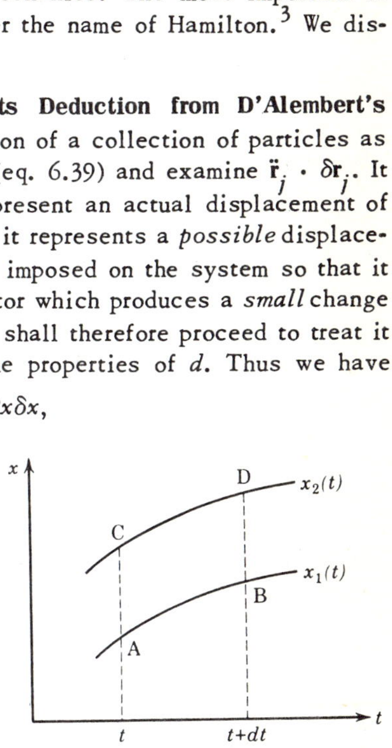

Fig. 8.2.

In order to distinguish the better between d and δ consider the accompanying figure (Fig. 8.2) which shows the x coordinate of a particle plotted as a function of t. We shall suppose that $x_1(t)$ indicates the actual dependence of x on t for a given motion. Let $x_2(t)$ denote another motion compatible with the constraints, a *varied* motion, we may call it. To each point of the actual motion we shall associate the corresponding point of the varied motion at the same time. Thus C on $x_2(t)$ corresponds to A on $x_1(t)$,

[1]Cf. H. Hertz, *Principles of Mechanics* (Dover reprint).

[2]Cf. E. T. Whittaker, *Analytical Dynamics*, (New York, Dover, 1944), pp. 254 ff.

[3]Sir W. R. Hamilton, *Phil. Trans. Roy. Soc. London*, 124, 307, 1834.

etc. We may denote A by (x, t). Then C will be $(x + \delta x, t)$. B will be $(x + dx, t + dt)$, and D will be $(x + dx + \delta(x + dx), t + dt)$ or $(x + \delta x + d(x + \delta x), t + dt)$. The fact that D can be represented in these two ways entails that

$$x + dx + \delta(x + dx) = x + \delta x + d(x + \delta x). \tag{8.1}$$

If we assume that δ like d obeys the distributive law, we obtain from (8.1)

$$\delta(dx) = d(\delta x), \tag{8.2}$$

i.e., the two operators commute. This means further that

$$\delta(\dot{x}) = \frac{d}{dt}(\delta x). \tag{8.3}$$

Since C represents a possible position of the particle it follows that we can treat δx as a possible or virtual displacement of the particle, i.e., one that we can *imagine* to take place, even if it does not actually take place. It thus satisfies the condition of D'Alembert's principle and the same will be true of $\delta \mathbf{r}_j$ in general. Moreover we shall also have from (8.3)

$$\delta(\dot{\mathbf{r}}_j) = \frac{d}{dt}(\delta \mathbf{r}_j). \tag{8.4}$$

Reverting to eq. (6.39) we see that

$$\ddot{\mathbf{r}}_j \cdot \delta \mathbf{r}_j = \frac{d}{dt}(\dot{\mathbf{r}}_j \cdot \delta \mathbf{r}_j) - 1/2 \cdot \delta(\dot{\mathbf{r}}_j \cdot \dot{\mathbf{r}}_j), \tag{8.5}$$

where we have employed (8.4). Noting that $\dot{\mathbf{r}}_j \cdot \dot{\mathbf{r}}_j = v_j^2$, or the square of the speed of the jth particle, we multiply (8.5) by m_j and sum over all the particles. The result is

$$\frac{d}{dt}\sum_{j=1}^{n} m_j \dot{\mathbf{r}}_j \cdot \delta \mathbf{r}_j = \sum_{j=1}^{n} m_j \ddot{\mathbf{r}}_j \cdot \delta \mathbf{r}_j + \frac{1}{2}\sum_{j=1}^{n} m_j \delta(v_j^2). \tag{8.6}$$

Next we integrate with respect to the time from t_1 to t_2, getting

$$\left.\sum_{j=1}^{n} m_j \dot{\mathbf{r}}_j \cdot \delta \mathbf{r}_j\right]_{t_1}^{t_2} = \int_{t_1}^{t_2}\left(\delta K + \sum_{j=1}^{n} m_j \ddot{\mathbf{r}}_j \cdot \delta \mathbf{r}_j\right) dt \tag{8.7}$$

$$\frac{1}{2}\sum_{j=1}^{n} m_j \delta(v_j^2) = \delta K, \tag{8.8}$$

or the variation in the total kinetic energy. If now we consider only those varied paths of the system for which the initial and final positions are the same, we have

$$(\delta \mathbf{r}_j)_{t_2} = (\delta \mathbf{r}_j)_{t_1} = 0, \tag{8.9}$$

i.e., every path begins at the same position as the actual path at $t = t_1$ and ends at the same position at $t = t_2$. On this condition (8.7) becomes (employing D'Alembert's principle)

$$\int_{t_1}^{t_2} \left(\delta K + \sum_{j=1}^{n} \mathbf{F}_{je} \cdot \delta \mathbf{r}_j \right) dt = 0. \tag{8.10}$$

In words, this says that the time integral of the variation in the total kinetic energy is equal to the negative of the time integral of the total virtual work done by the impressed forces during the same time interval. It is here understood that the variation in all quantities vanishes at t_1 and t_2. This result is known as the general form of *Hamilton's principle*.

As a simple illustration consider the simple harmonic damped oscillator. Here the motion is in one direction and $\mathbf{F}_{je}$ reduces to

$$F = -kx - R\dot{x}, \tag{8.11}$$

with k and R the stiffness and damping coefficient respectively as usual. Since

$$\int_{t_1}^{t_2} \delta K \, dt = \int_{t_1}^{t_2} m\dot{x}\delta(\dot{x}) \, dt = \int_{t_1}^{t_2} m\dot{x}\frac{d}{dt}(\delta x) \, dt, \tag{8.12}$$

we may integrate by parts and get

$$\int_{t_1}^{t_2} \delta K \, dt = -\int_{t_1}^{t_2} m\ddot{x}\delta x \, dt, \tag{8.13}$$

where the integrated part vanishes at the limits by the conditions laid down above. The principle then says that

$$\int_{t_1}^{t_2} (-m\ddot{x} - kx - R\dot{x})\delta x \, dt = 0. \tag{8.14}$$

Since δx although small, is arbitrary save at the end points of the path, the only way for the integral to vanish is for us to have

$$m\ddot{x} + R\dot{x} + kx = 0. \tag{8.15}$$

We are thus led immediately to the equation of motion of the oscillator.

The reader may object that if all Hamilton's principle can do is to provide the ordinary Newtonian equation of motion from a knowledge of the force it is a singularly complicated way of achieving a very simple result. From the ordinary elementary practical standpoint this judgment is sound, but from a methodological point of view it is extremely short-sighted. Of course we have illustrated the principle with a rather trivial example, but the point to keep in mind is that we have *deduced* an equation of motion (in the Newtonian form) from a more fundamental postulate and indeed one in which the concept of *energy* is paramount. We now have a recipe for solving mechanical problems in which it is unnecessary to start with Newton's laws as postulates. The significance of this becomes more marked if we restrict our attention to conservative systems.

Let us now introduce the assumption that a function $V(x, y, z)$ exists such that

$$\mathbf{F}_{je} = -\left[\mathbf{i}\frac{\partial V}{\partial x_j} + \mathbf{j}\frac{\partial V}{\partial y_j} + \mathbf{k}\frac{\partial V}{\partial z_j}\right] = -\nabla_j V. \tag{8.16}$$

Then (8.10) becomes

$$\int_{t_1}^{t_2} (\delta K - \delta V)\, dt = 0, \tag{8.17}$$

or from the distributive law

$$\int_{t_1}^{t_2} \delta(K - V)\, dt = 0. \tag{8.18}$$

It seems plausible to assume that the integral of the variation of $K - V$ is the variation of the integral. So we shall also write (8.18) in the form

$$\delta \int_{t_1}^{t_2} (K - V)\, dt = 0. \tag{8.19}$$

Closer mathematical investigation shows that this step is justified for the functions usually encountered in mechanics. Eqs. (8.17), (8.18), or (8.19) constitute the expression of Hamilton's principle for a *conservative system.* The quantity $K - V$ is usually called the Lagrangian function and denoted by L. In this formulation the physical content of the principle is as follows: If we form the time integral of the Lagrangian function between any two instants t_1 and t_2 for the *actual* path followed by the system, i.e., for the actual dependence of $\mathbf{r}_i$ on t, and then carry out the same sort of integration for any other possible alternative dynamical path (i.e., an altered dependence of $\mathbf{r}_j$ on t) which however has the same end points as the actual path, we shall find that for the actual path the integral is stationary with respect to its value for any of the alternative paths, i.e., it is either greater than that for any of the alternative paths or less than that for any of the alternative. (The integral might of course also be a mini-max.) Careful investigation shows that the integral for most dynamical behavior encountered in practice is a minimum and the statement of Hamilton's principle is often made in this fashion, although it is not strictly correct.[1] Nevertheless the relation of Hamilton's principle to the minimal principles mentioned in Sec. 8.1 is sufficiently close to justify treating it in the same category. Since

$$\int_{t_1}^{t_2} K\, dt = (t_2 - t_1)\bar{K}, \tag{8.20}$$

where $\bar{K}$ is the average kinetic energy over the indicated time interval, we can interpret the principle in its minimal sense as follows. The actual motion of a conservative dynamical system is that which makes the difference between the average kinetic energy and the average potential energy over any time interval a minimum. Actually the more correct statement would replace "a minimum" by "stationary" with respect to all possible motions of the system in same interval and having the same initial and final configurations.

It is of course clear that in the application of the principle, it is only the vanishing of the first variation of the time integral of the Lagrangian function that really matters. Nevertheless the mind persists in looking

[1]For the conditions under which the integral *is* a minimum, cf. Whittaker, *loc. cit,* pp. 250 ff.

for a more fundamental physical significance even at the expense of a certain haziness in accuracy.

The fact that the fundamental principles of mechanics can be expressed in terms of either the differential equations of motion of the Newtonian formulation on the one hand, or the integral stationarization or minimization of the Hamiltonian formulation on the other, may be characterized methodologically from two points of view. One may in the first place consider the two formulations merely as different mathematical descriptions, which are however essentially equivalent in the sense that either may be deduced from the other by purely formal means without introducing additional physical interpretation. From this standpoing it is meaningless to express a preference save from the angle of mathematical convenience: one may say that for certain problems the Newtonian formulation is more efficient while for others the Hamiltonian leads to the answer more readily. There is, however, another possible attitude somewhat more appealing to physicists who prefer to attach deeper philosophical significance to physical constructs. When one inserts a particular force function in the Newtonian equation of motion and then solves the resulting differential equation to find the actual dependence of the spatial coordinates on the time, it is hard to resist the feeling that one is determining the *effect* of a given *cause*. In this respect we may justly say that the Newtonian formulation stresses the causality aspect of mechanics: a particular force always produces a definite type of motion. If now we look at the Hamiltonian method with the idea of seeking in similar fashion for philosophical meaning, we note at once that the form of the principle emphasizes that nature forever tries to achieve a certain purpose or end and that the resulting motion realizes this end. There is little doubt, for example, that this teleological aspect was in the mind of Maupertuis when he introduced the principle of least action. All the minimal principles mentioned in Sec. 8.1 exemplify it.

It is scarcely necessary to add that no physicist stands in need of either philosophical view in the plying of his trade. Nevertheless it is also true that all physicists develop such views sooner or later and that in the historical development of the subject they have obviously played an important role. It is also of interest to realize that two such philosophically diverse points of view express themselves physically in mathematical formulations which turn out to be fundamentally equivalent. Presumably the student of methodology might study this fact with profit.

8.3. Generalized Coordinates and Lagrange's Equations. The equations of motion in Newtonian form have one disadvantage: they change their form with a change in the position coordinates. It would seem to be an obvious advantage to find a way of writing the fundamental equations of mechanics which does not suffer from this drawback. This problem was solved by Lagrange through the introduction of *generalized coordinates*. His method puts the energy in the position of preferred concept in place of force and has materially extended the range of applicability of mechanics to other branches of physics where the motion of particles is not directly observed.

Suppose we are trying to describe a physical phenomenon (e.g., acoustical or electrical) of the kind just mentioned. We carry out the description in terms of some parameters which vary with the time and serve to specify what we call the *state* of the system. Let us denote these parameters by $q_1 \cdots q_n$, the minimum number n necessary for a complete description of the behavior of the system being referred to as the number of *degrees of freedom* of the system. If the behavior of the system is susceptible of mechanical explanation we must assume that the q's are somehow related to the position coordinates of a system of particles which are really responsible for what the system is doing, but remain concealed from direct observation. The parameters $q_1 \cdots q_n$ were called by Lagrange the *generalized coordinates*. We shall express the rectangular coordinates of the equivalent concealed particles in terms of the q's as follows:

$$x_j = x_j(q_1 \cdots q_n); \quad y_j = y_j(q_1 \cdots q_n);$$
$$z_j = z_j(q_1 \cdots q_n). \tag{8.21}$$

Therefore

$$\dot{x}_j = \sum_{k=1}^{n} \frac{\partial x_j}{\partial q_k} \dot{q}_k, \tag{8.22}$$

with similar expressions for $\dot{y}_j$ and $\dot{z}_j$. We shall call the $\dot{q}_k$ the *generalized velocities*. The kinetic energy of the system can be expressed in terms of the q's and $\dot{q}$'s by substituting into the usual expression. The result is

$$K = \sum (1/2)m_j(\dot{x}_j^{\,2} + \dot{y}_j^{\,2} + \dot{z}_j^{\,2}) = \sum a_{ik}\dot{q}_i\dot{q}_k. \tag{8.23}$$

In words, the kinetic energy is a homogeneous quadratic function of the generalized velocities, with coefficients (the a_{ik}) which are functions of the generalized coordinates. If a potential energy function V exists it will be a function of the generalized coordinates alone. This will be the case for conservative systems, to which for the moment we shall restrict our attention.

Hamilton's principle for such a conservative system now takes the form

$$\int_{t_1}^{t_2} \delta\left[K(q, \dot{q}) - V(q)\right] dt = 0, \tag{8.24}$$

where q and $\dot{q}$ are abbreviations for $q_1 \cdots q_n$ and $\dot{q}_1 \cdots \dot{q}_n$, respectively. We carry out formally the variation, using $L(q, \dot{q})$ for $K - V$. The result is

$$\int_{t_1}^{t_2} \left[\sum_{j=1}^{n} \left(\frac{\partial L}{\partial q_j} \delta q_j + \frac{\partial L}{\partial \dot{q}_j} \frac{d}{dt}(\delta q_j)\right)\right] dt = 0. \tag{8.25}$$

Integration by parts transforms the second term of the integral as follows:

$$\begin{aligned} \int_{t_1}^{t_2} \sum \frac{\partial L}{\partial \dot{q}_j} \frac{d}{dt}(\delta q_j)\, dt &= \sum \frac{\partial L}{\partial \dot{q}_j} \delta q_j \Bigg]_{t_1}^{t_2} \\ &- \int_{t_1}^{t_2} \sum \frac{d}{dt}\left(\frac{\partial L}{\partial \dot{q}_j}\right) \delta q_j\, dt. \end{aligned} \tag{8.26}$$

The integrated part vanishes since $\delta q_j = 0$ for all j at $t = t_1$ and $t = t_2$. Hence the principle becomes

$$\int_{t_1}^{t_2} \sum \left[\frac{\partial L}{\partial q_j} - \frac{d}{dt}\left(\frac{\partial L}{\partial \dot{q}_j}\right)\right] \delta q_j\, dt = 0. \tag{8.27}$$

Since the δq_i although small are arbitrary, the only way the above integral can vanish identically is by having

$$\frac{d}{dt}\left(\frac{\partial L}{\partial \dot{q}_j}\right) - \frac{\partial L}{\partial q_j} = 0, \tag{8.28}$$

for all j running from 1 to n.

The n equations (8.28) are known as Lagrange's equations for the physical system. Each is a second order equation in terms of the time as independent variable, and hence its solution involves two arbitrary constants. The form of the type equation is independent of the coordinates used: the use of the equations demands only the ability to set up the expression for the Lagrangian function L as a function of the q's and $\dot{q}$'s. This is not always an easy task. Logically it is in the same category as setting up the force expression in ordinary Newtonian mechanics. This should remind us that it is of course possible to derive the Lagrangian equations directly from Newton's equations of motion without the agency of Hamilton's principle, merely by using the transformation equations (8.21). (Cf. Prob. 7 at the end of the chapter.) This was essentially the method of Lagrange[1] and, although algebraically complicated, is preferred by many to the deduction from Hamilton's principle. This is a matter of taste, which enters prominently into mechanics as with other human activities. Insofar as Hamilton's principle as a minimal principle constitutes an illuminating conceptual basis for the whole of mechanics, the readiness with which the Lagrangian equations evolve from it certainly gives it heuristic value and there is little sense in ignoring the derivation because of mathematical difficulties with the rigorous development of the calculus of variations.

No attempt is made here to discuss the details of the calculus of variations, although the reader who has made the acquaintance of this rather difficult branch of mathematics will recognize that the Lagrangian equations are really nothing but the Euler equations of the variation problem involved in Hamilton's principle. It is important to recall that these equations provide a *necessary* but not a *sufficient* condition for the existence of the required extreme value of the integral of the Lagrangian function. This possibly accounts for the distaste some physicists display for the use of Hamilton's principle in the deduction of the equations.

The relation of Lagrange's equations to the Newtonian form is well illustrated by the case of central motion of a single particle in a plane. Here we use polar coordinates for the q's with $q_1 = r$ and $q_2 = \theta$, so that

$$\begin{aligned} K &= \frac{1}{2}m(\dot{r}^2 + r^2\dot{\theta}^2), \\ V &= V(r). \end{aligned} \tag{8.29}$$

This is a dynamical system with two degrees of freedom and the two

[1]Cf. J. L. Lagrange, *Mechanique Analytique*, 3d Ed. (Paris, 1853), pp. 282ff, 301 (Part II, Sec. 4). Lagrange's deduction is based effectively on D'Alembert's principle with a liberal interpretation of the variation parameter δ. In reading his discussion due allowance must be made for the difference in his notation from that currently employed.

corresponding Lagrangian equations become, on carrying out the operations indicated in (8.28),

$$m\ddot{r} - mr\dot{\theta}^2 = -\frac{\partial V}{\partial r}, \tag{8.30}$$

$$2mr\dot{r}\dot{\theta} + mr^2\ddot{\theta} = 0. \tag{8.31}$$

The first equation is equivalent to Newton's second law for the r component of the motion and says that the resultant force in the r direction, which is the mass times the component acceleration in the r direction, viz., $\ddot{r} - r\dot{\theta}^2$ (cf. Sec. 2.5, eq. 2.40), is equal to $-\frac{\partial V}{\partial r}$. In similar fashion the second equation, (8.31), says that the resultant force in the direction perpendicular to the radius vector, which is the mass times the component acceleration perpendicular to the radius vector, viz., $2\dot{r}\dot{\theta} + r\ddot{\theta}$ (cf. again eq. 2.40), is equal to zero. This is of course consistent with the assumption of a *central* force.

We can also throw the eqs. (8.30) and (8.31) into another form equivalent to the Newtonian equations by recalling that the component *momentum* along the radius vector is

$$m\dot{r} = \frac{\partial L}{\partial \dot{r}}, \tag{8.32}$$

whereas the quantity

$$mr^2\dot{\theta} = \frac{\partial L}{\partial \dot{\theta}} \tag{8.33}$$

is the so-called moment of momentum or angular momentum (Sec. 6.1). If we agree to denominate these momenta by p_r and p_θ, respectively, we can write the Lagrangian equations as

$$\frac{dp_r}{dt} = \frac{\partial L}{\partial r} = mr\dot{\theta}^2 - \frac{\partial V}{\partial r}, \tag{8.34}$$

$$\frac{dp_\theta}{dt} = \frac{\partial L}{\partial \theta} = 0. \tag{8.35}$$

Here we have the equivalent Newtonian equations in momentum form, i.e., the time rate of change of momentum associated with each generalized coordinate is equal to the "force" associated with that coordinate. To be sure we must generalize the notions of momentum and force to achieve this result. In fact we decide to *define*

$$p = \frac{\partial L}{\partial \dot{q}} \tag{8.36}$$

as the *momentum conjugate* to the coordinate q. These quantities will not in general have the dimensions of mass times velocity, as we have defined momentum in our earlier treatment, nor will the time rate of change in general have the dimensions of force. But there is a certain advantage in using the term "generalized force" to denote the time rate of change of generalized or conjugate momentum. We can then write the Lagrangian equations for a conservative system in the form

$$\dot{p}_k = \frac{\partial L}{\partial q_k}. \tag{8.37}$$

The notion of generalized force is also useful for the writing of the Lagrangian equations for a nonconservative system. We revert to the general form of Hamilton's principle (8.10) and examine $\Sigma \mathbf{F}_{je} \cdot \delta \mathbf{r}_j$. With the usual decomposition into rectangular components this can be written

$$\sum_{j=1}^{n} \mathbf{F}_{je} \cdot \delta \mathbf{r}_j = \sum_{j=1}^{n} (F_{jex}\delta x_j + F_{jey}\delta y_j + F_{jez}\delta z_j). \tag{8.38}$$

Since the force components F_{jex}, etc., are in general functions of the x_j, y_j, z_j (as well as possibly explicitly of the time), we can transform (8.38) by means of the transformation equations (8.21) and ultimately obtain

$$\sum_{j=1}^{n} \mathbf{F}_{je} \cdot \delta \mathbf{r}_j = \sum_{j=1}^{n} Q_j \delta q_j. \tag{8.39}$$

Since the Q_j, which are functions of the generalized coordinates, when multiplied by the δq_j have the dimensions of work, it is appropriate to call them *generalized forces*, each associated with a particular generalized coordinate.

Application of the technique embodied in eqs. (8.24)--(8.28) produces from (8.10) the set of equations ($j = 1, 2, \cdots n$)

$$\frac{d}{dt}\left(\frac{\partial K}{\partial \dot{q}_j}\right) - \frac{\partial K}{\partial q_j} = Q_j, \tag{8.40}$$

which are then the more general form of Lagrange's equations, holding for nonconservative as well as conservative systems. Obviously if a part of Q_j is derivable from a potential function, i.e., if we can write

$$Q_j = Q_j' - \frac{\partial V}{\partial q_j}, \tag{8.41}$$

eq. (8.40) becomes

$$\frac{d}{dt}\left(\frac{\partial L}{\partial \dot{q}_j}\right) - \frac{\partial L}{\partial q_j} = Q_j', \tag{8.42}$$

still another way of writing the fundamental equations. Here, of course, we have used the fact that since V contains no $\dot{q}_j$,

$$\frac{\partial K}{\partial \dot{q}_j} = \frac{\partial L}{\partial \dot{q}_j}. \tag{8.43}$$

8.4. Illustration of Lagrange's Equations. We have already given a simple illustration of the use of Lagrange's equations in Sec. 8.3. Others are provided in the problems at the end of the chapter. Here we consider an illustration which is involved enough to bring out the value of the Lagrangian method. Fig. 8.3 depicts a wedge whose projection on the xy plane is ABC and whose mass is M. The wedge is able to slide freely on the horizontal plane (the xz plane in the diagram in which the z axis is not shown, but is supposed to be perpendicular to the plane of the figure).

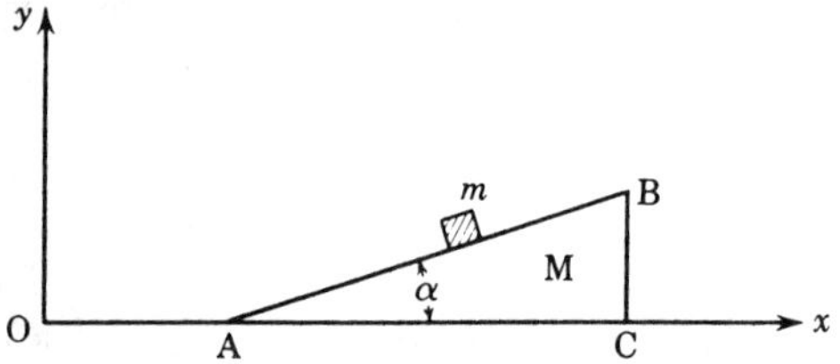

Fig. 8.3.

The perfectly smooth surface AB of the wedge makes the angle α with the horizontal. On this surface a particle of mass m is free to move in the xy plane. We wish to determine the motion of both m and M. The dynamical system here depicted obviously has two degrees of freedom. We are free to choose the two generalized coordinates in a variety of ways. If, for example, we take the xy axes fixed in the plane with origin O, we may denote OA as q_1 and take as q_2 the y coordinate of m. Inspection then shows that the Lagrangian function for the system is

$$L = \frac{1}{2}M\dot{q}_1^{\,2} + \frac{1}{2}m(\dot{q}_1 + \dot{q}_2/\tan\alpha)^2 + \frac{1}{2}m\dot{q}_2^{\,2} - mgq_2. \tag{8.44}$$

The Lagrangian equations become

$$M\ddot{q}_1 + m\ddot{q}_1 + m\ddot{q}_2/\tan\alpha = 0,$$
$$m\ddot{q}_2 + m\ddot{q}_1/\tan\alpha + m\ddot{q}_2/\tan^2\alpha + mg = 0. \tag{8.45}$$

The solution is left as an exercise. The problem should also be solved by taking as alternative coordinates the x and y coordinates of m. Obviously in any application of Lagrange's equations the aim should be so to choose the coordinates as to make the final equations as simple as possible. Note the interesting limiting case $m/M \to 0$.

8.5. Lagrangian Method in Oscillatory Phenomena. We may advantageously apply the Lagrangian equations to the problem of the small oscillations of a loaded string already treated in conventional fashion in Sec. 6.6. Here it is simplest to take the actual displacements of the n particles as the generalized coordinates. Then for the kinetic and potential energies we get respectively

$$K = \frac{m}{2}\sum_{j=1}^{n} \dot{q}_j^2,$$
$$V = \frac{T}{2a}\sum_{j=0}^{n} (q_{j+1} - q_j)^2. \tag{8.46}$$

The kinetic energy needs no explanation, but as usual V must be approached with some care. We reflect that the restoring force on each displaced particle is proportional to the *relative* displacement. Hence we expect that the total potential energy will be equal to one half the sum of the squares of the relative displacements each multiplied by the proper stiffness constant. The stiffness,is, however, the same for all and is equal to T/a. The Lagrangian equations based on (8.46) are indeed identical with eqs. (6.72).

If we examine the more general problem of the small oscillations of a dynamical system about a position of equilibrium, which Rayleigh[1] called the "main problem of Acoustics," we do not have much trouble convincing ourselves that the potential energy is a homogeneous quadratic function of the coordinates, if for the latter we take the differences between the values of the actual coordinates for the disturbed and undisturbed

[1] Cf. Lord Rayleigh, *Theory of Sound* (New York, Dover, 1945), Sec. 70 ff.

particles of the system. Calling these coordinates q's as usual, we can write

$$V = \sum_{i,j=1}^{n} b_{ij} q_i q_j, \tag{8.47}$$

where the b_{ij} must be such *constants* that V is positive for all real values of q_i and q_j. As usual the kinetic energy K is

$$K = \sum_{i,j=1}^{n} a_{ij} \dot{q}_i \dot{q}_j. \tag{8.48}$$

It will be noted that in the sum (8.47) there are two constants associated with each product $q_i q_j$, namely b_{ij} and b_{ji}. Since in the expansion which gives V the form it has there will actually be only one coefficient for $q_i q_j$, it is simplest to assume that $b_{ij} = b_{ji}$ so that the whole coefficient is simply $2b_{ij}$. We shall make the same assumption in (8.48), viz., $a_{ij} = a_{ji}$.

The Lagrangian equations now become

$$2\sum_{j=1}^{n} a_{kj} \ddot{q}_j + \sum_{i,j=1}^{n} \frac{\partial a_{ij}}{\partial q_k} \dot{q}_i \dot{q}_j + 2\sum_{j=1}^{n} b_{kj} q_j = 0, \tag{8.49}$$

as k runs from 1 to n. This of course neglects dissipation. To include this we must introduce dissipative forces and use the form of the equations appropriate to nonconservative systems. Actually all this involves is replacing the zero in (8.49) by Q_k, the kth dissipative force. If we assume that these forces are all linear functions of the $\dot{q}$'s we can follow Rayleigh and introduce a dissipation function $R(\dot{q}_1 \cdots \dot{q}_n)$ such that

$$R = \sum_{i,j=1}^{n} c_{ij} \dot{q}_i \dot{q}_j, \tag{8.50}$$

which of course must be carefully distinguished from the kinetic energy. The Lagrangian equations now take the form

$$\frac{d}{dt}\left(\frac{\partial K}{\partial \dot{q}_k}\right) - \frac{\partial K}{\partial q_k} + \frac{\partial V}{\partial q_k} = -\frac{\partial R}{\partial \dot{q}_k}. \tag{8.51}$$

The relation of R to the rate of dissipation of energy by the system should be noted. If there exist *external* driving forces in addition to the restoring forces represented by $\partial V/\partial q_k$, they must of course appear on the right-hand side of (8.51).

In Secs. 6.5 and 6.6 reference was made to the importance of *normal coordinates* in the description of vibrating systems and we emphasized the possibility of obtaining such coordinates for vibrating systems of arbitrarily great number of degrees of freedom. This can now be understood in terms of the treatment of the present section. For a dynamical system vibrating about a position of equilibrium the potential energy is always a homogeneous quadratic function of the generalized coordinates, as in eq. (8.47). Moreover the kinetic energy of any dynamical system is a homogeneous quadratic function of the time derivatives of the generalized coordinates. Now it is always possible[1] to find a linear transformation of coordinates which will reduce these homogeneous quadratic expressions to sums of squares. The resulting Lagrangian equations each involve but one of the new coordinates, leading to solutions each of which depends on a single frequency. It is clear that the new coordinates will be normal coordinates in the sense of Sec. 6.5. This turns out to be of great practical importance in connection with the vibrations of the constituent parts of molecules and of atoms in crystal lattices.[2]

8.6. Meaning of a Mechanical Theory of Physical Phenomena. Illustration from Acoustics. The key problem of classical mechanics is the determination of the motion of a collection of n particles acted on by arbitrary but known forces, subject to arbitrary but known constraints and possessing known initial positions and velocities in the primary inertial system. We have seen that the problem may be phrased in a variety of ways, e.g., the equations of Newton, the principle of D'Alembert, or the equations of Lagrange. The last named have indeed the advantage that they are invariant in form with respect to transformations of the coordinates. But the Lagrangian equations also possess two other fundamental merits which have powerfully advanced the possibility of using mechanical concepts in describing other physical phenomena in which the motion of particles is not observed.

In the first place the Lagrangian equations put the chief emphasis on energy and work rather than on force, and in the second place the use of generalized coordinates means that the q's do not have to refer directly

[1] Cf., e.g., Margenau and Murphy, *Mathematics of Physics and Chemistry*, (D. Van Nostrand, New York, 1943), p. 311.

[2] Cf., e.g., F. Seitz, *Modern Theory of Solids* (New York, McGraw-Hill, 1940), Chap. 3.

to the positions or displacements of actual particles. It is well known that there are many physical phenomena, e.g., those of acoustics, optics, and electricity, in which no particle motions are readily detectable, but in which the concepts of energy and work have proved of very great value. The suggestion is strong that by proper choice of Lagrangian function and suitable identification of coordinates with measurable quantities relating to the given phenomenon, it should be possible to describe it by means of a set of Lagrangian equations. If this can be done, we shall say that we have a mechanical theory of the phenomenon in question.

Let us give a simple illustration. The Helmholtz resonator, which has been referred to in one of the problems in Chapter 4, is a vessel of volume V with a small opening of cross-sectional area S and having a neck of length l. It is well-known that if a source of sound of suitable frequency is brought close to the opening of the resonator, it greatly amplifies the sound. Here then is a phenomenon in which no motions are directly observed (we assume that no one has as yet thought of inventing microphones for detecting small motions of air, etc.); it is a purely auditory phenomenon, which at first might seen quite outside the scope of mechanics. However, if we are impressed with the importance of energy as a general physical concept, as certainly was the case with nineteenth century physicists,[1] we feel convinced that this behavior of the resonator has something to do with energy and hence a mechanical description should be possible. If this is assumed to take the form of the actual motion of a plug of air back and forth in the neck of the resonator, we could look for a Lagrangian equation in which the coordinate q is the displacement of the air plug from equilibrium, and the kinetic and potential energies will have the form

$$K = \frac{1}{2}m\dot{q}^2; \qquad V = \frac{1}{2}fq^2, \tag{8.52}$$

where m is the equivalent mass and f the equivalent stiffness of the system. The Lagrangian equation is of course that of simple harmonic motion for q with frequency

$$\nu = \frac{1}{2\pi}\sqrt{f/m}. \tag{8.53}$$

It is only natural to interpret this as the resonance frequency of the resonator, which can be measured in terms of the dimensions of the resonator and the properties of air. We can then interpret f and m in terms of

[1]Cf., e.g., H. v. Helmholtz's essay on "The Conservation of Force" (meaning energy, of course) in *Popular Lectures on Scientific Subjects* (New York, Dover Publications), p. 317.

these measured quantities. Of course in this case it is simple to take m directly as ρSl, where ρ is the equilibrium density of the air. Since the measured frequency comes out to be

$$\nu = \frac{c}{2\pi}\sqrt{\frac{S}{lV}}, \tag{8.54}$$

where V is the volume of the resonator and c is the velocity of sound in air, we can at once interpret the stiffness f by equating the two expressions for the frequency. The result is

$$f = \rho c^2 S^2/V. \tag{8.55}$$

So far this is a purely formal process. However, it is important to note that we do not stop here, but strive to justify (8.55) by a direct examination of the mechanical effect which the resonator chamber would have on the motion of the plug of air if the latter really existed and were the cause of the observed behavior of the resonator. When this is done, (8.55) is shown to follow directly from Hooke's law of elasticity.

The example just given now seems to be a trivial one for the mechanical theory of acoustics has been well established for a long time. Nevertheless it should focus attention on the importance of the endeavor to interpret physical phenomena in general in terms of a mechanical model, for it is precisely in this way that we are stimulated to look more deeply into the phenomena in an organized fashion and thus learn more about it in an efficient way. It is no exaggeration to say that a large part of the development of physical science has taken place in this fashion.

In the next section we discuss an application of the Lagrangian method to electricity and also include some comments on its limitations.

8.7. Lagrange's Equations for Electric Currents. As a further illustration of the Lagrangian method we shall consider electric currents.

It will be recalled from elementary physics that the flux of magnetic induction through a closed circuit is directly proportional to the current, the coefficient being known as the *self-inductance* or, more simply, inductance. The flux of induction through one circuit due to current in a neighboring circuit is also proportional to the latter circuit, the coefficient being known as the *mutual inductance*. If then we have a set of circuits, n in number, with currents $i_1, i_2, \cdots i_n$ flowing in them, respectively (these being in general functions of the time), the flux of magnetic induction N_j in the jth circuit is given by the expression

$$N_j = \sum_{k=1}^{n} M_{jk}(x) i_k, \tag{8.57}$$

where $M_{jk}(x)$ is the mutual inductance between the jth and kth circuits and depends on the position coordinates of these circuits, represented in abbreviated form by x. M_{jj} is the self-inductance of the jth circuit. Moreover, from the very nature of mutual inductance $M_{jk} = M_{kj}$. The electrical energy of the system of current-carrying circuits, i.e., the work necessary to set up the currents in them, is given by

$$W_e = 1/2 \cdot \sum_{j=1}^{n} N_j i_j. \tag{8.58}$$

This will be established in Sec. 15.5 but may be accepted here as a hypothesis. With the combination of (8.57) and (8.58) there results

$$W_e = 1/2 \cdot \sum_{j,\,k} M_{jk}(x) i_k i_j. \tag{8.59}$$

This assumes, of course, that the circuits are fixed in space. If they are allowed to move freely, there will exist mechanical kinetic energy of the form

$$W_m = 1/2 \cdot \sum_{j,\,k} a_{jk}(x) \dot{x}_j \dot{x}_k, \tag{8.60}$$

where $\dot{x}_j$ is the time rate of change of the jth position coordinate. The total energy of the system of circuits is

$$E = W_e + W_m. \tag{8.61}$$

In order to apply the Lagrangian method it is necessary to decide how E is to be divided into kinetic and potential energy. This is one of the difficulties encountered in the application of Lagrange's equations to situations in which other things besides particle motions occur. In the present case it seems reasonably clear that if $i_j = \dot{q}_j$ is to be treated as analogous to a generalized velocity rather than to a generalized coordinate, E is entirely kinetic, and the potential energy is zero or at most a constant. This we shall assume, and for simplicity let us suppose further that the circuits are rigidly fastened, so that W_m vanishes. Then

$$L = K = 1/2 \cdot \sum_{j,k} M_{jk}(x)\dot{q}_j\dot{q}_k \tag{8.62}$$

Lagrange's equations in the form

$$\frac{d}{dt}\left(\frac{\partial K}{\partial \dot{q}_i}\right) - \frac{\partial K}{\partial q_i} = Q_i, \tag{8.63}$$

then become

$$\frac{d}{dt}\left[\sum_{j=1}^{n} M_{ij}(x)\dot{q}_j\right] = Q_i, \tag{8.64}$$

where we have utilized the fact that $M_{ij} = M_{ji}$. If we employ (8.57) we can rewrite (8.64) in the form

$$\frac{dN_i}{dt} = Q_i. \tag{8.65}$$

The question now arises as to the physical meaning of Q_i in this connection. It is a generalized force which when multiplied by the generalized coordinate gives energy or work. But in the present example the generalized coordinate is the charge. Now for an electric current, energy divided by charge is electromotive force. Hence it seems necessary to interpret Q_i as the electromotive force E_i in the ith circuit. With this interpretation (8.65) becomes *Faraday's law of electromagnetic induction.* Strictly speaking, if E_i is the external impressed electromotive force, eq. (8.65) applies only to a perfect (i.e., resistance-less) conductor. In this discussion capacity effects are also neglected, so it is rather idealized. For a rather elaborate discussion of the possible application of Lagrangian dynamics to physical phenomena in general, the treatise by J. J. Thomson, *Applications of Dynamics to Physics and Chemistry* (Macmillan, 1888), should be consulted. Both the advantages and limitations of the method are here treated at length.

As the foregoing simple illustration makes clear the Lagrangian method contains nothing in itself to decide how to separate the kinetic and potential energies of a system. Hence its use inevitably involves much judicious guesswork. More serious indeed is the difficulty any mechanical theory encounters when the number of degrees of freedom becomes so large that the coordinates are largely uncontrollable, i.e., initial conditions cannot effectively be set up. This situation comes to the fore particularly as soon as we enter the field of molecular and atomic theory, as, for example, the kinetic theory of gases. As we shall see in Chapter 10, the application of mechanical ideas to such fields demands the introduction of probability and statistical considerations which are quite foreign to the classical mechanics we have been developing up to this point.

8.8. The Principle of Least Action. In Sec. 8.1 we mentioned the principle of least action of Maupertuis and now wish to consider it briefly. The discussion here will be limited to a conservative system of n particles, of masses $m_1 \cdots m_n$ and position vectors $\mathbf{r}_1 \cdots \mathbf{r}_n$, for which the total momentum is $\sum_{j=1}^{n} m_j \dot{\mathbf{r}}_j$. The space integral of this quantity between any two positions of the system, P_1 and P_2, viz.,

$$\int_{P_1}^{P_2} \sum_{j=1}^{n} m_j \dot{\mathbf{r}}_j \cdot d\mathbf{r}_j, \tag{8.66}$$

is defined to be the *action* of the system for the path in question. The principle of least action states that the actual motion of the system from P_1 to P_2 takes place in such a way that the action has a stationary value with respect to all other possible paths between P_1 and P_2 corresponding to the *same total energy*. The difference between this principle and that of Hamilton should be carefully noted. In the latter all the varied paths are performed in the same time; in the former the total energy must remain constant but the times need not necessarily be the same. It is of interest to note that the space integral in (8.66) can be transformed to a time integral from the fact that

$$\sum_{j=1}^{n} m_j \dot{\mathbf{r}}_j \cdot d\mathbf{r}_j = \sum_{j=1}^{n} m_j v_j^2 \, dt = 2K \, dt, \tag{8.67}$$

where K is the total kinetic energy. We may therefore rewrite the content of the principle in the form

$$\delta \int_{t_1}^{t_2} 2K\, dt = 0. \tag{8.68}$$

In carrying out the variation we must be careful, however, to remember that the energy and not the time remains constant over the varied paths.

It is not our purpose to present a demonstration of the principle,[1] which has not actually had such an important influence on the development of classical mechanics as that of Hamilton. We mention it primarily because the concept of action in itself has played a very significant role in modern atomic physics. It would be tempting indeed to try to represent the motion of photons or light particles in terms of least action. This is what Maupertuis tried to do. If we denote the element of arc along the path of the light particle by ds and call the magnitude of its velocity v, the principle for a single particle takes the form

$$\delta \int_{P_1}^{P_2} mv\, ds = 0. \tag{8.69}$$

We do not know the mass of the light particle, but if it is constant, this should not interfere with the application of the principle. The trouble is that (8.69) is in direct contradiction to Fermat's principle of least time in accordance with which the v should be in the denominator of the integrand and not the numerator. But we know that Fermat's principle gives the correct laws of reflection and refraction of light. It would therefore appear that least action does not at all apply and this would be a blow to the hope of a mechanical theory of optical phenomena. The situation has been saved by the modern stratagem of redefining the momentum of a photon by means of the quantum theory. This is analyzed in detail in Chapter 11, but we here merely state the new definition, namely $h\nu/v$, where ν is the frequency of the light, v is the velocity of the light particle, and h is the fundamental constant of the quantum theory, the so-called action constant of Planck. It has the dimensions of action and the value 6.55×10^{-27} erg sec. Hence $h\nu/v$ has the dimensions of momentum. If we use this in the principle of least action, the latter reduces to the principle of Fermat and all is well. It should be remarked that this definition is not purely *ad hoc* but yields correct results in many experimental phenomena involving the interaction of light with matter.

[1] Cf., E. T. Whittaker, *Analytical Dynamics* (New York, Dover Publications, 1944), pp. 247 f.

PROBLEMS

1. Use Hamilton's principle to derive the equations of motion of a particle moving in a field of force in which the potential energy is $-C/r$, where r is the distance of the particle to the origin and C is a positive constant.

2. Show that Hamilton's principle for a conservative system can be stated in the following way: A conservative dynamical system moves in such a way that the difference between the time averages of kinetic and potential energies has a stationary value with respect to all paths performed in the same time and having the same end points.

3. A particle of mass m moves in a uniform field directed along the x axis and characterized by potential energy of magnitude mgx. Calculate the time integral of the Lagrangian function over the actual dynamical path $x = gt^2/2 + v_0t$ from $t = 0$ to $t = T$. Consider a varied path $x = gt^2/2 + v_0t + \lambda \sin 2\pi t/T$, which has the same end points as the dynamical path at $T = 0$ and $t = T$, where λ is a constant. Find the time integral of the Lagrangian function over the varied path from $t = 0$ to $t = T$ and compare with the previous result.

4. The propagation of light follows the principle of Fermat, i.e., the principle of least time of transit. How can this principle be made consistent with the principle of least action for the motion of photons (light particles)?

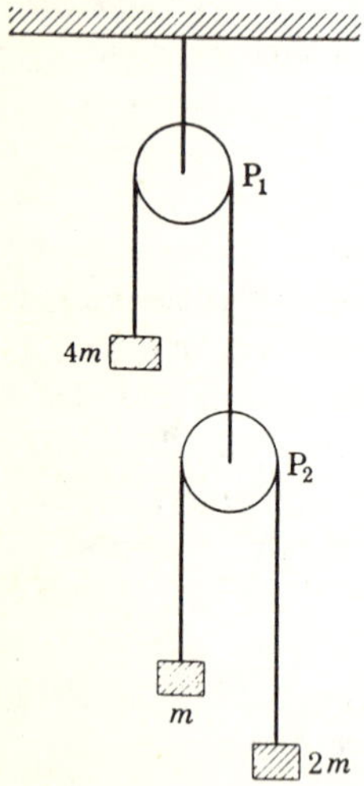

5. Deduce the equations of motion of a particle moving in a plane and subject to a uniform field along the x axis [i.e., $L = \frac{m}{2}(\dot{x}^2 + \dot{y}^2) - mgx$] by means of the principle of least action.

6. Determine by means of Lagrange's equations the motion of the masses in the frictionless pulley hook-up shown in the accompanying figure. The movable pulley P_2 has mass m.

7. A particle of mass m moves in a plane in a conservative field of force characterized by the potential function V, so that the cartesian equations of motion are

$$m\ddot{x} = -\frac{\partial V}{\partial x}; \qquad m\ddot{y} = -\frac{\partial V}{\partial y}.$$

Introduce the generalized coordinates $q_1 = q_1(x, y)$ and $q_2 = q_2(x, y)$ and show by direct substitution of the transformation that the above equations take on the Lagrangian form, i.e.,

$$\frac{d}{dt}\left(\frac{\partial T}{\partial \dot{q}_i}\right) - \frac{\partial T}{\partial q_i} = -\frac{\partial V}{\partial q_i} \; (i = 1, 2).$$

8. A particle of mass m moves in a central field of force characterized by the potential energy function $V = -C/r$ where C is a positive constant. Write Lagrange's equations for the motion using $q_1 = r$, $q_2 = \theta$, $q_3 = \phi$, i.e., the usual spherical coordinates. Where in the discussion does it come out that the motion takes place in a plane?

9. Two particles of mass m connected by a weightless rigid bar of length l are set moving in arbitrary fashion on a smooth horizontal plane. Set up and solve Lagrange's equations for this case, using as coordinates the rectangular coordinates of the center of the bar and the angle the bar makes with the x axis.

10. Use the method of Sec. 8.7 to determine the force of attraction between the two parallel straight current-carrying wires.

11. Find the equation of the curve connecting two points A and B in the xy plane which minimizes the area of the surface of revolution formed by rotating the curve about the x axis.

12. Deduce the laws of reflection and refraction of light by means of Fermat's principle of least time.

13. A particle of mass m moves in a plane subject to gravitational force of attraction toward the origin and in addition to a constant force perpendicular to the radius vector. Is this a conservative system of forces? Write the Lagrangian equations of motion for the particle.

14. A dynamical system of n degrees of freedom described by the generalized coordinates $q_1 \cdots q_n$ is also subject to constraints described by the following k $(k < n)$ equations connecting the q's and the time

$$\phi_1(q_1 \cdots q_n, t) = 0, \quad \phi_2(q_1 \cdots q_n, t) = 0,$$
$$\cdots \phi_k(q_1 \cdots q_n, t) = 0.$$

Show that the Lagrangian equations for the system may be written in the form

$$\frac{d}{dt}\left(\frac{\partial L}{\partial \dot{q}_i}\right) - \frac{\partial L}{\partial q_i} - \sum_{j=1}^{k} \lambda_j(t) \frac{\partial \phi_j}{\partial q_i} = 0, \qquad (i = 1 \cdots n)$$

where $\lambda_j(t)$ are initially undetermined functions of the time alone. These reduce to constants if the equations do not involve the time.

15. A smooth wire is shaped as an inverted semicircle of radius a and placed in a vertical plane. A bead of mass m slides freely on the wire under the influence of gravity. Determine its motion by the use of the Lagrangian equations for a system subject to constraints (cf. Prob. 14).

16. In a *non-holonomic* dynamical system the constraint conditions are nonintegrable, i.e., are expressible only by relations connecting the δq_i. Suppose that for a system of n degrees of freedom there are k linear equations connecting the δq_i of the form

$$\sum_{j=1}^{n} a_{1j}\, \delta q_j = \sum_{j=1}^{n} a_{2j}\, \delta q_j = \cdots = \sum_{j=1}^{n} a_{kj}\, \delta q_j = 0,$$

where the a_{ij} are functions of the q's. Show that the Lagrangian equations for the system take the form

$$\frac{d}{dt}\left(\frac{\partial L}{\partial \dot{q}_i}\right) - \frac{\partial L}{\partial q_i} = \sum_{j=1}^{k} \lambda_j a_{ji},$$

where λ_j are the initially undetermined parameters. Indicate how the λ_j may be determined.

17. Show that the principle of least action leads to the conservation of momentum in the elastic collision of two mass particles moving with initial velocities $\mathbf{v}_1$ and $\mathbf{v}_2$, respectively, along the same straight line.

18. Write the Lagrangian equations of motion for a particle moving freely on the surface of a sphere (zero impressed force). Show that the path is a great circle of the sphere.

19. A particle of mass m moves in a plane with potential energy

$$V = 1/2 \cdot k_1 x^2 + 1/2 \cdot k_2 y^2,$$

where k_1 and k_2 are positive constants. Set up and solve the Lagrangian equations of motion.

20. Solve eqs. (8.45).

CHAPTER IX

THE TRANSFORMATION THEORY OF MECHANICS. ATOMIC STRUCTURE.

9.1. The Canonical Equations of Hamilton. In Sec. 8.3 we introduced a new set of quantities called the generalized *momenta* (eq. 8.36). Thus, to each coordinate q_k we associated a *conjugate momentum* p_k defined by

$$p_k = \frac{\partial L}{\partial \dot{q}_k}. \tag{9.1}$$

Note that for the case of a single free particle, where $L = 1/2 \cdot m\dot{q}^2$, $p = m\dot{q}$ and is the usual momentum. In general, however, p_k will not have the dimensions of mass times velocity, although clearly $p_k\dot{q}_k$ will have the dimensions of energy, independently of the nature of the generalized coordinates. We now proceed to make use of these momenta.

Consider for a conservative dynamical system of n degrees of freedom the function

$$H = \sum_{k=1}^{n} p_k\dot{q}_k - L(q, \dot{q}), \tag{9.2}$$

where the q, $\dot{q}$ in $L(q, \dot{q})$ are an abbreviation for $q_1 \cdots q_n$ and $\dot{q}_1 \cdots \dot{q}_n$, respectively. The formation of the total differential of H leads at once to

$$dH = \sum_{k=1}^{n} \dot{q}_k \, dp_k - \sum_{k=1}^{n} \frac{\partial L}{\partial q_k} \, dq_k, \tag{9.3}$$

whereby it appears that H is really a function of the p's and q's and not a function of the $\dot{q}$'s at all. From (9.3) we immediately conclude that

$$\frac{\partial H}{\partial p_k} = \dot{q}_k, \tag{9.4}$$

$$\frac{\partial H}{\partial q_k} = -\frac{\partial L}{\partial q_k} = -\dot{p}_k, \tag{9.5}$$

where the last step in (9.5) comes from the use of the Lagrangian equations (8.28) and the definition of p_k in (9.1).

The $2n$ first-order partial differential equations (9.4) and (9.5) (obtained by letting k run from 1 to n) are usually called the *canonical equations* of Hamilton for the dynamical system, or sometimes Hamilton's equations for short. To understand their significance we must make clear the physical meaning of H. From (9.3) the time rate of change of H becomes

$$\dot{H} = \sum_{k=1}^{n} \dot{q}_k \dot{p}_k - \sum_{k=1}^{n} \dot{p}_k \dot{q}_k = 0. \tag{9.6}$$

Hence H is constant in time and we naturally suspect some connection with the energy. As a matter of fact since, for a conservative system, the potential energy contains no terms in the $\dot{q}$'s, we can write

$$H = \sum_{k=1}^{n} \frac{\partial K}{\partial \dot{q}_k} \dot{q}_k - L. \tag{9.7}$$

Since K is a homogeneous quadratic function of the $\dot{q}$'s (cf. eq. 8.23), Euler's theorem can be applied with the result

$$\sum_{k=1}^{n} \frac{\partial K}{\partial \dot{q}_k} \dot{q}_k = 2K. \tag{9.8}$$

The combination of (9.7) and (9.8) at once leads to

$$H = K + V = \text{total energy} = E. \tag{9.9}$$

The function H, called the *Hamiltonian function*, thus appears as the expression for the total energy of the system in terms of the p's and q's. As such it must, of course, be carefully distinguished from the ordinary expression for the total energy in terms of the q's and $\dot{q}$'s.

As a simple example, consider the simple harmonic oscillator with

$$E = 1/2 \cdot m\dot{x}^2 + 1/2 \cdot fx^2, \tag{9.10}$$

where we here denote the stiffness by f. The Hamiltonian for this system of one degree of freedom with $x = q$, $\dot{x} = \dot{q}$, $m\dot{x} = p$ is then

$$H = p^2/2m + fq^2/2. \tag{9.11}$$

The canonical equations (9.4) and (9.5) are now two in number, viz.,

$$\dot{p} = -fq,$$
$$\dot{q} = p/m.$$

It is seen at once that the first equation is really the ordinary Newtonian equation of motion of the oscillator, while the second is an identity connecting p with $\dot{q}$. The situation might seem to deprive the canonical equations of more than curiosity value. However in the next section their real utility in providing the basis for a more powerful technique in advanced dynamics will become clear.

9.2. Canonical Transformations. Consider the general transformation of coordinates and momenta q_k, p_k for a dynamical system of n degrees of freedom to a new set Q_k, P_k, viz.,

$$Q_k = Q_k(q, p); \quad P_k = P_k(q, p), \tag{9.12}$$
$$q_k = q_k(Q, P); \quad p_k = p_k(Q, P). \tag{9.13}$$

We shall not allow this to be completely arbitrary. Let us suppose there exist two functions, $L'(Q, \dot{Q})$ and $S(q, Q, t)$ such that the Lagrangian function in terms of the original coordinates is given by

$$L(q, \dot{q}) = L'(Q, \dot{Q}) + \frac{dS}{dt}. \tag{9.14}$$

We shall indeed define

$$P_k = \frac{\partial L'}{\partial \dot{Q}_k} \tag{9.15}$$

and shall further introduce the function $H'(P, Q)$ as follows:

$$H'(P, Q) = \sum P_k \dot{Q}_k - L'(Q, \dot{Q}). \tag{9.16}$$

Now from Hamilton's principle we have

$$\int_{t_1}^{t_2} \delta L \, dt = \int_{t_1}^{t_2} \delta L'(Q, \dot{Q}) \, dt + \int_{t_1}^{t_2} \delta\left(\frac{dS}{dt}\right) dt = 0. \tag{9.17}$$

But

$$\int_{t_1}^{t_2} \delta\left(\frac{dS}{dt}\right) dt = \int_{t_1}^{t_2} d(\delta S) = (\delta S)_2 - (\delta S)_1, \tag{9.18}$$

where $(\delta S)_1$ is the value of δS at the beginning of the path, etc. Now if the δq_j and the δQ_j are all zero at the extremities of the path, it is clear that the integral in (9.18) is zero and (9.17) becomes

$$\int_{t_1}^{t_2} \delta L'(Q, \dot{Q})\, dt = 0. \tag{9.19}$$

so that L' satisfies the Lagrangian equations, namely,

$$\frac{d}{dt}\left(\frac{\partial L'}{\partial \dot{Q}_j}\right) - \frac{\partial L'}{\partial Q_j} = 0. \tag{9.20}$$

Now applying the same analysis to $H'(P, Q)$, as defined in (9.16), that we used on $H(p, q)$ in (9.2) we see that

$$\dot{P}_k = -\frac{\partial H'}{\partial Q_k}, \quad \dot{Q}_k = \frac{\partial H'}{\partial P_k}. \tag{9.21}$$

Thus the transformed coordinates also satisfy the canonical equations. The transformation is therefore known as a canonical transformation. Eq. (9.14) defines the transformation. This can be put into more convenient form by the use of (9.2) and (9.16). Thus

$$\sum P_k \dot{Q}_k - H'(P, Q) + \frac{dS}{dt} = \sum p_k \dot{q}_k - H(p, q). \tag{9.22}$$

If we multiply through by dt, this becomes

$$\sum \frac{\partial S}{\partial q_k} dq_k + \sum \frac{\partial S}{\partial Q_k} dQ_k + \frac{\partial S}{\partial t} dt =$$

$$\sum p_k dq_k - \sum P_k dQ_k + (H' - H)\, dt. \tag{9.23}$$

Since this is an identity we must have ($k = 1, 2, \cdots n$)

$$p_k = \frac{\partial S}{\partial q_k}; \quad P_k = -\frac{\partial S}{\partial Q_k}; \quad H' = H + \frac{\partial S}{\partial t}. \tag{9.24}$$

Eqs. (9.24) show that S really defines the transformation, since the first $2n$ equations enable us to write the eqs. (9.12) or (9.13). The equation $H' = H + \partial S/\partial t$ enables us to find the new function H', which so to speak corresponds to H. In fact, if we choose S in such a way that it does not contain the time explicitly, $\partial S/\partial t = 0$ and $H' = H$; in this case H' is simply H with the p's and q's replaced by the corresponding P's and Q's from (9.13).

There are obviously other ways of introducing the transformation function S. It need not be a function of q_k, Q_k, t. It might equally well be a function of p_k, Q_k, t or q_k, P_k, t or p_k, P_k, t. For each of these cases eq. (9.16) must be altered appropriately. For example, let us write

$$S = S(q_k, P_k, t), \tag{9.25}$$

and then define $H'(P, Q)$ so that

$$H'(P, Q) = -\sum Q_k \dot{P}_k - L'(P, \dot{P}). \tag{9.26}$$

We choose $L'(P, \dot{P})$ so that the original Lagrangian

$$L(q, \dot{q}) = L'(P, \dot{P}) + \frac{dS}{dt} \tag{9.27}$$

and *define*

$$Q_k = -\frac{\partial L'}{\partial \dot{P}_k}. \tag{9.28}$$

Then, following through once more the earlier analysis, we arrive at the result that L' satisfies the Lagrangian equations in the form

$$\frac{d}{dt}\left(\frac{\partial L'}{\partial \dot{P}_k}\right) - \frac{\partial L'}{\partial P_k} = 0. \tag{9.29}$$

Eq. (9.26) yields

$$\sum \frac{\partial H'}{\partial P_k} dP_k + \sum \frac{\partial H'}{\partial Q_k} dQ_k =$$

$$- \sum Q_k d\dot{P}_k - \sum \dot{P}_k dQ_k - \sum \frac{\partial L'}{\partial P_k} dP_k - \sum \frac{\partial L'}{\partial \dot{P}_k} d\dot{P}_k. \quad (9.30)$$

But from (9.28) the first and last terms on the right cancel and we get by comparison of the corresponding remaining terms,

$$\dot{P}_k = -\frac{\partial H'}{\partial Q_k}, \quad \dot{Q}_k = \frac{\partial H'}{\partial P_k}, \quad (9.31)$$

i.e., the usual canonical equations. Eq. (9.27) now becomes

$$\sum \frac{\partial S}{\partial q_k} dq_k + \sum \frac{\partial S}{\partial P_k} dP_k + \frac{\partial S}{\partial t} dt$$

$$= \sum p_k dq_k + \sum Q_k dP_k + (H' - H)\, dt. \quad (9.32)$$

Hence (9.24) is replaced by

$$p_k = \frac{\partial S}{\partial q_k}, \quad Q_k = \frac{\partial S}{\partial P_k}, \quad H' = H + \frac{\partial S}{\partial t}. \quad (9.33)$$

This is occasionally a more convenient set-up.

It must be emphasized that every continuous differentiable function S can define a canonical transformation. The question now arises: Of what use is such a transformation? One possible answer is: by choosing an appropriate transformation the expression for the motion may be rendered much simpler in the new coordinates than the old. This might seem indeed a questionable gain, particularly if one has to proceed to locate desirable S or transformation functions by guesswork. It is therefore fortunate that a way exists for finding an S function which is guaranteed to be useful.

9.3. The Hamilton-Jacobi Equation. We have seen that the Hamiltonian function H is equal to the total energy expressed as a function of the p's and q's. We are therefore entitled to write the equation

$$H(p_1 \cdots p_n, \quad q_1 \cdots q_n) = E. \quad (9.34)$$

As it stands it is an algebraic equation connecting the p's and q's. But nothing prevents us from replacing p_k by $\partial S/\partial q_k$ from (9.24) and so obtaining

$$H\left(\frac{\partial S}{\partial q_1} \cdots \frac{\partial S}{\partial q_n}, \quad q_1 \cdots q_n\right) = E, \tag{9.35}$$

which has precisely the same physical significance as before, although mathematically it is now a partial differential equation of the first order for the function $S(q_1 \cdots q_n)$. Such an equation has, of course, in general an infinity of solutions, most of which are of no interest to us here. Let us suppose there exists a *complete* integral or *primitive* of the equation, which is a solution in the form of a relation connecting S with the n q's and containing n arbitrary constants of integration such that when these constants are eliminated from the integral and the n equations obtained from it by differentiating once with respect to each coordinate, the original differential equation results. As an illustration, consider the first-order equation

$$\frac{\partial z}{\partial x} \cdot \frac{\partial z}{\partial y} = 4xy.$$

This has a complete integral in the form

$$z = \frac{x^2}{a} + ay^2 + b,$$

as is readily clear by substitution. The a and b are the two arbitrary constants corresponding to the two independent variables x and y.

If the eq. (9.35), usually known as the Hamilton-Jacobi equation, has a complete integral, let it be written in the form

$$S = S(q_1 \cdots q_n, \quad E, \quad \alpha_1 \cdots \alpha_n), \tag{9.36}$$

where the n arbitrary constants of integration are $\alpha_1 \cdots \alpha_n$, and E is included since it is a fundamental parameter in the original equation. Note that there is no explicit time dependence. Let us now choose this S as a transformation function defining a transformation from q, p to Q, P, where in line with the discussion at the beginning of Sec. 9.2 we naturally take the constants as our new Q's. Specifically we set

$$Q_1 = E, \quad Q_2 = \alpha_2, \cdots, \quad Q_n = \alpha_n. \tag{9.37}$$

To the question what happens to α_1, we reply that one constant in a complete integral is always additive (witness b in the example given just above.) We shall let α_1 be this additive constant and not worry about it further. The equations $P_k = -\partial S/\partial Q_k$ in (9.24) now become

$$P_1 = -\frac{\partial S}{\partial E}, \quad P_k = -\frac{\partial S}{\partial \alpha_k} (k = 2, 3, \cdots n). \tag{9.38}$$

But now $H' = H = E$ and hence the canonical equations become

$$\dot{P}_1 = -\frac{\partial H'}{\partial E} = -\frac{\partial E}{\partial E} = -1, \tag{9.39}$$

$$\dot{P}_k = -\frac{\partial H'}{\partial \alpha_k} = -\frac{\partial E}{\partial \alpha_k} = 0 (k = 2, 3, \cdots n), \tag{9.40}$$

whence

$$P_1 = -t - \beta_1; \quad P_k = -\beta_k (k = 2, 3, \cdots n), \tag{9.41}$$

where the $\beta_1, \cdots, \beta_n$ are new constants of integration. The combination of (9.41) and (9.38) yields

$$\frac{\partial S}{\partial E} = t + \beta_1; \quad \frac{\partial S}{\partial \alpha_k} = \beta_k (k = 2, 3, \cdots n). \tag{9.42}$$

The eqs. (9.42) provide a means for expressing the n q's in terms of the time. But this is the solution of the dynamical problem of the motion of the system whose Hamiltonian is given in (9.34). If we like we may consider the set of $n-1$ equations $\partial S/\partial \alpha_k = \beta_k (k = 2 \cdots n)$ as the geometrical equations of the *path* of the system, whereas the first equation $\partial S/\partial E = t + \beta_1$ is the *time* equation, i.e., it shows how the path is traversed in time.

The method of Jacobi, using the Hamilton–Jacobi equation, is apt to appear rather mysterious. Hence it will be appropriate to see how it works in a couple of examples. The first is the very simple one of a single free particle of mass m, having three degrees of freedom. Then $q_1 = x$, $q_2 = y$, $q_3 = z$, if we choose to use rectangular coordinates. The Hamiltonian function becomes

$$H = \frac{1}{2m} (p_x{}^2 + p_y{}^2 + p_z{}^2). \tag{9.43}$$

Hence the Hamilton–Jacobi equation is

$$\left(\frac{\partial S}{\partial x}\right)^2 + \left(\frac{\partial S}{\partial y}\right)^2 + \left(\frac{\partial S}{\partial z}\right)^2 = 2mE, \tag{9.44}$$

where S is, of course, a function of x, y, z. A standard method of finding a complete integral of such an equation is by separation of the variables. Thus we try to get a solution in the form

$$S = S_x(x) + S_y(y) + S_z(z), \tag{9.45}$$

where $S_x(x)$ is a function of x alone, etc. Substitution into eq. (9.44) yields

$$\left(\frac{dS_x}{dx}\right)^2 = 2mE - \left(\frac{dS_y}{dy}\right)^2 - \left(\frac{dS_z}{dz}\right)^2. \tag{9.46}$$

Now the left-hand side is a function of x alone, while the right-hand side contains only y and z. The only way in which both sides can be identically equal is for each to be equal to a constant. Let the constant be α_2^2. Then

$$\frac{dS_x}{dx} = \alpha_2; \quad S_x = \alpha_2 x + C_1. \tag{9.47}$$

(We choose the positive sign before the square root, although this is not of fundamental significance.) Now repeat the process, obtaining

$$S_y = \alpha_3 y + C_2, \tag{9.48}$$

and finally

$$S_z = \sqrt{2mE - \alpha_2^2 - \alpha_3^2}\, z + C_3, \tag{9.49}$$

where α_3, C_1, C_2, and C_3 are constants. The solution (9.45) then appears as

$$S = \alpha_2 x + \alpha_3 y + \sqrt{2mE - \alpha_2^2 - \alpha_3^2}\, z + \alpha_1, \tag{9.50}$$

where we have put $C_1 + C_2 + C_3 = \alpha_1$, the additive constant. This solution satisfies the requirement of the complete integral, as it has the requisite number of arbitrary constants. The solution of the problem is now inherent in eqs. (9.42) which become

$$\begin{aligned} \frac{mz}{\sqrt{2mE - \alpha_2^2 - \alpha_3^2}} &= t + \beta_1. \\ x - \frac{\alpha_2 z}{\sqrt{2mE - \alpha_2^2 - \alpha_3^2}} &= \beta_2, \\ y - \frac{\alpha_3 z}{\sqrt{2mE - \alpha_2^2 - \alpha_3^2}} &= \beta_3. \end{aligned} \tag{9.51}$$

In fact the second and third equations by themselves represent planes and taken together represent the straight line which is the intersection of these planes. In parametric form the solution may be written

$$x = \frac{\alpha_2}{m}t + \gamma_1,$$

$$y = \frac{\alpha_3}{m}t + \gamma_2, \tag{9.52}$$

$$z = \frac{\sqrt{2mE - \alpha_2^2 - \alpha_3^2}}{m}t + \gamma_3,$$

where the $\gamma_1, \gamma_2, \gamma_3$ are the arbitrary constants in a new form. These are the kinematic equations for motion along a straight line with constant velocity having the rectangular components

$$\frac{\alpha_2}{m}, \quad \frac{\alpha_3}{m}, \quad \frac{\sqrt{2mE - \alpha_2^2 - \alpha_3^2}}{m}, \tag{9.53}$$

and starting at the point $(\gamma_1, \gamma_2, \gamma_3)$. The resultant speed is, of course,

$$\sqrt{\frac{\alpha_2^2}{m^2} + \frac{\alpha_3^2}{m^2} + \frac{2mE}{m^2} - \frac{\alpha_2^2}{m^2} - \frac{\alpha_3^2}{m^2}} = \sqrt{\frac{2E}{m}}, \tag{9.54}$$

as it should be.

This is obviously not a convenient way of studying the motion of a free particle. Our reason for introducing the illustration is to make the method clear. Nevertheless the methodological significance of a method which is based wholly on the form of the Hamiltonian function and nowhere involves the conventional Newtonian equations of motion should be evident.

9.4. Central Field Motion by the Hamilton – Jacobi Method. The power of the method becomes more convincing if we apply it to central motion in an inverse square field. Here we shall find it convenient to employ polar coordinates r and θ in the plane of the motion. To be specific we shall also suppose the moving particle is an electron of negative charge e moving about a stationary nucleus with positive charge Ze where Z is an integer denoting the so-called atomic number. Denoting the conjugate momenta along the r and θ directions respectively by p_r and p_θ the Hamiltonian function becomes

$$H = \frac{1}{2m}(p_r^2 + p_\theta^2/r^2) - e^2Z/r. \tag{9.55}$$

Hence the Hamiltonian – Jacobi equation takes the form

$$\left(\frac{\partial S}{\partial r}\right)^2 + \frac{1}{r^2}\left(\frac{\partial S}{\partial \theta}\right)^2 = 2mE + \frac{2me^2Z}{r}, \tag{9.56}$$

where E is the total energy. Following the usual procedure of separating the variables we set

$$S = S_r(r) + S_\theta(\theta), \tag{9.57}$$

and obtain

$$\left(\frac{dS_\theta}{d\theta}\right)^2 = r^2\left[2mE + 2\frac{me^2Z}{r} - \left(\frac{dS_r}{dr}\right)^2\right]. \tag{9.58}$$

It follows that

$$\frac{dS_\theta}{d\theta} = \alpha_2 = \text{constant}, \tag{9.59}$$

and

$$\frac{dS_r}{dr} = \sqrt{2mE + 2\frac{me^2Z}{r} - \frac{\alpha_2^2}{r^2}}. \tag{9.60}$$

The transformation function then becomes

$$S = \alpha_2\theta + \int\sqrt{2mE + \frac{2me^2Z}{r} - \frac{\alpha_2^2}{r^2}}\,dr + \alpha_1. \tag{9.61}$$

The integrated equations of motion are given as usual in the form

$$\frac{\partial S}{\partial E} = t + \beta_1; \quad \frac{\partial S}{\partial \alpha_2} = \beta_2. \tag{9.62}$$

The second of these provides the equation of the orbit. This is therefore

$$\beta_2 = \theta - \int\frac{\alpha_2 dr}{r^2\sqrt{2mE + 2me^2Z/r - \alpha_2^2/r^2}}. \tag{9.63}$$

The integral is evaluated by the use of Peirce's Tables (No. 183) with the final result that r and θ are related by the equation

$$\frac{1}{r}=\frac{me^2Z}{\alpha_2^{\,2}}-\sqrt{\frac{m^2e^4Z^2}{\alpha_2^{\,4}}+\frac{2mE}{\alpha_2^{\,2}}}\cdot\cos\,(\theta-\beta_2'), \tag{9.64}$$

with $\beta_2' = \beta_2 + \pi/2$. This can further be written

$$r=\frac{\epsilon l}{1-\epsilon\cos\,(\theta-\beta_2')}, \tag{9.65}$$

where

$$\epsilon l = \alpha_2^{\,2}/me^2Z, \tag{9.66}$$

and

$$\epsilon = \sqrt{1+2E\alpha_2^{\,2}/me^4Z^2}: \tag{9.67}$$

It will be recalled from eq. (5.41) that (9.65) is the equation of a conic with eccentricity ϵ. If $E < 0$, the orbit is an ellipse, if $E = 0$, it is a parabola, and if $E > 0$, it is a hyperbola. Incidentally we can establish an important relation between the major axis of the conic, when it is an ellipse, and the energy. From (9.65) we can at once infer that

$$2a = 2\epsilon l/(1-\epsilon^2), \tag{9.68}$$

whence by the use of (9.67) it follows that

$$E=-\frac{e^2Z}{2a}. \tag{9.69}$$

9.5. Application of the Hamilton – Jacobi Method to Bohr Atomic Theory. In Sec. 9.4 the application of the Hamilton – Jacobi method to the motion of an electron around the nucleus of an atom was purely classical. There is nothing to restrict the values of ϵ and a and hence nothing to restrict either the shape or size of the elliptic orbits in the case of bounded motion. Moreover from (9.69) the energy may have any value. However, in Bohr's quantum theory certain nonclassical restrictions are arbitrarily imposed and these serve to select certain orbits and energy values as actually realized in the so-called stationary states of the atom.

It will perhaps be worth while to recapitulate here the fundamental postulates of the Bohr theory. These are: (1) Among the classically possible states of an atom the only stable ones in which the atom can exist without gaining or losing energy are certain ones called *stationary* states. All changes in the atom involve transitions from one stationary state to another. Each stationary state is characterized by a definite value of the energy. (2) Whenever a transition takes place from one stationary state to another with absorption or emission of radiation, the frequency of the radiation is equal to the gain or loss of energy divided by a universal constant h, having the dimensions of action and equal to 6.55×10^{-27} erg sec. This is called the Planck constant of action. It is fundamental to the quantum theory. We shall recur to these postulates, which are of very general character, in Chapter 11. At the moment we are interested principally in Bohr's method of picking out the stationary states in an atom like hydrogen whose model is the single electron moving in an inverse square force field about the positively charged nucleus. If the nucleus is considered at rest, the simple assumption we employ here, the energy of the atom resides in the electron. Bohr picks out the possible states of motion of the electron by certain quantum conditions which may be stated as follows:

$$\oint p_\theta d\theta = n_1 h, \tag{9.70}$$

$$\oint p_r dr = n_2 h. \tag{9.71}$$

In these equations n_1 and n_2 are positive integers, p_θ and p_r are the moment of momentum and the radial momentum of the electron in its path, respectively. In each case the integral is to be taken about a complete cycle of the corresponding variable. In the first integral this means that θ is to be allowed to run from 0 to 2π. In the second integral r runs from r_{max} to r_{min} and back to r_{max}, thus completing a period.

Let us now work out the integrals in (9.70) and (9.71) and see what restrictions they place on the electron orbits in hydrogen. We begin with the integral over θ. From (9.59) we have

$$p_\theta = \frac{dS_\theta}{d\theta} = \alpha_2 = \text{constant.}$$

Hence (9.70) yields at once

$$\alpha_2 = n_1 h/2\pi. \tag{9.72}$$

If we use (9.60) and (9.72) we can write (9.71) in the form

$$\oint \sqrt{2mE + 2me^2Z/r - n_1{}^2h^2/4\pi^2r^2}\, dr = n_2h. \tag{9.73}$$

It must be emphasized that in evaluating the integral on the left we use the positive sign in front of the radical as r goes from r_{min} to r_{max} (i.e., when dr is positive) and the negative sign as r goes from r_{max} to r_{min}. Hence the equation becomes

$$2\int_{r_{min}}^{r_{max}} \sqrt{2mE + 2me^2Z/r - n_1{}^2h^2/4\pi^2r^2}\, dr = n_2h. \tag{9.74}$$

We also note that r_{min} and r_{max} are the two roots of the quadratic in r in the radiand. With these facts in mind we can use No. 187 of Peirce's Tables of Integrals to get

$$2\pi me^2Z/\sqrt{-2mE} - n_1h = n_2h, \tag{9.75}$$

whence solving for E yields at once

$$E = -\frac{2\pi^2me^4Z^2}{(n_1 + n_2)^2h^2}. \tag{9.76}$$

for the allowed or quantized energy values. For a particular single electron atom characterized by the atomic number Z (this is equal to 1 for normal hydrogen, 2 for singly ionized helium, 3 for doubly ionized lithium, etc.) the quantized energy states are given by the integers n_1 and n_2 called quantum numbers, n_1 the so-called *azimuthal* quantum number, and n_2 the *radial* quantum number. Actually the energy depends only on the sum of these numbers and hence on a single integer, usually called the *principal* quantum number. The same is true of the semi-major axis of the corresponding orbit, as is clear from (9.69). Thus we have

$$a = \frac{(n_1 + n_2)^2h^2}{4\pi^2me^2Z}. \tag{9.77}$$

The smallest value of a that is possible in hydrogen is then

$$a_0 = h^2/4\pi^2me^2 \doteq 0.53 \times 10^{-8}\ cm, \tag{9.78}$$

if we substitute the relevant values. It is usually taken as the unit of distance in atomic problems. The energy of the electron in the orbit is

$$E_0 = -e^2/2a_0 \doteq -21.5 \times 10^{-12} \text{ erg.} \tag{9.79}$$

As a matter of fact the erg is not a satisfactory unit of energy in the atomic domain. Much better is the *electron volt*, which is the energy gained by an electron in being accelerated through a potential difference of 1 volt. It turns out to be 1.6×10^{-12} erg. Hence $E_0 = 13.5$ electron volts. This is called the ionization energy or potential of hydrogen. It represents the energy with which the electron is bound to the hydrogen nucleus and thereby the energy which must be given to the atom in order to separate the electron from it, i.e., ionize it.

Although the energy and semi-major axis associated with each elliptical electron orbit depend only on the sum of n_1 and n_2, a glance at eq. (9.67) shows that the eccentricity and thereby the shape of the orbit depends on n_1 as well as n. In fact, we have on substitution into (9.67)

$$\epsilon = \sqrt{1 - \left(\frac{n_1}{n_1 + n_2}\right)^2}. \tag{9.80}$$

It is therefore clear that we can have in general several different orbits for each value of $n = n_1 + n_2$. Thus for $n_1 + n_2 = 1$, we can have $n_1 = 1$, $n_2 = 0$, or $n_1 = 0$, $n_2 = 1$. The first yields $\epsilon = 0$, i.e., a circular orbit, while the second gives $\epsilon = 1$, i.e., an ellipse degenerated to a straight line. This has been ruled out by the Bohr theory on obvious physical grounds and in general $n_1 = 0$ is forbidden. The possibilities in the case of larger values of n are obvious and are left to the reader to explore.

The time of revolution in the orbit can be found by use of the law of areas. Reverting to eqs. (5.29) and (5.41) we see that the constant of areas C becomes

$$C = \alpha_2/m = n_1 h/2\pi m. \tag{9.81}$$

But C is equal to twice the rate at which the position vector sweeps out area. Since the area of an ellipse is πab, where b is the semi-major axis, we have

$$n_1 h/2\pi m = 2\pi ab/P, \tag{9.82}$$

where P is the period or time of revolution. Now from analytic geometry

$$b = a\sqrt{1 - \epsilon^2}. \tag{9.83}$$

Hence

$$P = \frac{4\pi^2 ma^2 \sqrt{1-\epsilon^2}}{n_1 h}. \tag{9.84}$$

But from eq. (9.68),

$$\sqrt{1-\epsilon^2} = \sqrt{\epsilon l/a}. \tag{9.85}$$

Substitution shows that P varies as $a^{3/2}$, which is the content of Kepler's third law of elliptical motion in an inverse square force field. If we make the necessary substitutions into (9.84) we ultimately obtain

$$P = \frac{(n_1 + n_2)^3 h^3}{4\pi m e^4 Z^2}. \tag{9.86}$$

As a matter of fact P is not so important in the Bohr atomic model as the corresponding quantity in the motion of a planet around the sun.

9.6. Angle and Action Variables. The importance of the quantum conditions (9.70) and (9.71) in the Bohr theory suggests the introduction of certain new variables in the Hamilton – Jacobi method. Suppose that the Hamilton – Jacobi equation has been solved and the complete integral found. The momentum conjugate to q_k then is found to be

$$p_k = \frac{\partial S}{\partial q_k} = \frac{dS_k}{dq_k} = f_k(q_k, E, \alpha_2 \cdots \alpha_n). \tag{9.87}$$

Now introduce the set of quantities

$$J_k = \oint p_k \, dq_k = \oint f_k(q_k, E, \alpha_2 \cdots \alpha_n) \, dq_k. \tag{9.88}$$

The integral is known as a *phase* integral. It has a meaning only if each q_k is confined to a finite range of values running from $q_{k,\,min}$ to $q_{k,\,max}$. The integration is then conducted over the interval from the minimum value to the maximum and back again, i.e., over a complete cycle of values of q_k. The result is a definite integral from which q_k has disappeared and which depends on E and the α's only. If q_k is a coordinate that does not appear in the Hamiltonian, it is termed a *cyclic* variable. (An example is the longitude ϕ in the Hamiltonian for a free particle.) The momentum p_k conjugate to a cyclic variable is evidently constant (cf. eq. 9.5). In such a case the p_k in (9.88) comes outside the integral sign and the range of integration by definition runs from 0 to 2π.

After integrating we then get

$$J_k = J_k(E, \alpha_2 \cdots \alpha_n), \tag{9.89}$$

or rather n expressions of this kind as k goes from 1 to n. Ideally we can solve these for E and the $(n - 1)$ α's in terms of the J's. Let us substitute the resulting expressions into the expression for S and obtain

$$S = S(q_1 \cdots q_n, \quad J_1 \cdots J_n). \tag{9.90}$$

We now have a transformation function defining a transformation from the coordinates $q_1 \cdots q_n$ to the coordinates $J_1 \cdots J_n$, in accordance with Sec. 9.2. As a matter of fact it is more common to consider the J's as new momenta and employ the situation exemplified in eqs. (9.33). Thus we now have

$$Q_k = \frac{\partial S}{\partial J_k} = w_k, \tag{9.91}$$

since it is customary to refer to the new coordinates in this specific case as w_k. The canonical equations become

$$\dot{w}_k = \frac{\partial H'}{\partial J_k}, \quad \dot{J}_k = -\frac{\partial H'}{\partial w_k}, \tag{9.92}$$

where H' is the transformed Hamiltonian. But in the new coordinates, since $H' = E$, it is a function of the J's only and contains no w's. Therefore

$$\dot{J}_k = 0 \text{ and } \dot{w}_k = \omega_k = \text{constant}. \tag{9.93}$$

These equations say simply that the J_k are constants, whereas the w_k are linear functions of the time. We summarize these results as follows

$$J_k = \text{constant}; \quad w_k = \omega_k t + \gamma_k, \tag{9.94}$$

where the γ_k are also constant in time. Since the w_k are linear functions of the time they have received the name *angle variables*. On the other hand, from their method of construction the J_k are naturally called *action variables*. Indeed, no matter what the nature of the coordinates may be, the J_k as defined in (9.88) always have the physical dimensions of action.

The physical significance of the ω_k can be seen as follows. Consider the change in w_k associated with a complete cycle of values of q_j, which may be written

$$\Delta w_k = \oint \frac{\partial w_k}{\partial q_j}\, dq_j. \tag{9.95}$$

But from (9.91) this takes the form

$$\Delta w_k = \oint \frac{\partial}{\partial q_j}\left(\frac{\partial S}{\partial J_k}\right) dq_j. \tag{9.96}$$

We shall assume S to be of such mathematical form that we may safely interchange the order of differentiation and write

$$\Delta w_k = \frac{\partial}{\partial J_k} \oint \frac{\partial S}{\partial q_j}\, dq_j. \tag{9.97}$$

From (9.88) it follows finally that

$$\begin{aligned}\Delta w_k = \frac{\partial J_j}{\partial J_k} &= 0 \text{ for } j \neq k,\\ &= 1 \text{ for } j = k.\end{aligned} \tag{9.98}$$

Thus the change in w_k when q_k goes through a complete cycle is unity, whereas the change in w_k when any other q goes through its cycle is zero. Hence q_k is a periodic function of w_k with a period equal to unity and therefore periodic in time with period equal to $1/\omega_k$.

As an example we apply the method of angle and action variables to elliptical inverse square field motion. From (9.59) and (9.60) we obtain

$$J_\theta = \oint \alpha_2\, d\theta = 2\pi\alpha_2, \tag{9.99}$$

$$J_r = \oint \sqrt{2mE + 2me^2Z/r - \alpha_2{}^2/r^2}\, dr. \tag{9.100}$$

We have already encountered the integral in J_r in (9.74). Using (9.75)

$$J_r = \frac{2\pi me^2 Z}{\sqrt{-2mE}} - 2\pi\alpha_2. \tag{9.101}$$

Elimination of α_2 between (9.99) and (9.101) gives

$$J_r + J_\theta = 2\pi m e^2 Z/\sqrt{-2mE}, \tag{9.102}$$

whence

$$H'(J_r, J_\theta) = E = -\frac{2\pi^2 m e^4 Z^2}{(J_r + J_\theta)^2}. \tag{9.103}$$

From (9.92) the fundamental frequencies ω_r and ω_θ are obtained as follows

$$\begin{aligned} \omega_\theta = \dot{w}_\theta &= \frac{\partial H'}{\partial J_\theta} = \frac{4\pi^2 m e^4 Z^2}{(J_\theta + J_r)^3}, \\ \omega_r = \dot{w}_r &= \frac{\partial H'}{\partial J_r} = \frac{4\pi^2 m e^4 Z^2}{(J_\theta + J_r)^3}. \end{aligned} \tag{9.104}$$

Since these frequencies are the same the problem is said to be *degenerate.* The motion involves really only one frequency.

The eccentricity of the orbit (from eq. 9.67) can now be written in the form

$$\epsilon = \sqrt{1 - J_\theta^2/(J_\theta + J_r)^2}, \tag{9.105}$$

and the major axis becomes

$$2a = \frac{(J_\theta + J_r)^2}{2\pi^2 m e^2 Z}. \tag{9.106}$$

The degeneracy is again indicated by the dependence on the sum of J_θ and J_r. Comparison of the Bohr quantum conditions (9.70) and (9.71) with the defining equations for the action variables J_k in eq. (9.88) indicates that the use of action variables is particularly helpful in atomic structure problems involving the Bohr theory. With

$$J_\theta = n_1 h, \quad J_r = n_2 h, \tag{9.107}$$

the energy expression (9.103) becomes precisely that of the quantized states of the one electron atom model. The relation between (9.105) and (9.80) under similar circumstances is very striking as is also the relation between (9.106) and (9.77). The reader will also see the similarity between (9.104) and (9.86). The method of angle and action variables, although suggested long before the advent of the Bohr theory, turns out to be admirably fitted to its needs.

All the characteristics of elliptical inverse square field motion are inherent in the analysis given above. But we can go further and find the dependence of r on w_r and hence on the time. Thus from (9.61)

$$w_r = \frac{\partial S}{\partial J_r} = \int \frac{\omega_r mr\, dr}{\sqrt{2mEr^2 + 2me^2Zr - \alpha_2^2}}. \tag{9.107}$$

From the property of the ellipse this becomes

$$w_r = \frac{\omega_r m}{\sqrt{-2mE}} \int \frac{r\, dr}{\sqrt{[r - a(1+\epsilon)][-r + a(1-\epsilon)]}}. \tag{9.108}$$

If we introduce the eccentric anomaly u (cf. Lindsay's *Physical Mechanics*, (2d ed.), page 81, where it is denoted by E) we get

$$w_r = \frac{m\omega_r a}{\sqrt{-2mE}} \int (1 - \epsilon \cos u)\, du = \frac{1}{2\pi}(u - \epsilon \sin u) + C, \tag{9.109}$$

where for convenience we take the constant $C = 0$. Since from the definition of the eccentric anomaly

$$r = a(1 - \epsilon \cos u), \tag{9.110}$$

and we know that r is a periodic function of w_r with a period equal to unity, we can expand r in a Fourier series in w_r. Thus, noting that r is an even function

$$\frac{r}{a} = \frac{B_0}{2} + \sum_{s=1}^{s=\infty} B_s \cos 2\pi s w_r. \tag{9.111}$$

The coefficients are calculated in the usual way, viz.,

$$B_f = \frac{4}{a}\int_0^{1/2} r \cos 2\pi f w_r\, dw_r. \tag{9.112}$$

By the use of (9.109) and (9.110) the integrals may be readily calculated. Thus

$$B_0 = 2 + \epsilon^2, \tag{9.113}$$

whereas by the use of trigonometry, and integration by parts, for $f > 1$,

$$B_f = \frac{\epsilon}{\pi f}\left\{\int_0^\pi \cos\,[(f+1)u - f\epsilon \sin u]\,du - \int_0^\pi \cos\,[(f-1)u - f\epsilon \sin u]\,du\right\}. \tag{9.114}$$

Since it is well known that Bessel's function of order f is given by[1]

$$J_f(z) = \frac{1}{\pi}\int_0^\pi \cos\,[fu - z \sin u]\,du, \tag{9.115}$$

we can express r finally in the form

$$\frac{r}{a} = 1 + \frac{\epsilon^2}{2} + \epsilon\sum_{f=1}^{f=\infty}\frac{1}{f}[J_{f+1}(f\epsilon) - J_{f-1}(f\epsilon)]\cos 2\pi f(\omega_r t + \gamma_r) \tag{9.116}$$

which enables r to be computed for every value of t. This method of expressing r has proved of value in both celestial mechanics and Bohr atomic mechanics.

9.7. The Bohr Theory of Polyelectronic Atoms. We have already considered in Section 9.5 the application of Hamiltonian mechanics to the Bohr theory of the single electron atom. The same method has also been applied with success to obtain approximate order of magnitude information about atoms with more than one extra nuclear electron. This scheme has, of course, been since replaced by quantum mechanics (Chapter 11) but it has made fundamental contributions to the latter which justify a brief survey here.

A polyelectronic atom is a collection of electrons each acted on by the Coulomb inverse square force of attraction to the nucleus and the inverse square force of repulsion by every other electron. The problem of the motion of these electrons is then an n-body problem of the type discussed in Chapter 6 and mathematically insoluble in exact form for $n > 2$

[1]Cf. Margenau and Murphy, *The Mathematics of Physics and Chemistry* (New York, Van Nostrand, 1943), p. 113.

(where the nucleus is of course included as one of the particles). Nevertheless some progress can be made toward finding the *approximate* quantized energy states of the collection by the following simplifying assumptions.

We may assume in spite of the complicated field of force in which each electron actually moves that, if the number is considerable, the field is approximately central. This is equivalent to assuming that each electron will affect all the others as if it were a continuous spherical distribution of charge about the nucleus. It is most natural to assume that the density of this distribution is such that the magnitude of the total charge within any spherical shell about the nucleus is equal to that fraction of the charge e on the electron corresponding to the time which the electron spends in this shell during its revolution about the nucleus.

If we knew the actual orbit for each electron in the atom we could find the above average charge density distribution by knowing how the radius vector r depends on the time t and hence how much time is spent by the electron in going from r to $r + dr$. The knowledge of the average charge density enables one to compute the potential energy of an electron in the field of such a continuous distribution. This to be sure is a problem in field physics to be discussed in Chapter 12. It is an illustration of the fact that when particle physics gets complicated because of the large number of particles involved we must inevitably try some other consideration and in this case indeed the effect of a continuous distribution.

Having the potential energy V as a function of r (usually not a function in closed form but only a table of numerical values of V for an appropriate range of r) we can at once apply the radial quantum condition. Thus the Hamiltonian is in general

$$H = \frac{1}{2m}(p_r^{\,2} + p_\theta^{\,2}/r^2) + V(r) = E. \qquad (9.117)$$

Hence

$$\oint p_r\, dr = \oint \sqrt{2mE - 2mV(r) - n_2^{\,2}h^2/4\pi^2\, r^2}\, dr = n_1 h, \qquad (9.118)$$

where the azimuthal quantum condition has already been imposed. The insertion of the tabular function $V(r)$ into the phase integral makes possible its evaluation by numerical or graphical integration. The value of E must then be adjusted until the integration yields the correct integral value of n_1 for the type of electron orbit being considered.

The reader may object that the above procedure is actually impossible to carry out since we do not know the orbits of the various electrons to begin with and hence have no basis for computing the equivalent continuous distribution of charge on which the determination of $V(r)$ depends. The answer obviously is that we must proceed by a trial-and-error scheme We must begin by assuming what appears to be a reasonable set of orbits for the given atom or alternatively a reasonable initial $V(r)$ field for each electron to move in. Then after completing the process of quantization indicated in (9.118) for each type of electron (characterized by the usual quantum numbers n_1 and n_2), we can use the appropriate energy E to determine a transformation function S by the equation

$$S = \frac{n_2 h}{2\pi}\theta + \int \sqrt{2mE - 2mV(r) - n_2{}^2 h^2/4\pi^2 r^2}\, dr. \tag{9.119}$$

From this a corresponding orbit can be plotted using eqs. (9.62). From the orbits of all the type electrons a new charge distribution can be calculated. This must be compared with the original charge distribution. It will in general disagree. From the nature of the disagreement we can guess what a better approximation to the actual charge distribution will be and proceed to repeat the process, continuing until the distribution *resulting* from the quantization agrees within previously specified limits with the distribution used in the quantization. We then have achieved consistency and may call the final $V(r)$ field for each electron a *self-consistent* field.[1]

The theory just described yields estimates for the quantized energy values of the electrons in the ground (lowest energy) state of any atom as well as the size of the atom in this state as measured by r_{max} for the outermost orbit. The reader is referred to the article mentioned in the footnote for details in the case of the alkalies (lithium, sodium, etc.). States of higher quantization or excited states (i.e., with larger n_1 and n_2 values) can also be computed and thence from the Bohr frequency condition mentioned in Sec. 9.4 the frequencies of emitted or absorbed spectral lines can be estimated.

[1]So far as the author is aware the idea of the use of the self-consistent field in Bohr polyelectronic atoms was first proposed by him in a paper on the atomic models of the alkali metals in *Journal of Mathematics and Physics (M.I.T.)*, 3, 191, 1924. A similar idea was applied in 1928 by D. R. Hartree to the wave mechanical theory of polyelectronic atoms (*Proc. Camb. Phil. Soc.*, 24, 89, 1928) and has proved very successful.

PROBLEMS

1. Prove that the Hamiltonian eqs. (9.4) and (9.5) are invariant in form under a "point" transformation, i.e., one such that

$$q_j = q_j(Q_1 \cdots Q_n),$$
$$Q_j = Q_j(q_1 \cdots q_n),$$

$j = 1, 2, 3, \cdots n$.

2. A particle of mass m moves in the xy plane under the impressed force

$$\mathbf{F} = \mathbf{i}(-k_1 x) + \mathbf{j}(-k_2 y),$$

where k_1 and k_2 are different positive constants. Write the Hamilton–Jacobi equation and find the complete integral. Set up the action variables and express the energy in terms of them.

3. An electron moving in the central field of force of a nucleus of charge Ne is also acted on by a uniform electric field of intensity F acting in the z direction, so that the potential energy has the form

$$V = -\frac{Ne^2}{r} + eFz.$$

Express the Hamiltonian function in terms of parabolic coordinates, using the transformation eqs. (2.5). Write the Hamilton–Jacobi equation and show that it is separable in these coordinates. (This is the basis of the Bohr theory method of handling the Stark effect.)

4. In relativistic mechanics the kinetic energy is found to have the form

$$K = m_0 c^2 \left(\frac{1}{\sqrt{1 - \beta^2}} - 1 \right),$$

(cf. Chapter 17, eq. 17.48), where $\beta = v/c$, c being the velocity of light and $\mathbf{v}$ the velocity of the particle. The rest mass or limiting mass as $v \to 0$ is m_0. Find the relativistic Hamiltonian function for a particle moving in an inverse square central force field (i.e., $V = -Ne^2/r$).

5. A simple harmonic oscillator of one degree of freedom has mass m and stiffness k. Consider the canonical transformation defined by the transformation function

$$S = \frac{\sqrt{mk}}{2} q^2 \cot Q,$$

where q and Q have the significance of Sec. 9.2. Find the equations defining the transformation in the form

$$q = q(P, Q), \quad p = p(P, Q),$$

Find the canonical equations in the new coordinates and draw the appropriate physical conclusions.

6. Prove that for a conservative system the action function

$$\int \Sigma m_j \dot{\mathbf{r}}_j \cdot d\mathbf{r}_j$$

satisfies the Hamilton – Jacobi equation, i.e., reduces to the transformation function $S(q_1 \cdots q_n, \alpha_1 \cdots \alpha_n, E)$ plus an arbitrary constant.

7. If $g(q_1 \cdots q_n, p_1 \cdots p_n)$ and $h(q_1 \cdots q_n, p_1 \cdots p_n)$ are any two continuous and differentiable functions of the generalized coordinates and conjugate momenta of a dynamical system of n degrees of freedom, the so-called Poisson bracket for g, h is defined as follows

$$\{g, h\} = \sum_{j=1}^{n} \left(\frac{\partial g}{\partial p_j} \frac{\partial h}{\partial q_j} - \frac{\partial g}{\partial q_j} \frac{\partial h}{\partial p_j} \right).$$

Write the canonical equations of Hamilton in Poisson bracket form. Find the values of the following brackets:

$$\begin{array}{llll} \{p_j, p_k\} & \text{for} & j = k, & j \neq k, \\ \{q_j, q_k\} & '' & '' & '' , \\ \{p_j, q_k\} & '' & '' & '' , \\ \{H, t\}. & & & \end{array}$$

In the last H is the Hamiltonian function. Recalling that p_k and q_k are canonically conjugate, what can be said of H and t?

Consider the canonical transformation of Prob. 5. Find $\{P, Q\}$ and comment on its physical significance. What plausible generalization suggests itself?

8. Solve the problem of the three-dimensional harmonic oscillator with stiffness coefficients k_x, k_y, k_z, respectively, in the three coordinate directions, by the Hamilton – Jacobi method.

9. Show that the time mean value of the radius vector r (over a single period of revolution) of an electron moving in a Bohr elliptical orbit in hydrogen is given by

$$\bar{r} = n^2 a_0 \left[1 + \frac{1}{2}\left(1 - \frac{n_\theta{}^2}{n^2}\right)\right],$$

where n_θ = azimuthal quantum number, n = principal quantum number, and a_0 = radius of Bohr orbit of lowest energy in hydrogen. Comment on the physical significance of the above result for various values of n_θ and n.

10. Carry out the expansion for r/a in eq. (9.116) to five terms for the special case of the Bohr orbit in hydrogen for which $n = 2$ and $n_1 = 1$. What is ω_r in this case?

11. Quantize the three-dimensional harmonic oscillator in the Bohr theory by means of the method of angle and action variables.

CHAPTER X

STATISTICAL MECHANICS AND THERMODYNAMICS

10.1. The Statistical Method in Particle Physics. The reader will have noted from the discussion in Chapter 6 that classical mechanics can provide only results of very general character for dynamical systems containing a large number of particles. As soon as we ask detailed information about the behavior of the component particles of such systems we find ourselves up against insoluble mathematical problems to say nothing of our inability to specify initial conditions of motion even if we could solve the differential equations. It is true that we are able to make helpful statements by the introduction of the concept of center of mass, and the energy equation for an aggregate is also useful. But the lack of ability to prove more detailed theorems seems a serious handicap in the development of the mechanical theory of aggregates. It appears evident that progress here demands a change in method. Two routes are open to us. If the number of particles per unit volume is very large indeed, we can achieve a certain measure of success by treating the collection as a so-called *continuous* medium. We have already encountered one illustration of this in the rigid body. This is a very special case since the component particles are allowed no relative motion. The more general case will be that of deformable solids and fluids. We shall postpone consideration of these to Part Two, which deals with field physics and introduces a new set of physical concepts appropriate for continuous media.

The second route retains the particle concept as fundamental but deliberately foregoes the attempt to describe what each particle of the collection is doing at every instant and contents itself with describing the *average* behavior of each. This means it averages certain important properties (e.g., energy) of the collection over long periods of time and strives to find relations connecting these average values. The hope is that when the system is examined experimentally it is just such average values that are measured; the average relations just mentioned can then be tested to see whether they are laws descriptive of the behavior of the collection as a whole.

When we concern ourselves in this way with the average behavior of a collection, we are using a method commonly called statistical. Help in understanding this method is provided by the concept of the *virial*, which we now proceed to examine as an introduction to elementary kinetic theory of gases, itself an admirable simple illustration of statistical reasoning.

10.2. The Virial Theorem and Elementary Kinetic Theory of Gases. Let us suppose we have a system of n particles of masses m_1, m_2, m_3, $\cdots$ m_n and position vectors $\mathbf{r}_1$, $\mathbf{r}_2$, $\cdots$ $\mathbf{r}_n$, respectively. We proceed to form the arbitrary scalar function

$$Q = \frac{1}{2}\sum_{i=1}^{n} m_i \mathbf{r}_i \cdot \mathbf{r}_i. \tag{10.1}$$

If this is differentiated twice with respect to the time the result is

$$\ddot{Q} = \sum_{i=1}^{n} m_i \dot{\mathbf{r}}_i \cdot \dot{\mathbf{r}}_i + \sum_{i=1}^{n} m_i \mathbf{r}_i \cdot \ddot{\mathbf{r}}_i. \tag{10.2}$$

Now form the time average of $\ddot{Q}$ over the interval τ and obtain

$$\overline{\ddot{Q}} = \frac{1}{\tau}(\dot{Q}_\tau - \dot{Q}_0), \tag{10.3}$$

the bar, as usual, denoting time average, where

$$\dot{Q} = \sum_{i=1}^{n} m_i \mathbf{r}_i \cdot \dot{\mathbf{r}}_i, \tag{10.4}$$

and $\dot{Q}_\tau$ and $\dot{Q}_0$ denote the values of $\dot{Q}$ at times τ and 0, respectively. If the particles are assumed to move always in a bounded region, $\mathbf{r}_i$ is bounded for all i. We shall likewise assume that $\dot{\mathbf{r}}_i$ is bounded, thus denying the possibility of infinitely great velocities. Hence $\dot{Q}$ is bounded and as τ increases

$$\overline{\ddot{Q}} \to 0. \tag{10.5}$$

Therefore from (10.2) it follows that

$$\frac{1}{2}\overline{\sum_{i=1}^{n} m_i \dot{\mathbf{r}}_i \cdot \dot{\mathbf{r}}_i} = -\frac{1}{2}\overline{\sum_{i=1}^{n} m_i \ddot{\mathbf{r}}_i \cdot \mathbf{r}_i}. \tag{10.6}$$

The left-hand side of (10.6) is the time average of the total kinetic energy of the system. If we recall that $m_i\ddot{\mathbf{r}}_i$ is the *resultant* force $\mathbf{F}_i$ acting on the ith particle, we may rewrite (10.6) in the form

$$\frac{1}{2}\overline{\sum_{i=1}^{n} m_i v_i^2} = -\frac{1}{2}\overline{\sum_{i=1}^{n} \mathbf{r}_i \cdot \mathbf{F}_i}. \tag{10.7}$$

The quantity on the right was called by Clausius the *virial* of the system (usually denoted by Ω). Eq. (10.7) then expresses the result that the average kinetic energy of the collection of particles moving subject to the imposed conditions is equal to its virial. This is the *virial theorem.*

A simple illustration is provided by a collection of particles in which the only forces acting are central forces between the individual particles varying in magnitude as the nth power of the distance of separation, where n can be negative as well as positive. We can apply the virial theorem to show that if $\overline{K}$ is the average total kinetic energy of the collection and $\overline{V}$ the average total potential energy

$$\overline{K} = \frac{n+1}{2} \cdot \overline{V}. \tag{10.8}$$

The demonstration is left for the reader. A result of this sort is of interest in connection with atomic systems, particularly when $n = -2$.

The virial theorem can be used as the starting point for the kinetic theory of gases. It will be remembered from elementary physics that according to this theory a gas is assumed to be composed of a large number of material particles called molecules. These are supposed to be in ceaseless motion with velocities varying in direction and magnitude over a wide range. When the gas is confined in a vessel the molecules collide frequently with the walls and these collisions are held to be responsible for the *pressure* which the gas is observed to exert on the walls. The mean kinetic energy per molecule, on the other hand, is interpreted as the theoretical explanation of the *temperature* of the gas. Experimental observation establishes for every gas a relation connecting volume, pressure, and temperature which is usually called the equation of state. For a so-called ideal gas, approximated to by gases like hydrogen, oxygen, and nitrogen in the neighborhood of room temperature, this equation has the well-known form

$$pV = RT, \tag{10.9}$$

where p, V, and T stand for pressure, volume, and absolute temperature, respectively, and R is called the gas constant.

It is one obvious task of kinetic theory to provide a theoretical basis for eq. (10.9), i.e., to show that it follows as a natural deduction from the fundamental principles of the theory. This can be done by applying the virial theorem.

The forces acting on the molecules of a gas arise presumably from several sources: collisions of molecules with each other, collisions of molecules with the walls, and direct attraction or repulsion of molecules by gravitational or electrical or similar action. Hence the picture, mechanically speaking, must be considered rather complicated. We therefore proceed at once to simplify by the usual process of idealization and assume that we can describe certain gases at any rate by taking into account only those forces which arise from collisions with the walls. We assume in other words that in calculating the virial of the collection of gas molecules we may neglect all forces save those exerted by the surface of the vessel on the molecules.

We need not restrict the discussion to any particular shape of vessel. Consider an element of surface which we may represent by the vector area $\mathbf{n}dS$, where dS is the numerical area and $\mathbf{n}$ is the unit outward drawn normal. The time average force exerted by this surface on the molecules which collide with it can be represented as the product of the scalar pressure p (magnitude of force per unit area) and the vector surface area. We are here falling back on the well-known behavior of fluids at rest in exerting pressure at right angles to any surface they touch. Thus we are really introducing the additional assumption that the gas, phenomenologically considered, behaves like a fluid. Strictly speaking this should be also derivable as a prediction of kinetic theory, but we are here concerned with an elementary, approximate treatment. The averaged sum on the right-hand side of (10.7) now becomes the integral

$$\frac{p}{2}\int \mathbf{r} \cdot \mathbf{n}dS, \tag{10.10}$$

where $\mathbf{r}$ is the position vector of the center of the surface element. Note that the minus sign in (10.7) has now disappeared, since $\mathbf{n}$ is the *outward* drawn normal and the force due to the pressure is directed *inward*. The integral is taken over the whole closed surface of the vessel. It is most simply evaluated by a theorem called the divergence theorem, derived in Sec. 12.3. This changes the surface integral into a volume integral as follows

$$\frac{p}{2}\int \mathbf{r}\cdot \mathbf{n}dS = \frac{p}{2}\int \nabla\cdot \mathbf{r}dV, \tag{10.11}$$

where

$$\nabla\cdot\mathbf{r} = \left(\mathbf{i}\frac{\partial}{\partial x} + \mathbf{j}\frac{\partial}{\partial y} + \mathbf{k}\frac{\partial}{\partial z}\right)\cdot(\mathbf{i}x + \mathbf{j}y + \mathbf{k}z) = 3. \tag{10.12}$$

Hence the virial theorem (10.7) takes the form

$$\frac{1}{2}\overline{\Sigma m_i v_i^{\,2}} = \frac{3}{2}pV, \tag{10.13}$$

where V is the total volume of the vessel. This is a very suggestive result. On the right-hand side we have the product of two *operationally* defined quantities characteristic of the gas, namely pressure and volume. On the left-hand side we have a quantity constructed from the *mechanical* behavior of the assumed constituents of the gas. To turn this equation into a *law* of gases it is necessary to identify the left-hand side with some other quantity descriptive of the gas. Let us therefore make the additional assumption that the average total kinetic energy of all the molecules of the gas is proportional to the absolute temperature and write

$$\frac{1}{2}\overline{\Sigma m_i v_i^{\,2}} = \frac{3}{2}RT. \tag{10.14}$$

With this identification (10.13) becomes precisely the equation of state. Moreover if all the molecules have the same mass m and their number is N, we can write (10.13) in the form, familiar from elementary physics,

$$p = \frac{1}{3}\frac{Nmv_m^{\,2}}{V}, \tag{10.15}$$

where

$$v_m^{\,2} = \overline{\Sigma v_i^{\,2}/N} \tag{10.16}$$

is therefore the *mean square velocity* for each molecule. Moreover we can now rewrite (10.14) in the form

$$\frac{1}{2}mv_m^{\,2} = \frac{3}{2}kT, \tag{10.17}$$

where

$$k = R/N \tag{10.18}$$

may be called the effective gas constant per molecule, and the kinetic theory definition of absolute or Kelvin temperature appears as the average kinetic energy per molecule divided by $3k/2$.

This identification of *operationally* defined quantities like pressure and temperature of a gas with *theoretically* defined averages of purely mechanical quantities like molecular velocity is a characteristic feature of the statistical method. We may legitimately refer to the kinetic theory as a *mechanical* theory in the sense of Chapter 8, but it is more appropriately referred to as a *statistical mechanical* theory, for while it operates with mechanical quantities it uses average values of the latter rather than instantaneous values.

The above elementary treatment of the kinetic theory of an ideal gas is intended to be illustrative merely. It is clear that to get a better approximation to the equation of state of a real gas, we should have to take into account the contribution to the virial due to other forces besides those connected with impact on the walls, e.g., forces of interaction between individual molecules. Moreover, attention would have to be paid to the collisions of molecules.

It should be clear, at any rate, that in the use of the statistical method as exemplified in the kinetic theory illustration just treated we give up the calculation of precise results from given initial conditions, so characteristic of classical mechanics as we have so far developed it. All average quantities calculated on the basis of the statistical method have *probable* validity only, and we may expect to find fluctuations from them when we compare with the experimentally determined analogies. As a matter of fact all averages calculated from experimental measurements by some selected theory of errors show the same sort of fluctuations from individual measurements. In a sense therefore the statistical method appears ideally indicated as appropriate for describing most types of experimental physical phenomena.[1]

10.3. The Canonical Distribution. We now wish to develop the statistical method in physics somewhat more systematically. For this it is necessary to introduce the concept of a statistical distribution. We confine our attention once more to a mechanical aggregate, i.e., a very large number of particles all having the same mass and enclosed in a finite vessel. These particles are continually bumping into each other and the

[1]Cf. Lindsay, *Physical Statistics* (Dover), Chap. I, for a discussion of the methodological differences between the statistical method and classical dynamics.

walls of the vessel. We shall assume for simplicity that the only forces acting on the particles are those arising on their collision. We can learn much about the behavior of the aggregate by endowing the particles with kinetic energy only. The kinetic energy for a particular particle can change on collision and we can therefore think of a more or less continual interchange among the particles as time goes on. However if no external forces act on the system as a whole the total energy must remain constant (Sec. 6.2).

How may we proceed to use the idea of statistical distribution in the evaluation of the average quantities which we hope will correspond to the measured values characteristic of the gas or other system we are trying to describe? We have already agreed that it is impossible to specify the motion of every particle in the system. However, if we are able in some reasonable way to specify *how many* particles *on the average* have energy lying within a given energy interval, it is not difficult to calculate the average energy per particle and from this ultimately to deduce the average values of quantities characteristic of the system as a whole. This specification of the average number of particles in each energy region is what we mean by a statistical distribution.

Let us be more pictorial and at the same time more precise. Any particle may have energy running all the way from zero up to some maximum value. We shall find it convenient to think of the various possible energy values as associated with n energy *boxes*, where n is a very large but finite integer. Let the energy E_j be associated with the jth box in the sense that any particle in this box will have the energy E_j. This is equivalent to introducing a set of energy intervals all of the same size in the neighborhood of the energy values $E_1, E_2, E_3, \cdots E_n$. Actually the energy values may be continuous, but it will be more convenient for us to consider them as forming a discrete set. But, if we choose n large enough, i.e., make the intervals or boxes small enough, we can simulate a continuous distribution as closely as we please for physical purposes.

We now ask ourselves the question: In how many different and independent ways can we distribute the N particles among the n boxes so that there will be precisely N_1 particles in the first box (corresponding to energy E_1), N_2 in the second, etc., until we reach the nth box, which is to have N_n? The use of the words "different and independent" implies that we are not interested in the *order* in which the particles get into the boxes. Now we can choose the first particle to go into the first box in N ways, the second in $N - 1$ ways, the third in $N - 2$ ways, etc., and finally the N_1th in $(N - N_1 + 1)$ ways. Hence the total number of ways in which N_1 particles can be put into box 1 is

$$N(N-1)(N-2)\cdots(N-N_1+1).$$

But since we are not interested in the order in which these particles are put into the box (or more properly put themselves in by collisions) we must reduce the above number to the proper number of different independent ways by dividing by the number of ways in which the same N_1 particles can be rearranged in order. This is just the number of permutations of the N_1 particles N_1 at a time. This is $N_1!$ or $1\cdot 2\cdot 3\cdot 4\cdot 5\cdots N_1$. Hence the total number of ways sought above is really

$$\frac{N(N-1)\cdots(N-N_1+1)}{N_1!}=\frac{N!}{N_1!(N-N_1)!}, \tag{10.19}$$

where the algebraic transformation to give the form on the right should be sufficiently clear from the definition of the factorial.

We now repeat the process for the N_2 particles which are to be chosen out of the remaining $N-N_1$ to go into box 2 and have energy E_2. This number is clearly

$$\frac{(N-N_1)!}{N_2!(N-N_1-N_2)!}. \tag{10.20}$$

Repeating the scheme for each set of particles we finally arrive at the required *total* number of ways to distribute the N particles so that there are N_j in the jth box in the form

$$\frac{N!}{N_1!(N-N_1)!}\cdot\frac{(N-N_1)!}{N_2!(N-N_1-N_2)!}\cdots\frac{(N-N_1-N_2\cdots-N_n)!}{N_n!}$$

$$=\frac{N!}{N_1!N_2!\cdots N_n!}. \tag{10.21}$$

We shall call this quantity w. It is usually referred to as the statistical probability. Its magnitude will of course depend on the N_j values. Thus if $N_1=N$ and $N_2=N_3=\cdots=N_n=0$, i.e., if all the particles are to be in the first box, $w=1$; there is only *one* way to achieve this. On the other hand, if $N_1=N_2=\cdots=N_n=N/n$, i.e., if the particles are equally divided among the boxes, w is much larger than unity. As a matter of fact it becomes *approximately*, if N is very large,

$$w=(2\pi N)^{(1-n)/2}\cdot n^{N+(n/2)}. \tag{10.22}$$

The only point we are trying to bring out now is that w differs considerably for different distributions. It seems natural to suppose that the larger w is, the greater is the likelihood that the corresponding distribution will actually take place. This tacitly assumes that each way of realizing a given distribution is equally likely, an assumption few will question if we assign the same probability to each energy box.

If we can find the distribution in which w has the largest possible value consistent with the total energy of the aggregate and the total number of particles remaining constant, we shall presumably have found the distribution most likely to occur in nature for a system of constant energy and constant number of particles. The mathematical problem facing us therefore is to make w a maximum subject to the conditions that

$$\sum_{j=1}^{n} N_j E_j = E = \text{constant},$$
$$\sum_{j=1}^{n} N_j = N = \text{constant}. \qquad (10.23)$$

It proves more convenient to work with log w. Thus

$$\log w = \log N! - \sum_{j=1}^{n} \log N_j!. \qquad (10.24)$$

Let us suppose that N_j changes by an amount δN_j, which is an integer considerably smaller than N_j. Therefore the change in log w associated therewith is

$$\delta \log w = -\sum_{j=1}^{n} \delta \log N_j!$$
$$= -\sum_{j=1}^{n} [\log (N_j + \delta N_j)! - \log N_j!]. \qquad (10.25)$$

But

$$\log (N_j + \delta N_j)! - \log N_j! = \log \frac{(N_j + \delta N_j)!}{N_j!}$$
$$= \log (N_j + 1)(N_j + 2)\cdots(N_j + \delta N_j). \qquad (10.26)$$

Hence

$$\delta \log w = -\sum_{j=1}^{n} [\log (N_j + 1) + \log (N_j + 2) + \cdots + \log (N_j + \delta N_j)]. \tag{10.27}$$

Now since $\delta N_j \ll N_j$, it follows that $N_j \gg 1$, etc. We may therefore approximate safely and write

$$\delta \log w \doteq -\sum_{j=1}^{n} \delta N_j \log N_j. \tag{10.28}$$

This approximation may be expected to work better as N_j grows larger. Actually the same sort of approximation results from the use of Stirling's formula[1] for $N!$ If $\log w$ is to be a maximum (or have a stationary value which may be maximum, minimum, or mini-max), its variation for a *small* variation in N must be zero. Hence the condition for our problem is

$$\delta \log w = -\sum_{j=1}^{n} \delta N_j \log N_j = 0. \tag{10.29}$$

But the δN_j are not all independent since conditions (10.23) must be satisfied. Since the E_j are treated as constants, these conditions yield the relations

$$\sum_{j=1}^{n} E_j \delta N_j = 0, \qquad \sum_{j=1}^{n} \delta N_j = 0. \tag{10.30}$$

We now employ Lagrange's method of undetermined multipliers to find the w which satisfies (10.29) subject to (10.30). If λ_1 and λ_2 are two initially undetermined parameters, it is clear that we can write (no matter what their values are)

$$\sum_{j=1}^{n} (-\log N_j + \lambda_1 + \lambda_2 E_j) \delta N_j = 0. \tag{10.31}$$

[1] Cf., e.g., Lindsay, *Physical Statistics*, p. 19.

Now let us *fix* λ_1 and λ_2 so that

$$\begin{aligned} -\log N_1 + \lambda_1 + \lambda_2 E_1 &= 0, \\ -\log N_2 + \lambda_1 + \lambda_2 E_2 &= 0. \end{aligned} \tag{10.32}$$

Then (10.31) becomes

$$\sum_{j=3}^{n} (-\log N_j + \lambda_1 + \lambda_2 E_j)\delta N_j = 0. \tag{10.33}$$

Now the variations $\delta N_3, \delta N_4, \cdots \delta N_n$ occurring in (10.33) are *all* arbitrary, since only two of the δN's are fixed in terms of the rest by the conditions (10.30) and we have chosen those as δN_1 and δN_2. They are eliminated from the condition (10.31) by the choice of λ_1 and λ_2 in (10.32). But the only way for (10.33) to hold identically is to have the coefficient of every independently varying δN_j vanish identically. This leads to the result

$$N_j = e^{\lambda_1 + \lambda_2 E_j}, \tag{10.34}$$

valid for all j. For $j = 1, 2$ this comes from (10.32), while for $j = 2, 3, 4, \cdots n$, it comes from (10.33). This gives us the required maximal distribution. We are still left with the necessity of providing physical significance for λ_1 and λ_2. For this purpose it is more convenient to replace these by two other parameters Θ and ψ, where

$$\lambda_1 = \psi/\Theta + \log Nn; \quad \lambda_2 = -1/\Theta. \tag{10.35}$$

With this choice (10.34) becomes

$$N_j = Nne^{\psi/\Theta} e^{-E_j/\Theta}. \tag{10.36}$$

We have tacitly assumed that Θ is positive. As a matter of fact, if it were negative, we should have the result that as E_j becomes indefinitely large (and positive), N_j likewise increases indefinitely. This is not physically reasonable.

We naturally wish to give some physical meaning to the parameters ψ and Θ. Recalling (10.23) we have

$$\psi = -\Theta \log \Sigma n e^{-E_j/\Theta}. \tag{10.37}$$

Also

$$\Sigma N_j E_j = E = Nne^{\psi/\Theta} \cdot \Sigma E_j e^{-E_j/\Theta}, \tag{10.38}$$

whence from (10.36)

$$E = N \frac{\Sigma E_j e^{-E_j/\Theta}}{\Sigma e^{-E_j/\Theta}}. \tag{10.39}$$

The quantity E/N is the average energy per particle and will be denoted by $\bar{E}$. Hence

$$\bar{E} = \frac{\Sigma E_j e^{-E_j/\Theta}}{\Sigma e^{-E_j/\Theta}}. \tag{10.40}$$

Eq. (10.39) is a transcendental equation from which Θ may be determined once the values of E, N, and the E_j are known. Then from (10.37) ψ may be determined. The distribution given by N_j in (10.36) is often referred to as the *canonical distribution*, and ψ and Θ are known as *distribution moduli*. From (10.36) and (10.37) we can write

$$N_j = \frac{Ne^{-E_j/\Theta}}{\Sigma e^{-E_j/\Theta}}. \tag{10.41}$$

In the special case for which all the E_j are equal to the same constant energy ϵ, (10.41) yields

$$N_j = \frac{Ne^{-\epsilon/\Theta}}{ne^{-\epsilon/\Theta}} = \frac{N}{n}. \tag{10.42}$$

In this case the canonical distribution, i.e., that corresponding to a maximum likelihood of occurrence, is a uniform distribution in which the number of particles in each energy box is the same.

The discussion of this section will undoubtedly appear rather mysterious to one who has mastered the method of classical mechanics as set forth in the earlier chapters of this book and has grown accustomed to its deterministic precision. The idea of the most likely distribution of entities with respect to some property is of course a statistical notion by no means restricted to mechanical applications. In fact it is much more commonly met with in biological and sociological problems. Its success in the treatment of a mechanical system containing a large number of particles must be tested of course by the degree to which the concrete results obtained from it agree with the observed large-scale behavior of the system. In other words, the statistical method applied to mechanics constitutes a physical theory of the same logical character as that of classical mechanics but with a different postulational approach. This should become clearer to the thoughtful reader as he proceeds with the study of the subject.

10.4. General Canonical Distribution and Kinetic Theory. The canonical distribution (10.41) is a special case of a more general type. It has been introduced as an illustration of the technique of the statistical method, and its relative analytical simplicity gives it heuristic value. However we must generalize it in order to show its real value for kinetic theory.

It will be recalled that in deriving (10.41) we assumed that each energy box has the same probability in the sense that it is equally likely that a particle shall reside in any box. It was on this basis that we justified the choice of

$$w = \frac{N!}{N_1!N_2! \cdots N_n!}$$

as a measure of the likelihood of occurrence of a distribution in which N_1 particles are in the first box, etc. But now it is necessary to be more general and suppose that each box has its own *a priori* probability which for the jth box we shall denote by g_j. The simplest interpretation of this is that it is a proper fraction denoting the relative frequency with which a particle distributed at random falls into the jth box.

A simple example is provided by the *spatial* distribution of N identical molecules in a volume V, divided into subvolumes as actual boxes of volumes $V_1, V_2, \cdots V_n$, where

$$\sum_{i=1}^{n} V_i = V.$$

If we grant that the molecules distribute themselves at random, which is just another way of saying that we know of no reason why a particular molecule should occupy any one point inside the volume V rather than any other one, it follows that the frequency with which a molecule finds itself in a given volume V_j is directly proportional to the volume. Hence V_j/V is a measure of the relative frequency and may be taken as equal to the *a priori* probability to be attached to the volume V_j. The same principle may be applied to energy boxes.[1] If the *a priori* probability is

[1]Cf., Lindsay and Margenau, *Foundations of Physics*, pp. 159 ff., for a discussion of the frequency theory of probability as used in physics.

the same for all energy boxes, $g_j = 1/n$ for every j. In general, however, we must suppose that $g_i \neq g_j$, if $i \neq j$. Since a particle must always turn up in some box, the g_j must satisfy the relation

$$\Sigma g_j = 1. \tag{10.43}$$

Now from elementary probability theory[1] the probability of a distribution in which N_1 particles are in box 1, N_2 in box 2, etc., would be

$$g_1^{N_1} g_2^{N_2} \cdots g_n^{N_n}, \tag{10.44}$$

if there were only one way of realizing the distribution. Since there are actually the number of ways given in (10.21), the total probability for the distribution is

$$P = \frac{N!}{N_1!N_2! \cdots N_n!} g_1^{N_1} \cdot g_2^{N_2} \cdots g_n^{N_n}. \tag{10.45}$$

The generalized canonical distribution is then found by making log P a maximum subject to the conditions (10.23). The method is identical with that followed in Sec. 10.3 and need not be repeated here. The result is that (10.36) is replaced by

$$N_j = g_j Nne^{\psi/\Theta} e^{-E_j/\Theta}, \tag{10.46}$$

where now we have

$$\psi = -\Theta \log \Sigma ng_j e^{-E_j/\Theta}. \tag{10.47}$$

The distribution may also be rewritten in the form

$$N_j = \frac{ng_j Ne^{-E_j/\Theta}}{\Sigma ng_j e^{-E_j/\Theta}}. \tag{10.48}$$

If g_j = constant = $1/n$, (10.48) reduces to the special case (10.41). It is customary to refer to the denominator in (10.48) as the *partition* function and denote it by Z. Thus

$$Z = \Sigma ng_j e^{-E_j/\Theta}. \tag{10.49}$$

Z might equally well be called the *distribution* function since its evaluation for any particular system is a necessary preliminary to determining the canonical distribution N_j in (10.48). It has other interesting uses which will also appear in the course of the discussion.

[1] Cf., for instance, H. B. Fine, *College Algebra* (New York, Dover Publications), p. 409. Also A. C. Aitkin, *Statistical Mathematics*, (New York, Interscience Publishers, 1942), pp. 12 ff.

As an illustration let us calculate the partition function for a collection of particles having kinetic energy only, i.e.,

$$E_j = \frac{1}{2} m v_j^2. \tag{10.50}$$

This should throw some light on the expected average distribution of the particles with respect to their velocities. Actually it proves more convenient to use the momentum p_j in place of the velocity and write for the cartesian coordinate components

$$p_{jx} = m v_{jx}, \text{ etc.} \tag{10.51}$$

$$E_j = \frac{p_{jx}^2 + p_{jy}^2 + p_{jz}^2}{2m}. \tag{10.52}$$

In evaluating Z the question is: How shall we choose g_j in the most plausible manner?

We find it desirable to be somewhat abstract and formal here by describing the behavior of a particle of the system in terms of not merely its three rectangular position coordinates but its three component momentum values as well. The six quantities $x_j, y_j, z_j, p_{jx}, p_{jy}, p_{jz}$ are said to define the *phase* of the particle. For a given particle of mass m, if we know its phase at any instant then we know or can find every other mechanical property of the particle at that instant. We can give a straightforward geometrical interpretation of this by setting up a six-dimensional cartesian space called a phase space in which the quantities $x_j, y_j, z_j, p_{jx}, p_{jy}, p_{jz}$ fix the position of a point. As the actual particle moves in ordinary three-dimensional space its corresponding phase point moves in phase space in an orbit which cannot be entirely arbitrary since the motion of the particle is governed by the Hamiltonian equations of motion

$$\frac{\partial H}{\partial q_j} = -\dot{p}_j, \quad \frac{\partial H}{\partial p_j} = \dot{q}_j, \tag{10.53}$$

where the q_j's here are of course x_j, y_j, z_j, etc.

It appears plausible to connect the *a priori* probabilities g_j somehow with the phase space. In fact the most natural assumption is to assume that g_j is directly proportional to the element of phase volume in the neighborhood of the phase coordinates corresponding to energy E_j. The complete justification of this assumption demands a more detailed study of statistical mechanics than we are prepared to give here.[1] We therefore shall be content to accept it on a plausibility basis. We are going to suppose further that the phase space is not continuous but has a cellular structure with the phase volume of the fundamental cell, inside which we are unable to distinguish phase values, equal to Δ. Let us also postulate that the phase space is so bounded as to contain only n_c cells; the total phase volume is then $n_c\Delta$. We shall now take

$$g_j = \frac{dx_j\, dy_j\, dz_j\, dp_{jx}\, dp_{jy}\, dp_{jz}}{n_c\Delta}. \tag{10.54}$$

From (10.49) we therefore can write for the partition function

$$Z = \frac{n}{n_c\Delta}\Sigma e^{-(p_{jx}^2 + p_{jy}^2 + p_{jz}^2)/2m\Theta}\, dx_j\, dy_j\, dz_j\, dp_{jx}\, dp_{jy}\, dp_{jz}. \tag{10.55}$$

For the practical evaluation of this sum it is desirable to employ an approximation and treat the sum as an integral by imagining that the contiguous summands differ from each other infinitesimally. This allows us to write approximately

$$Z = \frac{n}{n_c\Delta}\int_{-\infty}^{+\infty} e^{-(p_x^2 + p_y^2 + p_z^2)/2m\Theta}\, dx\, dy\, dz\, dp_x\, dp_y\, dp_z. \tag{10.56}$$

The infinite limits apply only to p_x, p_y, p_z, and not to x, y, z. In fact we have

$$\int dx\, dy\, dz = V, \tag{10.57}$$

where V is the volume of the physical space in which the particles are confined. This puts Z in the form

$$Z = \frac{nV}{n_c\Delta}\int_{-\infty}^{+\infty} e^{-(p_x^2 + p_y^2 + p_z^2)/2m\Theta}\, dp_x\, dp_y\, dp_z. \tag{10.58}$$

The integral is strictly a triple one, written as a single one for simplicity. Its evaluation can be found in any book on advanced calculus and leads to the final answer for Z in terms of Θ

[1] Cf. Lindsay, *Physical Statistics*, Chap. VI.

$$Z = \frac{nV}{n_c\Delta} \cdot (2\pi m\Theta)^{3/2}. \qquad (10.59)$$

We are now prepared to write the canonical distribution (10.48) for a collection of free particles in the form

$$dN = \frac{Ne^{-(p_x^2 + p_y^2 + p_z^2)/2m\Theta}}{V(2\pi m\Theta)^{3/2}} \cdot dx\, dy\, dz\, dp_x\, dp_y\, dp_z. \qquad (10.60)$$

This is the expression for the number of particles whose phase lies in the phase interval $dx\, dy\, dz\, dp_x\, dp_y\, dp_z$ in the neighborhood of the component momentum values p_x, p_y, p_z and the space coordinates x, y, z. Actually dN does not depend explicitly on the spatial coordinates since the energy is kinetic only and not a function of position. If we are not interested in the spatial distribution of the particles but only in the distribution with respect to momentum or velocity, we integrate dN over dx dy dz and get for the desired distribution number

$$(dN)_p = \frac{Ne^{-(p_x^2 + p_y^2 + p_z^2)/2m\Theta}}{(2\pi m\Theta)^{3/2}} \cdot dp_x\, dp_y\, dp_z. \qquad (10.61)$$

Finally we may wish to find the number of particles which, independently of their position in the confining vessel, have their *velocity* components lying in the interval $dv_x\, dv_y\, dv_z$ in the neighborhood of the values v_x, v_y, v_z. Then (using eq. 10.51) we have

$$(dN)_v = N\left(\frac{m}{2\pi\Theta}\right)^{3/2} e^{-m(v_x^2 + v_y^2 + v_z^2)/2\Theta} dv_x\, dv_y\, dv_z. \qquad (10.62)$$

With the appropriate identification of the distribution modulus this is the famous Maxwell velocity distribution formula, whose physical significance we shall now examine. The physical meaning of Θ will appear presently.

First let us notice that we can use (10.62) to calculate the *average* velocity of each particle. This will be done by treating $(dN)_v/N$ as the probability of finding a particle with velocity lying in the interval v, $v + dv$ or components lying in the interval v_x, $v_x + dv_x$, v_y, $v_y + dv_y$, v_z, $v_z + dv_z$. Note that $v = \sqrt{v_x^2 + v_y^2 + v_z^2}$ is the magnitude of a vector whose components are v_x, v_y, v_z or as usual

$$\mathbf{v} = \mathbf{i}v_x + \mathbf{j}v_y + \mathbf{k}v_z.$$

Consequently the average resultant velocity

$$\bar{\mathbf{v}} = \mathbf{i}\bar{v}_x + \mathbf{j}\bar{v}_y + \mathbf{k}\bar{v}_z. \tag{10.63}$$

But

$$\begin{aligned}\bar{v}_x &= \int v_x \frac{(dN)_v}{N} \\ &= \left(\frac{m}{2\pi\Theta}\right)^{3/2} \int_{-\infty}^{+\infty} v_x e^{-m(v_x^2 + v_y^2 + v_z^2)/2\Theta} dv_x\, dv_y\, dv_z,\end{aligned} \tag{10.64}$$

with similar expressions for $\bar{v}_y$ and $\bar{v}_z$. We are forming the average of v_x by multiplying it by the probability that the x component velocity shall lie in the prescribed interval about this value and integrating over all values of the velocity components. This is essentially the definition of a *weighted* average; thus each v_x value carries a weight proportional to the probability that a particle shall have this v_x value. From the distribution formula (10.62) we see that small magnitudes of $v_x^2 + v_y^2 + v_z^2$ correspond to larger probability values than larger magnitudes.

Now we do not need actually to carry out the integration in (10.64) to see that

$$\bar{v}_x = \bar{v}_y = \bar{v}_z = 0, \tag{10.65}$$

since $e^{-m(v_x^2 + v_y^2 + v_z^2)/2\Theta}$ is an even function of v_x, v_y, v_z, while the factor v_x is an odd function. We reach the conclusion that insofar as (10.62) represents the most likely distribution of particles in the confining vessel with respect to their velocities, the average resultant velocity per particle is zero. What does this mean physically? If a little demon were to take his stand at any point in the vessel and examine the particles that passed him going along any particular straight line, he would find that in any given interval of time the number of particles which passed him with given speed going in one direction along this line would be equal to the number of particles with the same speed going in the opposite direction.

The reader might well object that it is hardly essential to go through all the work leading to the Maxwell distribution to achieve this result. This is true! But we can now go on to deduce from (10.62) other results which are not so intuitively obvious.

In particular let us next try to find the fractional number of molecules having their *speeds* in the range v to $v + dv$. This disregards the *direction* of motion. We imagine now a space in which each point represents a possible momentum value mv for a molecule. The points corresponding to molecules with momentum *magnitudes* lying between mv and $mv + mdv$ will lie in a spherical shell of radius mv and thickness mdv. This has the volume $4\pi m^3 v^2 dv$. If we call the number of molecules corresponding to this volume $(dN)'_p$, it is clear that the relation between this and $(dN)_p$ in (10.61) is

$$\frac{(dN)'_p}{(dN)_p} = \frac{4\pi m^3 v^2 dv}{dp_x\, dp_y\, dp_z}. \tag{10.66}$$

Substitution from (10.61) yields

$$(dN)'_p = N \cdot 4\pi v^2 (m/2\pi\Theta)^{3/2} e^{-mv^2/2\Theta}\, dv. \tag{10.67}$$

If we plot the coefficient of Ndv as a function of v we get the curve indicated in Fig. 10.1, another form of the Maxwell distribution. It is at

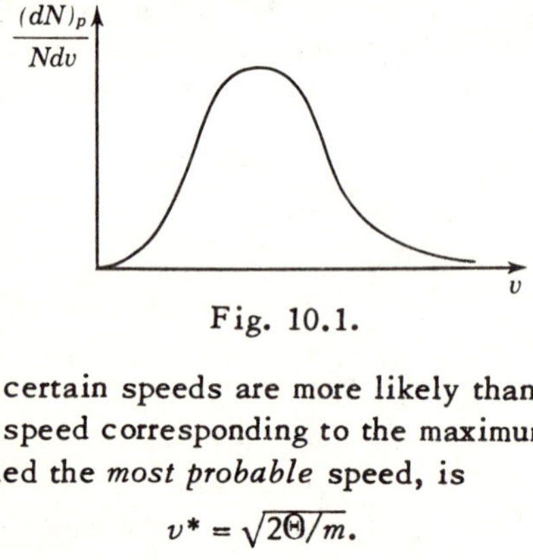

Fig. 10.1.

once evident that certain speeds are more likely than others. In fact it turns out that the speed corresponding to the maximum of the curve, which may be called the *most probable* speed, is

$$v^* = \sqrt{2\Theta/m}. \tag{10.68}$$

More important is the square root of the mean square speed (r.m.s. speed). Carrying out the average process by the same method as in (10.64) we are led to

$$\sqrt{\overline{v^2}} = \sqrt{3\Theta/m}, \tag{10.69}$$

of the same order of magnitude as v^* but nevertheless significantly different. This difference in precise value depending on the precise type of average is one of the penalties we must pay for the pleasure of using statistical reasoning.

It is now time to give some physical meaning to the parameter Θ. If the analysis of the present section is to bear a consistent relation to that of Sec. 10.2, where we applied the virial theorem to the motion of a confined aggregate of particles, we surmise that $\sqrt{\overline{v^2}}$ calculated in (10.69) should be the same as the v_m calculated by the virial theorem and presented in (10.15) in terms of the pressure of the corresponding gas as

$$v_m = \sqrt{\frac{3pV}{Nm}} = \sqrt{\frac{3NkT}{Nm}}, \tag{10.70}$$

where we have replaced pV by the value NkT to which it is equal for an ideal gas, k being the universal constant of Boltzmann (the gas constant per molecule and equal to 1.37×10^{-16} ergs/°C). Equating v_m to $\sqrt{\overline{v^2}}$ at once yields

$$\Theta = kT. \tag{10.71}$$

It is seen that if the statistical distribution theory of the present section is to provide a description of the behavior of an ideal gas, it is essential that the distribution modulus Θ be representative of and indeed proportional to the absolute temperature of the gas. We find it advantageous to go even further and use Θ to provide a *statistical definition* of absolute temperature. This means that from the standpoint of statistical mechanics the absolute temperature of a system is equal to two thirds the mean kinetic energy per molecule divided by the Boltzmann constant k. It remains necessary to identify this definition with the quantity operationally determined as temperature by means of a thermometer. If we could always deal with an ideal gas, this would not be an essentially difficult matter: we should merely have to associate with whatever thermometer we happen to use such values of T as make Θ/k satisfy the fundamental statistical equation (10.13). We must, of course, face the difficulty that different thermometers (e.g., thermocouples, resistance thermometers, mercury-in-glass thermometers, etc.) may demand the assignment of somewhat different values of T to systems that so far as one can tell are in the same state. It is a matter of practical importance to establish a scale of temperature which is independent of the particular thermometer used. We shall return to this in our study of thermodynamics (Sec. 10.7).

In the foregoing paragraphs of this section we have at our disposal the essential apparatus for constructing a theory of the behavior of gases based on the hypothesis of molecular constitution coupled with statistical analysis. Before we proceed further it will be wise to introduce a few more concepts.

10.5. Further Properties of the Canonical Distribution. Let us return to the general canonical distribution (10.48) and calculate the total energy in the usual way. Thus with Θ put equal to kT

$$E = \Sigma N_j E_j = \frac{N\Sigma n g_j E_j e^{-E_j/kT}}{\Sigma n g_j e^{-E_j/kT}} . \tag{10.72}$$

Now from (10.45) form $\log n^N P_c$, where P_c is the value of P for the canonical distribution. It turns out by substitution that

$$\log n^N P_c = \frac{E - N\psi}{kT} . \tag{10.73}$$

We shall find it advantageous to refer to $n^N P_c$ as the *canonical statistical probability* and denote it by w_c, keeping w for the corresponding general statistical probability $n^N P$. We then rewrite (10.73) in the form

$$\log w_c = \frac{E - N\psi}{kT} . \tag{10.74}$$

This is an important relation. From (10.47) and (10.49) we may write it in still another form, viz.,

$$\log w_c = E/kT + N \log Z. \tag{10.75}$$

It is to be emphasized that this holds only when the system is canonically distributed. We recall that this is the state of maximum probability and hence presumably the one toward which the system tends to move if left by itself.

Now from the definition of Z, it is clear that it depends on the energy values E_j assigned to each box as well as on the g_j and T. The dependence on the E_j and g_j can be thought of as equivalent to dependence on certain parameters $\xi_1 \cdots \xi_k$ which control the relation between the system and its external environment. Of these the most important is the volume of the vessel containing the system. If we change this volume we shall certainly change the E_i somehow. In this process E itself will in general alter. Let us now ask ourselves the question: How will $\log w_c$ change when the external environment of the system changes along with the temperature T? It is understood of course that all through this change the system remains canonically distributed. Forming the differential of $\log w_c$ in (10.75) we have[1]

$$d \log w_c = dE/kT - EdT/kT^2 + N \frac{\partial \log Z}{\partial T} dT + N \sum_{i=1}^{k} \frac{\partial \log Z}{\partial \xi_i} d\xi_i. \tag{10.76}$$

Reference to the definition of Z and to the expression for E in (10.72) shows at once that

$$-E/kT^2 + N \frac{\partial \log Z}{\partial T} = 0. \tag{10.77}$$

Therefore (10.76) reduces to

$$kTd \log w_c = dE + NkT \sum_{i=1}^{k} \frac{\partial \log Z}{\partial \xi_i} d\xi_i. \tag{10.78}$$

Let us now examine the second term on the right. This becomes

$$\sum \frac{\partial \log Z}{\partial \xi_i} d\xi_i = \sum_{i=1}^{k} \frac{\partial}{\partial \xi_i} \left(\log \sum_{j=1}^{n} ng_j e^{-E_j/kT} \right) d\xi_i$$
$$= -\frac{1}{ZkT} \sum_i \sum_j ng_j e^{-E_j/kT} \frac{\partial E_j}{\partial \xi_i} d\xi_i. \tag{10.79}$$

Now when the external coordinates ξ_j vary and with them the E_j values, work is done either on the system or by the system (e.g., the case of a gas expanding or being compressed). If a particular particle is in the jth energy state and ξ_i is changed by $d\xi_i$ the work done by the particle is $-\frac{\partial E_j}{\partial \xi_i} d\xi_i$, since this is the negative of the change in energy associated with $d\xi_i$(if $\partial E_j/\partial \xi_i$ is positive it means that the particle gains energy

[1]The use of the differential operator d rather than the variation operator δ reflects the fact that the change in question is to be carried out so that the system remains canonically distributed.

when ξ_i is changed or work is done *on* the particle). Therefore the work done by all the particles (N_j in number) in the jth state when ξ_i changes is

$$-N_j \frac{\partial E_j}{\partial \xi_i} d\xi_i.$$

Hence the total work done by the whole system when all the external coordinates vary is

$$dW = -\sum_i \sum_j N_j \frac{\partial E_j}{\partial \xi_i} d\xi_i. \tag{10.80}$$

But from eqs. (10.48) and (10.79) it is clear that

$$dW = NkT \sum_i \frac{\partial \log Z}{\partial \xi_i} d\xi_i. \tag{10.81}$$

Hence eq. (10.78) reduces finally to

$$kTd \log w_c = dE + dW. \tag{10.82}$$

In words, this says that when T and the external environment of the system are altered in such a way that the distribution of the aggregate of particles composing the system remains canonical, i.e., the system remains at every instant in a state of equilibrium, the sum of the change in the internal energy of the system and the work done *by* the system is equal to kT times the change in the natural logarithm of the canonical statistical probability. For some purposes it is more significant to consider the equation in the form

$$dE = kTd \log w_c - dW, \tag{10.82a}$$

where $-dW$ is the work done *on* the system by the changes in the external environment. The importance of this equation lies in its expression of the fact that the change in the total internal energy of the system is not in general equal simply to the work done on it by the change in the external environment. Rather there is an additional contribution $kTd \log w_c$. It therefore becomes necessary to seek a physical explanation for the existence of this latter term.

As we critically examine physical phenomena in general it appears that the only way in which the internal energy of a physical system can be changed other than by doing mechanical work on it is by the communication of *heat* to it.[1] We shall therefore proceed at once to interpret the term $kTd \log w_c$ as the quantity of heat communicated to the system in the change under consideration and write

$$dQ = kTd \log w_c. \tag{10.83}$$

This is the first time in our discussion that the idea of heat has entered, although its phenomenological significance has already been implied by the introduction of the temperature T. The natural methodological attitude appropriate to the present treatment is to use (10.83) as the *definition* of a *new* concept in physics and then to seek to identify it with something in physical experience. The identification with heat as measured phenomenologically turns (10.82) into an equation serving as the basis for a dynamical theory of heat. This theory is called *thermodynamics*.

10.6. The Principles of Thermodynamics. In the classical theory of thermodynamics we encounter a physical theory which employs certain dynamical concepts but in a fashion different from the way they are used in the particle dynamics we have been discussing so far in this book. Instead of defining the state of a physical system in terms of the dynamical specification of the particles composing it, classical thermodynamics specifies a state in terms of operationally defined concepts like pressure, volume, and temperature relating to large-scale objects. Moreover the state so specified is assumed always to be one in which the object, if left to itself, will stay forever, i.e., a state of *equilibrium*. The purpose of thermodynamics is to describe the behavior of large-scale objects when undergoing changes from one equilibrium state to another, especially those changes in which change of temperature, i.e., heating or cooling, is involved. The theory recognizes that in most such changes, called technically *thermodynamic processes*, work will be done, e.g., in expanding or compressing a gas (or a solid or liquid, for that matter). It also recognizes that mechanical work can often be performed on an object or thermodynamical system without the appearance of the corresponding amount of kinetic and potential energy of the system as a whole, which would be expected from the principles of particle dynamics. It comforts itself in this apparent violation of the dynamical principle of the

[1] We are apparently excluding from our consideration here the presence of chemical reactions or electromagnetic effects in the system. But we may readily generalize the statement to include these effects by imagining the external coordinates ξ_j to be related to chemical activity and the presence of electric and magnetic fields.

conservation of energy by the observation that in all such cases there is the phenomenological appearance of heat. To treat this heat as a form of energy is an obvious next step, justified by the careful experiments first performed by Joule (1840) and others since, in which the apparent disappearance of a certain amount of mechanical work is always associated with a definite amount of heat as operationally defined (the mechanical equivalent of heat).

The classical theory of thermodynamics further recognizes that the processes it describes have a tendency to proceed in certain directions rather than others, e.g., heat (as operationally defined) always tends of itself to flow from a hotter to a cooler body. It proceeds to build principles embodying these empirical results. These are the "laws" of thermodynamics. The rest is largely a matter of mathematical exploitation of the laws and application to specific examples, many of which are familiar to the reader from his study of elementary physics, e.g., the expansion of gases, changes from gaseous to liquid or solid states, etc. The power of the method becomes more obvious when it is applied to phenomena in which magnetic, electric, and chemical changes take place.

In this book we shall not develop thermodynamics from the classical standpoint briefly reviewed above. Rather we shall treat it as the logical consequence of the application of the statistical method to the behavior of large aggregates of particles with appropriate identification of the statistical averages. Thus combining (10.83) with (10.82) we may write

$$dQ = dE + dW, \tag{10.84}$$

which may be interpreted as follows: the quantity of heat dQ added to a system of particles, called hereafter a thermodynamic system, is equal to the change dE in the internal energy of the system, i.e., the total energy of the particles making up the system, plus the work dW done by the system. This is known as the first "law" of thermodynamics. In classical thermodynamics it is one of the fundamental postulates of the theory. Here it appears as a direct deduction from the treatment of the aggregate of particles as a statistical distribution with maximum probability. Some comments are necessary.

We have already mentioned the concept of equilibrium state. Quantities descriptive of the system which depend only on the state and not on the process by which the system got into that state, and which therefore return to their original values when the system undergoes a process which takes it back into its original state (a so-called *cyclic* process) are called *state variables*. The internal energy of the system E is such a variable, since each equilibrium state is characterized by a definite canonical distribution function N_j (from (10.48)), and with E_j fixed for each substate, $E = \Sigma N_j E_j$ is completely determined for each state. On the other hand the work done by the system in a change from one equilibrium state to another cannot be considered a state variable. It is perfectly possible for quite different amounts of work to be done in the transition between two particular states, depending on how T and the ξ_i are varied. If the system, for example, is an ideal gas which is first heated to a given temperature at constant volume and then allowed to expand at constant temperature to double its original volume, the work done is not the same as that which would be done if the gas were first allowed to double its original volume at constant temperature. We commonly say that the work done by a gas depends on the nature of the process it undergoes. Hence W is not a state variable.

It therefore follows from (10.84) that Q is not a state variable either, since if it were, in any cyclic process both dQ and dE would have to be zero, and hence dW would have to vanish also. But we have just seen that dW will not necessarily vanish in a cyclic process. Hence neither will dQ *in general*, although we can visualize processes in which this may be true. In fact they are processes known as *adiabatic* and are of considerable practical importance.

In eq. (10.84) all quantities are measured in energy units. However, in order to use the equation in practical experience dQ must be interpreted calorimetrically, and this leads to the necessity for an energy equivalent of heat, in accordance with which 4.2 joules is equivalent to 1 calorie.

If we revert to eq. (10.82) we see that, although $kTd \log w_c$, which has been *defined* as dQ, is not the change in a state variable, $d \log w_c$ itself must be considered the change in a state variable. The reason is fundamental: the value of the statistical probability w_c depends solely on the equilibrium state of the system, since it defines this state by its own maximum value, subject to the constancy of N and E. Hence it has nothing to do with the process whereby the system got into the state. It must again be emphasized that w_c is defined only for equilibrium states, i.e., those of maximum probability. Any change in $\log w_c$ denoted by $d \log w_c$ refers only to the change from one equilibrium state to another.

It follows that

$$\frac{dQ}{T} = d(k \log w_c) \qquad (10.85)$$

is the change in a state variable, as far as equilibrium states are concerned, and those are the only states considered in our discussion. The state variable $k \log w_c$ is an interesting one, since for an isolated system, i.e., one not allowed to interact with the environment, it will not in general decrease as time goes on. The reason is of course associated with the nature of the statistical probability w_c, which is a maximum for a canonical distribution. If there were a state of greater probability than that for the canonical distribution the system would naturally tend to change until it attained it and in the process the statistical probability would increase. If $k \log w_c$ were to decrease for an isolated system, it would mean that the system was moving to a state of lower probability. This is not impossible in the framework of statistical theory but obviously improbable. The approach of a system to equilibrium (i.e., to a canonical distribution) therefore appears from the statistical point of view to be essentially a one-way or irreversible process.

As a measure of this irreversibility of approach toward equilibrium of a system with fixed energy we may well choose $k \log w_c$ itself, possibly plus a properly chosen constant. We shall indeed write

$$S = k \log w_c + C, \tag{10.86}$$

and call S the *entropy* of the system. It turns out to be advantageous to choose $C = -k \log N!$ so that

$$S = k \log \frac{w_c}{N!}. \tag{10.87}$$

For an isolated system S will not in general decrease since this would involve passage to a state of smaller probability, which is from the very nature of probability, highly improbable. Of course it is not impossible since fluctuations are allowable in statistical considerations. Therefore we are not entitled to say that S must *never* decrease for such a system; merely that it will be highly unlikely that it does.

The above statement about the entropy of an isolated system is one way of expressing what has come to be called the *second law of thermodynamics*. It purports to say something about the *direction* of thermodynamic processes, about which the first law says nothing, since the latter merely emphasizes the essential equivalence of heat and mechanical energy independent of the direction of the process. It is well known that experiment indicates that many natural processes appear to be one directional. Thus the flow of heat always appears to take place naturally from a body at a higher temperature to one at lower temperature. This does not mean, of course, that it is impossible for heat to flow from a cooler to a hotter body – this happens every day in a refrigerator, but we all realize that it involves the expenditure of energy and costs money; it is not a process that goes by itself.

It may be worthwhile to analyze this situation further from the statistical standpoint. Let us suppose we have a thermodynamical system in a state characterized by the statistical probability w. We shall not assume that the system is in equilibrium and hence w is not the canonical statistical probability w_c. Suppose in a second state which can be reached by the system the statistical probability is $w + \Delta w$. In proceeding from the first state to the second the gain in statistical probability is Δw. A reasonable measure of the tendency of the system to proceed from state 1 to state 2 is $(w + \Delta w)/w = z$. We may then write

$$\log \frac{w + \Delta w}{w} = \log z = \Delta \log w, \tag{10.88}$$

or

$$z = e^{\Delta \log w}. \tag{10.89}$$

Now it is difficult in general to calculate $\Delta \log w$. However there are cases in which we can treat states 1 and 2 as if they were *equilibrium* states and hence replace $\Delta \log w$ by $\Delta \log w_c$. This enables us at once to calculate z from (10.85), i.e., by computing

$$\frac{\Delta S}{k} = \frac{\Delta Q}{kT}.$$

As a simple example, consider a gas originally confined in a closed vessel of volume V_1 which forms part of a larger vessel of volume V_2 (cf. Fig. 10.2). We assume that the gas in V_1 is in a state of equilibrium. However, it is easily possible to upset this equilibrium by opening a small door in the wall separating the gas from the rest of the chamber. At the instant the door is opened the gas is no longer in a state of equilibrium. In fact we know from experience that it will stream out of the opening until the whole volume V_2 has been occupied, when it will again settle down to a state of equilibrium. We speak of this as a *natural* process. It is certainly more natural than the reverse process in which the gas in equilibrium in V_2 gradually decreases its volume and boxes itself up in V_1. The latter process is conceivable but we never see it occur. We say it is more *natural* for a gas to expand if left to its own devices than to contract, but this is only because we have observed this sort of thing so often that we believe it extremely unlikely that it will ever be contravened.

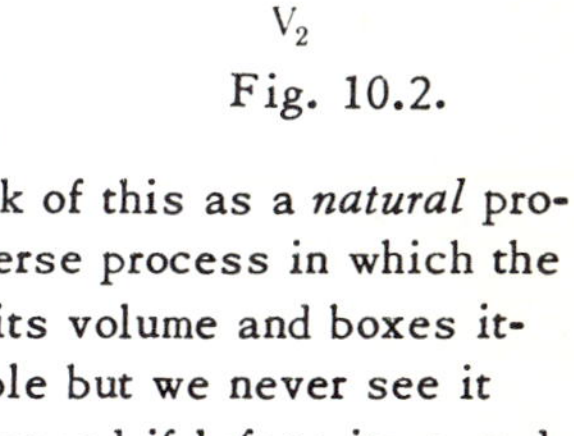

Fig. 10.2.

Now we can treat the process above described as if it took the gas from volume V_1 at temperature T to volume V_2 at the same temperature. Some question might be raised as to how we know that the temperature will remain the same. For this we must have recourse to experience which in the case in question indicates that the change in temperature is so small that we may safely neglect it to a first approximation. It is possible to imagine the change from volume V_1 to volume V_2 at constant temperature to take place in such a way that the gas remains in equilibrium at each stage of the process; or, as we may say, passes in succession through a continuous series of equilibrium states. It is true that this will be difficult to realize precisely in practice, but we can approach it the more nearly the more slowly we permit the expansion to take place. Such ideal processes in which the system is in equilibrium at every stage are called *reversible* processes, for they can be made to go in either direction by an infinitesimal change in the parameters controlling the change. Thus in the case under consideration, the expansion can be changed to contraction by an infinitesimal increase in the *pressure* on the gas.

The great theoretical advantage in dealing with reversible processes is that we may employ in their description equations like the so-called equation of state, which for an ideal gas takes the form (10.9). Such equations involve quantities which are defined only for equilibrium states and hence describe processes which are a continuous succession of such states.

This at once enables us to calculate $\Delta Q/kT$ for the reversible process which has the same initial and final states as those contemplated in the example under consideration. We revert to (10.84) and assume that the isothermal character of the process (negligible temperature change) means that the change in internal energy of the gas in expanding from volume V_1 to volume V_2 is negligible. As a matter of fact this follows from the assumption that the internal energy is due wholly to the kinetic energy of the molecules. The discussion in Sec. 10.4 showed that the mean square molecular velocity then depends only on the distribution modulus Θ (i.e., kT) and hence the kinetic energy must also depend only on Θ. We therefore reach the conclusion that the heat required to carry through the reversible isothermal transformation in question is simply equal to the work done by the expanding gas during the expansion.

By a very simple consideration of the force exerted by the gas on a movable piston we conclude that the work done when the piston moves through such a distance as to change the volume of the gas by dV is

$$dW = p\, dV. \tag{10.90}$$

In the case considered here

$$p = NkT/V, \tag{10.91}$$

where we have replaced the gas constant R by Nk, as usual. Hence the work done is

$$\Delta W = \int_{V_1}^{V_2} \frac{NkT\, dV}{V} = NkT \log \frac{V_2}{V_1}. \tag{10.92}$$

Therefore

$$\Delta \log w_c = \frac{\Delta W}{kT} = N \log \frac{V_2}{V_1}, \tag{10.93}$$

and from (10.89) z becomes

$$z = e^{N \log \frac{V_2}{V_1}} \tag{10.94}$$

We are now ready to estimate the tendency of the system to move from the state 1 (volume V_1) to state 2 (volume V_2) by the magnitude of z. In the trivial case where $V_2 = V_1$, $z = 1$, as should obviously be true. If the amount of gas is a mole, N = Avogadro's number. Suppose $V_2 = 2V_1$. Then

$$z = e^{6.03 \times 10^{23} \log 2}$$

This is a tremendously large number which the reader may express as a power of 10 if he wishes. It is clear that the "gain" in statistical probability in the expansion of one mole of an ideal gas to double its volume is enormously large. This will serve to give some idea of the measure of the irreversibility of natural processes supplied by the statistical theory of such processes.

But we must notice one thing: no matter how large N is, we can imagine such a change in volume that V_2/V_1 differs from unity by a vanishingly small quantity. In this case it turns out that z can become very close to unity, indicating that the probability of such a change is not very great. What does this mean? Evidently the smaller the change in volume, the less decisive is the statistical likelihood of the change going in one direction rather than the other. We note that the same sort of thing is implied when the number of molecules in question is very small. For very small changes involving small numbers of molecules it is, in this picture, impossible to predict with great assurance the direction of physical processes. Fluctuations from the expected change will occur. As the number of entities involved and the magnitude of the change become greater the probability of large fluctuations becomes less. It is possible to generalize this point of view to make it apply to all physical phenomena and to use it as the basis for the existence of observed regularities, i.e., the possibility of physical laws.[1]

10.7. Thermodynamic Scale of Temperature. At the end of Sec. 10.4 we introduced the statistical mechanical definition of temperature in terms of the statistical distribution modulus Θ for a statistical aggregate corresponding to an ideal gas. We pointed out there that, although this is a perfectly satisfactory theoretical definition, it leaves open the problem of setting up a temperature scale in terms of which practical thermometers can be calibrated. We now discuss this question in terms of the principles of thermodynamics.

It will be recalled from elementary physics that the efficiency of a heat engine working in a cyclic process in which the working substance takes in heat Q_1 at absolute temperature T_1 and ejects heat Q_2 at absolute temperature T_2 is given by $(Q_1 - Q_2)/Q_1$. According to the Carnot theorem proved in many elementary texts[2] this efficiency is the same for all reversible engines working between the same two temperatures, and no irreversible engine working between these two temperatures can have a greater efficiency. Lord Kelvin seized on this theorem to define a temperature *scale* independent of the properties of any particular substance. Since $(Q_1 - Q_2)/Q_1$ is dependent only on the temperatures T_1 and T_2 it is tempting to define the new scale so that

$$T_1/T_2 = Q_1/Q_2. \tag{10.95}$$

If then we draw the usual Carnot cycle we can set up the thermodynamic or Kelvin scale of temperature by dividing the area cut off by any two adiabatics between the isothermal for melting ice at atmospheric pressure and the isothermal for steam at atmospheric pressure into 100 equal parts by the introduction of 99 isothermal curves. If the isothermal for melting ice is labeled 0°C, the next in order will correspond to 1°C, etc. Then the temperature associated with the isothermal in which heat Q is absorbed will turn out to be

$$T = \frac{Q}{\dfrac{Q_{100} - Q_0}{100}} \tag{10.96}$$

[1]Cf., E. Schrödinger, *What is Life* (Cambridge Univ. Press, 1945), Chap. I.
[2]Cf., e.g., Lindsay, *General Physics*, (New York, John Wiley, 1939), pp. 216 ff.

on the centigrade scale.

It will be shown in Sec. 10.9 that a constant pressure gas thermometer (in which the volume of the gas is the thermometric variable) using an ideal gas yields temperatures agreeing with the Kelvin thermodynamic scale. This will complete the identification of the statistical definition of temperature with the operational one.

10.8. Some Applications of Thermodynamics. Free Energy and the Equation of State. We have already introduced the internal energy E and the entropy S in addition to volume, temperature, and pressure as thermodynamic variables. It now remains to bring in other state variables which can be formed from these and investigate their usefulness.

The function

$$A = E - TS \tag{10.97}$$

is usually called the Helmholtz free energy. Let us form its total differential with respect to changes in the independent variables, e.g., pressure and volume. Then

$$dA = dE - TdS - SdT. \tag{10.98}$$

Now if the only external coordinate ξ_i (cf. Sec. 10.5) of interest is the volume V, and this will cover a great many important cases, e.g., simple gases, we can write the first law in the form

$$dE = T\,dS - p\,dV. \tag{10.99}$$

Hence

$$dA = -p\,dV - S\,dT. \tag{10.100}$$

In an *isothermal* process, $dT = 0$ and (10.100) becomes

$$-\left(\frac{\partial A}{\partial V}\right)_T = p. \tag{10.101}$$

If the free energy A can be found as a function of T and V, this equation relates p, V, and T for the system and hence serves as the equation of state.

To revert for a moment to the statistical point of view let us use (10.87) for S and write for A in the case of a gas with N particles

$$A = E - kT\log\frac{w_c}{N!} = E - kT\log w_c + kT\log N! \tag{10.102}$$

whence from (10.74) we obtain

$$A = N\psi + kT \log N! \tag{10.103}$$

or to use the partition function Z (eq. 10.75)

$$A = -NkT \log Z + kT \log N!$$

$$= -NkT \left(\log \frac{Z}{N} + 1\right), \tag{10.104}$$

wbere the last step follows the use of Stirling's formula for $N!$ If therefore we are able to evaluate the partition function for a particular system, we can deduce A from (10.104) and then from (10.101) derive the equation of state. This shows the power of the statistical method. We can illustrate by deriving the equation of state of an ideal gas from the partition function already worked out in Sec. 10.4 and given by eq. (10.59). Rewriting this with $\Theta = kT$ gives for the free energy

$$A = -NkT\left[1 + \log\left(\frac{nV}{n_c N\Delta}\right) + \frac{3}{2}\log 2\pi mkT\right]. \tag{10.105}$$

Consequently

$$p = -\left(\frac{\partial A}{\partial V}\right)_T = \frac{NkT}{V}, \tag{10.106}$$

which is the required equation of state. Actually for more complicated real gases Z is very difficult to determine and the derivation of the state equation is by no means as simple as this ideal case might suggest.

Let us go back to (10.100) and note that if the right side is a perfect differential we must have not only (10.101) but also

$$S = -\left(\frac{\partial A}{\partial T}\right)_V. \tag{10.107}$$

This enables us to write

$$A = E + T\left(\frac{\partial A}{\partial T}\right)_V, \tag{10.108}$$

which is often referred to as the Gibbs – Helmholtz equation. For the case of an ideal gas, use of (10.105) with (10.108) gives simply

$$E = \frac{3}{2}NkT, \tag{10.109}$$

indicating that the internal energy of an ideal gas depends solely on the temperature. Of course we have previously derived this result on the basis of kinetic theory.[1] In many chemical reactions the free energy varies so slightly with temperature that $T(\partial A/\partial T)_V$ may be considered negligible compared with E. In such cases

$$A \doteq E, \tag{10.110}$$

and the free energy becomes practically equivalent to the total internal energy.

A standard application of (10.108) is to a reversible electric cell, e.g., a storage cell. For this the first law takes the form

$$dE = T\,dS - \epsilon\,dq. \tag{10.111}$$

in which ϵ is the electromotive force of the cell and q denotes the charge which passes. $\epsilon\,dq$ is the work done *by* the cell in the passage of charge dq around the circuit containing the cell. Eq. (10.100) now becomes

$$dA = -\epsilon\,dq - S\,dT, \tag{10.112}$$

whence

$$\left(\frac{\partial A}{\partial q}\right)_T = -\epsilon. \tag{10.113}$$

Therefore

$$A = E + T\left(\frac{\partial A}{\partial T}\right)_q. \tag{10.114}$$

Now differentiate both sides partially with respect to q, keeping T constant. The result is

$$\left(\frac{\partial A}{\partial q}\right)_T = -\epsilon = \left(\frac{\partial E}{\partial q}\right)_T - T\left(\frac{\partial \epsilon}{\partial T}\right)_q \tag{10.115}$$

or

$$\left(\frac{\partial E}{\partial q}\right)_T = T\left(\frac{\partial \epsilon}{\partial T}\right)_q - \epsilon. \tag{10.116}$$

This is the Gibbs–Helmholtz equation applied to the cell. The left-hand side is the change in internal energy of the cell associated with the passage of unit charge through it at constant temperature. This appears in general as the heat of reaction of the chemical change taking

[1] Actually we defined the temperature statistically so that this result follows.

place in the cell. Hence if the electromotive force and its temperature coefficient are accurately known we have here a precise method of measuring the heat of chemical reaction.

10.9. Other Thermodynamic Functions. Maxwell's Relations. Specific Heats. We have introduced in Sec. 10.8 the Helmholtz free energy as an important thermodynamic function. Two others have also proved of use. The first is the so-called *enthalpy* or H function defined by

$$H = E + pV. \tag{10.117}$$

The second is the Gibbs free energy F, given by

$$F = E + pV - TS = H - TS = A + pV. \tag{10.118}$$

Since both H and F are simple functions of state variables they are themselves state variables and their differentials are therefore perfect differentials. From this fact and the corresponding property of dE and dA, we may immediately deduce the following four partial differential relations

$$\left(\frac{\partial S}{\partial V}\right)_T = \left(\frac{\partial p}{\partial T}\right)_V; \quad \left(\frac{\partial T}{\partial V}\right)_S = -\left(\frac{\partial p}{\partial S}\right)_V. \tag{10.119}$$

$$\left(\frac{\partial T}{\partial p}\right)_S = \left(\frac{\partial V}{\partial S}\right)_p; \quad \left(\frac{\partial S}{\partial p}\right)_T = -\left(\frac{\partial V}{\partial T}\right)_p. \tag{10.120}$$

These are called Maxwell's thermodynamic relations. Note that they may be written in even more general form if we replace V by the general environmental parameter ξ_j (e.g., q in the case of the electric cell already discussed) and p by the corresponding dynamical parameter (e.g., ϵ, the emf, in the electrical case). The reader will recognize several familiar quantities in (10.119) and (10.120). Thus

$$\frac{1}{V}\left(\frac{\partial V}{\partial T}\right)_p = \alpha_p, \tag{10.121}$$

or the coefficient of volume expansion at constant pressure. Moreover

$$\frac{1}{p}\left(\frac{\partial p}{\partial T}\right)_V = \beta_V, \tag{10.122}$$

the coefficient of pressure change at constant volume. For an ideal gas, it will be recalled that these are both equal to $1/T$. Of course this is not the coefficient as ordinarily defined in elementary physics. To get the latter we must multiply by $T/273$, since the coefficient of expansion is usually referred to conditions holding at 273°K.

The above relations take on more significance if we introduce the heat involved in the various changes which a system can undergo. Even though this is not a thermodynamic variable (cf. Sec. 10.6), it is extremely useful since it is so readily measured in many cases. It is always possible to choose two thermodynamic variables as the independent ones and express the change in Q in terms of the changes in these with appropriate coefficients. Thus choosing the independent variables in pairs

$$dQ = C_V dT + L_T dV = C_p dT + \Lambda_T dp = l_V dp + \lambda_p dV. \quad (10.123)$$

The coefficients are named in accordance with their physical significance Thus

$$C_V = \left(\frac{\partial Q}{\partial T}\right)_V = \text{heat capacity at constant volume},$$

$$C_p = \left(\frac{\partial Q}{\partial T}\right)_p = \text{heat capacity at constant pressure},$$

$$L_T = \left(\frac{\partial Q}{\partial V}\right)_T = \text{latent heat of change of volume},$$

$$\Lambda_T = \left(\frac{\partial Q}{\partial p}\right)_T = \text{latent heat of change of pressure},$$

$$l_V = \left(\frac{\partial Q}{\partial p}\right)_V = \text{heat of compression at constant volume},$$

$$\lambda_p = \left(\frac{\partial Q}{\partial V}\right)_p = \text{heat of expansion at constant pressure}.$$

Of the above the first two are the most familiar. When divided by the mass m of the system they become the usual specific heats at constant volume and constant pressure respectively.

It should be observed that, although we have used the standard differential and derivative notation in connection with Q, this does not imply that Q is a state variable. In other words, dQ is not a perfect differential of T and V in (10.123). However, with this warning the notation should not prove misleading.

A great variety of relations can be derived from (10.123). As an example, consider

$$(C_p - C_V)\, dT = L_T dV - \Lambda_T dp.$$

Now suppose p is constant so that $dp = 0$. It follows that

$$C_p - C_v = L_T\left(\frac{\partial V}{\partial T}\right)_p. \qquad (10.124)$$

This takes on more direct physical meaning if we reflect that the first relation in (10.123) is actually the first law of thermodynamics (10.84) for an ideal gas. Hence at once in this case

$$L_T = p. \qquad (10.125)$$

For an ideal gas, however, the equation of state yields

$$\left(\frac{\partial V}{\partial T}\right)_p = \frac{Nk}{p}.$$

Therefore (10.124) becomes

$$C_p - C_V = Nk, \qquad (10.126)$$

a well-known result in elementary physics.

The use of the enthalpy function H is well exemplified by the so-called porous plug experiment, which in turn enables one to determine experimentally the absolute thermodynamic temperature of any environment. Fig. 10.3 shows schematically the arrangement in which unit mass of gas is forced through the plug by motion of the pistons 1 and 2 to the right in such a way that the pressure on the left of the plug is maintained at p_1 and that on the right is maintained at p_2. To the left of the plug the volume of the unit mass is changed from V_1 to zero by pressure p_1 while to the right the volume is increased from zero to V_2 under pressure p_2. Hence work $p_1 V_1$ is done *on* the gas and work $p_2 V_2$ is done by the gas. There is then a net increase in the internal energy of the gas (no heat being exchanged) equal to

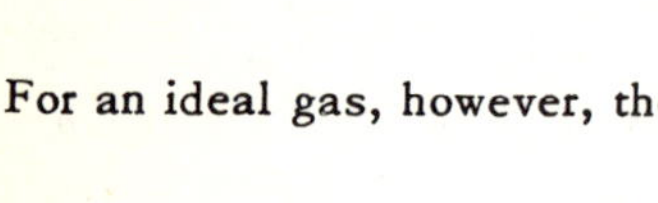

Fig. 10.3.

$$E_2 - E_1 = p_1 V_1 - p_2 V_2,$$

whence in the process

$$p_1 V_1 + E_1 = p_2 V_2 + E_2 = \text{constant}. \qquad (10.127)$$

From eq. (10.117) it then appears that the process takes place at constant enthalpy. It therefore follows from the first law that for the process,

$$T\,dS + V\,dp = 0. \tag{10.128}$$

This may be rewritten in the form

$$T\left(\frac{\partial S}{\partial T}\right)_p dT + \left[T\left(\frac{\partial S}{\partial p}\right)_T + V\right] dp = 0. \tag{10.129}$$

From the fact that

$$T\left(\frac{\partial S}{\partial T}\right)_p = C_p, \tag{10.130}$$

and by the use of Maxwell's relation

$$\left(\frac{\partial S}{\partial p}\right)_T = -\left(\frac{\partial V}{\partial T}\right)_p,$$

we modify (10.129) still further to the form

$$C_p\,dT + \left[V - T\left(\frac{\partial V}{\partial T}\right)_p\right] dp = 0. \tag{10.131}$$

Therefore it follows that

$$T\left(\frac{\partial V}{\partial T}\right)_p = V + C_p\left(\frac{\partial T}{\partial p}\right)_H. \tag{10.132}$$

The term $\left(\frac{\partial T}{\partial p}\right)_H$ represents the rate of change in temperature with pressure as the gas flows through the plug. This can be measured experimentally. For most gases it is a small negative quantity at room temperature (hydrogen being an exception). The much larger cooling at low temperatures is used in the liquefaction of gases.

If we employ (10.121) we can put (10.132) in the form

$$T = \frac{1}{\alpha_p}\left[1 + \frac{C_p}{V}\left(\frac{\partial T}{\partial p}\right)_H\right]. \tag{10.133}$$

Measurement of the quantities on the right serves to fix T for any particular environment. In particular we recall that the constant pressure gas thermometer measures essentially $1/\alpha_p$, or the corresponding quantity on the centigrade scale. If the gas is an ideal one $\left(\frac{\partial T}{\partial p}\right)_H$ vanishes (eq. 10.132) and $T = 1/\alpha_p$ precisely, so that the constant volume gas thermometer using an ideal gas gives exactly the thermodynamic temperature. For a real gas the correction provided by the second term in the bracket in (10.133) is necessary.

10.10. Gibbs Free Energy and the Clausius – Clapeyron Equation. In the previous section we defined the Gibbs' free energy F as

$$F = E + pV - TS.$$

If we form the total differential of both sides we get

$$dF = dE + pdV + Vdp - TdS - SdT. \tag{10.134}$$

But using the first law given in the form (10.99) we can write this more simply as

$$dF = V\,dp - S\,dT. \tag{10.135}$$

The importance of this lies in its implication that if the process in question proceeds without change of pressure or temperature $dF = 0$, i.e., there is no change in the Gibbs free energy. The Gibbs free energy remains unchanged in any reversible, isothermal, isobaric process. Changes in phase from liquid to vapor, etc., approximately satisfy this criterion. We can utilize this result to establish an important relation governing simple phase changes.

Suppose a system consists of a certain mass of substance in phase 1 and that at temperature T and pressure p, it changes reversibly to phase 2. The free energy constancy may be expressed as

$$F_1 = F_2. \tag{10.136}$$

If the temperature and pressure were $T + dT$ and $p + dp$, respectively, the Gibbs free energy in each phase would change but the equality in (10.136) would still hold. We may now write it as

$$F_1 + dF_1 = F_2 + dF_2. \tag{10.137}$$

Hence by subtraction and use of (10.135)

$$V_1 dp - S_1 dT = V_2 dp - S_2 dT,$$

or

$$\frac{dp}{dT} = \frac{S_2 - S_1}{V_2 - V_1}. \qquad (10.138)$$

Here $V_2 - V_1$ represents the change in volume associated with the change in phase and $S_2 - S_1$ the corresponding change in entropy. Since the process takes place reversibly

$$S_2 - S_1 = L/T, \qquad (10.139)$$

where L is the quantity of heat associated with the change. We usually term this the latent heat of change of phase. If the mass of material is one gram, eq. (10.138) becomes

$$\frac{dp}{dT} = \frac{L_m}{T(v_2 - v_1)}, \qquad (10.140)$$

where L_m is the latent heat per gram and v_2, v_1 are the respective specific volumes. Eq. (10.140) is called the Clausius – Clapeyron equation. It yields the relation between pressure and temperature for a reversible change of phase. Thus if $v_2 > v_1$, i.e., the specific volume increases in the change of phase and L_m is positive, i.e., heat is absorbed in the change, $dp/dT > 0$ and increase in pressure increases the temperature at which the change takes place. The application to the vaporization of water is clear. For the change of phase from liquid water to ice, $v_2 > v_1$ but L_m is negative. Hence $dp/dT < 0$ in this case, leading to the well-known lowering of the melting point of ice with pressure.

The application of thermodynamics to the thermal phenomena associated with change in phase leads at once to the consideration of so-called heterogeneous systems, in which several phases of different substances exist together in equilibrium. Here in many problems the Gibbs' free energy provides the key to the solution of how the equilibrium is maintained. In particular it yields the famous Gibbs phase rule for the connection between the number of chemical constituents in the system and the number of coexisting phases for which equilibrium will be maintained. If c represents the former number and n the latter, the *variance* of the system is

$$\delta = c - n + 2. \qquad (10.141)$$

The variance is the excess of the total number of variables defining the system (i.e., temperature, pressure, and the fractional amounts of each constituent) over the number of equations specifying the equilibrium. A zero value of δ means that equilibrium is maintained for a unique value of each variable defining the system. This is, for example, the case for the equilibrium of ordinary water for the three coexisting solid, liquid, and gas phases ($c = 1$, $n = 3$). If δ is not zero, equilibrium can be maintained for a variety of values of the variables. The case of liquid water in equilibrium with its vapor at once occurs as a simple illustration. It must be emphasized that the phase rule as stated does not contemplate chemical reactions taking place in the equilibrium.[1]

10.11. Thermodynamics of Radiation. Although our discussion of the applications of thermodynamics has been limited largely to gases, it should be clear that the method of this discipline is very general and therefore application should be possible to all physical phenomena. We shall illustrate this by a brief discussion of black body thermal radiation. Imagine a cavity composed of any material whatever, provided only it is able to stand high temperature, and imagine that it communicates with the outside by a hole of dimensions small compared with the size of the cavity. When heated so that the interior walls of the cavity are all at the same temperature the radiation emerging from the opening is known as black body radiation. Let us assume that the radiation inside the cavity may be likened to a gas in equilibrium in the sense that if the temperature is maintained constant the average rate of emission *from* the walls equals the average rate of absorption *by* the walls. We shall think moreover of the radiation in the cavity as possessing a constant energy which depends only on the temperature and the volume. It appears reasonable to suppose that this energy is directly proportional to the volume and hence the energy per unit volume or energy density is dependent on the temperature alone. At any rate, let us make this additional assumption.

If the radiation in the cavity acts like a gas, it should exert a pressure on the surrounding surface. This pressure has indeed been observed and measured.[2] Moreover if the radiation pressure is at all like the pressure exerted by ordinary gas molecules we should expect it to have a relation to the average energy density. Recall in this connection that the pressure in an ideal gas is two thirds the mean total energy of the molecules divided by the volume. One might therefore be tempted to assign to the radiation pressure the value two thirds of the average energy density. However this would be stretching the analogy too far. We are by no means justified in assuming that the radiation behaves like an *ideal* non-atomic gas in which the molecules have kinetic energy of translation only.

[1]For further discussion of the material of this section, cf. M. W. Zemansky, *Heat and Thermodynamics* (New York, McGraw–Hill, 2nd Ed. 1943), Chaps. XV and XIX.

[2]Cf. P. Lebedew, *Ann. der Phys.* 6, 433, 1901. E. F. Nichols and G. F. Hull, *ibid.*, 12, 225, 1903.

Actually it turns out that the correct expression for the radiation pressure is

$$p = E/3V. \tag{10.142}$$

The application of thermodynamic reasoning to the above situation now leads to some interesting consequences. Let us think of the entropy S of the radiation gas as a function of volume and temperature. Then

$$T\,dS = T\left(\frac{\partial S}{\partial V}\right)_T dV + T\left(\frac{\partial S}{\partial T}\right)_V dT. \tag{10.143}$$

But from the first law

$$T\,dS = dE + pdV.$$

Hence

$$dE = T\left(\frac{\partial S}{\partial T}\right)_V dT + \left[T\left(\frac{\partial S}{\partial V}\right)_T - p\right] dV. \tag{10.144}$$

But since from the property of a total differential

$$dE = \left(\frac{\partial E}{\partial T}\right)_V dT + \left(\frac{\partial E}{\partial V}\right)_T dV, \tag{10.145}$$

we must conclude on comparing identical terms in (10.144) and (10.145) that

$$\left(\frac{\partial E}{\partial V}\right)_T = T\left(\frac{\partial S}{\partial V}\right)_T - p, \tag{10.146}$$

and

$$\left(\frac{\partial E}{\partial T}\right)_V = T\left(\frac{\partial S}{\partial T}\right)_V = C_V. \tag{10.147}$$

The last equality follows from the definition of specific heat at constant volume. Here we are more interested in (10.146) which we shall rewrite by the use of the first Maxwell relation in (10.119), namely

$$\left(\frac{\partial S}{\partial V}\right)_T = \left(\frac{\partial p}{\partial T}\right)_V,$$

getting

$$\left(\frac{\partial E}{\partial V}\right)_T = T\left(\frac{\partial p}{\partial T}\right)_V - p. \tag{10.148}$$

Now for the radiation gas

$$\left(\frac{\partial p}{\partial T}\right)_V = \frac{1}{3}\frac{d(E/V)}{dT}.$$

Hence (10.148) becomes

$$E/V = \frac{T}{3}\frac{d(E/V)}{dT} - \frac{1}{3}E/V, \tag{10.149}$$

which when integrated gives

$$E/V = aT^4, \tag{10.150}$$

where a is a constant. It seems reasonable to suppose that the mean energy density of radiation in the cavity is directly proportional to the rate at which energy is being radiated by the walls. More careful study shows that this is so. If we denote the rate of radiation per second per unit area of the black body by $\dot{u}$, we then have from (10.150)

$$\dot{u} = bT^4, \tag{10.151}$$

where b is an appropriately chosen constant. This result is actually the well-known law of Stefan – Boltzmann and has been experimentally verified. It is useful in the determination of the temperature of glowing solids.

It is important to note the role of thermodynamic reasoning in the above illustration for the light it casts on thermodynamic reasoning in general. Before we can apply thermodynamics to the study of radiation, it is necessary to postulate that the radiation is a system characterized by the sort of quantities with which thermodynamics operates. This immediately necessitates *some* specific hypotheses regarding the nature of radiation, at any rate sufficient to enable us to write an equation of state like (10.142). Thermodynamics alone is powerless to help us here although, by suggesting the analogy with a gas composed of particles, it provides a stimulus to further research and indicates hopeful lines of investigation. The ability of thermodynamic reasoning to predict the temperature dependence of black body radiation once the pressure-energy density relation (10.142) is assumed is rather typical. We recall that, although thermodynamics is unable to derive the equation of state of a gas without prior knowledge of the partition function, it is competent to derive many useful consequences of any particular equation of state, e.g., specific heat relations, etc. Moreover, precisely because its concepts are not restricted to any specific system it provides a very helpful method of seeing fundamental analogies between what might be otherwise considered as wholly unrelated phenomena. Its generalizing value can therefore scarcely be overestimated.[1]

PROBLEMS

1. Calculate the virial for an ideal gas due to the impact forces on the walls in the special case in which the gas is confined to a sphere of radius R (i.e., do not use the divergence theorem, but work out the expression directly).

2. Derive eq. (10.8).

3. Show that for a real gas in which the intermolecular forces are attractive but decrease very sharply with increase in intermolecular distance, the contribution of these forces to the virial varies directly as the product of the volume and the square of the molecular concentration.

4. Find the average energy per particle in a collection of independent linear simple harmonic oscillators all of frequency ν and having possible energy values given by $E_j = (j - 1/2)\, h\nu$, where h is Planck's constant and $j = 1, 2, 3, \cdots$. Assume $g_j = 1/n$ for all j.

5. Calculate the partition function for the collection in Prob. 4.

6. Derive the expressions for the Helmholtz free energy and the entropy of the collection in Prob. 4.

7. Generalize the Maxwell distribution formula (10.62) to apply to a collection of independent particles in a uniform gravitational field. Apply the result to the determination of the variation of density with height in the earth's atmosphere.

8. In a canonical distribution of N free particles, find the expression for the number of particles with kinetic energy in the range from E to $E + dE$. Evaluate this number for the special cases where $E = \overline{E} = 3\Theta/2$ and $E = (1.01)\overline{E}$. Take the ratio of the two results and interpret it physically.

9. If α is the volume coefficient of thermal expansion and κ is the isothermal compressibility of a fluid, prove that

[1]Cf. J. W. Gibbs, "On the Equilibrium of Heterogeneous Substances" in Vol. I of *The Scientific Papers of J. Willard Gibbs* (Dover Publications), p. 55, especially the introductory section. For a classic statement of the relation of statistical reasoning to thermodynamics, cf. also the preface to Gibbs' "Elementary Principles in Statistical Mechanics" in Vol. II of *Collected Works*.

$$\left(\frac{\partial \alpha}{\partial p}\right)_T = -\left(\frac{\partial \kappa}{\partial T}\right)_p .$$

How does the compressibility of an ideal gas vary with the temperature? Answer the same question for a Van der Waals gas.

10. Prove that for any reversible cycle the integral of dQ/T around the cycle is equal to zero.

11. Calculate the change in entropy of a mass m of ideal gas in expanding irreversibly to double its volume, it being assumed that there is no change in temperature.

12. If the specific volume of water at 100°C is 1 cm^3/g and that of steam is 1686 cm^3/g, find the change in the boiling point of water produced by lowering the pressure by 10 cm of Hg in the neighborhood of standard atmospheric pressure.

If the difference in specific volume between the liquid and solid phases of water at 0°C is 0.1 cm^3/g, find the depression of the freezing point of water associated with an increase in pressure of one atmosphere.

13. Given one mole of an ideal gas under standard conditions of temperature and pressure. Find the change in free energy if the volume is doubled at constant temperature. Find the change in free energy if the temperature is raised by 1°C while the volume is kept constant.

14. Consider one mole of hydrogen gas under standard conditions (0°C and 760 mm of mercury pressure). Find the number of molecules with speeds between 0 and 100 meters/sec; 100 and 200 meters/sec, and so on up to 4000 meters/sec. Plot the results in the form of a curve and indicate on it the positions of the root-mean-square molecular speed and the most probable speed.

15. Let the effective potential energy of a single molecule in an uncompressed fluid of volume V_0 in equilibrium be denoted by E_0. Then when the volume of the fluid is compressed to V, the corresponding energy can be written

$$E = E_0 + (V - V_0)\left(\frac{dE}{dV}\right)_0 + \frac{(V - V_0)^2}{2} \cdot \left(\frac{d^2E}{dV^2}\right)_0 + \cdots .$$

Note that $\left(\frac{dE}{dV}\right)_0$ is the derivative relating to the equilibrium condition. If equilibrium corresponds to minimum molecular potential energy, what value must be assigned to the second term on the right? Derive the following expression for the bulk modulus of the fluid

$$B = V_0 N \left(\frac{d^2E}{dV^2}\right)_0,$$

where N is the number of molecules in volume V_0. What is the effective value of $\left(\frac{d^2E}{dV^2}\right)_0$ for an ideal gas? Comment on its physical significance in this case.

16. The critical temperature of a substance may plausibly be defined from the kinetic theory point of view as the temperature for which the average kinetic energy per molecule equals the average potential or binding energy. Use the result of Prob. 15 to get an estimate of the order of magnitude of the critical temperature of water on this assumption.

17. The average number of collisions suffered per second by a spherical molecule of diameter D may be calculated to be

$$Z_c = \pi D^2 n \bar{v}_r,$$

where $\bar{v}_r$ is the average *relative* velocity of the molecules and n is the number of molecules per unit volume. Show that on simple kinetic theory in which all the molecules are assumed to move with the *same* speed v_m, one gets

$$Z_c = 4\pi/3 \cdot D^2 n v_m.$$

Show that if the Maxwell distribution equation (10.62) is used one obtains the somewhat altered result

$$Z_c = 4\sqrt{\pi/3} D^2 n v_m.$$

18. Prove that the adiabatic reversible rate of change of temperature with increase of pressure for any substance is

$$\left(\frac{\partial T}{\partial p}\right)_S = \frac{TV\alpha_p}{C_p},$$

where the symbols have their usual meaning in the text. Find the change in temperature when one gram of water is compressed adiabatically by a change in pressure of 500 atmospheres at 0°C, for which $\alpha_p = -67 \times 10^{-6}$ per °C. Do the same for water at 5°C for which $\alpha_p = 15 \times 10^{-6}$ per

°C. Take C_p in the former case as 1.009 cal/g °C and in the latter 1.005 cal/g °C.

19. Calculate the Joule – Kelvin coefficient $\left(\frac{\partial T}{\partial p}\right)_H$ (eq. 10.132) for water at 1 atmosphere and at 0°C and 5°C respectively, using the data in Prob. 18.

20. Show that the equation of state of black body radiation considered as a thermodynamic system has the form

$$p = \frac{a}{3} T^4$$

where a is the constant in eq. (10.150). Use this to determine the equation for the change in entropy of black body radiation with temperature and volume, viz.,

$$dS = 4aVT^2 dT + \frac{4}{3} aT^3 dV.$$

Consider some special cases of this general result.

21. Use the canonical distribution equation (10.60) to obtain the average number of particles per unit volume irrespective of their momentum values.

CHAPTER XI

QUANTUM THEORY

11.1. Classical and Quantum Theory of Particles. In by far the larger part of the preceding chapters our applications of mechanics have been based on the fundamental postulates of Chapter 3. In accordance with these the position and velocity of every particle of a mechanical system (i.e., the *state* of the system) are completely specified once the initial conditions of the motion have been assigned. We have noted to be sure the difficulty of solving the general problem for systems containing more than two particles, but regardless of this, we have seen no reason for placing any restrictions on the nature or magnitude of the initial conditions. In particular, fundamental quantities like momentum, moment of momentum and energy have been allowed to take on *arbitrary* values. A simple pendulum, for example, can have a *continuous* range of energy values depending on its initial displacement and velocity.

We have already seen, however, in Chapter 9 that in the attempt to apply classical mechanics to the particles constituting atoms, difficulties are encountered which have necessitated the imposition of additional conditions on the motion of such particles. Thus the Bohr theory introduced certain *quantum* conditions which serve to define so-called stationary states of atoms (cf. eqs. 9.70 and 9.71). The key idea of these conditions is the existence of a basic constant of nature, called the Planck constant of action and denoted by h. Effectively we may view the existence of the constant h as imposing a certain discreteness on physical phenomena not contemplated in classical physics. We may express this by saying that the classical physics of particles treats the fundamental substance whose behavior is the goal of physical explanation, i.e., matter, as *atomic* in nature but does not associate atomicity with the *motions* of the particles. The quantum theory quantizes not only matter but also all the processes in which matter takes part.

The story of quantum theory is an interesting chapter in the history of physics and it will be worthwhile to review briefly the salient points about its development, since we are particularly concerned in this book with concepts and methods.

Although our principal reference to quantum theory so far has been in connection with the Bohr theory of atomic structure, actually quantum theory did not arise out of any difficulty in particle physics. The problem to which it owes its birth is one in heat radiation. In Sec. 10.11 we discussed briefly the contribution of thermodynamics to the deduction of the Stefan – Boltzmann law of black body radiation. This proceeds from rather general considerations based on the assumption of a particular relation between radiation pressure and energy density in a cavity. But there exist other interesting questions concerning heat radiation besides the relation between the total radiation and the absolute temperature. The most important is the distribution of energy in the spectrum of the radiation produced, e.g., when light from a black body is analyzed by a prism or grating. The energy per unit frequency interval rises from zero at very low frequencies to a maximum and then falls to zero again at very high frequencies. The ordinate for a given frequency rises as the absolute temperature increases, and the maximum for higher temperatures is shifted to higher frequencies. The form of the curve for given temperature is reminiscent of the Maxwell distribution of velocities for the molecules of an ideal gas in equilibrium (eq. 10.67 and Fig. 10.1). This indeed suggests the possibility of accounting for the spectral distribution by some statistical considerations similar to the kinetic theory of gases. Attempts in this direction led directly to the introduction of the quantum theory by Planck. The details are presented in the next section.

11.2. Classical Statistics of Radiation. Rayleigh – Jeans Law. The classical statistical treatment of the equilibrium of black body radiation in a cavity considers the radiation from the standpoint of wave motion in a continuous medium and in fact determines by the application of boundary conditions the number of possible modes of standing wave vibrations which can be set up in a given finite space with frequencies lying in a given interval. We shall investigate this kind of problem in Chapter 14 and shall merely quote the result here.

Imagine the cavity to be a cube of side l and volume V so that $l^3 = V$. If all waves constituting the radiation in the cavity have the velocity c of light in free space, it turns out that the number of allowed standing waves with frequencies lying between ν and $\nu + d\nu$ (if the frequency is as large as that corresponding to visible light) is given by[1]

$$dN = \frac{8\pi\nu^2 V\, d\nu}{c^3}. \tag{11.1}$$

[1] Cf. deduction in Sec. 11.9.

Since we want to calculate the average energy associated with the frequency interval in question, we must know how much average energy is associated with each mode. This was treated historically[1] by analogy with the corresponding problem of the assignment of average energy to molecules of a gas in the kinetic theory. Thus we can interpret eq. (10.17) as associating the average kinetic energy $3kT/2$ to each molecule of an ideal monatomic gas at temperature T. Since a molecule that has translational motion only has three degrees of freedom this is equivalent to assigning $kT/2$ energy per degree of freedom. That this is justified in the kinetic theory can be seen by a re-examination of the derivation of the virial of an ideal gas, where the significance of the factor 3 is made perfectly plain. Reasoning by analogy would then suggest that to each mode of radiation vibration in a cavity in equilibrium at temperature T there is associated the energy $kT/2$. This leaves out of account the possibility that to each mode there should be assigned more than one equivalent degree of freedom. Actually other methods of approach as well as the exigencies of experimental confirmation suggest that to each mode it is more proper to assign two degrees of freedom and hence the average energy kT. If this is done the average energy density associated with the radiation in the frequency interval from ν to $\nu + d\nu$ becomes

$$\frac{(dE)_\nu}{V} = \frac{8\pi\nu^2 kT}{c^3}\, d\nu. \tag{11.2}$$

Using wave length $\lambda = c/\nu$ (cf. Sec. 14.2) in place of frequency enables us to write this in the alternative form

$$\frac{(dE)_\lambda}{V} = \frac{8\pi kT d\lambda}{\lambda^4}. \tag{11.3}$$

This is known as the Rayleigh – Jeans law for the energy distribution in the spectrum of black body radiation. It agrees pretty well with experimental results for large values of λ, but it is clear it cannot agree with experiment for small λ, since as $\lambda \to 0$, it provides infinitely large radiation energy densities. To obtain a law of the type observed experimentally clearly demands a different kind of association of average energy per mode of vibration. This is discussed in the following section.

11.3. Quantum Statistics of Radiation. Planck's Distribution Law. Let us now suppose that black body radiation in a cavity is distributed with respect to energy like a collection of particles, effectively limitless

[1] Cf., e.g., Rayleigh, *Phil. Mag.* (5) 41, 539, 1900.

in number, which unlike the molecules of a material gas, cannot individually take on *all* possible energy values but are restricted to integral multiples of a certain fundamental energy. This fundamental energy is dependent on the frequency of the radiation ν, and in fact it is taken as $h\nu$, where h is a new universal constant of nature of the dimensions of *action*, or energy multiplied by time.

We shall still insist that the radiation gas be distributed canonically in the sense of Secs. 10.3 and 10.4. To find the average energy per light particle we have only to apply the method of eq. (10.38) to eq. (10.48) and get (with the assumption that the weights g_j are all equal)

$$\overline{E} = \frac{\Sigma E_j e^{-E_j/kT}}{\Sigma e^{-E_j/kT}}, \tag{11.4}$$

where we have already replaced the distribution modulus Θ by kT. We are to evaluate the sums in (11.4) by assuming that

$$E_j = (j - 1)h\nu, \tag{11.5}$$

where j runs from unity to a very large number, which we may take effectively as infinity. The sums then become infinite series and we can write

$$\Sigma e^{-(j-1)h\nu/kT} = \frac{1}{1 - e^{-h\nu/kT}} \tag{11.6}$$

from the binomial expansion. If now we differentiate both sides of the above equation with respect to T we obtain

$$\Sigma(j - 1)h\nu \cdot e^{-(j-1)h\nu/kT} = \frac{h\nu e^{-h\nu/kT}}{(1 - e^{-h\nu/kT})^2}. \tag{11.7}$$

Hence from (11.4) $\overline{E}$ is given by

$$\overline{E} = \frac{h\nu}{e^{h\nu/kT} - 1}. \tag{11.8}$$

This is now the average energy to be assigned to each light particle of frequency ν if the radiation is at absolute temperature T. It is seen that if $h\nu << kT$, i.e., at low frequencies or high temperatures

$$\overline{E} \doteq kT, \tag{11.9}$$

which is precisely the assignment suggested by elementary kinetic theory. If however we use the general average in (11.8) and combine it with (11.1), now interpreted to mean the average number of light particles in the cavity with frequency between ν and $\nu + d\nu$, we obtain for the resulting distribution law

$$\frac{(dE)_\nu}{V} = \frac{8\pi h\nu^3 d\nu}{c^3(e^{h\nu/kT} - 1)}. \tag{11.10}$$

This radiation distribution law was first obtained by Planck and is known by his name. It is seen by inspection to provide a radiation energy density which at given T goes to zero both for $\nu \to 0$ and $\nu \to \infty$ and attains a maximum in between at a frequency depending on T. As a matter of fact it fits the experimental distribution curve very accurately provided h is chosen to have the value

$$h = 6.55 \times 10^{-27} \text{ erg sec.}$$

This is the fundamental constant of the quantum theory.

11.4. Successes and Shortcomings of the Classical Quantum Theory. The New Point of View. The success of Planck's quantum point of view in deducing the correct formula for the energy distribution in black body radiation was at first regarded as rather formal because of the unusual and revolutionary character of the fundamental hypothesis. If the theory was to be regarded as really successful and not merely a happy accident, it was evidently necessary that it should predict other experimental laws which failed to fit into the classical picture. The photoelectric effect is a good illustration.

When light falls on a metal plate, it is well known that electrons are emitted. Experiment indicates that the *number* of electrons coming off per unit time increases with the *intensity* of the light, but that the *maximum kinetic energy* is wholly independent of the intensity and varies linearly with the *frequency* in accordance with the law

$$K_{max} = h\nu - h\nu_0, \tag{11.11}$$

where ν_0 is a minimum frequency or threshold below which the emission does not take place. This is the famous Einstein equation which along with the other facts just mentioned strongly suggests the existence of light particles of energy $h\nu$ corresponding to light of frequency ν. The value of h measured from the experimental results described by eq. (11.11) agrees with that given by the black body radiation law.

When X-rays are scattered by the electrons in material substances, a change in frequency is observed (the Compton effect) which is very difficult to explain on the wave theory of this kind of radiation. On the other hand, the quantum theory notion that the radiation consists of light particles of energy $h\nu$ provides a straightforward explanation. For we expect that such a particle in colliding with an electron must satisfy the principles of conservation of energy and momentum. Application of these shows that the energy of the scattered light particle becomes $h\nu'$ where $\nu' < \nu$, while the electron also recoils in a manner experimentally observed in a Wilson cloud chamber. The magnitude of the calculated change from ν to ν' also agrees with experiment. This renders very persuasive the assumption of the existence of light particles, now usually called *photons*, and provides further evidence of the power of the quantum theory.

It is true that the field or wave theory of radiation (cf. Chapter 16) has proved very successful in explaining many well-known properties of light such as diffraction, interference, and polarization, and it is rather puzzling to reconcile this success with the need for the quantum theory of photons exhibited by the photoelectric and Compton effects. All early versions of the quantum theory had to remain content with a dualistic view point in which radiation is treated as wave-like from certain aspects and particle-like from others. Only modern quantum mechanics has been able to remove this unfortunate dichotomy in physical theory.

Probably the most conspicuous of all the successes of the quantum point of view has been in the realm of atomic structure. We have already reviewed the Bohr theory in Chapter 9 and have noted its uncanny agreement with experimental measurements in spectroscopy, particularly with reference to observed emission and absorption frequencies in the spectrum of hydrogen. We need only recall here that Bohr introduced the quantum theory by means of the so-called quantum conditions in which certain action integrals are put equal to integral multiples of Planck's constant h. Here again the same value of h found in the radiation law as well as in the photoelectric effect is found to agree with the most precise spectroscopic data. Bohr was able to extend his theory to give approximate information about the intensity and polarization of spectral lines and hence strengthen confidence in the essential soundness of the quantum point of view.

At the same time, the attempt to extend Bohr's theory to the structure of atoms with more than one electron met with considerable difficulty. As has been indicated in Chapter 9, the broad outlines worked out approximately (self-consistent field), but precise agreement with experimentally determined spectral frequencies failed even in the presumably comparatively simple case of helium with two electrons moving around a nucleus. The fundamental postulates of the theory seemed cogent enough, but it became clear in the early 1920's that the attempt to implement them by means of quasi-mechanical models involving actual electron orbits was doomed to failure. At the same time other difficulties arose, notably in connection with dispersion and with the effect of a magnetic field on spectral lines (Zeeman effect). It is unnecessary to enter into details about these here.

A new approach to the application of quantum theory to atomic structure was made by Heisenberg in 1925. The important methodological characteristic of his theory was the complete abandonment of pictures or models of electrons moving in orbits and their replacement by a purely abstract mathematical structure in which the symbols represent quantities directly observed, e.g., spectral frequencies (or the corresponding atomic energies related to them by the Bohr frequency condition $E_1 - E_2 = h\nu$), and intensities. By making rather arbitrary postulates Heisenberg was able to solve certain atomic problems which the Bohr theory in its original form was unable to handle successfully. This was the origin of the new quantum mechanics. At about the same time the French physicist DeBroglie devised a still different approach stressing the analogy between particle and wave motion. With this he was able to predict successfully the behavior of electrons reflected from simple crystal metal surfaces or transmitted through thin metallic films. This behavior, discovered experimentally by Davisson and Germer and G. P. Thomson, seemed to demand the interpretation that under certain conditions electrons can be diffracted like waves. DeBroglie's theory, called *wave mechanics*, was actually able to assign an equivalent wave length to an electron in agreement with experimental results (cf. Sec. 14.11 and Sec. 17.8). At about the same time Schrödinger showed how to apply the wave mechanics to the determination of the allowed energy levels of atoms, thus solving perfectly generally the quantization problem of Bohr.

A synthesis of the viewpoints of Heisenberg, DeBroglie, and Schrödinger finally led in the hands of Dirac and others to what has come to be called quantum mechanics. We shall attempt in this book to give a brief survey of its methods with a few results. We here meet an interesting pedagogical problem which has methodological connotations. In studying a new discipline like quantum mechanics for the first time, is it the most expeditious plan to study the historical evolution of the theory in all its various phases, or is it more practical to examine the postulates and concepts from a purely formal point of view and show how the conclusions agree with experiment without inquiring too closely into the connection between the principles and those of so-called classical physics? It is possible to take the stand that no one can really understand a new theory without studying the background out of which it has developed. On the other hand, we rarely if ever do this with classical mechanics, probably because we feel that the student may become too readily confused over the changing points of view that finally culminated in the successful theory. Moreover if the purpose in studying a theory is to learn how to use it to get results, it is probably more efficient to study its formal structure and postpone closer inspection of its methodological relation with previous theories to later consideration.

In this book we shall present a survey of quantum mechanics from the formal point of view but shall at the same time try to make clear the distinction between its concepts and those of classical mechanics.

11.5. Concepts and Postulates of Quantum Mechanics. Quantum mechanics is a theory which applies to physical systems of the same kind as those considered in classical physics, namely collections of particles, charged or uncharged. Since classical mechanics with the use of statistics provides a satisfactory description of the behavior of large-scale bodies it would seem evident that quantum mechanics will find its chief field of application in the atomic domain. However, insofar as the purpose of atomic theory is to supply an explanation of many observed large-scale properties like electrical conductivity, elasticity, and magnetism, it is clear that quantum mechanics may be expected to supplement the ordinary statistical mechanics in the treatment of such properties of matter characterizing solid, liquid, and gaseous states.

Now the concepts and postulates of quantum mechanics are different from those of classical mechanics and certainly appear formal and abstract to the uninitiated. But the reader must bear in mind that the notions about falling bodies which Galileo set forth in his studies of mechanics seemed very strange and unfamiliar to his contemporaries. A new point of view demands study if one is to achieve familiarity with it.

Certain terms used in quantum mechanics have a different meaning from that associated with them in classical mechanics. It is essential that this difference be clearly grasped. Thus consider the term "state of a system." If we are talking in classical mechanics about the state of a collection of particles or dynamical system we mean the specification of the position and velocity of every particle in it. Classical theory postulates that from the knowledge of the state in this sense and its temporal changes all the knowable properties of the system can be deduced. We have seen how this works, for example, in celestial mechanics, in the vibrations of collections of particles and in thermodynamic systems considered from the statistical point of view.

Now, although quantum mechanics deals with the behavior of systems of particles, it attaches a quite different meaning to the term "state." It introduces for this purpose a function of the coordinates of the particles ψ (and more generally of the time explicitly, although we shall not consider this generalization here). The function is called a *state function*, in the sense that from the knowledge of ψ certain physical properties of the system may be deduced. For example, if the system is a single particle with three degrees of freedom, so that ψ appears as a function of the form

$$\psi(x, y, z),$$

then the quantity

$$\psi(x, y, z)\psi^*(x, y, z)\, dx\, dy\, dz$$

gives the probability of finding the particle in the volume interval $dx\, dy\, dz$ in the neighborhood of the point (x, y, z), where $\psi^*(x, y, z)$ is the complex conjugate of ψ(i.e., if $\psi = \psi_r + i\psi_i$, then $\psi^* = \psi_r - i\psi_i$). We see immediately that the kind of information which the quantum mechanical state function gives about the system it presumes to describe is quite different from that which the classical mechanics state provides. In particular we note the emphasis on probability: when a system is in a certain state, there exists no certainty that the various particles will be found at definite positions but in general only a certain probability of finding them here. For a system of n particles, of course, ψ becomes a function of the form

$$\psi(x_1, y_1, z_1 \cdots x_n, y_n, z_n),$$

and the analysis grows more elaborate.

Now there are many properties of a system about which we should wish the state function to give information: e.g., position, velocity, momentum, moment of momentum, energy. Such properties are called *observables*. Classical mechanics treats such observables in general as functions of the time to be determined at each instant from equations of motion or from principles like that of Hamilton in combination with arbitrary initial conditions. Quantum mechanics, on the other hand, merely seeks to find the *possible* values of any particular observable which a given system may take on and the probability that when it is in a given state the observable has values ranging throughout a specified interval. Such a program has little meaning in classical mechanics, which, although it frequently restricts a given observable to lie in a specified region, always allows it to take on all real values lying in the interval. For example, in considering a simple harmonic oscillator, if the amplitude (or energy) is given, the range of the displacement coordinate is restricted, but it is assumed that the coordinate at some time or other takes on *all* values of the displacement consistent with the restriction. Moreover, no restriction is imposed on the possible magnitude of the amplitude or energy, except by considerations extraneous to the system itself. Classical mechanics would see no meaning to the assertion that the energy of an oscillator of definite frequency can take on only a *discrete* set of values. But this assertion has a definite significance for quantum mechanics and indeed the application of quantum mechanics to an oscillator is precisely for the purpose of ascertaining the allowed values of its energy. To be more precise we should say this is the business of quantum *statics*. Quantum mechanics also has something to say about how the state of a system changes with time and how the allowed observable values change thereby. This is the special function of quantum dynamics—a more difficult branch of the subject which will not be discussed in this book.[1]

It is now necessary to elucidate the method by which the possible values of a particular observable are obtained. The plan which has been developed appears a somewhat mysterious one, but it has the merit that it works. To each observable is assigned an operator significance, that is a symbol which when placed to the left of the ψ function transforms it into another function of the same coordinates. Such operators may simply be algebraic multipliers or they may be differential operators. Thus the operator associated with the observable "position of a particle on the x axis" is simply the multiplier x itself. On the other hand, the operator corresponding to linear momentum in the x direction is $\dfrac{h}{2\pi i}\dfrac{\partial}{\partial x}$, i.e., a differential operator. Hence if we denote the momentum operator by P we have the identity

$$P\psi \equiv \frac{h}{2\pi i}\frac{\partial\psi}{\partial x}. \tag{11.12}$$

The operator corresponding to kinetic energy of a particle of mass m constrained to move along the x axis is

$$\frac{-h^2}{8\pi^2 m}\frac{\partial^2}{\partial x^2}.$$

The association of operators with observables is an important characteristic of quantum mechanics. Although it seems abstract, actually

[1] Cf. the introductory discussion in Lindsay and Margenau, *Foundations of Physics*, pp. 469 ff. See also Prob. 17 at the end of the chapter.

the notion of attributing physical significance to an operator is not foreign to classical mechanics. Newton's second law says that in order to get the resultant force per unit mass acting on particle in the x direction we must operate on the x coordinate with the differential operator $\frac{d^2}{dt^2}$. The difference of course comes in the fact that in classical mechanics we do not attribute *observable* significance to the operator $\frac{d^2}{dt^2}$.

The reader will next be puzzled over how we are to find the right operator for each observable. The only answer that can be given is the same that we gave to the question: How does one find the "right" forces to use in Newton's second law? We found that this was strictly a matter of hypothesis, of trial and error by which we build up a catalogue of useful functions to serve as forces in mechanical problems. Essentially the same is true of the choice of observable operators in quantum mechanics. We can indeed make use of one valuable device: having picked an operator for a simple observable, momentum, for example, we may be able to construct the operator for an observable which involves momentum, like energy, by observing the relation between the algebraic expressions for the two quantities. Since kinetic energy for *one* dimensional motion in terms of momentum is $p^2/2m$, if we assign $h/2\pi i \cdot \partial/\partial x$ to p, the suggestion is strong that

$$\left(\frac{h}{2\pi i}\right)^2 \cdot \frac{1}{2m} \cdot \left(\frac{\partial}{\partial x}\right)^2 = \frac{-h^2}{8\pi^2 m} \frac{\partial^2}{\partial x^2}$$

should be associated with kinetic energy. Of course the final test is always: Does it work?

The question now is: How does the association of an operator with an observable enable us to assign the *numerical* values of the observable which are allowed by the theory? This is done by the following "cookbook" recipe. If the observable is denoted by the symbol p and its associated operator by P, we write the equation

$$P\psi = p\psi, \tag{11.13}$$

in which $P\psi$ has the meaning already stated, while $p\psi$ is simply p multiplied by ψ. Thus, in general, (11.13) will be a partial differential equation for the state function ψ. As such it has many solutions involving arbitrary functions. Before one can pick out a solution having physical value it is necessary to impose conditions on ψ. These are mathematically similar to the boundary conditions imposed on the solution of the partial differential equations of classical physics like the wave equation for acoustical or electromagnetic waves (cf. Sec. 14.3). These matters will be examined thoroughly in Part II of the present book, but the reader need not have had an advanced course in wave theory to recall that, although waves traveling on an infinite string, for example, can have arbitrary shape, as soon as one restricts attention to a *finite* string fastened rigidly at the ends, the wave motion in the string is limited to the sum of a definite set of harmonic waves with discrete frequencies simply connected with the length of the string. These are the so-called harmonics of the stretched string. The imposition of boundary conditions, in other words, severely limits the possible solutions of the wave equation and moreover picks out of the infinite gamut of frequencies a discrete set (cf. Sec. 14.3). In similar fashion the imposition of restrictive conditions on ψ in (11.13) leads to the result that solutions satisfying these conditions do not in general exist for *all* values of the parameter p, but usually for only a discrete set. These are the so-called characteristic values of the observable, or in simpler terms, the *possible* values which alone can ever be observed. It has become customary to refer to them as the *eigenvalues* of the observables. The corresponding solutions for ψ are called *eigenfunctions* and the states which they describe are called *eigenstates.*

It is now necessary to say something about the nature of the fundamental conditions laid down on the ψ functions. They can be expressed in various forms, but perhaps the most inclusive formulation is to say that ψ must be: (1) a single-valued function of the coordinates, and (2) quadratically integrable, which means that the integral

$$\int \psi^* \psi d\tau$$

must exist, where $d\tau$ represents the element of volume in the configuration space of the system and the integration is taken over the whole configuration space occupied. Thus for a single particle $d\tau = dx\, dy\, dz$. If ψ is to satisfy these conditions, it will have to be suitably continuous and vanish properly at infinity. Moreover at a boundary separating two regions in which particles are subject to different types of forces (e.g., have different potential energies) we shall in general demand that ψ and its derivatives with respect to the coordinates (i.e., slope) be continuous. This is physically reasonable in view of the meaning assigned to $\psi\psi^*$ above.

In Sections 11.6 and 11.7 we shall give illustrations of the use of eq. (11.13) and the associated boundary conditions for the determination of eigenvalues, i.e., the so-called quantization problem.

11.6. Eigenstates and Eigenvalues of Momentum and Moment of Momentum. As the first illustration of eq. (11.13) suppose the observable in question is the ordinary linear momentum in the x direction. The operator associated therewith is $h/2\pi i \cdot \dfrac{d}{dx}$, in accordance with the arbitrary assignment already discussed in the previous section (eq. 11.12). We replace the partial differential operator by the ordinary differential operator, since ψ is here assumed to depend on x alone.

Eq. (11.13) then becomes in this special case the ordinary differential equation

$$\frac{d\psi}{dx} = \frac{2\pi i}{h} p\psi, \tag{11.14}$$

with the solution

$$\psi = Ce^{2\pi i\, px/h}, \tag{11.15}$$

where C is the constant of integration. The question now arises: Will this solution satisfy the conditions laid on state functions in the previous section for *all* real values of p (observables must be real quantities), or will it be necessary to select a certain discrete set for this purpose? Certainly the single-valuedness condition is met for every real p. In applying the integrability condition it is necessary to define the range of the coordinate x. Clearly if x is allowed to take on *all* negative and positive values, it is impossible to satisfy the condition for *any* value of p unless $C = 0$ identically. Hence to have a nonvanishing solution we must restrict the range of variation of x. Let us say that the particle is restricted to $-l \leqq x \leqq +l$. Then

$$\int_{-l}^{+l} \psi^*\psi\, dx = 2lC^*C, \tag{11.16}$$

independent of p. Hence we conclude that within the imposed restriction, all real values of p are possible. These are the eigenvalues of the momentum, which therefore form a *continuous* set. To each one there corresponds a single ψ function from (11.15).

Let us now pass on to another illustration in which the observable is moment of momentum. We shall take it about the z axis, so that classically the expression is $m_z = xp_y - yp_x$, where p_x and p_y denote the component momenta along the x and y axes, respectively. Consequently the suggested operator representation is

$$M_z = \frac{h}{2\pi i}\left(x\frac{\partial}{\partial y} - y\frac{\partial}{\partial x}\right). \tag{11.17}$$

Before writing the fundamental eq. (11.13) for this problem it is advantageous to transform (11.17) to polar coordinates, and express it in terms of the angle θ about the z axis. Using $x = r\cos\theta$, $y = r\sin\theta$, we show finally that

$$M_z = \frac{h}{2\pi i}\frac{d}{d\theta}, \tag{11.18}$$

whence the fundamental equation becomes again an ordinary differential equation, viz.,

$$\frac{d\psi}{d\theta} = \frac{2\pi i}{h} m_z \psi, \tag{11.19}$$

with solution

$$\psi = Ce^{2\pi i m_z \theta/h}, \tag{11.20}$$

of the same form as (11.15). However, the imposition of the basic conditions now leads to different consequences. Since the range of θ lies between 0 and 2π, no limitation is placed on m_z by the integrability condition. But the single valuedness of ψ requires that when θ increases by 2π, ψ shall return to the same value. Hence

$$\frac{2\pi m_z \cdot 2\pi}{h} = 2\pi n,$$

where n is integral. Therefore

$$m_z = \frac{nh}{2\pi}. \tag{11.21}$$

In other words, the moment of momentum about the z axis is restricted to the discrete set formed by integral multiples of $h/2\pi$. It will be recalled that this is precisely the Bohr quantum condition for angular momentum (cf. eq. 9.72). Bohr introduced it as a postulate, but here it appears as a deduction from quantum mechanics.

The corresponding eigenfunctions are

$$\psi = Ce^{in\theta}. \tag{11.22}$$

What can we say about C? Obviously it is arbitrary until further conditions are placed on ψ. It is customary to demand that state functions shall be normalized, viz.,

$$\int \psi^* \psi \, dt = 1. \tag{11.23}$$

In the present case this leads to

$$|C| = 1/\sqrt{2\pi}. \tag{11.24}$$

Hence the probability of finding the rotating or revolving particle in the interval $d\theta$ when it is in an eigenstate of moment of momentum is

$$\frac{d\theta}{2\pi}. \tag{11.25}$$

This is independent of θ, indicating that the probability in question is the same for every unit angular interval. One might indeed have intuitively reached this conclusion.

11.7. Eigenstates and Eigenvalues of Energy. A somewhat more important illustration of the fundamental eq. (11.13) is the quantization of energy. What shall we choose for the energy operator? We have already intimated that in the case of kinetic energy it is plausible to take $p^2/2m$ and replace p by $h/2\pi i \cdot \partial/\partial x$ as usual. We shall generalize this by using the Hamiltonian function H in terms of rectangular coordinates and component momenta and then replacing each quantity by its corresponding operator. The fundamental equation then becomes

$$H\psi = E\psi, \tag{11.26}$$

in which E is the energy and H the Hamiltonian operator. This is the so-called Schrödinger equation, from the inventor of wave mechanics (cf. Secs. 14.11 and 17.8). The solution of (11.26) subject to the conditions imposed on ψ yields the possible values of the energy of the system and the corresponding eigenfunctions. If the system is an atom, we have here the quantum mechanical equivalent of the quantization procedure of Bohr.

Let us take a simple illustration to make the method clear. Consider a single particle of mass m constrained to move along the x axis between $x = 0$ and $x = l$. We suppose the particle is subject to a force such that the potential energy is $V(x)$, and further specialize to the case where $V(x) = \text{constant} = 0$ for $0 < x < l$ while $V(x) = \infty$ for $x < 0$ and $x > l$. This means effectively that the particle is under the action of *no* force between 0 and l but would have to have infinite force exerted to get outside these limits. Looked at from the standpoint of classical mechanics, it is as if there were infinitely high potential barriers at each end of the interval, so that the particle can never get out of the interval, even though it moves quite freely therein.

The Hamiltonian function is

$$H = p^2/2m + V(x), \tag{11.27}$$

and the corresponding Schrödinger equation is the ordinary differential equation

$$\frac{d^2\psi}{dx^2} + \frac{8\pi^2 m}{h^2}[E - V(x)]\psi = 0. \tag{11.28}$$

It is convenient to solve this under the assumption that for $0 < x < l$, $V(x) = V_0$ and for $x < 0$ and $x > l$, $V = V_1$. Having once found the solution we shall let $V_0 = 0$ and $V_1 = \infty$ and ascertain the results of this procedure. Since $E > V_0$ inside the interval, the solution there is in the form

$$\psi = Ae^{ikx} + Be^{-ikx}, \tag{11.29}$$

where

$$k = \sqrt{8\pi^2 m(E - V_0)/h^2}, \tag{11.30}$$

with A and B arbitrary constants. Since $E < V_1$ outside the interval, the solution there is

$$\psi = A'e^{k'x} + B'e^{-k'x}, \tag{11.31}$$

with

$$k' = \sqrt{8\pi^2 m(V_1 - E)/h^2}. \tag{11.32}$$

We are now ready to apply the boundary conditions on ψ mentioned at the end of Sec. 11.5. Continuity of ψ and $\frac{d\psi}{dx}$ at $x = 0$ leads to

$$\begin{aligned} A + B &= A' + B', \\ ik(A - B) &= k'(A' - B'). \end{aligned} \tag{11.33}$$

Now as x becomes indefinitely large negatively the term $B'e^{-k'x}$ will go to infinity unless $B' = 0$. We therefore must put $B' = 0$ to avoid this calamity. But from the second equation in (11.33) we then have

$$ik(A - B) = k'A', \tag{11.34}$$

and if we set $V_1 = \infty$, k' also becomes infinite. The only way to assure the finiteness of the left-hand side of (11.34) is to set $A' = 0$. Consequently the first equation in (11.33) gives

$$A + B = 0, \tag{11.35}$$

which means that ψ must vanish at $x = 0$ to meet the boundary conditions imposed. Precisely the same kind of argument shows that ψ also vanishes at $x = l$.

The solution of the problem of determining the eigenvalues of E for the particle restricted to the interval 0 to l then reduces to (11.35) and the auxiliary condition at $x = l$, namely

$$Ae^{ikl} + Be^{-ikl} = 0. \tag{11.36}$$

This yields

$$\sin kl = 0,$$

or

$$kl = n\pi, \tag{11.37}$$

where n is any integer. The corresponding possible energy values are then

$$E = \frac{n^2h^2}{8ml^2} + V_0, \tag{11.38}$$

in which we are of course interested in the special case in which $V_0 = 0$.

The importance of this result is that the possible energy values form a discrete set, whereas in classical mechanics all values of the energy would be possible. As a matter of fact the very small value of h in c.g.s. units assures us that for m equal to or greater than the mass of the electron and l of ordinary macroscopic dimensions the difference between successive values of E is very small indeed, in fact negligibly so for all practical purposes. The larger l grows, the more nearly continuous the variation in E becomes. This illustrates the way in which quantum mechanical results merge into those of classical mechanics when dimensions approach macroscopic values. The big difference lies of course in the microscopic domain, i.e., when l becomes of the order of magnitude of atomic dimensions (10^{-8}cm).

The value and significance of the Schrödinger equation will become more evident in subsequent sections where further applications are considered.

11.8. Quantum Mechanical Averages and their Significance. The Indeterminacy Principle. We have seen that one of the important problems of quantum mechanics is the evaluation of the possible or allowed values of a given observable. Depending on the observable these may be either continuous or discrete, with the latter forming a characteristic feature o the theory (e.g., the discrete values of the energy). One would expect that the theory should also predict the *particular* value of an observable that results when a "measurement" of this observable is attempted. Certainly classical mechanics provides a method for this, for, in fact, its very definition of *state* implies a definite prediction of the observed values of the state variables. Presumably there should be something corresponding to this in quantum mechanics, and there is, but with this difference: whereas classical mechanics makes precise predictions based on precisely known initial conditions, quantum mechanics limits itself to predicting the *average* value of the observable which will result from a set of repeated observations. Thus its predictions are *statistical* in character. In particular suppose a system is in the state ψ. It is assume that the average value of an observable p, namely $\bar{p}$, will then come out to be

$$\bar{p} = \frac{\int \psi^* P \psi d\tau}{\int \psi^* \psi d\tau}. \tag{11.39}$$

In interpreting (11.39) note that $P\psi$ has the same meaning as in Sec. 11. i.e., it is the result of operating on ψ with the operator P associated wi the observable. Note also that we do *not* assume here that ψ is necessa ily an eigenstate of the observable p. It may well be an eigenstate of an other observable. If it *is* an eigenstate of p, say ψ_k corresponding to eigenvalue p_k, then by (11.13) we have

$$P\psi_k = p_k \psi_k,$$

and (11.39) becomes at once

$$\bar{p} = \frac{p_k \int \psi^* \psi d\tau}{\int \psi^* \psi d\tau} = p_k, \tag{11.40}$$

or the eigenvalue associated with state ψ_k. In this case the predicted value is a definite one. In general this will not be so. In fact, it can be proved that the average $\bar{p}$ is an average over all possible eigenvalues of the observable p with weights depending on the state of the system (cf. Prob. 15 at the end of the chapter). It is only if the state of the sys tem is an eigenstate of p that the average reduces to a precisely predicted value of p.

This raises the interesting question: Does it ever happen that an eigenstate of one observable is simultaneously an eigenstate of another one, so that the certain measurement of a definite value of one observab is always associated with the certain measurement of a definite value of the other? This turns out indeed to be the case provided only that the operators associated with the observables are *commutative*. Thus if P and Q are these operators, this demands

$$PQ = QP, \tag{11.41}$$

which means that if one operates first on a ψ function with Q and then operates on the result with P, the function obtained is the same that would ensue if the operations were performed in the reverse order. We shall not prove this theorem,[1] but shall find it interesting to observe the consequences when P and Q do *not* satisfy the condition (11.41). For example, if P is the operator for component momentum in the x direction and Q that for the x coordinate itself, we have at once

$$(PQ - QP)\psi = \frac{h}{2\pi i}\left[\frac{\partial}{\partial x}(x\psi) - \frac{x\partial\psi}{\partial x}\right] = \frac{h}{2\pi i}\psi. \tag{11.42}$$

Operators which behave in this way are said to be *canonically conjugate*.

As one might guess, if observables corresponding to operators which do not satisfy (11.41) are measured when the system is in a given state it turns out that certainty of measurement of one observable is accompanied by uncertainty in the measurement of the other. This is the celebrated principle of uncertainty or indeterminacy in quantum mechanics, usually associated with the name of Heisenberg. We shall give a demonstration of this important result.

As a measure of the indeterminacy of measurement of the observable p we take the mean square deviation, viz.,

$$\overline{(p - \bar{p})^2} = \overline{(\Delta p)^2} = \int \psi^*(P - \bar{p})^2\psi d\tau \tag{11.43}$$

from the definition (11.39). The precision in the measurement of p will then be proportional to the reciprocal of $\left[\overline{(\Delta p)^2}\right]^{1/2}$. We shall prove a relation between $\overline{(\Delta p)^2}$ and $\overline{(\Delta q)^2}$, where p and q are two observables whose corresponding operators P and Q do not commute.

It will be assumed that the operators P and Q satisfy the relations

$$\begin{aligned} \int\psi^*P\psi d\tau &= \int(P^*\psi^*)\psi d\tau, \\ \int\psi^*Q\psi d\tau &= \int(Q^*\psi^*)\psi d\tau, \end{aligned} \tag{11.44}$$

[1] Cf. Lindsay and Margenau, *Foundations of Physics*, p. 417.

where P^* means the complex conjugate operator to P. Thus if $P = \frac{h}{2\pi i}\frac{\partial}{\partial x}$ then $P^* = -\frac{h}{2\pi i}\frac{\partial}{\partial x}$, etc. Or if P is real, then clearly $P^* = P$. Operators which obey the condition (11.44) are said to be *Hermitian.*

If u and v are any arbitrary integrable functions of the coordinates on which ψ depends, we can form the integral over the whole configuration space

$$\int (\lambda u + v)(\lambda u^* + v^*)d\tau, \tag{11.45}$$

where λ is a real parameter that does not depend on the coordinates. Since

$$\lambda u^* + v^* = \text{complex conjugate of } \lambda u + v,$$

and the product of a complex quantity by its conjugate is always the square of the absolute value of the quantity, it follows that the integral in (11.45) is always greater than zero (or possibly equal to zero in the case in which $\lambda u = -v$, which is a trivial case excluded here). This means that

$$\lambda^2 \int u^*u d\tau + \lambda \int (u^*v + uv^*)d\tau + \int v^*v d\tau > 0 \tag{11.46}$$

for all real λ. But the left-hand side of this inequality is a quadratic form in λ and the condition that it have no real roots is

$$\int u^*u d\tau \cdot \int v^*v d\tau \geqq \frac{1}{4}\left[(u^*v + v^*u)d\tau\right]^2. \tag{11.47}$$

Suppose the system is in state ψ and define

$$u = (P - \bar{p})\psi; \quad v = i(Q - \bar{q})\psi. \tag{11.48}$$

On substitution the inequality (11.47) becomes

$$\int (P - \bar{p})^* \psi^*(P - \bar{p})\psi d\tau \cdot \int (Q - \bar{q})^*\psi^*(Q - \bar{q})\psi d\tau \geqq \tag{11.49}$$

$$\geqq \frac{1}{4}\left[i\int (P-\bar{p})^*\psi^*(Q-\bar{q})\psi d\tau - i\int (Q-\bar{q})^*\psi^*(P-\bar{p})\psi d\tau\right]^2. \qquad (11.49)$$

The employment of the Hermitian condition (11.44) enables us to write this in the form

$$\int \psi^*(P-\bar{p})^2\psi d\tau \cdot \int \psi^*(Q-\bar{q})^2\psi d\tau \geqq \frac{1}{4}\left[\left[\int \psi^*(PQ-QP)\psi d\tau\right]^2, \qquad (11.50)$$

or by virtue of (11.43) as

$$\overline{(\Delta p)^2}\cdot\overline{(\Delta q)^2} \geqq \frac{1}{4}\left[-\int \psi^*(PQ-QP)\psi d\tau\right]^2. \qquad (11.51)$$

It is thus clear that as long as $PQ - QP \neq 0$, i.e., the operators do not commute, we cannot make *both* $\overline{(\Delta p)^2}$ and $\overline{(\Delta q)^2}$ indefinitely small, i.e., we cannot expect to measure both p and q with the same degree of precision. On the other hand, if $PQ - QP = 0$, it is possible for $(\Delta p)^2$ and $(\Delta q)^2$ both to be zero. In particular let us assume that $PQ - QP = h/2\pi i$, i.e., the operators are canonically conjugate. Then (11.51) clearly becomes

$$\overline{(\Delta p)^2}\cdot\overline{(\Delta q)^2} \geqq h^2/16\pi^2, \qquad (11.52)$$

or

$$\sqrt{\overline{(\Delta p)^2}}\cdot\sqrt{\overline{(\Delta q)^2}} \geqq \frac{h}{4\pi}. \qquad (11.52a)$$

Let us examine for a moment the significance of this result. It means that if we measure the observable p when the system is in an eigenstate of p, so that $\overline{(\Delta p)^2}$ is zero, we are able to measure p with precision but q cannot be measured with any precision at all, and vice versa. It is scarcely necessary to stress how fundamentally different this result is from the assumptions of classical mechanics. In order, for example, to specify the state of a system classically at any time it is necessary to know the position and velocity (or momentum) of every particle at some one instant. According to the indeterminacy principle of quantum mechanics, since the position coordinate q and the conjugate momentum p

are canonically conjugate variables, it is impossible to obtain this knowledge with complete precision as to both position and momentum. In other words in quantum mechanics the absolutely precise assignment of initial conditions becomes impossible. Actually this result is of no great practical significance in systems of macroscopic size, for h is a very small number in c.g.s. units. However, it becomes of very considerable significance in atomic problems, wherein the atomic mass is a very small quantity in the same system of units.

The indeterminacy principle (11.52a) applies to any pair of canonically conjugate observables. Although the case of position and momentum is the standard illustration, we ought to note that it is shown in more elaborate treatments of quantum mechanics that the Hamiltonian function and the time are canonically conjugate observables. Hence it is possible to express a relation of the form

$$(\Delta E)_m \cdot (\Delta t)_m \geqq \frac{h}{4\pi}, \tag{11.52b}$$

where $(\Delta E)_m$ is the r.m.s. deviation of the energy of the system from its mean and $(\Delta t)_m$ is the r.m.s. time interval during which the system has the energy E, i.e., a measure of the lifetime of the energy state. This has an important consequence for the spectral lines which result from an atomic transition from one energy state to another. If the average lifetime of the two states involved in a particular transition is large, the $(\Delta E)_m$ values are correspondingly small. This means, from the Bohr frequency condition (Sec. 9.5) that the frequency of the corresponding spectral line is more precisely defined and the line correspondingly sharp. (Cf. Sec. 14.4 for a discussion of this matter from a different point of view.)

It should be pointed out that in many treatments of the indeterminacy principle, it is derived from a consideration of the disturbing influence of a measurement on the system under measurement and hence on the quantity being measured. The formulation presented in the foregoing treatment is more fundamental since it says nothing about the specific nature of the measurement. It is not in essential disagreement with the more conventional viewpoint, however.[1]

11.9. The Free Particle in a Rectangular Box. Distribution of Photons in a Cavity. An important generalization of the eigenvalue problem for the energy of a particle constrained to move in a given finite straight-line

[1]Cf. Lindsay and Margenau, *Foundations of Physics*, pp. 420 f. Also, H. Margenau, *The Nature of Physical Reality*, Chap. 18.

segment, although otherwise free, is obviously the equivalent quantization problem for the free particle in a box.

If the constant potential energy inside the box is V_0 we have for the Schrödinger equation (11.26)

$$\frac{\partial^2\psi}{\partial x^2} + \frac{\partial^2\psi}{\partial y^2} + \frac{\partial^2\psi}{\partial z^2} + k^2\psi = 0, \tag{11.53}$$

where

$$k = \sqrt{\frac{8\pi^2 m}{h^2}(E - V_0)}. \tag{11.54}$$

The solution of (11.53) proceeds by the separation of variables. Thus we assume that

$$\psi = X(x)Y(y)Z(z), \tag{11.55}$$

whence on substitution there results the ordinary differential equation

$$\frac{1}{X}\cdot\frac{d^2X}{dx^2} + \frac{1}{Y}\cdot\frac{d^2Y}{dy^2} + \frac{1}{Z}\cdot\frac{d^2Z}{dz^2} + k^2 = 0. \tag{11.56}$$

Note that the first term on the left is necessarily a function of x alone, the second a function of y alone, and the third a function of z alone; whereas k^2 is a constant. Therefore each term must be equal separately to a constant and the sum of these must equal $-k^2$. If we call the constants $-k_1^2$, $-k_2^2$ and $-k_3^2$, respectively, (11.56) becomes equivalent to the three equations

$$\frac{d^2X}{dx^2} + k_1^2X = 0, \quad \frac{d^2Y}{dy^2} + k_2^2Y = 0, \quad \frac{d^2Z}{dz^2} + k_3^2Z = 0, \tag{11.57}$$

with

$$k_1^2 + k_2^2 + k_3^2 = +k^2. \tag{11.58}$$

The solutions of (11.57) are

$$\begin{aligned} X &= A_x e^{ik_1x} + B_x e^{-ik_1x}, \\ Y &= A_y e^{ik_2y} + B_y e^{-ik_2y}, \\ Z &= A_z e^{ik_3z} + B_z e^{-ik_3z}. \end{aligned} \tag{11.59}$$

Application of the boundary conditions expressing the fact that the particle remains inside the box, assumed to be a rectangular parallepiped with sides l_1, l_2, l_3 and volume $\tau = l_1 l_2 l_3$, can be made in the manner shown to be cogent in Sec. 11.7. Thus we require that X, Y, Z vanish on the boundaries of the box, i.e., $X = 0$ for $x = 0, l$, etc. This gives us the conditions

$$A_x + B_x = 0, \quad A_x e^{ik_1 l_1} + B_x e^{-ik_1 l_1} = 0, \tag{11.60}$$

and two similar pairs of equations for the other two directions. Use of (11.60) for example gives

$$\sin k_1 l_1 = 0, \text{ or } k_1 l_1 = n_1 \pi, \tag{11.61}$$

where n_1 is any integer. The result is that we may not have *any* values of k_1, k_2, k_3 we please but rather that

$$k_1 = n_1 \pi / l_1, \quad k_2 = n_2 \pi / l_2, \quad k_3 = n_3 \pi / l_3, \tag{11.62}$$

where n_1, n_2, n_3 are any set of integers. But this immediately enables us to obtain the allowed or eigenvalues of the energy, viz.,

$$E_{n_1, n_2, n_3} = V_0 + \frac{h^2}{8m}\left[\left(\frac{n_1}{l_1}\right)^2 + \left(\frac{n_2}{l_2}\right)^2 + \left(\frac{n_3}{l_3}\right)^2\right]. \tag{11.63}$$

To each set of integers n_1, n_2, n_3 corresponds one and only one eigenvalue E_{n_1, n_2, n_3} of the energy. Moreover, it is clear that with each energy value there is associated an eigenfunction

$$\psi_{n_1, n_2, n_3} = c \sin\left(\frac{n_1 \pi x}{l_1}\right) \cdot \sin\left(\frac{n_2 \pi y}{l_2}\right) \cdot \sin\left(\frac{n_3 \pi z}{l_3}\right), \tag{11.64}$$

where the constant c is arbitrary but can be evaluated if we *normalize* by assuming that

$$\int \psi^* \psi \, dx \, dy \, dz = 1, \tag{11.65}$$

where the integration is taken over the whole box. This is equivalent to the statement that there is unit probability or certainty of finding the particle *somewhere* in the box. We get thereby

$$c = \sqrt{\frac{8}{\tau}}. \tag{11.66}$$

Although the allowed energy levels are discrete, it is clear that if l_1, l_2, l_3 are of macroscopic dimensions, the various E values are so close together as to be effectively continuous: For large regions, quantum mechanics merges with classical mechanics. The situation is far otherwise, however, when the region in question is of atomic size.

It is an interesting and important problem to find the number of allowed energy values lying between a given energy E and the slightly different value $E + dE$. This is in its general form strictly a problem in the theory of numbers and its precise solution is a difficult matter. However if we restrict our attention to cases in which the integers n_1, n_2, and n_3 are very large, we may safely employ an approximation method. We set up a three-dimensional rectangular lattice in which the sides of the unit cell are $1/l_1$, $1/l_2$, $1/l_3$, respectively, as indicated in Fig. 11.1. The coordinates of every point in the lattice are then in the form n_1/l_1, n_2/l_2, n_3/l_3, respectively, and the distance of the point from the origin of the space is

$$r = \sqrt{\left(\frac{n_1}{l_1}\right)^2 + \left(\frac{n_2}{l_2}\right)^2 + \left(\frac{n_3}{l_3}\right)^2}. \tag{11.67}$$

From (11.67) it is clear that to each point in the lattice there is associated an energy eigenvalue and the problem of finding the required number of energy values is that of counting the corresponding number of lattice points. Note that a change in sign of n_1, n_2, or n_3 will shift the corresponding lattice point from the first octant to another one, but the energy value will not change since the integers are squared. Hence the number of eigenvalues lying in the assigned energy interval is equal to the number of lattice points lying between the octant of the sphere of radius r corresponding to energy E and the octant of the sphere of radius $r + dr$ corresponding to energy $E + dE$. But ultimately each lattice point can be associated with a lattice cell and therefore the number of lattice points in question is also equal to the number of lattice cells contained in the lattice space. The latter is in turn equal

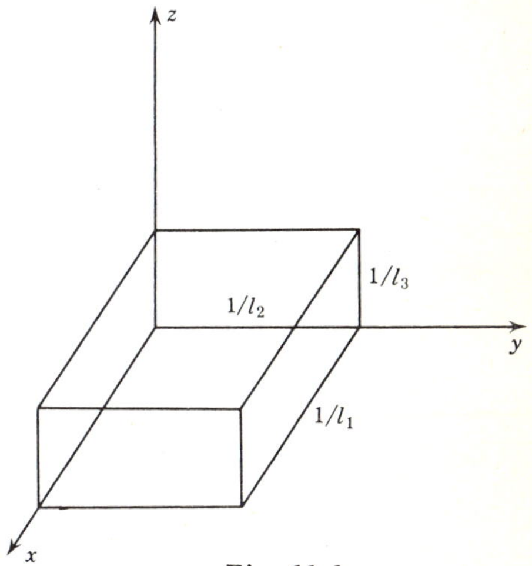

Fig. 11.1.

to the volume of the spherical shell octant divided by the volume of the unit lattice cell, namely, $1/\tau$. The larger τ is, the better is the approximation involved.

Now the lattice shell octant has the volume

$$\frac{4\pi}{8} \cdot r^2\, dr.$$

But (taking $V_0 = 0$ for convenience here) we have from (11.67)

$$r = \frac{2}{h}\sqrt{2mE}, \quad dr = \frac{1}{h}\sqrt{\frac{2m}{E}} \cdot dE. \tag{11.68}$$

Hence the number of energy eigenvalues we are looking for is given at once by

$$dN = \frac{2\pi\tau}{h^3}(2m)^{3/2}E^{1/2}dE. \tag{11.69}$$

This is an important formula. We may apply it, for example, to the evaluation of the number of light particles or *photons* residing in a given cavity with frequencies lying between ν and $\nu + d\nu$. It will be recalled that we needed this in the discussion of Secs. 11.2 and 11.3 leading to the deduction of the formula for the distribution of energy in the spectrum of black body radiation. Let us express (11.69) in terms of the momentum rather than the energy by means of the expression

$$E = p^2/2m. \tag{11.70}$$

Then the number of corresponding momentum values lying between p and $p + dp$ is

$$dN = \frac{4\pi\tau}{h^3}p^2dp. \tag{11.71}$$

Let us now associate with a photon of frequency ν the momentum $h\nu/c$, where c is the velocity of light in free space (cf. Sec. 8.8). Then

$$dN = \frac{4\pi\tau}{c^3}\nu^2d\nu, \tag{11.72}$$

which differs from (11.1) only by the factor 2. (We are using τ in place of V in 11.1.) If we then assume that there are two photons for each given energy value or frequency ν, we can obtain precisely the result used in the derivation of the distribution law. It is clear that our use of (11.69) does demand subsidiary hypotheses, e.g., the association of momentum $h\nu/c$ with a light particle, but these have been amply justified by other considerations (such as, for example, the Compton effect).

11.10. Quantum Mechanical Theory of Metals. An interesting application of the energy quantization of an otherwise free particle confined in a box, as exemplified by the eigenvalues in eq. (11.63), is provided by the modern electron theory of metals. It will be recalled that this attributes the conductivity of metals for heat and electricity to the presence of more or less free electrons which by their motion carry kinetic energy (heat conduction) and electric charge (flow of current). It is natural to try to explain these effects by a kinetic or statistical theory in which the electrons are treated as forming to a first approximation an ideal gas with a Maxwellian distribution of velocities and energies like that already discussed in Chapter 10. It was early discovered, however, that this assumption is in direct contradiction to certain well-known properties of metals. In particular, if the number of effectively free electrons per unit volume is chosen of sufficient size to account for the observed electrical and thermal conductivity, which means in general approximately at least one free electron per atom of the metal, the contribution of the electron gas to the specific heat of the metal will equal that of the atoms themselves (i.e., there will be a kind of superposition of *two* gases) and this will provide a total specific heat about double that actually observed, which seems to be due to the atoms alone. This difficulty remained a puzzle for a relatively long time. After the advent of quantum mechanics, a new form of quantum statistical distribution was developed independently by Fermi and Dirac.[1] This differs from the Maxwellian distribution in that at appropriately low values of the temperature the number of particles per unit energy interval remains practically constant with energy increase up to a certain energy value and then falls off to zero rather rapidly in a fashion approximately Maxwellian. It is then found that for an electron gas in a metal at temperatures less than 10,000°C this *degeneracy*, as it is called, is particularly marked. It means effectively that all the lowest energy states are occupied at ordinary room temperature and considerably above that as well, so that a change in temperature cannot be expected to bring about much change in the overall energy of the electrons. But this means that the contribution of the electrons in such a distribution to the total specific heat of the metal is negligible until the temperature reaches the neighborhood of several thousand degrees centigrade. This solves the specific heat problem just mentioned.

[1]This is discussed in Sec. 11.15. Also cf. Lindsay, *Physical Statistics*, Chap. Chap. VIII.

Another consequence of the Fermi–Dirac quantum statistical distribution of electrons in metals is the fact that even at the absolute zero of temperature the electrons in the metal must possess some energy. In fact we see from eq. (11.63) that, if the number of electrons in each energy state is to be constant (the number is usually chosen as two for reasons we cannot go into here,[1] only two electrons can have precisely zero energy. Those in the higher levels must contribute energy to the total. The result is the so-called *zero point energy* of the metal. We can use eq. (11.69) to calculate this. Since (11.69) gives the number of energy eigenvalues in the energy range dE about the value E, if all energy values from zero up to some upper limit E_{max} are each occupied by *two* electrons and the total number of electrons is N, the value of E_{max} is given by

$$\frac{4\pi\tau}{h^3}(2m)^{3/2}\int_0^{E_{max}} \sqrt{E}\,dE = N. \tag{11.73}$$

Integration yields for the maximum energy value possible for the assembly

$$E_{max} = \frac{h^2}{2m}\left(\frac{3N}{8\pi\tau}\right)^{2/3}. \tag{11.74}$$

Consequently the total energy of the whole degenerate collection of electrons becomes

$$\begin{aligned} E_0 &= \frac{4\pi\tau}{h^3}(2m)^{3/2}\int_0^{E_{max}} E^{3/2}dE \\ &= \frac{3}{10}\frac{Nh^2}{m}\cdot\left(\frac{3N}{8\pi\tau}\right)^{2/3}. \end{aligned} \tag{11.75}$$

This is the zero point energy, which thus appears to depend on the total number of free electrons in the aggregate as well as the number per unit volume.

11.11. The Transmission of Electrons Through Potential Barriers. The photoelectric and thermionic effects involve the transmission of electrons across the interfaces separating metals and the surrounding media. Since the electrons do not cross these surfaces spontaneously to any extent but demand some external agency, e.g., light or temperature increase, to drive them across in large numbers, it is natural to assume that every metal is

[1] Cf. Lindsay, *Physical Statistics*, pp. 188, 203.

surrounded by a barrier of electric potential whose tendency is to prevent the electrons from leaving the metal. This suggests the importance of the quantum mechanical problem of the transmission of electrons through such a barrier. We here present a very simple illustrative case.

Fig. 11.2 represents a potential barrier of height V_0 in the form of an abrupt discontinuity at $x = 0$. We assume that an electron moving in the region to the left of the origin is free and in a field of potential zero; while to the right of the origin it is still a free particle but moving in a field of potential V_0. The Schrödinger equation for the first region is (11.28) with $V(x) = 0$ and with solution (11.29), i.e.,

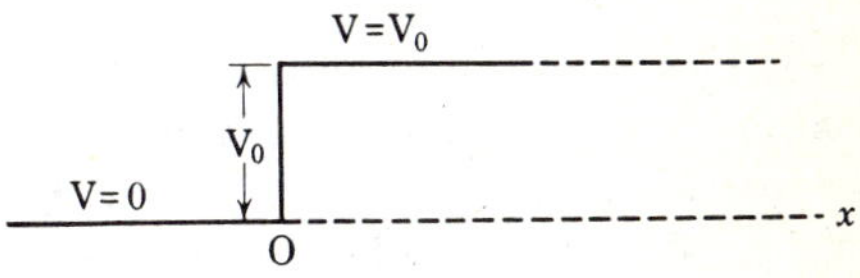

Fig. 11.2.

$$\psi = Ae^{ikx} + Be^{-ikx},$$

where $k = \sqrt{8\pi^2 mE/h^2}$. On the other hand, the Schrödinger equation for the region to the right of the origin is

$$\frac{d^2\psi}{dx^2} + \frac{8\pi^2 m}{h^2}(E - V_0)\psi = 0, \tag{11.76}$$

with solution

$$\psi = A'e^{ik'x} + B'e^{-ik'x}, \tag{11.77}$$

and

$$k' = \sqrt{\frac{8\pi^2 m(E - V_0)}{h^2}}. \tag{11.78}$$

As in Sec. 11.7 the constants A, B, A', B' cannot be completely arbitrary, since it is essential that the ψ functions in the two regions shall match at the barrier and moreover have the *same slope* there. This is demanded by the meaning associated with ψ in the physical problem under discussion (cf. Sec. 11.5). Hence we must satisfy the following equations:

$$A + B = A' + B', \tag{11.79}$$

$$k(A - B) = k'(A' - B'). \tag{11.80}$$

Before we proceed to solve equations (11.79) and (11.80) it is necessary to associate further specific meaning with the A and B parts of the ψ function. We shall assume that Ae^{ikx} is the part of the ψ function connected with electrons moving from left to right, while Be^{-ikx} similarly refers to electrons moving from right to left (reflected from the barrier, as it were). This is rendered somewhat reasonable from the fact that

$$k = 2\pi mv/h, \tag{11.81}$$

where v is the velocity of the electron, and positive k is associated with positive v and negative k with negative v. We shall see a more striking interpretation of (11.81) when we study wave motion in Part II.

Now if the barrier of height V_0 extends all the way to infinity there is no physical reason for the existence of electrons moving from right to left in the region to the right of $x = 0$, since there is nothing to reflect them. This will lead to $B' = 0$, and the equations (11.79) and (11.80) become

$$A + B = A', \quad A - B = k'/k \cdot A'.$$

We can eliminate B between the two and obtain

$$A'/A = 2k/(k + k') = \frac{2\sqrt{E}}{\sqrt{E} + \sqrt{E - V_0}}. \tag{11.82}$$

From the significance of $\psi^*\psi$ (Sec. 11.5) we conclude that

$$|A'|^2/|A|^2 = \frac{4E}{(\sqrt{E} + \sqrt{E - V_0})^2} \tag{11.83}$$

denotes the ratio of the probability of finding an electron to the right of the origin to that of finding it to the left. Now if $\psi^*\psi dx$ denotes the probability of finding an electron inside the interval dx in the neighborhood of given x we can also interpret $e\psi^*\psi/S$ as the *average charge density* if e is the electronic charge, and S is the area of cross section of the metal slab in question. If we multiply this by the velocity v, we obtain the average rate of flow of charge. Therefore $ev|A|^2$ represents the average rate of flow to the right in the region where $V = 0$ and $ev'|A'|^2$ the corresponding flow to the right in the region where $V = V_0$. Here

$$v = \sqrt{\frac{2E}{m}}, \quad v' = \sqrt{\frac{2(E - V_0)}{m}}. \tag{11.84}$$

We shall define

$$\frac{ev'|A'|^2}{ev|A|^2} = D = \frac{4\sqrt{E}\sqrt{E - V_0}}{(\sqrt{E} + \sqrt{E - V_0})^2} \tag{11.85}$$

as the *transmission coefficient* of the barrier. It is equal to unity for $V_0 = 0$ for all E. For $V_0 \neq 0$, D exists as a real quantity only for $E \gtreqless V_0$. As E increases the transmission increases from zero and approaches unity asymptotically. Strictly speaking we should investigate the case $E < V_0$ independently. It leads to a different form of solution from (11.77) above. In fact the reader should show that even though the average charge density is different from zero to the right of the origin there is nevertheless no average flow of charge there since the velocity v' is then imaginary.

The contrast between this quantum mechanical treatment of barrier transmission and that of classical mechanics is rather striking. We recall that in ordinary mechanics if the energy of a particle exceeds by ever so little the height of a potential barrier the particle gets over. In quantum mechanics this is no longer true: there is some "reflection" for all energies and complete transmission is attained only asymptotically for $E - V_0 \to \infty$.

The result just obtained has important consequences for thermionic emission, i.e., the emission of electrons from hot bodies, specifically metallic filaments. It may be recalled[1] that the thermionic current density is given by the Richardson equation and varies with the absolute temperature T of the emitter through an expression of the form

$$J = AT^2 e^{-W/kT}, \tag{11.86}$$

in which W is the important work function or work necessary to remove an electron from the metal. A is a constant which depends on the mass and charge of the electron as well as on h and k (where k here is the Boltzmann gas constant—cf. eq. 10.70 on page 201). The theory on which (11.86) is based neglects the existence of a potential barrier at the surface of the metal. Experimentally it is found that the predicted value of A is too high. The existence of the transmission coefficient D (11.85) leading to reflection of some of the electrons can thus serve as a partial explanation of this discrepancy. This view is further strengthened by the quantum mechanical solution of the transmission through a potential hill of finite width like that shown in Fig. 11.3. In decided contrast with the results of classical mechanics it here turns out that a certain fraction of electrons with energies lying between V_1 and V_0 will penetrate the barrier and give rise to transmission. This is known as the "tunnel" effect, and is useful in studying thermionic emission from filaments with metallic oxide coatings.

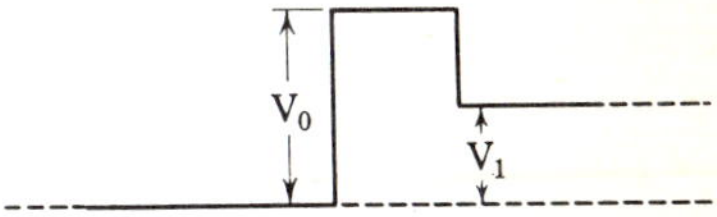

Fig. 11.3.

[1]Cf., e.g., Lindsay, *Physical Statistics*, p. 278.

11.12. Application of Quantum Mechanics to Atomic Structure. Energy Levels of the Hydrogen Atom. As has been mentioned at the end of Sec. 11.4, quantum mechanics was originally developed to provide a more consistent treatment of the problem of atomic structure, and in particular for the more precise prediction of the energy states of atoms, than was possible under the Bohr theory. The method is simply the solution of the fundamental Schrödinger equation $H\psi = E\psi$, using the Hamiltonian function appropriate for the atomic system in question and with the application of the appropriate boundary conditions. Unfortunately the mathematical complexities are considerable as soon as systems of several degrees of freedom are investigated. As an illustration of the method we shall consider a simple case: that of an electron moving along a straight line in an inverse square force field produced by a fixed positive charge. This might be termed the one-dimensional hydrogen type atom.

The Hamiltonian function for such a case is

$$p^2/2m - e^2Z/x, \tag{11.87}$$

where Ze is the charge on the force center and we are assuming attraction (Z is a positive quantity). The electron is assumed to move along the positive x axis. The corresponding Schrödinger equation is

$$-\frac{h^2}{8\pi^2 m}\cdot\frac{d^2\psi}{dx^2} - \frac{e^2 Z}{x}\psi = E\psi, \tag{11.88}$$

which may be more conveniently written

$$\frac{d^2\psi}{dx^2} + \frac{8\pi^2 m}{h^2}\left(E + \frac{e^2 Z}{x}\right)\psi = 0. \tag{11.89}$$

When x is very large, this reduces to

$$\frac{d^2\psi}{dx^2} + \frac{8\pi^2 mE\psi}{h^2} = 0, \tag{11.90}$$

which for positive values of E is an equation of the sort already discussed in Sec. 11.7. However, we must now investigate the situation for *negative* E, i.e., the case in which the electron is bound to the positively charged particle. Let $E = -E'$, where E' is positive. The solution of eq. (11.90) in this case is

$$\psi = Ae^{\sqrt{8\pi^2 mE'/h^2}\,x} + Be^{-\sqrt{8\pi^2 mE'/h^2}\,x}. \tag{11.91}$$

This may be looked upon as the limiting or asymptotic solution of the original equation (eq. 11.88). However, it must be noted that the term with coefficient A is inadmissible since for large x it fails to satisfy the quadratic integrability condition of Sec. 11.5. Hence our asymptotic solution becomes

$$\psi = Be^{-\sqrt{8\pi^2 mE'/h^2}\cdot x}. \tag{11.92}$$

This suggests that the solution of the original Schrödinger equation may be put into the form

$$\psi = ve^{-\alpha x}, \tag{11.93}$$

where

$$\alpha = \sqrt{8\pi^2 mE'/h^2}, \tag{11.94}$$

and v is a function of x. Direct substitution into (11.89) shows that v must satisfy the equation

$$\frac{d^2v}{dx^2} - 2\alpha\frac{dv}{dx} + \frac{8\pi^2 me^2 Z}{h^2 x}v = 0. \tag{11.95}$$

We try to solve this by expanding v in a power series in x. It is clear from the last term that the series must start with a term in x, i.e., be of the form

$$v = x\sum_{j=0} a_j x^j. \tag{11.96}$$

If we substitute (11.96) into (11.95) we find the following recursion formula for the a's:

$$a_{j+1} = \frac{\left[2\alpha(j+1) - \dfrac{8\pi^2 me^2 Z}{h^2}\right]}{(j+1)(j+2)}\cdot a_j. \tag{11.97}$$

Now closer mathematical study than we shall make here[1] indicates that if the series (11.96) is allowed to continue indefinitely it will approach $e^{2\alpha x}$ in its higher terms and hence from (11.93) ψ will approach $e^{\alpha x}$ and become unable to satisfy the quadratic integrability condition as x increases indefinitely. The only way to avoid this occurrence is to have

[1]Cf., e.g., Lindsay and Margenau, *Foundations of Physics*, pp. 438 f.

the series terminate in a polynomial. For this to happen a_{j+1} must vanish for some j. This leads at once to

$$\propto = \frac{4\pi^2 me^2 Z}{(j+1)h^2}, \tag{11.98}$$

or

$$E = -E' = \frac{-2\pi^2 me^4 Z^2}{(j+1)^2 h^2}, \tag{11.99}$$

for $j = 0, 1, 2, 3, \cdots$. These are then the energy eigenvalues for the one-dimensional hydrogen type atom. Comparison with eq. (9.76) shows that these values agree precisely with those obtained by the Bohr quantization method for the three-dimensional hydrogen type atom if $j + 1$ in (11.99) is assumed equivalent to $n_1 + n_2$ in (9.76). As a matter of fact the quantum mechanical treatment of the three-dimensional case, conducted fundamentally along the lines indicated above but necessarily involving greater mathematical complexity because of the additional degrees of freedom, also yields for the energy eigenvalues those in (11.99), although the physical significance of the quantum number j becomes altered. This is taken up in Sec. 11.13.

11.13. The Hydrogen-like Atom in Three Dimensions. The idealized problem of the preceding section gains increased physical significance by its generalization to three dimensions. We now allow the electron three degrees of freedom in the neighborhood of the fixed positive charge and employ the more general Hamiltonian function

$$(p_x{}^2 + p_y{}^2 + p_z{}^2)/2m - e^2 Z/r, \tag{11.100}$$

in which p_x, p_y, and p_z are the x, y, z components of the resultant momentum respectively, and r is the distance from positive charge to electron. The Schrödinger equation now becomes

$$\frac{\partial^2\psi}{\partial x^2} + \frac{\partial^2\psi}{\partial y^2} + \frac{\partial^2\psi}{\partial z^2} + \frac{8\pi^2 m}{h^2}\left(E + \frac{e^2 Z}{r}\right)\psi = 0. \tag{11.101}$$

This partial differential equation is most readily handled by transforming to spherical coordinates r, θ, and ϕ (Sec. 2.2).[1] The result is

[1]Cf. Chap. 12 and in particular Prob. 6 for the method of carrying out the transformation.

$$\frac{1}{r^2}\frac{\partial}{\partial r}\left(r^2\frac{\partial\psi}{\partial r}\right)+\frac{1}{r^2\sin\theta}\frac{\partial}{\partial\theta}\left(\sin\theta\frac{\partial\psi}{\partial\theta}\right)$$

$$+\frac{1}{r^2\sin^2\theta}\frac{\partial^2\psi}{\partial\phi^2}+\frac{8\pi^2 m}{h^2}\left(E+\frac{e^2 Z}{r}\right)\psi=0. \qquad (11.102)$$

As in Sec. 11.9 the solution proceeds by separation of the variables, i.e., we write

$$\psi(r,\theta,\phi)=R(r)\cdot\Theta(\theta)\cdot\Phi(\phi), \qquad (11.103)$$

where $R(r)$ is a function of r alone, etc. If we substitute this into eq. (11.102) and divide through both sides by ψ, we obtain

$$\frac{1}{R}\cdot\frac{d^2R}{dr^2}+\frac{2}{Rr}\cdot\frac{dR}{dr}+\frac{1}{r^2\Phi\sin^2\theta}\cdot\frac{d^2\Phi}{d\phi^2}+\frac{1}{r^2\Theta}\cdot\frac{d^2\Theta}{d\theta^2}$$

$$+\frac{\cot\theta}{r^2\Theta}\cdot\frac{d\Theta}{d\theta}+\frac{8\pi^2 m}{h^2}\left(E+\frac{e^2 Z}{r}\right)=0. \qquad (11.104)$$

If now we multiply through by r^2, we can put the result in the form

$$\frac{r^2}{R}\cdot\frac{d^2R}{dr^2}+\frac{2r}{R}\cdot\frac{dR}{dr}+\frac{8\pi^2 mr^2}{h^2}\left(E+\frac{e^2 Z}{r}\right)$$

$$=-\frac{1}{\Phi\sin^2\theta}\cdot\frac{d^2\Phi}{d\phi^2}-\frac{1}{\Theta}\cdot\frac{d^2\Theta}{d\theta^2}-\frac{\cot\theta}{\Theta}\cdot\frac{d\Theta}{d\theta}. \qquad (11.105)$$

Since this says that a function of r alone is identically equal to a function of θ and ϕ alone, it follows that both sides of the identity must equal the same *constant*. Let us call the constant β. We then have

$$\frac{1}{\Phi}\cdot\frac{d^2\Phi}{d\phi^2}+\frac{\sin^2\theta}{\Theta}\cdot\frac{d^2\Theta}{d\theta^2}+\frac{\sin\theta\cos\theta}{\Theta}\cdot\frac{d\Theta}{d\theta}+\beta\sin^2\theta=0. \qquad (11.106)$$

We notice that here we can separate out the ϕ dependent part from the θ dependent part in a fashion similar to that exemplified by eq. (11.105). The result is

$$\frac{d^2\Phi}{d\phi^2}=-m^2\Phi, \qquad (11.107)$$

$$\frac{d^2\Theta}{d\theta^2}+\cot\theta\cdot\frac{d\Theta}{d\theta}+(\beta-m^2/\sin^2\theta)\Theta=0, \qquad (11.108)$$

where m is a real constant (to be distinguished from the mass of the electron!). The general solution of (11.107) is, of course,

$$\Phi = Ae^{im\phi} + Be^{-im\phi}. \tag{11.109}$$

We recall the restriction of single-valuedness laid on the function (Sec. 11.5). This demands in (11.109) that m be an integer. Eq. (11.108) is clearly more complicated that (11.107). However it has been well known to mathematicians for many years, and the solutions for Θ which satisfy the conditions laid on the ψ function are called the associated Legendre functions[1] usually written in the form $P_l{}^m(u)$, where $u = \cos\theta$ and

$$P_l{}^m(u) = (1 - u^2)^{m/2} \frac{d^m}{du^m} P_l(u). \tag{11.110}$$

Here

$$P_l(u) = \frac{1}{2^l l!} \frac{d^l (u^2 - 1)^l}{du^l} \tag{11.111}$$

is known as the Legendre polynomial of degree l. It is also well known mathematically that the existence of the solution (11.110) demands that β have the value

$$\beta = l(l + 1), \tag{11.112}$$

where l is either zero or a positive integer. For the part of ψ in (11.103) which depends on the angles θ and ϕ we can therefore write

$$\Theta(\theta)\Phi(\phi) = CP_l{}^{|m|}(\cos\theta) e^{im\phi}, \tag{11.113}$$

where m is now permitted to take on negative as well as positive integral values. C is a constant which can be evaluated only by assigning some auxiliary condition.

The completion of the quantization process which enables us tó fix the allowed values of the energy of the atom demands consideration of the equation for $R(r)$, which has the form

$$\frac{d^2R}{dr^2} + \frac{2}{r} \cdot \frac{dR}{dr} + \left[\frac{8\pi^2 m}{h^2}\left(E + \frac{e^2 Z}{r}\right) - \frac{l(l+1)}{r^2}\right] R = 0. \tag{11.114}$$

The method of handling this equation proceeds precisely as in the preceding section for eq. (11.89). We shall not go through the details which are discussed in the standard mathematical texts,[2] but merely note that

[1] Cf., e.g., L. Pauling and E. B. Wilson, Jr., *Introduction to Quantum Mechanics*, (New York, McGraw—Hill, 1935), pp. 125 ff.

[2] Cf., e.g., A. Sommerfeld, *Partial Differential Equations in Physics*, (New York Acad. Press, 1949), pp. 202 f. Also, H. Margenau and Murphy, *Mathematics of Physics and Chemistry*, (New York, Van Nostrand, 1943).

solutions for R satisfying the conditions laid on the ψ function for negative E, corresponding to a bound electron, demand that we satisfy the relation

$$2\alpha(n_r + l + 1) = \frac{8\pi^2 me^2}{h^2}, \tag{11.115}$$

where α has precisely the significance of eq. (11.94) and n_r is zero or a positive integer. Hence the allowed energy values are given by

$$E = -\frac{2\pi^2 me^4 Z^2}{(n_r + l + 1)^2 h^2}. \tag{11.116}$$

This agrees with the result of the simple one-dimensional case discussed in Sec. (11.12) provided the integer j in the simple case is given by

$$j = n_r + l. \tag{11.117}$$

The numbers n_r and $l + 1$ correspond to the radial quantum number and azimuthal quantum number respectively of the old Bohr quantum theory of the hydrogen atom (Sec. 9.5). It is shown in treatises on differential equations[1] that the solution $R(r)$ of eq. (11.114) under the imposed conditions is of the form

$$R(\rho) = C\rho^l e^{-\rho/2} \cdot \frac{d^{2l+1}}{d\rho^{2l+1}} L_{n+l}(\rho), \tag{11.118}$$

where $\rho = 2\alpha r$ and $L_{n+l}(\rho)$ is the so-called Laguerre polynomial of order $n + l$; further $n = n_r + l + 1$ = the principal quantum number. The Laguerre polynomial can be written in terms of the formula

$$L_n(\rho) = e^\rho \frac{d^n}{d\rho^n}(\rho^n e^{-\rho}). \tag{11.119}$$

It is instructive to expand this and note the set of polynomials to which it leads. C, as usual, is a constant.

The complete ψ function for the hydrogen-like atom then becomes

$$\psi(\rho, \theta, \phi) = C\rho^l e^{-\rho/2} \frac{d^{2l+1}}{d\rho^{2l+1}} L_{n+l}(\rho) P_l^{|m|}(\cos\theta) e^{im\phi}. \tag{11.120}$$

The constant C can be evaluated by demanding that ψ be normalized to unity, i.e., that

[1] Or see Pauling and Wilson, *Introduction to Quantum Mechanics*, pp. 121 f.

$$\int \psi\psi^* r^2 \sin\theta\, dr\, d\theta\, d\phi = 1. \tag{11.121}$$

In treatises on quantum mechanics the expression (11.120) is examined in detail.[1] We shall be content to note here merely that for given principal quantum number n, i.e., given quantized energy value E, there exist *many* different ψ functions. In fact it is not hard to show that the number of different functions corresponding to the same energy is equal to n^2. This is called the degree of *degeneracy* of the problem.

It will be recalled from Sec. 11.5 that

$$\psi\psi^* r^2 \sin\theta\, dr\, d\theta\, d\phi$$

denotes the probability of finding the electron in the spherical coordinate volume element $r^2 \sin\theta\, dr\, d\theta\, d\phi$ in the neighborhood of r, θ, ϕ. Multiplied by the charge e this quantity then yields the *average* charge located in this volume. Density plots computed in this way are very useful in studying quantum mechanically the properties of atoms. A simple example will suffice here. Let us take $n = 1$, for which

$$E = -2\pi^2 me^4 Z^2/h^2 = -e^2 Z^2/2a_0,$$

where a_0 is the radius of the smallest Bohr orbit in hydrogen. There exists then but one corresponding function, which is said to describe the *ground* state of the atom. In this case $n_r = l = m = 0$ and consequently (11.120) reduces to

$$\psi = -Ce^{-\alpha r}, \tag{11.122}$$

where now $\alpha = Z/a_0$. Moreover, if we apply (11.121),

$$C = 1/\sqrt{\pi a_0{}^3}. \tag{11.123}$$

Consequently the ground state is characterized by the state function

$$\psi = \frac{-e^{-Zr/a_0}}{\sqrt{\pi a_0{}^3}}. \tag{11.124}$$

The associated average charge density is

$$P = \epsilon/\pi a_0{}^3 \cdot e^{-2Zr/a_0}, \tag{11.125}$$

in which ϵ is used here to denote the charge on the electron,(to avoid confusion with the Naperian base e). This attains its maximum at $r = 0$ and

[1]Cf., e.g., Pauling and Wilson, *Introduction to Quantum Mechanics*, pp. 132 ff.

falls off asymptotically to zero as r increases. The maximum at the origin at which the positive charge is located might appear to constitute a physical inconsistency in terms of any reasonable pictorial representation of quantum mechanics. However the result becomes more understandable if we inquire about the average charge content of a spherical shell of thickness dr at distance r from the origin. This becomes

$$4\pi r^2 P\, dr = 4\epsilon r^2/a_0{}^3 \cdot e^{-2Zr/a_0} \cdot dr. \tag{11.126}$$

When $4\pi r^2 P$ is plotted, we get Fig. 11.4, in which the maximum charge per unit dr comes, not at the origin, but at $r = a_0/Z$. It is interesting to note that this is precisely the radius of the smallest orbit (a circle, of course) in the hydrogen-like atom with nuclear charge Z, in the Bohr theory (Sec. 9.5).

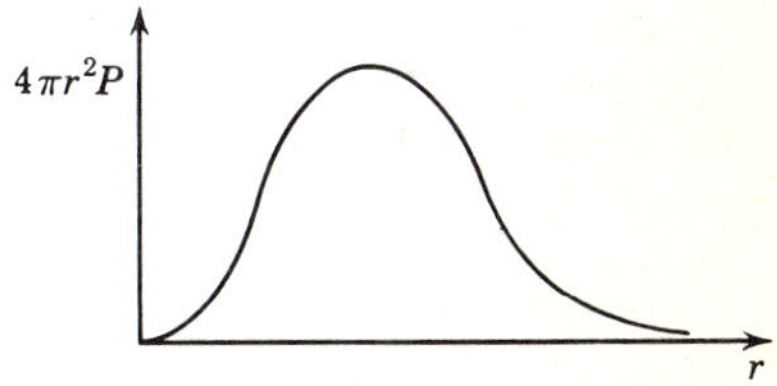

Fig. 11.4.

11.14. Other Illustrations of the Schrödinger Equation. To provide further indication of the physical utility of the Schrödinger equation, we shall find it worth while in this section to list the solutions and corresponding eigenvalues for certain other cases which have yielded to exact treatment. No attempt will be made to go through the analysis leading to the results.

(a) *Rigid rotator and rotator with two degrees of freedom.* The Hamiltonian of a mechanical system with moment of inertia I rotating about a fixed axis is

$$H = M^2/2I, \tag{11.127}$$

where M is the magnitude of the angular momentum or moment of momentum. The Schrödinger equation is

$$\frac{d^2\psi}{d\phi^2} + k^2\psi = 0, \tag{11.128}$$

where ϕ is the angle about the axis of rotation. This is a problem in one degree of freedom. We also have

$$k^2 = 8\pi^2 IE/h^2, \tag{11.129}$$

in which E is the rotator energy. The solution of the equation is

$$\psi = Ae^{ik\phi} + Be^{-ik\phi}, \tag{11.130}$$

where A and B are constants. The eigenvalues of the energy are

$$E_n = n^2h^2/8\pi^2I, \tag{11.131}$$

n being integral. These are the rotational energy levels which we should expect to find for a diatomic molecule. However, the result does not agree with experiment. Rather one needs

$$E_l = l(l+1)h^2/8\pi^2I, \tag{11.132}$$

where l is integral. As might be suspected from the treatment in Sec. (11.13), these are precisely the energy eigenvalues for the rotator with *two* degrees of freedom, i.e., not restricted to rotate about a fixed axis.

(b) *Harmonic oscillator.* For the linear harmonic oscillator of mass m and stiffness k the Hamiltonian is

$$H = p^2/2m + kx^2/2. \tag{11.133}$$

The associated Schrödinger equation readily reduces to

$$\frac{d^2\psi}{dx^2} + \frac{8\pi^2m}{h^2}\left(E - \frac{m\omega^2}{2}x^2\right)\psi = 0, \tag{11.134}$$

in which $\omega = 2\pi\nu$, and ν is the frequency of the oscillator (Sec. 4.2). The requried solution comes out in the form

$$\psi = CH_n(\xi)e^{-\xi^2/2}, \tag{11.135}$$

where $\xi = \sqrt{k/h\nu}\cdot x$, and $H_n(\xi)$ is the Hermite polynomial of order n, representable in the form

$$H_n(\xi) = (-1)^n e^{\xi^2}\frac{d^n(e^{-\xi^2})}{d\xi^n}. \tag{11.136}$$

The corresponding eigenvalues are

$$E_n = (n + 1/2)h\nu, \tag{11.137}$$

differing from the Bohr quantized energy values by the presence of the 1/2. This term indeed has an important physical significance, since it leads to the result that the minimum energy value for the quantized oscillator is not zero but rather $h\nu/2$.

It is easy to extend the preceding results to oscillators in two or three dimensions. Thus for the three-dimensional oscillator with frequencies ν_x, ν_y, ν_z in the three coordinate directions, respectively, the energy eigenvalues are

$$E_{n_x, n_y, n_z} = h\left[\left(n_x + \frac{1}{2}\right)\nu_x + \left(n_y + \frac{1}{2}\right)\nu_y + \left(n_z + \frac{1}{2}\right)\nu_z\right], \tag{11.138}$$

where n_x, n_y, and n_z are any three positive integers or zero, and the corresponding state functions are

$$\psi_{n_x, n_y, n_z} = Ne^{-\frac{1}{2}(\alpha_x x^2 + \alpha_y y^2 + \alpha_z z^2)} \cdot \tag{11.139}$$

$$H_{n_x}(\sqrt{\alpha_x}x)H_{n_y}(\sqrt{\alpha_y}y)H_{n_z}(\sqrt{\alpha_z}z). \tag{11.139}$$

Here

$$\alpha_x = k_x/h\nu_x, \quad \alpha_y = k_y/h\nu_y, \quad \alpha_z = k_z/h\nu_z, \tag{11.140}$$

where the k's are the stiffnesses in the three respective directions.

(c) *Particle in a spherical box.* We have already discussed (Sec. 11.9) the quantum mechanical treatment of a particle in a rectangular box. Here we list the result for a free particle of mass m in a spherical box. If the box is chosen with its center at the origin, the potential energy of the particle is a function of r only, and indeed we have $V(r) = \text{constant} = V_0$ for $r < a$ and $V(r) = 0$ for $r > a$, where a is the radius of the box. If V_0 is negative and equal to $-V_0'$ (V_0' positive) we can think of the box as equivalent to a kind of spherical "well" of finite radius and depth (somewhat difficult to picture, to be sure, in terms of an ordinary cylindrical well in the ground). Particles in the neighborhood of this "well" tend to be attracted to it, just as a particle in the neighborhood of an actual well runs the risk of falling into it. We consider only the case in which the particle has no angular momentum. Then $l = 0$, and the eigenfunctions become: for $r < a$,

$$\psi(r) = A \sin k_1 r + B \cos k_1 r, \tag{11.141}$$

with $k_1 = \sqrt{\dfrac{8\pi^2 m(V_0' - |E|)}{h^2}}$, while for $r > a$,

$$\psi(r) = Ce^{-k_2 r}, \tag{11.142}$$

with

$$k_2 = \sqrt{\frac{8\pi^2 m}{h^2}|E|}.$$

The eigenvalues E cannot be expressed explicitly but are given by the transcendental equation

$$k_1 \cot k_1 a = -k_2. \tag{11.143}$$

It then develops that there is no energy level unless

$$V_0' a^2 > h^2/32m. \tag{11.144}$$

In other words, for a "well" of given depth V_0', this sets a lower limit to the radius a in order that a particle may actually exist in the well.

This type of problem has proved of great value in the study of nuclear energy levels. For example, the energy levels of the deuteron (formed of a neutron and a proton) have been assessed approximately by the use of the spherical "well."[1]

(d) *Collection of non-interacting particles. Pauli exclusion principle.* If we have N identical, indistinguishable particles (e.g., electrons) forming an assembly in which the mutual separations are sufficiently great to permit the neglect of interaction, the Hamiltonian function will have the same form for all, although it will be expressed in terms of the coordinates appropriate for each. The Schrödinger equations for the various particles then become

$$H_j \psi_i(q_j) = E_i \psi_i(q_j), \quad j = 1, 2, 3, \cdots N. \tag{11.145}$$

The quantity q_j is symbolic for all the coordinates of the jth particle and H_j is the Hamiltonian operator for the jth particle. ψ_i is the ith state function for the particle, with E_i the associated eigenvalue. The Schrödinger equation for the whole collection of particles (neglecting interaction) is

$$(H_1 + H_2 + \cdots + H_N)\psi(q_1 \cdots q_N) = E\psi(q_1 \cdots q_N), \tag{11.146}$$

in which

$$E = \sum_{i=1}^{N} E_i. \tag{11.147}$$

It is clear that the number of ψ functions which satisfy (11.146) is very large. A possible one, for example, is evident at once on inspection as

$$\psi = \psi_{i_1}(q_1)\psi_{i_2}(q_2)\psi_{i_3}(q_3)\cdots\psi_{i_N}(q_N), \tag{11.148}$$

where $i_1, i_2, \cdots i_N$ are any *arbitrary* indices. From this, however, we can easily produce new state functions by interchanging particles in the assembly, and the sums of such functions will still be solutions and

[1] Cf., e.g., H. A. Bethe, *Elementary Nuclear Theory*, (New York, Wiley, 1947), pp. 31 ff.

hence state functions also. Pauli[1] introduced a principle to handle this somewhat embarrassing situation. He postulates that if the particles are elementary charged particles like electrons and protons, the only sums of ψ functions of the form (11.148) realized in nature are *antisymmetrical* with respect to an interchange of the coordinates of two particles, meaning that such an interchange produces a change in sign without changing the value. But examination discloses that there is only *one* such antisymmetrical sum and it can be written in the form of the determinant

$$\Psi = C \begin{vmatrix} \psi_{i_1}(q_1)\psi_{i_1}(q_2)\cdots\psi_{i_1}(q_N) \\ \psi_{i_2}(q_1)\psi_{i_2}(q_2)\cdots\psi_{i_2}(q_N) \\ \cdots\cdots\cdots\cdots\cdots \\ \psi_{i_N}(q_1)\psi_{i_N}(q_2)\cdots\psi_{i_N}(q_N) \end{vmatrix}, \tag{11.149}$$

where C is a constant. The interchange of two particles, which means the interchange of two q's yields the interchange of two columns of the determinant and this means merely a change in sign. On the other hand, if two of the particles were to be in the *same* state, it would mean that two of the i subscripts were the same, and this would make two rows of the determinant identical, leading to its vanishing. The Pauli *exclusion principle*, as it is commonly called, therefore implies that in such a collection as we are considering no two particles can be in the same state. This has important consequences for the quantum mechanical theory of atomic structure.[2]

One of these consequences is of such general importance that we ought to mention it briefly here: it is its implication of a type of interaction force between particles of quite different character from the standard forces of classical mechanics (Chapter 3). We have already noted in the foregoing analysis that the simplest state for a system of noninteracting particles (i.e., the solution of eq. 11.146) is represented by the simple product of state functions for the individual particles (eq. 11.148). Of course other solutions are possible by adding such products, since the Schrödinger equation is linear. However it is not difficult to show that only the simple products can represent states for which the particles are wholly independent of each other or what may be termed *uncorrelated*. A natural physical interpretation of absence of correlation in the case of any two particles is as follows.[3] If the ratio of the probability that particle 1 shall be at point P_1 to the probability that it shall be at point P_2

[1]Cf., W. Pauli, *Zs. F. Physik* 31, 765, 1925.

[2]Cf., e.g., Lindsay and Margenau, *Foundations of Physics*, p. 493.

[3]Cf., H. Margenau, *The Nature of Physical Reality*, (New York, McGraw–Hill, 1950), pp. 434 ff.

is independent of the position of particle 2, we may say that 1 and 2 are uncorrelated. On the other hand, if the ratio *is* dependent on the position of particle 2, we shall say that the particles are not independent but correlated. It is clear that if the state functions for the system composed of the pair of particles are simple products like that in eq. (11.148) the lack of correlation is assured. On the other hand, the use of algebraic sums of simple products at once leads to a violation of the criterion. But the Pauli exclusion principle demands the use of such a sum (the determinant in eq. 11.149). Hence the exclusion principle automatically requires that the individual particles shall exhibit correlated behavior even though the Schrödinger equation representing the collection employs a Hamiltonian operator which from classical mechanics corresponds to non-interacting particles. This inconsistency can be resolved in a simple way if we admit into the theory the existence of forces of different type from those which correspond to the Hamiltonian functions of classical mechanics. Closer inspection (cf. Margenau, *loc. cit.*) indeed verifies the validity of this viewpoint and establishes the significance of such forces in quantum mechanics. They are called "exchange" forces to distinguish them from the forces of classical mechanics.

There seems no way consistent with the principles of quantum mechanics of reducing exchange forces to the more usual physical forces. On the other hand, this does not preclude the possibility ultimately of accounting for *all* physical forces as exchange forces. This opens up a fundamental methodological problem which illustrates forcibly the fact that the logical structure of physics is by no means complete.

11.15. Quantum Statistics of Bose–Einstein and Fermi–Dirac. In Sec. 11.2 we pointed out that the statistical distribution of photons in a cavity with respect to their energy must be considered as different from that of the molecules of an ordinary gas, and in Sec. 11.10 we further indicated that the statistical distribution of electrons in an electron gas, as, e.g., in a metal at room temperature is not at all what would be expected from the classical statistics presented in Chapter 10. It will be worth while to consider briefly this interesting situation, which has given rise to the subject known as quantum statistics.

Let us consider first the photon distribution. If in eq. (11.10) we recall that the energy for each photon is $h\nu$, the *number* of photons per unit volume corresponding to the distribution is

$$\frac{dN}{\tau} = \frac{8\pi\nu^2 d\nu}{c^3(e^{h\nu/kT} - 1)}, \tag{11.150}$$

where we here are using τ to denote volume. Now we have already suggested in Sec. 8.8 that to make Fermat's principle of least time consistent with the principle of least action it is necessary to associate with a photon of frequency ν the momentum $h\nu/c$. Calling this p and reverting to E for $h\nu$, we may rewrite (11.150) in the form

$$\frac{dN}{\tau} = \frac{8\pi p^2 dp}{h^3(e^{E/kT} - 1)}. \qquad (11.151)$$

As a statistical distribution formula this certainly does not have the form of the Maxwellian as presented in eq. (10.67). If indeed we express the latter in terms of particle momentum, we have for the fractional number of particles with momentum magnitude lying between p and $p + dp$ and with kinetic energy E

$$\frac{dN}{N} = \frac{4\pi e^{-E/kT} p^2 dp}{(2mkT)^{3/2}}. \qquad (11.152)$$

Ignoring the appearance of h in (11.151), we see that the dependence on E is significantly different in the two formulas. Of course it is true that for $E \gg kT$, the energy dependence in the two cases becomes identical. This corresponds to high frequency or low temperature.

It has turned out that a distribution equation of the form of (11.151) is applicable not merely to photons but also to the molecules in ordinary gases under certain extreme conditions, e.g., very low temperature. The corresponding statistics is referred to as the *Bose – Einstein* statistics in order to distinguish it from the classical Maxwellian variety. As a matter of fact the relevant distribution formula can be derived by an analysis is very similar to that employed in Secs. 10.3 and 10.4 to obtain the Maxwell distribution equation. The expression for the statistical probability w is naturally quite different from that in Sec. 10.4. We shall not go into this here.[1] The discussion of this section is intended to be suggestive only.

In contrast to the Bose – Einstein statistics, that of *Fermi – Dirac* leads to a distribution formula of the form

$$\frac{dN}{\tau} = \frac{8\pi m}{h^3} \cdot \frac{p^2 dp}{Ae^{E/kT} + 1}, \qquad (11.153)$$

for particles of mass m. This should be compared with (11.151) and the significant difference in the sign of the unity in the denominator noted. It will be worth while to give a short derivation of (11.153). Suppose as in Sec. 10.4 we desire to distribute N identical particles in energy intervals or shells with N_j in the jth shell. Let us imagine that the jth shell has n_j

[1] Cf., Lindsay, *Physical Statistics*, Chap. VIII.

cells available for occupancy. In other words, the number of distinct possible energy values associated with the jth shell is n_j. In ordinary classical statistics there is *no* limit to the number of particles that may have a particular energy value. But we now assume that not more than *one* particle can be placed in a given cell. The number of ways in which N_j particles can be placed in n_j cells so that no cell contains more than one particle is equal to the number of ways in which the n_j cells can be divided into two groups, with N_j in one group (those containing *one* particle) and $n_j - N_j$ in the other (those containing *no* particle). This number is simply (Sec. 10.3)

$$\frac{n_j!}{N_j!(n_j - N_j)!}.$$

Consequently the total number of ways of distributing all N objects is

$$w = \Pi_j \frac{n_j!}{N_j!(n_j - N_j)!}, \tag{11.154}$$

where the symbolism means that we must take all products of the form indicated as j runs from 1 to the total number of energy shells. This is the new statistical probability. If we proceed to maximize it subject to the condition that the total number of particles and the total energy remain constant, we finally obtain for the average number of particles $\overline{N_j}$ in the jth shell

$$\overline{N_j} = \frac{n_j}{1 + e^{-\gamma_1} e^{E_j/kT}}. \tag{11.155}$$

where γ_1 is a statistical parameter like λ_1 in Sec. 10.3. If the particles are free, i.e., have kinetic energy only, we can logically take n_j as given by dN in eq. (11.69), and finally have

$$\overline{N_j} = \frac{4\pi\tau\sqrt{2mE_j}\,dE_j}{h^3\left(1 + e^{-\gamma_1} e^{E_j/kT}\right)} \tag{11.156}$$

When the particles are electrons, it appears necessary to assume that the maximum limit on the occupancy of a cell is *two* rather than *one*. In this case the factor 4 in (11.156) should be 8. With this understanding (11.156) can readily be thrown into the form (11.153) with $A = e^{-\gamma_1}$. This type of distribution has proved very successful in the study of an electron gas (as in a metal) at ordinary temperatures. At very high temperatures it reduces effectively to the Maxwellian form.

The derivation of the Fermi – Dirac statistical distribution as given above may well appear unpleasantly mysterious. It has a decided "cook book" air. Fortunately it has proved possible to show that both the Bose – Einstein and the Fermi – Dirac brands of statistics follow directly from the quantum mechanical treatment of assemblies of particles (Sec. 11.14, part d). Interestingly enough it develops that an aggregate of particles whose quantum mechanical state function is *symmetrical* obeys the Bose – Einstein statistics, whereas an aggregate whose state function is *antisymmetrical* follows the Fermi – Dirac statistics.[1] Thus the whole problem of statistical distribution has been integrated directly into the theory of quantum mechanics.

PROBLEMS

1. Use Planck's radiation law (eq. 11.10) to derive the total rate of radiation of all frequencies from a black body as a function of the temperature T. This is the Stefan – Boltzmann law.

2. Use Planck's radiation law to derive the Wien displacement law, i.e., the relation between the temperature T and the wave length λ_{max} for maximum rate of radiation.

3. Calculate $\overline{q}$ and $\overline{q^2}$, where q denotes the displacement from equilibrium of a simple harmonic oscillator, and the averages are defined by means of eq. 11.39. It is to be assumed that the oscillator is the lowest energy eigenstate. Compare these results with the corresponding time averages for the classical oscillator of the same frequency and energy.

4. Calculate $\overline{(\Delta q)^2}$ and $\overline{(\Delta p)^2}$ for the special case of the lowest energy eigenstate of a simple harmonic oscillator, where q and p denote the displacement and momentum observables, respectively (cf. eq. 11.43). Find their product and compare with the general form of the indeterminacy principle.

5. If q and p have their usual significance, find the observable canonically conjugate to q^2. Do the same for p^2 and $p^2 + q^2$.

6. Find the energy eigenvalues for a simple harmonic oscillator of frequency ν.

7. If ψ in (11.15) is multiplied by the factor $e^{i\omega t}$ where $\omega = 2\pi\nu$, and ν is frequency, it becomes a harmonic wave (Sec. 14.2) with wave length

$$\lambda = \frac{h}{p}.$$

[1]Cf., Lindsay, *Physical Statistics*, pp. 188 ff.

This is the celebrated DeBroglie wave mechanical relationship. Use it to derive a form of the indeterminacy principle (11.52).

8. Find the energy eigenvalues for a rigid body rotating about a fixed axis in terms of the moment of inertia about this axis. Discuss its physical inadequacy.

9. Calculate the zero point energies of the alkali metals per gram molecule on the assumption that they have one free electron per atom.

10. A cubical box of unit side with its center at the origin of a rectangular coordinate system contains one free particle. If the particle is in the energy eigenstate corresponding to $n_1 = n_2 = n_3 = 1$ (in eq. 11.64), what is the probability of finding the particle in the region bounded by the planes $x = 0, \frac{1}{4}$; $y = 0, \frac{1}{4}$; $z = 0, \frac{1}{4}$? Compare this with the probability for the region bounded by the planes $x = \frac{1}{4}, \frac{1}{2}$; $y = \frac{1}{4}, \frac{1}{2}$; $z = \frac{1}{4}, \frac{1}{2}$.

11. Calculate quantum mechanically the transmission through a potential hill like that shown in Fig. 11.3.

12. Show that the Planck radiation law reduces to the Rayleigh – Jeans law for sufficiently long wave lengths. Comment on the connection of the temperature with this result. What form does the Planck law take for very short wave lengths? This has been known as the Wien equation.

13. Calculate the total amount of radiation energy density for a black body at 10,000°K in the wave-length interval extending from λ_{max} (cf. Prob. 2) to infinite wave length.

14. The average temperature in interstellar space is now presumed to be about 3°K. What is the average radiation density in this region? From the Einstein mass-energy relation (Sec. 17.5) how much mass density does this correspond to? Compare this with the astronomically estimated average matter density in the universe. In a modern cosmological theory it is assumed that in the initial stages the universe had a temperature of 10^9 degrees K. What average radiation density then prevailed?

15. Let P be the operator corresponding to the observable p and suppose ψ_k is an eigenstate of P, i.e.,

$$P\psi_k = p_k\psi_k.$$

Prove that if ϕ is any other state (not necessarily an eigenstate of P) the average value of p measured in the state ϕ is

$$\bar{p} = \Sigma |a_k|^2 p_k,$$

where

$$a_k = \int \psi^*_k \phi d\tau.$$

Hint. Assume we can expand ϕ in the series

$$\phi = \Sigma a_k \psi_k,$$

and use orthogonal property of ψ_k, i.e.,

$$\int \psi_k \psi_j^* d\tau = 0 \text{ for } k \neq j, = 1 \text{ for } k = j.$$

16. In Prob. 15 $|a_k|^2$ appears as the *probability* of measuring the eigenvalue p_k of the observable p when the quantum mechanical system is in the state ϕ. Confine attention to one degree of freedom (e.g., motion of a particle along the x axis) and assume

$$\phi(x) = Ae^{-(x/l)^2},$$

where A and l are fixed parameters. From the fundamental physical meaning assigned to $\phi(x)\phi^*(x)$ in Sec. 11.5, what is the physical significance of l? Find the probability of measuring the precise value p_k of the *momentum* of the particle when it is in the state $\phi(x)$ written above. Plot the probability distribution as a function of p and comment on its physical significance.

17. The fundamental equation of quantum dynamics that treats state functions which depend on the time is

$$H\psi = -\frac{h}{2\pi i}\frac{\partial \psi}{\partial t},$$

sometimes called the Schrödinger *time* equation. Show that the solution of this equation which reduces to

$$\psi = Ae^{-(x/l)^2}$$

for $t = 0$, becomes in general as a function of t

$$\psi(x, t) = \frac{A}{\sqrt{1 - \dfrac{ht}{2\pi i m l^2}}} \cdot e^{-\dfrac{x^2}{2\left(l^2 - \dfrac{ht}{2\pi m}\right)}}$$

where m is the mass of the particle. Comment on the behavior of the plot of ψ as a function of x as t increases.

18. Prove that the number of different state functions for hydrogen asociated with the eigenvalue corresponding to quantum number n is n^2 (cf. Sec. 11.13).

19. Apply eq. (11.143) to the interaction of a proton and neutron to form a deuteron. Assume $a = 2.8 \times 10^{-13}$ cm and take $|E| = 2.18 \times 10^6$ electron volts. Calculate V_0^{I} to two significant figures. Compare with the value estimated from (11.144).

Part II

FIELD PHYSICS

CHAPTER XII

SCALAR AND VECTOR FIELDS

12.1. Concept of the Continuum. In Part I we studied physical phenomena from the standpoint of the motion of particles subject to certain forces. Even large-scale bodies have been treated essentially as aggregates of particles held together by interaction forces. This is really a heritage of the atomic theory of the Greeks. Now we wish to introduce an alternative point of view which is just as old as the atomic theory and which possesses wide applicability: the concept of the *continuum.* Historically it is interesting to note that many of the early Greek philosophers did not believe in the existence of atoms, but held that observed matter is continuous in the sense that it can be subdivided into parts indefinitely. Many common phenomena associated with solids, liquids, and gases certainly suggest that this view has much to commend it.

Let us therefore imagine a medium or region in which every volume element (save possibly for a discrete set), no matter how small, contains points of the region or material of the medium. This is our physical continuum, and we shall suppose that it occupies a three-dimensional Euclidean space.

What sort of physical quantities are necessary to describe the behavior of a continuum? In the mechanics of particles, the state of the system is determined once the position coordinates are known as functions of the time. Even if generalized coordinates are employed, it is always supposed that these can be expressed in terms of the position coordinates of a system of particles, which may indeed be concealed. The time is therefore the sole independent variable and the Newtonian equations of particle mechanics are fundamentally ordinary in character. This is essentially true also of the Lagrangian and Hamiltonian equations, in spite of the fact that they are written as partial differential equations. The Hamilton-Jacobi equation is another thing again, although the end result is the same: the q's as functions of t.

The situation is different for a continuum in that we must be able to express what is happening at every point at every instant of time. Therefore all physical quantities descriptive of the behavior of a continuum must be functions of space as well as time. As illustrations we can cite the *density* of the medium, the *flow velocity* (i.e., the volume of material flowing across unit area normal to the flow per second), the *pressure* in the case of a liquid, and the *electric and magnetic field intensities* in the case of an electromagnetic field, which is also a continuum, albeit a nonmaterial one. The fundamental equations for the behavior of continua will therefore always be partial differential equations.

12.2. Scalar and Vector Fields. A continuum need not necessarily be a material medium like a fluid or solid. For many purposes it is convenient to treat the region surrounding a mass particle, an electrically charged particle or a magnet as a kind of continuum in the sense that another particle placed at any point therein will be influenced by the first to an extent which in general varies from point to point. Such a region is termed a *field*. It is characterized by quantities varying both in space and time. The most important classification of these quantities is into *scalars* and *vectors*. A vector field is one characterized by a certain vector which is defined for every point of the field at every instant of time. As an illustration consider a gravitational field, due to a distribution of material particles, in which the characterizing quantity is the *field intensity* (or force on a unit mass). On the other hand, this same field may be considered equally well a scalar field with the potential (potential energy per unit mass) as the characterizing quantity. Other illustrations of vector fields are provided by the electro-magnetic field and the flow of fluids, heat, and electrons. Illustrations of scalar fields will be found in the velocity potential of fluids, in temperature, fluid pressure, and density distributions.

The relation between the intensity and potential of a field is a particularly significant one. If we denote the intensity by the vector $\mathbf{F}$, the corresponding potential V (if one exists) is defined by the relation

$$\mathbf{F} = -\left(\mathbf{i}\frac{\partial V}{\partial x} + \mathbf{j}\frac{\partial V}{\partial y} + \mathbf{k}\frac{\partial V}{\partial z}\right). \tag{12.1}$$

A special vector notation has been devised for this by introducing the *vector operator* ∇ (pronounced "del"), where

$$\nabla \equiv \mathbf{i}\frac{\partial}{\partial x} + \mathbf{j}\frac{\partial}{\partial y} + \mathbf{k}\frac{\partial}{\partial z}. \tag{12.2}$$

We therefore write (12.1) in the form

$$\mathbf{F} = -\nabla V. \tag{12.3}$$

∇V is often called the gradient of V or grad V for short. There is an important interpretation of this from differential geometry. Since V is a function of x, y, z, the equation

$$V(x,y,z) = C, \tag{12.4}$$

where C is a constant, defines a family of surfaces, one for each value of C. The direction cosines of the normal to the surface are

$$\frac{\dfrac{\partial V}{\partial x}}{\sqrt{\left(\dfrac{\partial V}{\partial x}\right)^2 + \left(\dfrac{\partial V}{\partial y}\right)^2 + \left(\dfrac{\partial V}{\partial z}\right)^2}}, \quad \frac{\dfrac{\partial V}{\partial y}}{\sqrt{}}, \quad \frac{\dfrac{\partial V}{\partial z}}{\sqrt{}}, \tag{12.5}$$

where the radical is the same in each term. Hence ∇V is a vector directed along the normal to the surface. Moreover, if we denote the change in the position vector $\mathbf{r}$ of the point (x,y,z) by $d\mathbf{r}$ as usual, we have

$$\nabla V \cdot d\mathbf{r} = \frac{\partial V}{\partial x}\, dx + \frac{\partial V}{\partial y}\, dy + \frac{\partial V}{\partial z}\, dz = dV, \tag{12.6}$$

dV being the change in V associated with the displacement $d\mathbf{r}$. If we call dn the magnitude of the element of distance along the normal, we can write

$$\nabla V \cdot \frac{d\mathbf{r}}{dn} = \frac{dV}{dn}. \tag{12.7}$$

But by definition

$$\frac{dx}{dn} = \frac{\partial V}{\partial x} \Big/ \sqrt{}, \text{ etc.} \tag{12.8}$$

Hence ultimately

$$\frac{dV}{dn} = \sqrt{\left(\frac{\partial V}{\partial x}\right)^2 + \left(\frac{\partial V}{\partial y}\right)^2 + \left(\frac{\partial V}{\partial z}\right)^2} = |\nabla V| \tag{12.9}$$

Therefore the magnitude of ∇V is the rate of change of V in the direction of the normal to the surface.

To return to (12.3) and (12.6) we can write

$$\int_{\mathbf{r}_0}^{\mathbf{r}_1} \mathbf{F} \cdot d\mathbf{r} = V_0 - V_1. \tag{12.10}$$

The difference of potential between any two points is thus the work done in the motion of unit mass (or charge in the electrical case) from the one point to the other. Since this depends only on the end points of the path, it is independent of the path. We say that the field defined by V is a *conservative field*. The mathematical condition for the existence of a potential for a given field will be discussed in Sec. 14.7.

12.3. Divergence and the Divergence Theorem. Let $\mathbf{Q}$ be a vector field quantity defined at every point of space and for every instant of time. An important concept is the integral of the normal component of $\mathbf{Q}$ over a closed surface. This is formed by taking the normal component of $\mathbf{Q}$, i.e., Q_n, at each point of the surface, multiplying it by the element of surface at this point and then summing all the products so obtained for the whole surface. This is called the normal flux of $\mathbf{Q}$ through the surface. It is written as

$$\int Q_n dS = \int \mathbf{Q} \cdot d\mathbf{S}. \tag{12.11}$$

Here $d\mathbf{S} = \mathbf{n} dS$, where $\mathbf{n}$ is the unit outward drawn vector normal to the surface and dS the magnitude of the surface element, and of course $Q_n = \mathbf{Q} \cdot \mathbf{n}$, by definition of the scalar product. If $\mathbf{Q}$ is a field intensity, the integral in (12.11) is referred to as a force flux, whereas if $\mathbf{Q}$ is the velocity of flow of a fluid medium, the integral is the net efflux per unit time, i.e., the excess of volume of fluid leaving the volume enclosed by the surface per unit time over that entering the volume.

Let us calculate the normal $\mathbf{Q}$ flux through the surface of the rectangular parallelepiped $\Delta x \; \Delta y \; \Delta z$ shown in Fig. 12.1. The normal flux through the face $OCDG$ is by definition

$$-Q_x \, \Delta y \, \Delta z \, ,$$

if by Q_x we mean the *average* value of the normal component of $\mathbf{Q}$ over the whole face. The minus sign is due to the fact that the normal to the face is directed along the negative x axis. The flux through the parallel face *ABEF* is similarly, neglecting changes in Q_x of order higher than the first,

$$\left(Q_x + \frac{\partial Q_x}{\partial x} \cdot \Delta x\right) \Delta y \, \Delta z.$$

The *net* outward flux through *OCDG* and *ABEF* is therefore

$$\frac{\partial Q_x}{\partial x} \cdot \Delta x \, \Delta y \, \Delta z.$$

Similarly the net normal flux through the faces *OABC* and *GFED* is

$$\frac{\partial Q_y}{\partial y} \cdot \Delta x \, \Delta y \, \Delta z$$

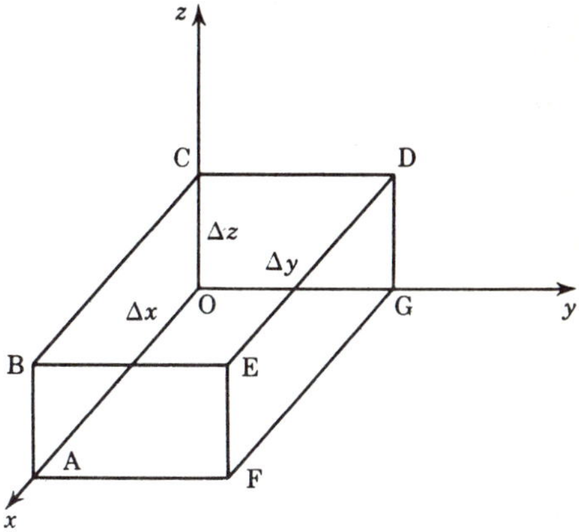

Fig. 12.1

and through the faces *OAFG* and *CBED* is

$$\frac{\partial Q_z}{\partial z} \cdot \Delta x \, \Delta y \, \Delta z \, .$$

Consequently the total net normal flux through all six faces of the volume element is

$$\left(\frac{\partial Q_x}{\partial x} + \frac{\partial Q_y}{\partial y} + \frac{\partial Q_z}{\partial z}\right) \Delta x \, \Delta y \, \Delta z. \qquad (12.12)$$

Referring to (12.2) we see that formally, at any rate,

$$\frac{\partial Q_x}{\partial x} + \frac{\partial Q_y}{\partial y} + \frac{\partial Q_z}{\partial z} = \nabla \cdot \mathbf{Q}. \qquad (12.13)$$

We shall call this expression the *divergence* of **Q** and sometimes write it div **Q.** It is conveniently called "del dot **Q.**" Div **Q** thus appears as the net normal flux of **Q** per unit volume through the surface enclosing the unit volume at any point. If now we take a volume of macroscopic size and divide it into a number of volume elements like that in Fig. 12.1, the integral

$$\int \nabla \cdot \mathbf{Q} d\tau \qquad (12.14)$$

over the whole volume will be the sum of the quantities (12.12) for all the individual volume elements, if $d\tau$ represents the general volume element at the point where $\nabla \cdot \mathbf{Q}$ is taken. Taking into account the fact that the efflux from one face of one volume element is the influx into the interface separating this from an adjacent volume element, we conclude that the integral (12.14) must represent the net normal flux through the whole surface bounding the given volume. But the latter is by definition the surface integral (12.11) over the surface bounding the volume. We therefore have

$$\int \nabla \cdot \mathbf{Q} d\tau = \int \mathbf{Q} \cdot \mathbf{n} dS. \tag{12.15}$$

This identification of the surface integral of the normal component of $\mathbf{Q}$ over the surface bounding a given volume with the volume integral of $\nabla \cdot \mathbf{Q}$ over the volume bounded by the surface is called the *divergence theorem*. The above intuitive discussion cannot be considered a proof of the theorem, but it does show clearly the physical significance of the result, which is confirmed by more careful mathematical reasoning.[1]

Applications of (12.15) are very numerous in theoretical physics. Consider, for example, a continuous fluid medium which is indestructible and uncreatable. Let $\mathbf{Q} = \rho \mathbf{v}$, where ρ is the density and $\mathbf{v}$ the fluid velocity. Here $\mathbf{Q}$ represents the rate at which fluid mass flows per second through a unit area placed normal to the direction of flow. Then the surface integral on the right-hand side of (12.15) represents physically the excess of efflux of fluid mass out of over influx into the region surrounded by the surface. Since the fluid is assumed to be continuous, indestructible, and uncreatable, this excess must equal the rate at which mass is decreasing in the region. This in turn is measured by the rate of decrease of density, since the volume is constant. The result of this physical consideration can be expressed by the relation

$$\int (\rho \mathbf{v}) \cdot \mathbf{n} dS = -\int \dot{\rho} d\tau. \tag{12.16}$$

[1]Cf. P. Franklin, *Treatise on Advanced Calculus* (Dover), p. 384. (Here the theorem is called Green's theorem.) Also Gillespie, *Integration* (University Math. Texts), p. 64. (Here the theorem is called Gauss' theorem.)

The combination of (12.15) and (12.16) leads to

$$\nabla \cdot (\rho \mathbf{v}) = -\dot{\rho}, \tag{12.17}$$

which is called the *equation of continuity* for a fluid. It may be looked upon as an expression of the *conservation of mass* for the fluid. For an *incompressible* homogeneous fluid it takes the special form

$$\nabla \cdot \mathbf{v} = 0. \tag{12.18}$$

Finally, if there exists a velocity potential $\phi(x, y, z, t)$ such that

$$\mathbf{v} = \nabla \phi, \tag{12.19}$$

eq. (12.18) becomes

$$\nabla^2 \phi = 0. \tag{12.20}$$

This is known as *Laplace's equation* and will be encountered again in what follows.

In another application of (12.15) $\mathbf{Q}$ is the intensity of the gravitational field due to a collection of material particles of masses m_1, m_2, . . . m_k. If we reflect that the intensity of the field at distance $\mathbf{r}$ from m_j ($\mathbf{r}$ being directed from m_j to the point in question) is

$$\mathbf{F}_j = -\frac{Gm_j}{r^3}\mathbf{r}, \tag{12.21}$$

and form the surface integral in (12.15) over a surface completely surrounding the mass particles, it follows that

$$\int \mathbf{F} \cdot \mathbf{n} dS = -4\pi G M, \tag{12.22}$$

where

$$M = \sum m_j = \text{total mass within the surface.}$$

This is often termed the *Gauss force flux* theorem. If the distribution of mass is continuous, we can write the mass in terms of the density thus

$$M = \int \rho d\tau. \tag{12.23}$$

Then the divergence theorem (12.15) leads to the differential equation

$$\nabla \cdot \mathbf{F} = -4\pi G\rho. \tag{12.24}$$

Since the field in this instance is conservative we may write **F** in the form (12.3), and (12.24) becomes

$$\nabla^2 V = 4\pi G\rho, \tag{12.25}$$

which is called Poisson's equation. A similar equation holds for a distribution of electrical charge. Of course, in the region in which $\rho = 0$, Poisson's equation reduces to Laplace's equation. It should be emphasized that the Gauss force flux theorem is essentially a geometrical consequence of the inverse square law and hence possesses a generality transcending particular physical phenomena.

12.4. Potential Calculations. Applications and Properties of Laplace's and Poisson's Equations. A simple application of Laplace's equation is to the field due to a single material or charged particle. For convenience let us discuss the former, although everything deduced applies with little change to an electrical charge. Now $\nabla^2 V = 0$ is a partial differential equation of the second order with an infinite number of solutions involving arbitrary functions. It would seem to be a rather difficult matter to track down just the right solution for the problem at hand. We must introduce some purely physical assumptions. The first is that the field about the mass is spherically symmetrical. Hence V is a function of r alone. Laplace's equation then becomes

$$\frac{d}{dr}\left(r^2 \frac{dV}{dr}\right) = 0, \tag{12.26}$$

with the solution

$$r^2 \frac{dV}{dr} = C_1, \tag{12.27}$$

where C_1 is an arbitrary constant which must be evaluated to satisfy the physical conditions of the problem. In this case we know that $-\nabla V$ is the intensity of the gravitational field at distance r from the particle of mass m. Hence from the Newtonian law of gravitation,

$$-\frac{dV}{dr} = -|\nabla V| = -\frac{GM}{r^2}.$$

Consequently from (12.27) we must have $C_1 = GM$. Another integration yields

$$V = -\frac{GM}{r} + C_2. \tag{12.28}$$

The second constant of integration must remain arbitrary until we have specified the potential at some point. It is customary here to set $V = 0$ at $r = \infty$. Hence $C_2 = 0$.

The next problem is the evaluation of the potential in the field of a spherical shell of radius R and mass M. We first determine the potential at a point P *outside* the shell and distant r from the center. Once more we go back to eq. (12.27) and try to evaluate C_1 to fit the conditions of the problem. The intensity of the field about the shell is most simply determined by the use of Gauss' force flux theorem (12.22). Utilizing the spherical symmetry property and taking as the Gaussian surface a sphere with center at the center of the shell and passing through the point P, we arrive once more at

$$-\frac{dV}{dr} = -\frac{GM}{r^2}, \tag{12.29}$$

and hence

$$V = -\frac{GM}{r}, \tag{12.30}$$

putting $C_2 = 0$ again by the same method as before. The potential at a point outside a shell is the same as if all the mass were concentrated at the center.

Now consider an *inside* point. Here the use of Gauss' theorem leads to $C_1 = 0$ in (12.27), or the intensity of the gravitational field at any point inside a shell vanishes. Hence further integration yields

$$V = C_2, \tag{12.31}$$

i.e., the potential at any point inside the shell is constant. From the requirement of continuity we conclude that $C_2 = -GM/R$, which is the value of the potential on the surface of the shell.

The reader is advised to apply similar reasoning to the evaluation of the potential at points inside and outside a solid homogeneous sphere or one in which the density is a function of r only. It develops that the potential at a point outside such a solid sphere is that produced by a single particle with mass equal to that of the sphere and placed at the center of the sphere. Historically this is one of the most celebrated problems of theoretical physics. It was necessary to confirm this result of the Newtonian law of gravitation in order to justify Newton's original (1666) theoretical calculation of the acceleration of gravity on the surface of the earth from the application of the inverse square law to the observed motion of the moon. It was not until 1685 that he was able to prove the theorem in question and hence feel assured that his previous calculation was logically sound.

A question immediately arises with respect to problems of the type just considered. How do we know that the solutions of Laplace's equation we have found are unique? Might there not be other solutions equally valid which would lead to different values of the potential at the same distance from the center of the shell, for example? Physically this does not appear reasonable, but it is satisfying, at any rate, to be able to prove that if a solution of Laplace's equation can be found which satisfies the imposed boundary conditions, this solution is *unique*. Suppose that two different solutions V_1 and V_2 were to exist satisfying the same boundary conditions. This means that

$$\nabla^2 V_1 = \nabla^2 V_2 = 0$$

everywhere in the given region where $\rho = 0$, and that $V_1 = V_2$ on all the surfaces of the masses distributed in this region. Let $V = V_1 - V_2$. Since

$$\nabla \cdot (V\nabla V) = (\nabla V) \cdot (\nabla V) + V\nabla^2 V \qquad (12.32)$$

by direct expansion, it follows that

$$\int (\nabla V) \cdot (\nabla V) d\tau = \int \nabla \cdot (V\nabla V) d\tau - \int V\nabla^2 V d\tau, \qquad (12.33)$$

where the volume integration is taken over the whole region in question. Since $\nabla^2 V = 0$ throughout this region, the second integral on the right vanishes. By the use of the divergence theorem the first integral on the right can be transformed to a surface integral over all surfaces bounding the region. Hence (12.33) becomes

$$\int (\nabla V) \cdot (\nabla V) d\tau = \int (V\nabla V) \cdot \mathbf{n} dS. \qquad (12.34)$$

But $V = 0$ over all these surfaces by fundamental assumption. Hence

$$\int (\nabla V) \cdot (\nabla V) d\tau = 0. \tag{12.35}$$

Since $(\nabla V) \cdot (\nabla V)$ is a positive quantity everywhere, this can be true only if identically

$$\left(\frac{\partial V}{\partial x}\right)^2 + \left(\frac{\partial V}{\partial y}\right)^2 + \left(\frac{\partial V}{\partial z}\right)^2 = 0, \tag{12.36}$$

which in turn demands that

$$\frac{\partial V}{\partial x} = \frac{\partial V}{\partial y} = \frac{\partial V}{\partial z} = 0. \tag{12.37}$$

Hence V is a constant in space. Therefore, since it vanishes on all the surfaces it must vanish everywhere. Hence $V_1 = V_2$ everywhere and the two solutions really are identical.

We now give a little consideration to Poisson's equation. One of the most interesting applications is to the space charge-limited current due to the flow of electrons in a vacuum tube. By the use of Coulomb's law the procedure followed above in eqs. (12.21) to (12.25) leads, for distributions of electric charge, to Poisson's law in the form

$$\nabla^2 V = -4\pi\rho, \tag{12.38}$$

where ρ is here the effective charge density. Since electrons are discrete charges, it might be thought that a continuum or field equation would not apply. However, the electron is so small and the number per cm^3 under the conditions considered here so great that we are justified in interpreting the charge as continuously distributed. To simplify the problem, let us assume that the electrons are allowed to move in one direction only (the x axis) in a cylinder of cross-sectional area A. We further assume that the electrons are emitted by a cathode (at potential zero because it is grounded) and move toward the anode distant l from the cathode and maintained at a positive potential V_0. Let the potential at distance x from the cathode be V. Then the potential energy there is $-eV$, if e is the *magnitude* of the electronic charge. If v_0 is the average velocity of the electrons as they are emitted from the cathode and v the corresponding velocity at x, conservation of energy demands

$$\frac{1}{2} m v_0^2 = \frac{1}{2} m v^2 - eV, \tag{12.39}$$

or

$$v = \sqrt{\frac{2eV}{m} + v_0^2}. \tag{12.40}$$

Now if n is the average number of electrons per unit volume at x, we have $\rho = -ne$. At the same time the current density is

$$j = nev. \tag{12.41}$$

Consequently Poisson's equation (12.38) becomes

$$\frac{d^2V}{dx^2} = 4\pi j\left(\frac{2eV}{m} + v_0^2\right)^{-\frac{1}{2}}. \tag{12.42}$$

This is in readily integrable form. If j is constant one integration yields

$$\frac{1}{2}\left(\frac{dV}{dx}\right)^2 = \frac{1}{2}\left(\frac{dV}{dx}\right)_0^2 + \frac{4\pi jm}{e}\left[\left(v_0^2 + \frac{2eV}{m}\right)^{\frac{1}{2}} - v_0\right]. \tag{12.43}$$

Here $-\dfrac{dV}{dx}$ is the field intensity at x and $-\left(\dfrac{dV}{dx}\right)_0$ is that at the cathode. The second integration of (12.43) to find V as a function of x is difficult. However, a reasonable approximation can be obtained by assuming that $\left(\dfrac{dV}{dx}\right)_0 = 0$ very close to the cathode and that $v_0^2 \ll \dfrac{2eV}{m}$. The equation then takes the approximate form

$$\frac{dV}{dx} = \sqrt{\frac{8\pi jm}{e}}\left(\frac{2eV}{m}\right)^{\frac{1}{4}}, \tag{12.44}$$

and integration gives

$$V = (9\pi j)^{2/3}\left(\frac{m}{2e}\right)^{1/3} x^{4/3}, \tag{12.45}$$

for the variation of potential along the tube. This satisfies the condition that $\left(\dfrac{dV}{dx}\right)_0 = 0$. Since practically $V = V_0$ for $x = l$, we can solve (12.45) for j and obtain

$$j = \frac{\sqrt{2e/m}}{9\pi} \cdot \frac{V_0^{3/2}}{l^2}, \tag{12.46}$$

which is the Langmuir-Child equation for the current-voltage relation in a vacuum tube. The departure from Ohm's law is noteworthy. In practical applications the cathode is a cylindrical wire. Poisson's equation must then be written in cylindrical coordinates. However, the analysis in essence proceeds similarly to that just given (cf. Prob. 13).

12.5. Poisson's Equation in Atomic Structure. Thomas-Fermi Model. An interesting modern application of Poisson's equation arises in connection with the attempt to calculate approximately the potential of the field due to the aggregate of electrons surrounding the nucleus of a neutral atom. Although the number of such electrons does not exceed 100 (Uranium has 92), considerable success has attended the statistical treatment of the aggregate as similar to a collection of gas molecules. As indicated in Sec. 11.15 the proper statistics to be employed is that of Fermi-Dirac. It develops that in this case the average number of electrons n per unit volume is the following function of the potential v of the field due to the electrons:[1]

$$n = \frac{8}{3\sqrt{\pi}h^3} \cdot (2\pi me)^{3/2} v^{3/2}. \tag{12.47}$$

If we treat this as an effectively continuous distribution of charge density $\rho = -ne$, where e is the charge on the electron, Poisson's equation becomes

$$\nabla^2 v = \frac{32\sqrt{\pi}}{3h^3} \cdot (2\pi me)^{3/2} e v^{3/2} = C v^{3/2}. \tag{12.48}$$

It is customary to assume spherical symmetry, whence eq. (12.48) becomes

$$\frac{d^2 v}{dr^2} + \frac{2}{r}\frac{dv}{dr} = C v^{3/2}. \tag{12.49}$$

This must be solved subject to the boundary conditions

$$\lim_{r\to 0} v = Ze/r; \qquad \int n d\tau = Z, \tag{12.50}$$

[1] Cf. Lindsay, *Physical Statistics*, pp. 226f.

where Z is the atomic number of the nucleus. It is assumed of course that the atom is neutral. The solution of (12.49) subject to these conditions has been obtained numerically by Fermi.[1] We shall not discuss the solution here. It may be pointed out, however, that the resulting charge distribution (n or ρ) is very useful in applications of atomic structure to solid state physics.

12.6. Green's Theorem. Let us apply the divergence theorem (12.15) to the vector function

$$\mathbf{Q} = U\nabla V, \tag{12.51}$$

where U and V are mathematically well-behaved scalar point functions (e.g., potential functions). We get (cf. eq. 12.33)

$$\int \nabla U \cdot \nabla V \, d\tau + \int U\nabla^2 V \, d\tau = \int U \frac{\partial V}{\partial n} dS. \tag{12.52}$$

This form of the divergence theorem is usually known as a form of Green's theorem.[2] It possesses many interesting applications. Thus suppose the continuum being described is a homogeneous incompressible fluid with a velocity potential $\phi(x,y,z,t)$ (cf. eq. 12.19) and assume that

$$U = V = \phi. \tag{12.53}$$

For such a fluid $\nabla^2\phi = 0$ everywhere. Hence over any finite volume of the fluid we have from (12.52)

$$\int \nabla\phi \cdot \nabla\phi d\tau = \int \phi \frac{\partial \phi}{\partial n} dS. \tag{12.54}$$

Multiplying both sides by $\rho/2$, where ρ is the constant density of the fluid, we get on the left the total kinetic energy K of the fluid enclosed in the given volume. Hence

$$K = \frac{\rho}{2} \int \phi v_n dS, \tag{12.55}$$

[1] Cf. *Physical Statistics*, p. 228.

[2] Cf. H. Margenau and G. M. Murphy, *Mathematics of Physics and Chemistry* (New York, Van Nostrand, 1943), p. 156.

where v_n is the normal component of the flow velocity at the surface. Now in hydrodynamics[1] it is shown that $\rho\phi$ represents the impulse of the pressure tending to set up motion in the fluid. In fact, $\rho\phi$ has precisely the dimensions dyne sec/cm^2 or impulse per unit area. We see from (12.55) that the impulse per unit area multiplied by the normal velocity at the surface when integrated over the whole surface yields twice the resulting kinetic energy of the fluid enclosed by the surface. The right-hand integral in (12.55) is really the work done by the impulse per unit area at the surface. In a sense we can look upon (12.55) as a kind of work-kinetic energy theorem for an incompressible fluid.

We shall meet other similar applications of Green's theorem in electricity, heat flow, and wave motion. Another useful form of the theorem is obtained by interchanging U and V in (12.52), whence there results on subtraction

$$\int (U\nabla^2 V - V\nabla^2 U)d\tau = \int \left(U\frac{\partial V}{\partial n} - V\frac{\partial U}{\partial n}\right) dS. \tag{12.56}$$

Clearly if both U and V satisfy Laplace's equation in the given region, this leads to

$$\int \left(U\frac{\partial V}{\partial n} - V\frac{\partial U}{\partial n}\right) dS = 0. \tag{12.57}$$

Suppose U is some arbitrary differentiable function of space satisfying Laplace's equation, and let $V = c/r$, where r is the distance from some fixed point $P(x,y,z)$, which may or may not be within the surface over which the integration is conducted in (12.57). Draw a sphere of radius r' (where r' is less than the shortest distance from P to the surface) about P as center. Then carry out the integration indicated in (12.57) over both the outer surface and the inner sphere, for which the area element will be denoted by dS'. This gives

$$\int \left[U\frac{\partial}{\partial n}\left(\frac{1}{r}\right) - \frac{1}{r}\frac{\partial U}{\partial n}\right] dS + \int \left[U\frac{\partial}{\partial n}\left(\frac{1}{r}\right) - \frac{1}{r}\frac{\partial U}{\partial n}\right] dS' = 0. \tag{12.58}$$

[1] Cf., e.g., H. Lamb, *Hydrodynamics* (New York, Dover Publications, 1945), para. 18.

By the use of the divergence theorem we can show that

$$\int \frac{1}{r}\frac{\partial U}{\partial n}\,dS' = 0, \tag{12.59}$$

and that

$$\int U\frac{\partial}{\partial n}\left(\frac{1}{r}\right)dS' = \int U\,d\omega, \tag{12.60}$$

where $d\omega$ is the solid angle subtended at P by dS'. In the limit as $r' \to 0$ and the sphere about P shrinks to the center, the right-hand side of (12.60) approaches $4\pi U(P)$. Hence we have shown that

$$U(P) = \frac{1}{4\pi}\int\left[\frac{1}{r}\cdot\frac{\partial U}{\partial n} - U\frac{\partial}{\partial n}\left(\frac{1}{r}\right)\right]dS, \tag{12.61}$$

if P lies within the closed surface over which the integration is taken. It is not difficult to show that if P lies outside the surface, the integral in (12.61) vanishes. The importance of (12.61) resides in the fact that it gives U at the point P in terms of the values of U and its derivative at the boundary of the surface. This result of Green's theorem is extremely useful in wave propagation problems. The reader will find it instructive to apply this to the special case of the potential due to a spherical shell (sec. 12.4).

We can also get an interesting result of (12.56) when $\nabla^2 U \neq 0$. In this case the reader may show (with $V = c/r$ as before) that (12.61) is replaced by

$$U(P) = \frac{1}{4\pi}\int\left[\frac{1}{r}\cdot\frac{\partial U}{\partial n} - U\frac{\partial}{\partial n}\left(\frac{1}{r}\right)\right]dS - \frac{1}{4\pi}\int\frac{\nabla^2 U}{r}\,d\tau. \tag{12.62}$$

Here $U(P)$ is determined not merely by its value and that of its derivative on the surrounding surface but also by the value of $\nabla^2 U$ throughout space. A special case of interest is provided by letting the surface over which the integration is conducted in (12.62) recede to infinity. As long as U falls off with r as fast as $1/r$, the surface integral vanishes and

$$U(P) = -\frac{1}{4\pi}\int \frac{\nabla^2 U}{r}\,d\tau, \tag{12.63}$$

so that U at the point P now depends on $\nabla^2 U$ throughout space. In particular, if U satisfies Poisson's equation (12.38), we get

$$U(P) = \int \frac{\rho\,d\tau}{r}, \tag{12.64}$$

which thus appears as a solution of Poisson's equation. This provides, of course, an alternative method of solving potential problems like those discussed at the beginning of Sec. 12.4.

PROBLEMS

1. Find ∇r^n where n is any integer.

2. Calculate $\nabla \cdot \mathbf{r}$, where $\mathbf{r}$ is the position vector of a point. Use the divergence theorem to establish the connection between the surface and volume of a sphere.

3. Apply the equation of continuity to the special case of the steady flow of an incompressible fluid through a cylindrical tube containing a constriction. In particular find the relation between the fluid velocity in the constriction and the rest of the tube.

4. Write the equation of continuity for a compressible fluid in terms of the condensation s, defined in terms of the density ρ by

$$\rho = \rho_0(1 + s),$$

where ρ_0 is the constant equilibrium density. What approximate form does the equation take when $s \ll 1$? How can this approximate form be written if a velocity potential exists?

5. Specialize the result of Prob. 4 to the case in which the motion of the compressible fluid is confined to the x axis and the condensation and fluid velocity are functions of x and t alone. Hence derive the expression for the condensation in terms of the particle displacement ξ of the fluid. (ξ is the time integral of the fluid velocity). Comment with a diagram on the physical meaning of the result.

6. Use the divergence theorem to deduce the expression for the Laplacian in spherical coordinates.

7. Prove that $\nabla^2(1/r) = 0$.

8. Given a spherically symmetrical distribution of attracting matter with the density obeying the law $\rho = Ce^{-\alpha r}$, where C and α are constants and r is the distance from the center of symmetry. Find C if the total mass of the distribution is M. Find the potential and field intensity produced by the distribution at a point distant r from the center. Find the total mutual potential energy of the distribution.

9. Show that if $\mathbf{F} = -\nabla V$, it follows that $\nabla \times \mathbf{F} = 0$, where $\nabla \times \mathbf{F} =$

$$\begin{vmatrix} \mathbf{i} & \mathbf{j} & \mathbf{k} \\ \dfrac{\partial}{\partial x} & \dfrac{\partial}{\partial y} & \dfrac{\partial}{\partial z} \\ F_x & F_y & F_z \end{vmatrix}$$

10. If the intensity of a field is given by

$$\mathbf{F} = \mathbf{i}ay - \mathbf{j}ax,$$

is the field conservative? (N.B., a is a constant.) Find the work done when a unit mass is moved from $(-a, 0)$ to $(+a, 0)$ along the x axis. Do the same when the motion is along the semi circle of radius a. Comment on the result.

11. Use Gauss' force flux theorem to find the potential of the gravitational field due to an infinitely long cylinder of mass m per unit length and radius a, at an external point distant r from the axis. Do the same for an internal point.

12. Use the divergence theorem to deduce the fundamental equation of heat conduction in the form

$$\dot{\theta} = \frac{K}{c\rho} \nabla^2 \theta,$$

where θ is the temperature, K the thermal conductivity, c the specific heat and ρ the density (both the latter assumed constant).

13. Write Poisson's equation in cylindrical coordinates and apply to the case of a vacuum tube with cylindrical cathode and anode. Carry the result as far as the equation corresponding to (12.42) in the text.

14. A sphere of radius a contains matter distributed throughout its volume with density depending on the radius r alone in accordance with the formula

$$\rho = k(a^2 - r^2).$$

Find k if the total mass is m. Find the magnitude of the gravitational intensity and potential at points inside and outside the sphere.

15. Show that the divergence of the vector $\mathbf{Q}$ in terms of the orthogonal curvilinear coordinates q_1, q_2, q_3 may be written in the form

$$\nabla \cdot \mathbf{Q} = \frac{1}{h_1 h_2 h_3}\left[\frac{\partial}{\partial q_1}(h_2 h_3 Q_1) + \frac{\partial}{\partial q_2}(h_3 h_1 Q_2) + \frac{\partial}{\partial q_3}(h_1 h_2 Q_3)\right]$$

where Q_1, Q_2, Q_3 are the components of $\mathbf{Q}$ in the directions associated with increasing q_1, q_2, q_3, respectively, and

$$h_1 = \frac{ds_1}{dq_1}, \text{ etc.,}$$

where ds_1 is the actual spatial displacement associated with dq_1. Specialize the above general result to cylindrical and spherical coordinates.

16. If V is any scalar show that in terms of the orthogonal curvilinear coordinates q_1, q_2, q_3

$$\nabla V = \frac{\mathbf{i}_1}{h_1}\frac{\partial V}{\partial q_1} + \frac{\mathbf{i}_2}{h_2}\frac{\partial V}{\partial q_2} + \frac{\mathbf{i}_3}{h_3}\frac{\partial V}{\partial q_3},$$

where $\mathbf{i}_1, \mathbf{i}_2, \mathbf{i}_3$ are the unit vectors in the directions of increasing q_1, q_2, q_3, respectively.

17. Consulting Prob. 9, show that if the vector $\mathbf{Q} = \nabla \times \mathbf{F}$, $\nabla \cdot \mathbf{Q} = 0$ identically.

18. Find the gravitational potential at a point external to a sphere of radius a containing matter of mass m distributed so that its density is a function of r only, where r is the distance from the center of the sphere.

CHAPTER XIII

DEFORMABLE MATERIAL MEDIA AND TENSOR FIELDS

13.1. Deformability and Strain. One of the characteristics of any finite portion of a continuous material medium is its tendency to suffer changes in size and shape under the influence of appropriate external forces. All known bodies are thus deformable, although for certain ones the deformability may be so small except under extremely large forces that for practical purposes they may be considered rigid (cf. Chapter 7). In this section we shall be concerned with the study of small deformations from equilibrium size and shape. It will first be desirable to classify the possible types of deformation as simply as possible.

Consider Fig. 13.1 which shows two nearby points in the medium P and Q, where the coordinates of P are x, y, z and those of Q are $x + dx$, $y + dy$, $z + dz$. The separation of P and Q has the magnitude ds, where

$$ds^2 = dx^2 + dy^2 + dz^2. \quad (13.1)$$

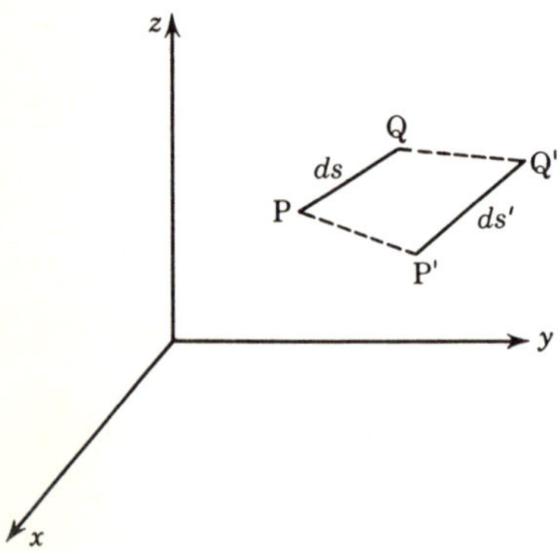

Fig. 13.1

Now let the medium be deformed in such a way that P is displaced to P' $(x + \delta x, y + \delta y, z + \delta z)$ and Q is displaced to $Q'[x + dx + \delta(x + dx), y + dy + \delta(y + dy), z + dz + \delta(z + dz)]$. The separation between P' and Q' is ds', and the difference between ds' and ds, which we may denote by $\delta(ds)$, is a measure of the deformation. Clearly we may get at this by finding $\delta(ds)^2 = 2ds \cdot \delta(ds)$. Thus

$$\delta(ds)^2 = 2\left[\left(\frac{\partial(\delta x)}{\partial x}dx^2 + \frac{\partial(\delta x)}{\partial y}dy\,dx + \frac{\partial(\delta x)}{\partial z}dz\,dx\right)\right.$$

$$+\left(\frac{\partial(\delta y)}{\partial x}dx\,dy + \frac{\partial(\delta y)}{\partial y}dy^2 + \frac{\partial(\delta y)}{\partial z}dz\,dy\right) +$$

$$\left.+\left(\frac{\partial(\delta z)}{\partial x}dx\,dz + \frac{\partial(\delta z)}{\partial y}dy\,dz + \frac{\partial(\delta z)}{\partial z}dz^2\right)\right]. \quad (13.2)$$

In writing (13.2) we have utilized the fact that

$$\delta(dx)^2 = 2dx\ \delta(dx) = 2dx\ d(\delta x)$$

$$= 2dx\left[\frac{\partial(\delta x)}{\partial x}dx + \frac{\partial(\delta x)}{\partial y}dy + \frac{\partial(\delta x)}{\partial z}dz\right]. \qquad (13.3)$$

Inspection shows that we may write (13.2) in the compact form

$$\frac{1}{2}\delta(ds)^2 = \frac{1}{2}\sum_{j,k=1}^{3}\left[\frac{\partial(\delta x_j)}{\partial x_k} + \frac{\partial(\delta x_k)}{\partial x_j}\right]dx_k\ dx_j, \qquad (13.4)$$

if we set $x = x_1$, $y = x_2$, $z = x_3$. It will now be convenient to consider δx_1, δx_2, δx_3 as *displacement coordinates* and denote them by ξ, η, ζ, respectively. Clearly the deformation will depend on the array of quantities or matrix

$$\mathbf{D} = \left\|\begin{matrix} \frac{\partial \xi}{\partial x}, & \frac{1}{2}\left(\frac{\partial \xi}{\partial y} + \frac{\partial \eta}{\partial x}\right), & \frac{1}{2}\left(\frac{\partial \xi}{\partial z} + \frac{\partial \zeta}{\partial x}\right) \\ \frac{1}{2}\left(\frac{\partial \eta}{\partial x} + \frac{\partial \xi}{\partial y}\right), & \frac{\partial \eta}{\partial y}, & \frac{1}{2}\left(\frac{\partial \eta}{\partial z} + \frac{\partial \zeta}{\partial y}\right) \\ \frac{1}{2}\left(\frac{\partial \zeta}{\partial x} + \frac{\partial \xi}{\partial z}\right), & \frac{1}{2}\left(\frac{\partial \zeta}{\partial y} + \frac{\partial \eta}{\partial z}\right), & \frac{\partial \zeta}{\partial z} \end{matrix}\right\| \qquad (13.5)$$

This matrix (the lines merely imply that the various quantities are considered as forming an array and do not signify any mathematical operation connecting the symbols) is called the *deformation or strain tensor,* and the nine quantities forming the array constitute the components of the tensor. The mathematical importance of the array comes from the transformation properties of its components, just as the mathematical significance of a vector comes from the transformation properties of *its* components (Sec. 2.3). The physical importance of **D** must now be investigated.

Consider the volume element in the shape of a rectangular parallelepiped $dx\,dy\,dz$, as in Fig. 13.2 and suppose that the displacement coordinates at 0 are ξ, η, ζ. At A we then expect the displacement in the x direction to be

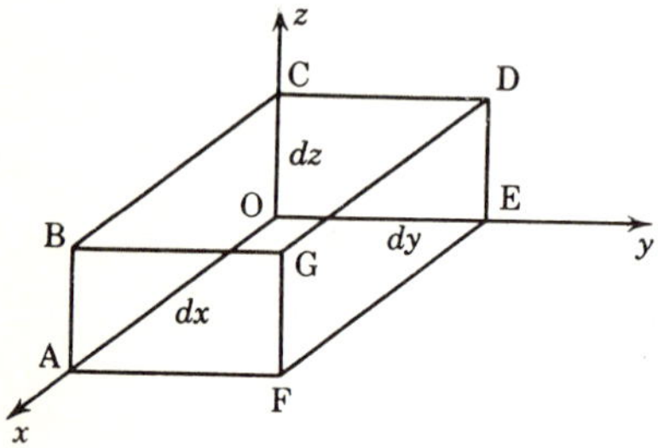

Fig. 13.2

$$\xi + \frac{\partial \xi}{\partial x} dx \tag{13.6}$$

out to terms involving first order differentials. In the present analysis we deliberately ignore higher order differentials since we are dealing with small deformations. The net change in length of the volume element in the x direction is thus $\partial\xi/\partial x \cdot dx$ and the *change in length per unit length* is

$$\frac{\partial \xi}{\partial x}, \tag{13.7}$$

which is called the *linear strain* in the x direction. It then follows that the linear strains in the y and z directions are

$$\frac{\partial \eta}{\partial y}, \frac{\partial \zeta}{\partial z}, \tag{13.8}$$

respectively. This provides the physical significance of the diagonal terms in the aggregate **D.** We introduce the notation

$$\delta_{xx} = \frac{\partial \xi}{\partial x}, \ \delta_{yy} = \frac{\partial \eta}{\partial y}, \ \delta_{zz} = \frac{\partial \zeta}{\partial z}.$$

To understand the significance of the other components of **D**, consider the section $OABC$ of the element in Fig. 13.2 and imagine that the rectangular section is deformed into the (cf. Fig. 13.3) parallelogram $OAB'C'$ in which OA remains fixed but CB slides over to $C'B'$, and other layers between OA and CB slide over distances proportional to their distance from OA. Such a deformation is called a *shear.* The simplest measure of this is the magnitude of the displacement CC' divided by OC (or dz). But since $CC' = \partial\xi/\partial z \cdot dz$, the *shearing strain* is therefore defined as

$$\frac{\partial \xi}{\partial z}. \tag{13.9}$$

It is accordingly the displacement parallel to one pair of faces of the element per unit length at right angles to these faces. It will be noted that both linear and shearing strains are nondimensional quantities. In the shear just described there is no deformation displacement along the y axis. Another, and indeed the only other, shear for which this is true is that which takes $OABC$ into $OA'B''C$. Clearly the shearing strain for this is $\partial\zeta/\partial x$. Hence we may call the total shearing strain about the y axis

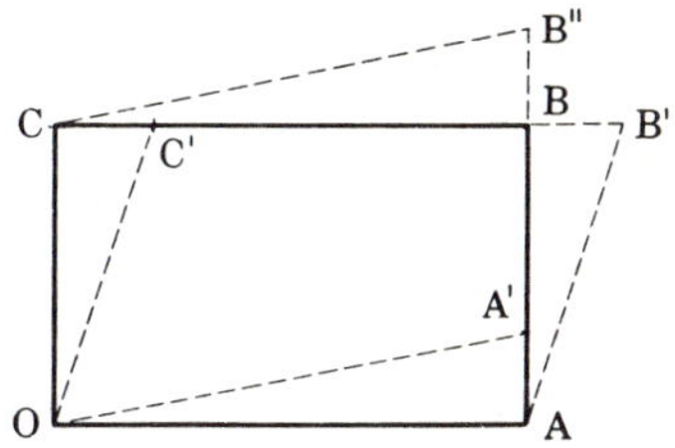

Fig. 13.3

$$\frac{\partial\xi}{\partial z} + \frac{\partial\zeta}{\partial x}. \tag{13.10}$$

For convenience we shall introduce the notation

$$\delta_{xz} = \frac{1}{2}\left(\frac{\partial\xi}{\partial z} + \frac{\partial\zeta}{\partial x}\right) = \delta_{zx}. \tag{13.11}$$

The total shearing strain about the x axis is likewise given by $\frac{\partial\eta}{\partial z} + \frac{\partial\zeta}{\partial y}$ and we write

$$\delta_{yz} = \frac{1}{2}\left(\frac{\partial\eta}{\partial z} + \frac{\partial\zeta}{\partial y}\right) = \delta_{zy}. \tag{13.12}$$

Finally the total shearing strain about the z axis is given by

$$\left(\frac{\delta\xi}{\partial y} + \frac{\partial\eta}{\partial x}\right), \text{ where}$$

$$\delta_{xy} = \frac{1}{2}\left(\frac{\partial\xi}{\partial y} + \frac{\partial\eta}{\partial x}\right) = \delta_{yx}. \tag{13.13}$$

These shears being invariant with respect to changes in order of the two terms in the sum, we are able to write $\delta_{xy} = \delta_{yx}$, etc. The aggregate **D** now takes the simple form

$$\mathbf{D} = \left\| \begin{matrix} \delta_{xx} & \delta_{xy} & \delta_{xz} \\ \delta_{yx} & \delta_{yy} & \delta_{yz} \\ \delta_{zx} & \delta_{zy} & \delta_{zz} \end{matrix} \right\| \tag{13.14}$$

An alternative view may serve to clarify the meaning of strain still further, and provide a reason for introducing another important vector operator. Let us form the vector whose components are the displacement coordinates ξ, η, ζ. We call this the *displacement vector* and write for the point P

$$\boldsymbol{\Delta} = \mathbf{i}\xi + \mathbf{j}\eta + \mathbf{k}\zeta. \tag{13.15}$$

For the nearby point Q the displacement vector is $\boldsymbol{\Delta} + d\boldsymbol{\Delta}$ where

$$\begin{aligned} d\boldsymbol{\Delta} = \mathbf{i}&\left(\frac{\partial\xi}{\partial x}dx + \frac{\partial\xi}{\partial y}dy + \frac{\partial\xi}{\partial z}dz\right) \\ + \mathbf{j}&\left(\frac{\partial\eta}{\partial x}dx + \frac{\partial\eta}{\partial y}dy + \frac{\partial\eta}{\partial z}dz\right) \\ + \mathbf{k}&\left(\frac{\partial\zeta}{\partial x}dx + \frac{\partial\zeta}{\partial y}dy + \frac{\partial\zeta}{\partial z}dz\right), \end{aligned} \tag{13.16}$$

and also serves as a measure of the deformation of the medium. Associated with $d\boldsymbol{\Delta}$ we have the array

$$\begin{Vmatrix} \dfrac{\partial\xi}{\partial x} & \dfrac{\partial\eta}{\partial x} & \dfrac{\partial\zeta}{\partial x} \\ \dfrac{\partial\xi}{\partial y} & \dfrac{\partial\eta}{\partial y} & \dfrac{\partial\zeta}{\partial y} \\ \dfrac{\partial\xi}{\partial z} & \dfrac{\partial\eta}{\partial z} & \dfrac{\partial\zeta}{\partial z} \end{Vmatrix} \tag{13.17}$$

If we assign to such an array the matrix properties of addition and subtraction defined[1] so that the sum of two arrays is an array in which each element is the sum of the corresponding elements of the arrays, we can readily write (13.17) in the form

$$\begin{Vmatrix} \dfrac{\partial\xi}{\partial x} & \dfrac{\partial\eta}{\partial x} & \dfrac{\partial\zeta}{\partial x} \\ \dfrac{\partial\xi}{\partial y} & \dfrac{\partial\eta}{\partial y} & \dfrac{\partial\zeta}{\partial y} \\ \dfrac{\partial\xi}{\partial z} & \dfrac{\partial\eta}{\partial z} & \dfrac{\partial\zeta}{\partial z} \end{Vmatrix} = \begin{Vmatrix} \dfrac{\partial\xi}{\partial x}, & \dfrac{1}{2}\left(\dfrac{\partial\eta}{\partial x} + \dfrac{\partial\xi}{\partial y}\right), & \dfrac{1}{2}\left(\dfrac{\partial\zeta}{\partial x} + \dfrac{\partial\xi}{\partial z}\right) \\ \dfrac{1}{2}\left(\dfrac{\partial\xi}{\partial y} + \dfrac{\partial\eta}{\partial x}\right), & \dfrac{\partial\eta}{\partial y}, & \dfrac{1}{2}\left(\dfrac{\partial\zeta}{\partial y} + \dfrac{\partial\eta}{\partial z}\right) \\ \dfrac{1}{2}\left(\dfrac{\partial\xi}{\partial z} + \dfrac{\partial\zeta}{\partial x}\right), & \dfrac{1}{2}\left(\dfrac{\partial\eta}{\partial z} + \dfrac{\partial\zeta}{\partial y}\right), & \dfrac{\partial\zeta}{\partial z} \end{Vmatrix} +$$

[1]Cf. H. Margenau and G. M. Murphy, *Mathematics of Physics and Chemistry* (New York, Van Nostrand, 1943), Chap. 10.

$$+\left\|\begin{matrix} 0, & \frac{1}{2}\left(\frac{\partial\eta}{\partial x}-\frac{\partial\xi}{\partial y}\right), & \frac{1}{2}\left(\frac{\partial\zeta}{\partial x}-\frac{\partial\xi}{\partial z}\right) \\ \frac{1}{2}\left(\frac{\partial\xi}{\partial y}-\frac{\partial\eta}{\partial x}\right), & 0, & \frac{1}{2}\left(\frac{\partial\zeta}{\partial y}-\frac{\partial\eta}{\partial z}\right) \\ \frac{1}{2}\left(\frac{\partial\xi}{\partial z}-\frac{\partial\zeta}{\partial x}\right), & \frac{1}{2}\left(\frac{\partial\eta}{\partial z}-\frac{\partial\zeta}{\partial y}\right), & 0 \end{matrix}\right\| \tag{13.18}$$

Inspection of (13.18) shows that the first array on the right is a symmetrical one in the sense that corresponding elements on either side of the principal diagonal are equal. On the other hand, the second array on the right is anti-symmetrical, since the corresponding terms on either side of the principal diagonal are the same in magnitude but differ in sign. It is seen that the first or symmetrical array is precisely **D** in (13.14). It remains to be seen whether a physical interpretation can be given to the anti-symmetrical array.

Corresponding to the array identity (13.18) we can write the deformation $d\Delta$ (13.16) in the form

$$\begin{aligned} d\Delta = \mathbf{i}&\left[\frac{\partial\xi}{\partial x}dx+\frac{1}{2}\left(\frac{\partial\eta}{\partial x}+\frac{\partial\xi}{\partial y}\right)dy+\frac{1}{2}\left(\frac{\partial\zeta}{\partial x}+\frac{\partial\xi}{\partial z}\right)dz\right] \\ +\mathbf{j}&\left[\frac{1}{2}\left(\frac{\partial\xi}{\partial y}+\frac{\partial\eta}{\partial x}\right)dx+\frac{\partial\eta}{\partial y}dy+\frac{1}{2}\left(\frac{\partial\zeta}{\partial y}+\frac{\partial\eta}{\partial z}\right)dz\right] \\ +\mathbf{k}&\left[\frac{1}{2}\left(\frac{\partial\xi}{\partial z}+\frac{\partial\zeta}{\partial x}\right)dx+\frac{1}{2}\left(\frac{\partial\eta}{\partial z}+\frac{\partial\zeta}{\partial y}\right)dy+\frac{\partial\zeta}{\partial z}dz\right] \\ +\mathbf{i}&\left[\frac{1}{2}\left(\frac{\partial\xi}{\partial y}-\frac{\partial\eta}{\partial x}\right)dy+\frac{1}{2}\left(\frac{\partial\xi}{\partial z}-\frac{\partial\zeta}{\partial x}\right)dz\right]. \\ +\mathbf{j}&\left[\frac{1}{2}\left(\frac{\partial\eta}{\partial x}-\frac{\partial\xi}{\partial y}\right)dx+\frac{1}{2}\left(\frac{\partial\eta}{\partial z}-\frac{\partial\zeta}{\partial y}\right)dz\right] \\ +\mathbf{k}&\left[\frac{1}{2}\left(\frac{\partial\zeta}{\partial x}-\frac{\partial\xi}{\partial z}\right)dx+\frac{1}{2}\left(\frac{\partial\zeta}{\partial y}-\frac{\partial\eta}{\partial z}\right)dy\right]. \end{aligned} \tag{13.19}$$

To understand the significance of the second set of **i, j, k** components in (13.19) we find it advisable to introduce another type of vector operator derived from ∇ (Sec. 12.2). Let **F** be an arbitrary vector. Then we define

$$\text{curl}\,\mathbf{F} = \nabla \times \mathbf{F} = \begin{vmatrix} \mathbf{i} & \mathbf{j} & \mathbf{k} \\ \dfrac{\partial}{\partial x} & \dfrac{\partial}{\partial y} & \dfrac{\partial}{\partial z} \\ F_x & F_y & F_z \end{vmatrix} . \tag{13.20}$$

It thus develops that the curl of the vector **F** (often referred to as "del cross **F**") is a vector which when written out in terms of its components becomes

$$\nabla \times \mathbf{F} = \mathbf{i}\left(\frac{\partial F_z}{\partial y} - \frac{\partial F_y}{\partial z}\right) + \mathbf{j}\left(\frac{\partial F_x}{\partial z} - \frac{\partial F_z}{\partial x}\right) + \mathbf{k}\left(\frac{\partial F_y}{\partial x} - \frac{\partial F_x}{\partial y}\right) \tag{13.21}$$

This proves to be of considerable use in field physics as the rest of this chapter and the succeeding ones will indicate.

Let us form $\nabla \times \Delta$ from (13.15). The result is

$$\nabla \times \Delta = \mathbf{i}\left(\frac{\partial \zeta}{\partial y} - \frac{\partial \eta}{\partial z}\right) + \mathbf{j}\left(\frac{\partial \xi}{\partial z} - \frac{\partial \zeta}{\partial x}\right) + \mathbf{k}\left(\frac{\partial \eta}{\partial x} - \frac{\partial \xi}{\partial y}\right). \tag{13.22}$$

The connection between the components of $\nabla \times \Delta$, the elements of the anti-symmetrical array in (13.18) and the coefficients of dx, dy, dz, in the second part of (13.19) is very suggestive. We proceed to form the vector product of $\frac{1}{2}\nabla \times \Delta$ with $d\mathbf{s}$, where

$$d\mathbf{s} = \mathbf{i}\,dx + \mathbf{j}\,dy + \mathbf{k}\,dz \tag{13.23}$$

is the vector separation of P and Q in Fig. 13.1. Thus

$$\frac{1}{2}(\nabla \times \Delta) \times d\mathbf{s} = \frac{1}{2}\begin{vmatrix} \mathbf{i} & \mathbf{j} & \mathbf{k} \\ \dfrac{\partial \zeta}{\partial y} - \dfrac{\partial \eta}{\partial z}, & \dfrac{\partial \xi}{\partial z} - \dfrac{\partial \zeta}{\partial x}, & \dfrac{\partial \eta}{\partial x} - \dfrac{\partial \xi}{\partial y} \\ dx & dy & dz \end{vmatrix}$$

$$= \frac{1}{2}\left[\mathbf{i}\left\{\left(\frac{\partial \xi}{\partial z} - \frac{\partial \zeta}{\partial x}\right)dz - \left(\frac{\partial \eta}{\partial x} - \frac{\partial \xi}{\partial y}\right)dy\right\} + \mathbf{j}\left\{\left(\frac{\partial \eta}{\partial x} - \frac{\partial \xi}{\partial y}\right)dx\right.\right.$$

$$\left.\left. - \left(\frac{\partial \zeta}{\partial y} - \frac{\partial \eta}{\partial z}\right)dz\right\} + \mathbf{k}\left\{\left(\frac{\partial \zeta}{\partial y} - \frac{\partial \eta}{\partial z}\right)dy - \left(\frac{\partial \xi}{\partial z} - \frac{\partial \zeta}{\partial x}\right)dx\right\}\right]. \tag{13.24}$$

Examination discloses that (13.24) is precisely the second part of (13.19). It is therefore a part of the deformation vector $d\boldsymbol{\Delta}$. What is its physical significance? We recall that $\frac{1}{2}(\nabla \times \boldsymbol{\Delta}) \times d\mathbf{s}$ by virtue of the definition of the cross product is perpendicular to $d\mathbf{s}$. It therefore cannot correspond to a change in $d\mathbf{s}$ which involves a change in the length ds. It is consequently not a genuine deformation. It can only correspond to a *rotation* of Q about P keeping their separation constant. This result indeed receives confirmation from the fact that if the motion of Q relative to P is a *rotation* about some instantaneous axis through a *small* angle $d\theta$, according to eq. (7.3) the resulting displacement $\boldsymbol{\Delta}'$ is given by

$$\boldsymbol{\Delta}' = d\boldsymbol{\theta} \times \mathbf{s}, \tag{13.25}$$

where now we take $\mathbf{s}$ as the position vector of Q relative to P, i.e., take P as origin. Hence

$$\nabla \times \boldsymbol{\Delta}' = \nabla \times (d\boldsymbol{\theta} \times \mathbf{s}). \tag{13.26}$$

If we carry out the indicated operation in the usual fashion we find

$$\nabla \times \boldsymbol{\Delta}' = d\boldsymbol{\theta}\, \nabla \cdot \mathbf{s} - (d\boldsymbol{\theta} \cdot \nabla)\mathbf{s}. \tag{13.27}$$

But since

$$\mathbf{s} = \mathbf{i}x + \mathbf{j}y + \mathbf{k}z,$$

we have

$$\nabla \cdot \mathbf{s} = 3,$$

and

$$(d\boldsymbol{\theta} \cdot \nabla)\mathbf{s} = \left[(d\boldsymbol{\theta})_x \frac{\partial}{\partial x} + (d\theta)_y \frac{\partial}{\partial y} + (d\theta)_z \frac{\partial}{\partial z}\right](\mathbf{i}x + \mathbf{j}y + \mathbf{k})$$

$$= \mathbf{i}(d\theta)_x + \mathbf{j}(d\theta)_y + \mathbf{k}(d\theta)_z = d\boldsymbol{\theta}. \tag{13.28}$$

Consequently

$$\nabla \times \boldsymbol{\Delta}' = 2d\boldsymbol{\theta}. \tag{13.29}$$

This says that if $\boldsymbol{\Delta}'$ is the displacement due to rotation alone $\frac{1}{2}\nabla \times \boldsymbol{\Delta}'$ is precisely the resulting infinitesimal rotation $d\boldsymbol{\theta}$. This crossed into $d\mathbf{s}$ therefore gives a displacement corresponding to rotation alone. The total deformation vector $d\boldsymbol{\Delta}$ is thus resolved into a part which involves rotation without strain and another part which therefore must represent the effective strain.

13.2. The Strain Tensor. We have referred to the aggregate (13.5) or its equivalent array (13.14) as the strain tensor. It now becomes necessary to investigate the mathematical properties of such a quantity more closely. It will be recalled that we found our most significant definition of a *vector* in terms of its behavior under a rotation of axes in a rectangular coordinate system. We proceed to study the 9-term array in similar fashion.

Let us introduce an arbitrary transformation of coordinates from x_1, x_2, x_3 to x_1', x_2', x_3', which we may define by

$$x_j = x_j(x_1', x_2', x_3'), \quad (j = 1, 2, 3). \tag{13.30}$$

Then

$$dx_j = \sum_{i=1}^{3} \frac{\partial x_j}{\partial x_i'} dx_i'; \quad dx_k = \sum_{l=1}^{3} \frac{\partial x_k}{\partial x_l'} \cdot dx_l'. \tag{13.31}$$

We substitute this transformation into the quadratic form (13.4) rewritten for brevity as

$$\frac{1}{2}\delta(ds)^2 = \sum_{j,k=1}^{3} a_{kj}\, dx_k\, dx_j. \tag{13.32}$$

The result is

$$\sum_{j,k=1}^{3} a_{kj}\, dx_k\, dx_j = \sum_{j,k=1}^{3} \sum_{i,l=1}^{3} a_{kj} \frac{\partial x_k}{\partial x_l'} \frac{\partial x_j}{\partial x_i'} dx_l'\, dx_i'. \tag{13.33}$$

Since $a_{jk} = a_{kj}$ we can write

$$\sum_{j,k=1}^{3} \sum_{i,l=1}^{3} a_{kj} \frac{\partial x_k}{\partial x_l'} \frac{\partial x_j}{\partial x_i'} dx_l'\, dx_i' =$$

$$= \sum_{j,k=1}^{3} \sum_{i,l=1}^{3} a_{il} \frac{\partial x_i}{\partial x_k'} \frac{\partial x_l}{\partial x_j'} dx_k' \, dx_j' . \tag{13.34}$$

It therefore follows that after the transformation the expression which in the new quadratic form replaces the a_{kj} of the original quadratic form is

$$(a_{kj}) = \sum_{i,l=1}^{3} a_{il} \frac{\partial x_i}{\partial x_k'} \frac{\partial x_l}{\partial x_j'} . \tag{13.35}$$

We shall say that the transformation changes the aggregate of quantities a_{kj} into the aggregate (a_{kj}) as given by (13.35). Any aggregate of quantities a_{kj} which transforms according to this rule is called a *covariant tensor* of the second order. The meaning of (13.35) must be carefully noted. Since ξ, η, ζ are functions of x, y, z, it is also true that the terms of the aggregate **D**, namely $\delta_{x_j x_k} = a_{jk}$, etc., are functions of x, y, z or x_1, x_2, x_3. When we make the transformation to x_1', x_2', x_3' given by (13.30) the quantities a_{jk} will be different functions of the primed coordinates than they were of the original unprimed coordinates. This, however, is not the kind of transformation which is meant here. We mean that whenever we wish to use a_{jk} in the primed coordinates, what we have to employ is (a_{jk}) as defined in (13.35). Of course this usage insures that the quadratic form $\frac{1}{2}\delta(ds)^2$ maintains the *same form* in the primed coordinates as in the unprimed coordinates. The analogy with a vector as defined in eqs. (2.9) and (2.10) of Chapter 2 is fairly clear: any set of quantities transforming as the coordinates of a point under rotation of axes we called a three-vector. In fact, we may consider such a vector as a kind of tensor of the first order. The components of the vector $d\mathbf{r}$ representing the separation between any two neighboring points are dx_1, dx_2, dx_3 (using the above convention regarding subscripts). After rotation of axes, we get dx_1', dx_2', dx_3', where

$$dx_k' = (dx_k) = \sum_{j=1}^{3} \frac{\partial x_k'}{\partial x_j} dx_j . \tag{13.36}$$

Here the primed coordinate is in the numerator, in contrast to (13.35). We shall speak of the three-vector as a *contravariant* tensor of the first order.

The reader should show that the gradient of a scalar is a covariant tensor of the first order.

In the study of deformability it appears that the covariant tensor is the important one. We may look upon the deformable medium as a covariant tensor field. As a matter of fact the transformation of greatest importance in deformability theory is the rotation of axes. It is not difficult to show that in this case

$$\frac{\partial x_j}{\partial x_i'} = \frac{\partial x_i'}{\partial x_j},$$

since the quantities in question are direction cosines of the same angle. Hence it follows that we can also write (13.35) in the form

$$(a_{kj}) = \sum_{i,l=1}^{3} a_{il} \frac{\partial x_k'}{\partial x_i} \frac{\partial x_j'}{\partial x_l}. \tag{13.37}$$

Here the aggregate a_{kj} appears as a *contravariant* tensor of the second order. As far as rotation of rectangular axes is concerned there is no difference between a covariant and a contravariant tensor.

Considerable simplification in the use of tensors can be achieved by means of a summation convention. If a subscript is repeated on the same side of the equation the quantity written down is to be summed over the whole range of variation of the subscript or index. In this notation (13.36) takes the form

$$(dx_k) = \frac{\partial x_k'}{\partial x_j} dx_j = \frac{\partial x_k'}{\partial x_i} dx_i, \tag{13.36a}$$

and (13.35) can be written

$$(a_{kj}) = a_{il} \frac{\partial x_i}{\partial x_k'} \frac{\partial x_l}{\partial x_j'} = a_{mn} \frac{\partial x_m}{\partial x_k'} \frac{\partial x_n}{\partial x_j'}. \tag{13.35a}$$

Note that the index over which summation is to be conducted is a dummy one and can be replaced by another if it proves more convenient.

It is not our purpose here to discuss in detail the mathematical theory of tensors and the various notations which have been invented to facilitate their use. Our intention is rather to call attention to the importance of tensors in certain physical problems.[1] Certain properties of tensors are discussed in the problems at the end of the chapter.

13.3. The Stress Tensor. It is now in order to study the forces associated with the deformation of the medium. From elementary physics we recall that it is customary to associate with the deformation *stresses*, or forces per unit area. These forces may be directed either normally or tangentially to the surfaces on which they act. To refer to Fig. 13.2 we suppose that on the face *OCDE* there is a stress X_{xx} acting along the x axis. Since this tends to elongate the volume element it is called a tensile stress. On the same face we shall imagine that forces may act along the y and z axes, respectively. The corresponding forces per unit area are called shearing stresses and will be denoted by X_{xy} and X_{xz}, respectively. The corresponding stresses on the face *OABC* will be denoted by X_{yx}, X_{yy}, X_{yz} respectively, the first and last being shearing stresses in the x and z directions, respectively, and the second being the tensile stress in the y direction. Finally the stresses on *OAFE* are X_{zx}, X_{zy}, X_{zz}, respectively. It seems plausible to assume that the complete stress specification of the medium is inherent in the array of nine quantities which we shall label **S** and arrange as follows:

$$\mathbf{S} = \left\| \begin{matrix} X_{xx} & X_{xy} & X_{xz} \\ X_{yx} & X_{yy} & X_{yz} \\ X_{zx} & X_{zy} & X_{zz} \end{matrix} \right\| \tag{13.38}$$

The suggestion is strong that this array like **D** is also a tensor of the second order. But we must first show that the transformation properties of $X_{x_i x_j}$ are the same as those of the $\delta_{x_i x_j}$ in the strain tensor. We shall limit the transformations to rotation of axes.[2]

Consider the face normal to the x_1 axis. The component of the stress $X_{x_1 x_1}$ along the x_1' axis is $X_{x_1 x_1} \dfrac{\partial x_1}{\partial x_1'}$ since $\dfrac{\partial x_1}{\partial x_1'}$ is the cosine of the angle between the two axes. Hence the stress directed along the x_1' axis

[1] Cf. for the mathematical treatment: H. Jeffreys, *Cartesian Tensors* (Cambridge Univ. Press, 1931).

[2] This restriction is not essential. Cf. for the more general treatment H. Thirring, *Phys. Zs.*, 26, 518, 1925.

on the face normal to the x_1 axis due to stresses $X_{x_1x_1}$, $X_{x_1x_2}$ and $X_{x_1}X_{x_3}$ is

$$X_{x_1x_1}\frac{\partial x}{\partial x_1{}'} + X_{x_1x_2}\frac{\partial x_2}{\partial x_1{}'} + X_{x_1x_3}\frac{\partial x_3}{\partial x_1{}'}\,.$$

But we want the contribution of this stress on unit area normal to the $x_1{}'$ axis. This is

$$\left(X_{x_1x_1}\frac{\partial x_1}{\partial x_1{}'} + X_{x_1x_2}\frac{\partial x_2}{\partial x_1{}'} + X_{x_1x_3}\frac{\partial x_3}{\partial x_1{}'}\right)\cdot\frac{\partial x_1}{\partial x_1{}'}\,,$$

where $\partial x_1/\partial x_1{}'$ is here a geometrical factor relating the faces normal to the x_1 and $x_1{}'$ axes, respectively. By applying the same reasoning to the stress components $X_{x_2x_1}$, $X_{x_2x_2}$, $X_{x_2x_3}$, etc.; we arrive at the *total* stress in the $x_1{}'$ direction on the face perpendicular to the $x_1{}'$ axis in the form

$$(X_{x_1x_1}) = \sum_{j,l=1}^{3} X_{x_jx_l}\frac{\partial x_j}{\partial x_1{}'}\frac{\partial x_l}{\partial x_1{}'}\,. \tag{13.39}$$

In general we have for the total stress in the $x_i{}'$ direction on the face perpendicular to the $x_k{}'$ axis

$$(X_{x_ix_k}) = \sum_{j,l=1}^{3} X_{x_jx_l}\frac{\partial x_j}{\partial x_i{}'}\frac{\partial x_l}{\partial x_k{}'}\,. \tag{13.40}$$

The new stresses are thus obtained from the old by the same rule as in the case of the strain tensor (cf. eq. 13.35). Hence we are justified in considering the aggregate **S** as a covariant tensor of the second order. As has already been suggested in Sec. 13.2, the tensor (13.40) can also be considered contravariant. However, this will not affect the comparison of **D** with **S.**

It has already been emphasized that the strain tensor **D** (eq. 13.5) is a symmetrical one. The question immediately arises: Is the stress tensor **S** likewise? In other words, do we have $X_{xy} = X_{yx}$, etc.? This is indeed true and we can see it from the fact that

$$(X_{xy} - X_{yx})dx\,dy\,dz$$

is twice the stress moment on the volume element $dx\,dy\,dz$ about an axis through the center of mass parallel to the z axis, and is therefore equal to the moment of inertia about the z axis multiplied by the angular acceleration about the z axis. The moment of inertia contains terms in dx^2 and dy^2 in addition to $dx\,dy\,dz$. Now as the volume of the rectangular element is decreased, both sides of the equation must approach zero at the same rate. This can only happen if $X_{yx} = X_{xy}$ identically. By a similar argument $X_{xz} = X_{zx}$, $X_{yz} = X_{zy}$. Consequently **S** is a symmetrical tensor.

13.4. Relation between Strain and Stress in an Isotropic Medium. Hooke's Law. Experiment indicates that in any deformable medium there exists a relation between stress and the associated strain. This relation may be expected to depend on the complexity of the medium. For example, for a crystalline medium whose properties differ in different directions or for a nonhomogeneous medium whose properties vary from point to point, the relation may well be very complicated. Likewise if the strains are large fractions of unity, we also expect difficulty in setting up any simple mathematical equations of deformability. However, it was discovered in the seventeenth century by Hooke and others that for small enough strains in homogeneous isotropic bodies the relation between stress and strain is a relatively simple one. Media that follow this law, commonly known by the name of Hooke, are said to be *elastic*. In what follows we shall confine our attention to elastic media, and in general to homogeneous, isotropic ones.

Reverting to Fig. 13.2 imagine that the tensile stress X_{xx} acts along the x axis. According to Hooke's law there will be associated with this a linear strain in the x direction equal to δ_{xx}, where

$$\delta_{xx} = X_{xx}/Y. \tag{13.41}$$

Here Y is a constant typical of the material undergoing strain, and is the well-known Young's modulus. Now it is an experimental fact that associated with the given tensile stress X_{xx} there is not only the linear strain along x but also a lateral strain at right angles to x involving a shrinking of all lengths perpendicular to x. The magnitude of the lateral strain is directly proportional to the linear strain, the coefficient of proportionality being called *Poisson's ratio*, denoted by σ. If the same tensile stress X (say) acts on the volume element along all axes simultaneously, it follows that the resultant linear strains along the three axes, respectively, are

$$\delta_{xx} = \delta_{yy} = \delta_{zz} = \frac{X}{Y}(1 - 2\sigma). \qquad (13.42)$$

The application of X to all faces of the parallelepiped simultaneously acts to produce a change in volume. The volume strain associated with this is defined as the change in volume per unit volume, and we may denote it by δ_v. The law of Hooke states here that the volume strain under uniform surface stress is directly proportional to the stress, the coefficient being known as the *bulk modulus* (or modulus of volume elasticity).[1] Thus

$$\delta_v = X/B, \qquad (13.43)$$

if we use B to denote the bulk modulus. It can readily be shown that the volume strain δ_v (assumed small) is equivalent to three mutually perpendicular linear strains each equal in magnitude to $\delta_v/3$. Hence we can write (13.43) in the form

$$X/3B = X(1 - 2\sigma)/Y, \qquad (13.44)$$

whence there results the important relation connecting the moduli Y and B

$$Y = 3B(1 - 2\sigma). \qquad (13.45)$$

Since for all known elastic media Y and B are positive it follows from (13.45) that σ cannot exceed ½.

We next look for a relation involving shearing strain. If we have a pure shear as in Fig. 13.3, we define the ratio of the shearing stress to the corresponding shearing strain as the shear modulus μ, which of course is defined only if the stress and strain are small enough to satisfy Hooke's law. It can now be shown that a shearing strain, e.g., δ_{xz}, is equivalent to two linear strains of opposite sign and of magnitude $\delta_{xz}/2$ at right angles to each other and each at 45° to the sheared planes. Suppose a positive tensile stress X acts along the z axis and a negative tensile stress $-X$ acts along the x axis. The result will be a shearing strain at an angle of 45° to the linear strains corresponding to these stresses, and the magnitude will be

$$\delta_{xz} = \frac{2X}{Y}(1 + \sigma). \qquad (13.46)$$

[1] If the surface stress acts inward on all faces, it is called a pressure.

But the same two tensile stresses are equivalent to a shearing stress of magnitude X acting at an angle of 45° to both the tensile stresses. This yields

$$\delta_{xz} = X/\mu, \tag{13.47}$$

which must be equal to (13.46). Consequently we deduce the relation

$$Y = 2\mu(1 + \sigma), \tag{13.48}$$

whence it follows that σ cannot be less than -1. The combination of (13.45) and (13.48) yields the equation connecting Y, μ, and B,

$$Y = \frac{9B\mu}{\mu + 3B}. \tag{13.49}$$

Two independent elastic moduli are thus sufficient to describe the stress-strain relations of an isotropic, homogeneous elastic medium.

Going back to Fig. 13.2, suppose that stresses X_{xx}, X_{yy}, and X_{zz} act along their respective axes. The associated linear strains are then

$$\begin{aligned} \delta_{xx} &= X_{xx}/Y - \sigma(X_{yy} + X_{zz})/Y, \\ \delta_{yy} &= X_{yy}/Y - \sigma(X_{xx} + X_{zz})/Y, \\ \delta_{zz} &= X_{zz}/Y - \sigma(X_{xx} + X_{yy})/Y. \end{aligned} \tag{13.50}$$

When these three equations are solved for X_{xx}, X_{yy}, and X_{zz} (most simply by first adding them) we obtain (after utilizing eqs. 13.45 and 13.48)

$$\begin{aligned} X_{xx} &= (B - 2\mu/3) \cdot \Theta + 2\mu\delta_{xx}, \\ X_{yy} &= (B - 2\mu/3) \cdot \Theta + 2\mu\delta_{yy}, \\ X_{zz} &= (B - 2\mu/3) \cdot \Theta + 2\mu\delta_{zz}. \end{aligned} \tag{13.51}$$

Here we have written

$$\Theta = \delta_{xx} + \delta_{yy} + \delta_{zz}, \tag{13.52}$$

called the "total dilatation" of the element. The three equations must be accompanied by the corresponding shear equations

$$X_{xy} = X_{yx} = 2\mu\delta_{xy},$$
$$X_{zx} = X_{xz} = 2\mu\delta_{xz}, \tag{13.53}$$
$$X_{yz} = X_{zy} = 2\mu\delta_{yz}.$$

If we set $B - 2\mu/3 = \lambda$ for convenience, we may replace equations (13.51) and (13.53) by the single tensor equation

$$\mathbf{S} = \left\| \begin{matrix} \lambda\Theta + 2\mu\delta_{xx}, & 2\mu\delta_{xy}, & 2\mu\delta_{xz} \\ 2\mu\delta_{yx}, & \lambda\Theta + 2\mu\delta_{yy}, & 2\mu\delta_{yz} \\ 2\mu\delta_{zx}, & 2\mu\delta_{zy}, & \lambda\Theta + 2\mu\delta_{zz} \end{matrix} \right\|, \tag{13.54}$$

the left-hand side of which in turn may be rewritten as the sum of two tensors, so that

$$\mathbf{S} = 2\mu\mathbf{D} + \lambda\mathbf{D}', \tag{13.55}$$

wherein $\mathbf{D}'$ is the diagonal tensor

$$\mathbf{D}' = \left\| \begin{matrix} \Theta & 0 & 0 \\ 0 & \Theta & 0 \\ 0 & 0 & \Theta \end{matrix} \right\| . \tag{13.56}$$

The tensor equation (13.55) is the generalized form of Hooke's law for a homogeneous, isotropic elastic medium.

It is worth while pausing to note that the fundamental physical significance of (13.55) is to be found in the fact that, under an arbitrary transformation of coordinates, the two sides of the equation transform in the same way and hence the equation will remain true in the new coordinates. The importance of this consideration in the special case of vectors has already been noted. If the equations of physics are to represent physical situations which are independent of the coordinate systems used in their description, it is clear that both sides of each equation must transform the same way under any arbitrary transformation of coordinates. The precise behavior under transformation is not so important as the existence of quantities which follow the same transformation rule. Perhaps it would be fairer to say that the physicist builds such quantities in order to insure the consequence just stated. An understanding of the properties of tensors therefore becomes of importance in the development of new quantities for the description of more complicated physical phenomena. This is well illustrated by the theory of relativity.

Although we have introduced tensors for the sake of the convenient representation of elastic medium behavior, the most elaborate use of tensors has occurred in general relativity.

13.5. The Motion of an Elastic Solid. The Wave Equation. An elastic solid can move in translation and rotation and for the study of this we need no more than the concepts of rigid body motion developed in Chapter 7. However, since the parts of an elastic medium can also move with respect to each other, it becomes of importance to pay special attention to such motions which are associated with the stresses applied to the medium.

Let us set up the equations of motion of an isotropic, homogeneous elastic medium under the influence of stress forces. The total force components due to stress acting on the volume element $dx\,dy\,dz$ in the x, y, z directions respectively are

$$\left(\frac{\partial X_{xx}}{\partial x} + \frac{\partial X_{yx}}{\partial y} + \frac{\partial X_{zx}}{\partial z}\right) dx\,dy\,dz,$$

$$\left(\frac{\partial X_{xy}}{\partial x} + \frac{\partial X_{yy}}{\partial y} + \frac{\partial X_{zy}}{\partial z}\right) dx\,dy\,dz, \tag{13.57}$$

$$\left(\frac{\partial X_{xz}}{\partial x} + \frac{\partial X_{yz}}{\partial y} + \frac{\partial X_{zz}}{\partial z}\right) dx\,dy\,dz,$$

the terms in the parentheses denoting the corresponding forces per unit volume. If we employ (13.51) and (13.53) we are led to

$$\frac{\partial X_{xx}}{\partial x} + \frac{\partial X_{yx}}{\partial y} + \frac{\partial X_{zx}}{\partial z} = (B + \mu/3)\frac{\partial \Theta}{\partial x} + \mu\nabla^2\xi,$$

$$\frac{\partial X_{xy}}{\partial x} + \frac{\partial X_{yy}}{\partial y} + \frac{\partial X_{zy}}{\partial z} = (B + \mu/3)\frac{\partial \Theta}{\partial y} + \mu\nabla^2\eta, \tag{13.58}$$

$$\frac{\partial X_{xz}}{\partial x} + \frac{\partial X_{yz}}{\partial y} + \frac{\partial X_{zz}}{\partial z} = (B + \mu/3)\frac{\partial \Theta}{\partial z} + \mu\nabla^2\zeta.$$

The component equations of motion then become

$$\rho\ddot{\xi} = (B + \mu/3)\frac{\partial\Theta}{\partial x} + \mu\nabla^2\xi,$$

$$\rho\ddot{\eta} = (B + \mu/3)\frac{\partial\Theta}{\partial y} + \mu\nabla^2\eta, \tag{13.59}$$

$$\rho\ddot{\zeta} = (B + \mu/3)\frac{\partial\Theta}{\partial z} + \mu\nabla^2\zeta.$$

if ρ is the density of the medium. Some question indeed arises as to the precise significance of $\ddot{\xi}$, etc. As written here it refers to the acceleration of deformation at a point and neglects the motion of the volume element during deformation. For most applications to solids this is a good approximation. The case of fluids is treated in Chapter 14.

From

$$\boldsymbol{\Delta} = \mathbf{i}\xi + \mathbf{j}\eta + \mathbf{k}\zeta,$$

we obtain

$$\nabla \cdot \boldsymbol{\Delta} = \frac{\partial\xi}{\partial x} + \frac{\partial\eta}{\partial y} + \frac{\partial\zeta}{\partial z} = \Theta. \tag{13.60}$$

With the introduction of $\boldsymbol{\Delta}$, the three equations (13.59) can be written as a single vector equation (cf. eq. 13.21),

$$\rho\ddot{\boldsymbol{\Delta}} = (B + 4\mu/3)\nabla\nabla \cdot \boldsymbol{\Delta} - \mu\nabla \times \nabla \times \boldsymbol{\Delta}. \tag{13.61}$$

There are two important special cases of the equation of motion (13.61). Let us first assume that the curl of $\boldsymbol{\Delta}$ is equal to zero. This means

$$\nabla \times \boldsymbol{\Delta} = 0. \tag{13.62}$$

Such a vector is called *irrotational.* The second term on the right of (13.61) now vanishes. Moreover we have

$$\nabla \times \nabla \times \boldsymbol{\Delta} = \nabla\nabla \cdot \boldsymbol{\Delta} - \nabla \cdot \nabla\boldsymbol{\Delta}, \tag{13.63}$$

where

$$\nabla \cdot \nabla\boldsymbol{\Delta} = \mathbf{i}\nabla^2\xi + \mathbf{j}\nabla^2\eta + \mathbf{k}\nabla^2\zeta, \tag{13.64}$$

and will be denoted by $\nabla^2\boldsymbol{\Delta}$ for convenience, although care must be taken in distinguishing this from the Laplacian of a scalar. Therefore if $\boldsymbol{\Delta}$ is irrotational the equation of motion (13.61) becomes

$$\rho\ddot{\boldsymbol{\Delta}} = (B + 4\mu/3) \cdot \nabla^2\boldsymbol{\Delta}, \tag{13.65}$$

one of the most famous of all the partial differential equations of physics. Before we discuss its physical meaning let us consider the second special case of (13.61), namely that in which the divergence of Δ vanishes, i.e.,

$$\nabla \cdot \Delta = 0. \tag{13.66}$$

In this case Δ is termed a *solenoidal* vector. Now the first term on the right of (13.61) vanishes and the second term becomes $\mu\nabla^2\Delta$, so that the equation of motion has the form

$$\rho\ddot{\Delta} = \mu\nabla^2\Delta, \tag{13.67}$$

i.e., the same form as (13.65), although we may expect some physical difference in the interpretation because of the difference in the nature of Δ.

Reverting to (13.65), we introduce the simplification that ξ, η, ζ are independent of y and z and dependent on x and t only. Then eq. (13.65) becomes

$$\begin{aligned} \rho\ddot{\xi} &= (B + 4\mu/3)\frac{\partial^2\xi}{\partial x^2}, \\ \rho\ddot{\eta} &= (B + 4\mu/3)\frac{\partial^2\eta}{\partial x^2}, \\ \rho\ddot{\zeta} &= (B + 4\mu/3)\frac{\partial^2\zeta}{\partial x^2}. \end{aligned} \tag{13.68}$$

Equations of this type were presumably first discussed and solved by D'Alembert, who showed that a solution of the first, for example, can be had in the form

$$\xi = \xi_1(x - ct) + \xi_2(x + ct), \tag{13.69}$$

where

$$c = \sqrt{\frac{B + 4\mu/3}{\rho}}, \tag{13.70}$$

and ξ_1 and ξ_2 are arbitrary differentiable functions of the arguments $x - ct$ and $x + ct$ respectively, with continuous second derivatives over some range of the argument. Later it was shown that the solution (13.69) is general. Direct substitution of (13.69) into (13.68) verifies the solution sufficiently for our present purpose. A little examination shows that $\xi_1(x - ct)$ represents a function which moves along the positive x axis with velocity c. It can therefore represent a disturbance, in this case the displacement ξ, moving with velocity c in the positive x direction. Such a moving disturbance in a medium is called a *wave*. Similarly $\xi_2(x + ct)$ represents a wave in the negative x direction.

The motion of the medium described by eqs. (13.68) therefore is wave motion with wave velocity given by eq. (13.70) Since the displacement vector $\mathbf{\Delta}$ is independent of y and z, at any instant Δ is the same everywhere on a plane perpendicular to the x axis. In general the surface which is the locus of points for which the *phase* of the wave disturbance, i.e., the argument $x - ct$ of the wave function in (13.69) is the same, is called the *wave front*. In the present case the wave front is a plane and the wave is a plane wave.

Since for (13.68) $\nabla \times \Delta = 0$ holds, we have

$$\frac{\partial \zeta}{\partial y} = \frac{\partial \eta}{\partial z}, \quad \frac{\partial \xi}{\partial z} = \frac{\partial \zeta}{\partial x}, \quad \frac{\partial \eta}{\partial x} = \frac{\partial \xi}{\partial y}.$$

Moreover, since we assumed that ξ, η, ζ are functions of x and t only, it follows that $\dfrac{\partial \eta}{\partial x} = \dfrac{\partial \zeta}{\partial x} = 0$. Hence η and ζ are constants in space and are not propagated as waves. The only propagated displacement is ξ which is directed along the direction of propagation. Hence the wave is called *longitudinal*. An irrotational plane wave is thus longitudinal. It must be observed that this would not be true in general for any type of wave front.

The type of wave under discussion is usually known as *dilatational* since no shears are involved in the propagated displacement, even though the shear modulus enters into the expression for the wave velocity.

We next consider (13.67) which also presumably represents wave motion, since if we assume that the wave is plane by taking Δ as dependent on x and t only, it becomes

$$\rho\ddot{\Delta} = \mu \frac{\partial^2 \Delta}{\partial x^2}. \tag{13.71}$$

The velocity of this type of wave is

$$c = \sqrt{\mu/\rho}. \tag{13.72}$$

Let us look more closely at the nature of this wave. Since $\nabla \cdot \Delta = 0$ here, we necessarily conclude that $\partial\xi/\partial x = 0$ and ξ is constant in space. Evidently only η and ζ can change with both x and t. Thus the propagated displacement is at right angles to the direction of propagation and the wave is called *transverse*. We must again be careful not to conclude that all elastic waves for which $\nabla \cdot \Delta = 0$ have the resultant propagated displacement perpendicular to the direction of propagation. We have just shown this to be true of a *plane* wave only. Hence a plane solenoidal wave is transverse and therefore a *shear* wave.

We may expect to find in general both dilatational and shear waves when a small disturbance from equilibrium is created in a homogeneous isotropic elastic medium. It is clear that the dilatational wave will always have the larger velocity. The torsional wave in a solid rod is a good illustration of (13.67) (cf. Prob. 14). The dilatational or extensional wave in a solid rod does not, however, have the velocity (13.70). Rather we have in this case

$$c = \sqrt{Y/\rho}, \tag{13.73}$$

which for a good many solids is about 10% smaller than (13.70). The reader should work out (13.73) for himself.

PROBLEMS

1. Let (x, y, z) be a point P in a deformable medium and let l, m, n be the direction cosines of a line through this point. A point Q on the line distant s from the point P has what coordinates? If after small deformation the particle at (x, y, z) moves to $(x + \xi, y + \eta, z + \zeta)$, to what point does the particle originally at Q move? If the distance s changes after deformation to s', show that, neglecting squares and products of such small quantities as

$$\frac{\partial \xi}{\partial x}, \frac{\partial \eta}{\partial y} \text{ etc.,}$$

we have

$$s' = s(1 + \delta_{xx}l^2 + \delta_{yy}m^2 + \delta_{zz}n^2 + 2\delta_{yz}mn + 2\delta_{xz}ln + 2\delta_{xy}lm).$$

2. A solid cube has all its faces acted on by the same compressive stress p directed inward. Prove that the normal stress on any plane through the cube is equal to p.

3. Writing $\mathbf{S}$ in the schematic (dyadic) form

$$\begin{aligned}\mathbf{S} = \mathbf{ii}X_{xx} + \mathbf{ij}X_{xy} + \mathbf{ik}X_{xz}\\ + \mathbf{ji}X_{yx} + \mathbf{jj}X_{yy} + \mathbf{jk}X_{yz}\\ + \mathbf{ki}X_{zx} + \mathbf{kj}X_{zy} + \mathbf{kk}X_{zz}\end{aligned}$$

and similarly for $\nabla\Delta$, etc., show that the tensor eq. (13.55) can be written in the form

$$\mathbf{S} = 2\mu(\nabla\Delta + \mathbf{D}^{\text{п}}) + \lambda\nabla \cdot \Delta(\mathbf{I})$$

where $\mathbf{I}$ = unit diagonal matrix. Find the dyadic form that $\mathbf{D}^{\text{п}}$ must have. Find $\nabla \cdot \mathbf{S}$ and show that

$$\nabla \cdot \mathbf{S} = (2\mu + \lambda)\nabla(\nabla \cdot \Delta) - \mu\nabla \times \nabla \times \Delta.$$

4. In a widely used notation in elasticity[1] the quantities

$$e_{xx} = \delta_{xx},\ e_{yy} = \delta_{yy},\ e_{zz} = \delta_{zz},$$
$$e_{xy} = 2\delta_{xy},\ e_{yz} = 2\delta_{yz},\ e_{zx} = 2\delta_{zx},$$

are called the "strain components." Moreover, the stress components are written in terms of the notation in Sec. 13.3

$$X_x = X_{xx},\ Y_y = X_{yy},\ Z_z = X_{zz},$$
$$X_y = X_{yx},\ X_z = X_{zx},\ Y_z = X_{zy}$$
$$(\text{with } X_y = Y_x = X_{xy}, \text{ etc.}).$$

Hooke's law in its most *general* form is then written as follows

$$e_{xx} = s_{11}X_x + s_{12}Y_y + s_{13}Z_z + s_{14}Y_z + s_{15}Z_x + s_{16}X_y,$$
$$e_{yy} = s_{21}X_x + s_{22}Y_y + s_{23}Z_z + s_{24}Y_z + s_{25}Z_x + s_{26}X_y,$$
$$\cdots\cdots\cdots\cdots\cdots\cdots\cdots\cdots\cdots\cdots\cdots\cdots$$
$$e_{xy} = s_{61}X_x + s_{62}Y_y + s_{63}Z_z + s_{64}Y_z + s_{65}Z_x + s_{66}X_y,$$

where s_{11}, s_{12} (36 in number) are called elastic *constants*. These may be solved for the stress components in terms of the strain components, i.e.,

$$X_x = c_{11}e_{xx} + c_{12}e_{yy} + c_{13}e_{zz} + c_{14}e_{yz} + c_{15}e_{zx} + c_{16}e_{xy}$$
$$Y_y = c_{21}e_{xx} + c_{22}e_{yy} + c_{23}e_{zz} + c_{24}e_{yz} + c_{25}e_{zx} + c_{26}e_{xy}$$
$$\cdots\cdots\cdots\cdots\cdots\cdots\cdots\cdots\cdots\cdots\cdots\cdots$$
$$Xy = c_{61}e_{xx} + c_{62}e_{yy} + c_{63}e_{zz} + c_{64}e_{yz} + c_{65}e_{zx} + c_{66}e_{xy},$$

[1]Cf., e.g., C. Zener, *Elasticity and Anelasticity of Metals* (Univ. of Chicago Press, 1948) pp. 7ff.

where the c_{ij} are called the elastic *coefficients*. Prove that the matrix or array of the c's is the "reciprocal" of the matrix of the s's, i.e., their product equals the unit matrix. Thus

$$\|s\| \cdot \|c\| = \begin{Vmatrix} 1 & 0 & 0 & 0 & 0 & 0 \\ 0 & 1 & 0 & 0 & 0 & 0 \\ 0 & 0 & 1 & 0 & 0 & 0 \\ 0 & 0 & 0 & 1 & 0 & 0 \\ 0 & 0 & 0 & 0 & 1 & 0 \\ 0 & 0 & 0 & 0 & 0 & 1 \end{Vmatrix}$$

5. In Prob. 4, prove that the c's and s's are not all independent, i.e., show that $c_{ij} = c_{ji}$ and $s_{ij} = s_{ji}$.

6. Apply the general equations in Prob. 4 to an isotropic body and find the c's and s's in terms of the bulk and shear moduli.

7. For a quartz fiber Young's modulus is 5.179×10^{11} dynes/cm^2 and the shear modulus is 2.88×10^{11} dynes/cm^2. Calculate Poisson's ratio and comment on any interesting physical significance of the result. Calculate the c's and s's for this material.

8. For a crystal of cubic symmetry like the metal gold, e.g., the c matrix reduces to the form

$$\begin{Vmatrix} c_{11} & c_{12} & c_{12} & 0 & 0 & 0 \\ c_{12} & c_{11} & c_{12} & 0 & 0 & 0 \\ c_{12} & c_{12} & c_{11} & 0 & 0 & 0 \\ 0 & 0 & 0 & c_{44} & 0 & 0 \\ 0 & 0 & 0 & 0 & c_{44} & 0 \\ 0 & 0 & 0 & 0 & 0 & c_{44} \end{Vmatrix}$$

and the s matrix has precisely the same form with s replacing c. Prove that for a cubic crystal

$$c_{11} = \frac{s_{11} + s_{12}}{(s_{11} - s_{12})(s_{11} + 2s_{12})}$$

$$c_{12} = \frac{-s_{12}}{(s_{11} - s_{12})(s_{11} + 2s_{12})}$$

$$c_{44} = \frac{1}{s_{44}},$$

and solve also for the s's in terms of the c's. Show that the bulk modulus is given by

$$B = \frac{c_{11} + 2c_{12}}{3}.$$

For crystalline gold, $s_{11} = 2.33 \times 10^{-12} \text{cm}^2/\text{dyne}$, $s_{12} = -1.07 \times 10^{-12}$ cm^2/dyne and $s_{44} = 2.38 \times 10^{-12}$ cm^2/dyne. Calculate the bulk modulus and compressibility and compare with the values for polycrystalline gold.

9. A possible modification of Hooke's law to take account of the "relaxation time" effect (cf. Sec. 4.6) may be written in the form

$$X = A\delta + B\dot{\delta},$$

where X is the stress and δ the corresponding strain, and A and B are constants. Suppose $X = 0$ for $0 < t < t_0$, $X = X_0 =$ constant for $t_0 < t < t_1$ and $X = 0$ for $t > t_1$. Solve to find the dependence of δ on the time. Give a plot of δ versus time, using reasonable values for A and B.

10. Prove that the gradient of a scalar function of space is a covariant tensor of the first order. Is any three-vector a covariant tensor of the first order?

11. A tensor of the second order of which the type term is K_{ij} is symmetrical if $K_{ij} = K_{ji}$ and antisymmetrical if $K_{ij} = -K_{ji}$ (note that this implies $K_{ii} = 0$ for any i). Prove that any tensor of the second order can be expressed as the sum of a symmetrical and an antisymmetrical tensor (cf. eq. 13.18).

12. Show that if K_{ij} (all real constants) is a symmetrical tensor the equation

$$\sum_{i,j=1}^{3} K_{ij} x_i x_j = C,$$

where C is a constant, represents a central quadric surface with center at the origin.

13. Using the result of Prob. 12, prove that by suitable transformation of axes any symmetrical tensor may be reduced to diagonal form (cf. eq. 13.56).

14. Show that if a torsional or twist disturbance travels along a hollow tube of inner radius r and thickness $\Delta r(\ll r)$, the wave equation becomes

$$\ddot{\theta} = \frac{\mu}{\rho} \cdot \frac{d^2\theta}{dx^2},$$

where θ is the angular displacement of any section distant x from the origin. Note the wave velocity and compare with Sec. 13.5.

15. Can an elastic solid in general propagate a shear wave without at the same time propagating a dilatational wave? Discuss some special (even if idealized) cases.

16. Given the symmetric tensor of the second order with type term K_{ij}. The divergence of this tensor is a vector defined as follows

$$(\text{div } K)_j = \frac{\partial K_{ij}}{\partial x_i},$$

where we are employing the summation convention described in Sec. 13.2, and the left-hand side represents symbolically the jth component of the divergence of the tensor K (i and j run from 1 to 3). Prove the divergence theorem for tensors, i.e.,

$$\int (\text{div } K)_j d\tau = \int K_{jk} dS_k = \int (K_{j1} dS_1 + K_{j2} dS_2 + K_{j3} dS_3),$$

Here the expression on the left is the volume integral of the jth component of the divergence of K. In the surface integral dS_1 is the element of area perpendicular to the x axis, etc.

CHAPTER XIV

WAVE MOTION IN SOLID AND FLUID MEDIA

14.1. Waves in an Elastic String and Fluid in a Tube. In Sec. 13.5 we encountered for the first time the notion of wave propagation. It is sufficiently important to deserve a whole chapter. We begin it by considering what is probably the simplest illustration of wave motion, that of small transverse displacements in an infinite perfectly flexible elastic string of line density ρ_l. With reference to Fig. 14.1 let the string be suspended vertically from a rigid ceiling at point O. The disturbance whose motion along the string constitutes the wave is the horizontal displacement ξ, which will be assumed to be so small compared with the disturbed part of the string at any instant that the element of length of the disturbed part ds is approximately equal to dz. It is assumed that the linear tension of the string is T and that this is unaffected appreciably by the disturbance. The equation of motion of the element ds under the resultant elastic force is then

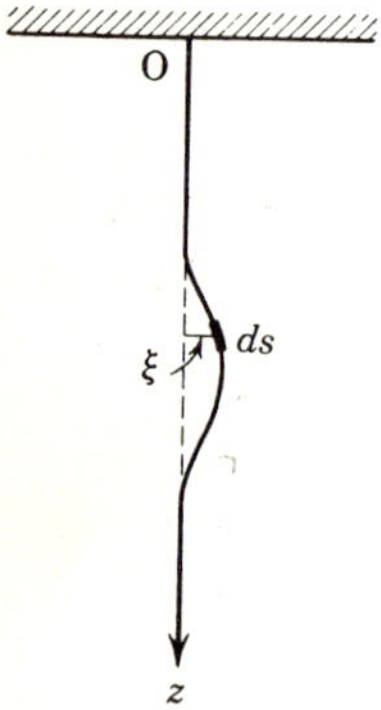

Fig. 14.1

$$T\left[\frac{\partial\xi}{\partial z}+\frac{\partial}{\partial z}\left(\frac{\partial\xi}{\partial z}\right)dz\right]-T\frac{\partial\xi}{\partial z}=\rho_l\ddot{\xi}dz. \tag{14.1}$$

This reduces to

$$\ddot{\xi}=T/\rho_l\cdot\frac{\partial^2\xi}{\partial z^2}, \tag{14.2}$$

an equation of precisely the same form as those in (13.68). Consequently the general solution is

$$\xi=\xi_1(z-ct)+\xi_2(z+ct), \tag{14.3}$$

corresponding to wave motion up and down the string with velocity c given by

$$c=\sqrt{T/\rho_l}. \tag{14.4}$$

The discerning reader will observe some crudities in the preceding derivation. The neglect of higher order derivatives is justified by the approximation which treats $\left|\frac{\partial \xi}{\partial z}\right| << 1$, and moreover insists on the continuity of ξ and the first derivative, etc. Somewhat more serious would appear to be the physical error involved in writing the equation of motion of the element as if it moves as a whole, whereas actually the element is deformed during its motion. Actually it turns out that this does not have a serious affect on the result, provided that the displacement is small enough. We can see most clearly what is involved here by discussing the corresponding problem of the propagation of a one-dimensional compressional disturbance in a fluid medium[1] confined to a long horizontal rigid tube of constant cross section S.

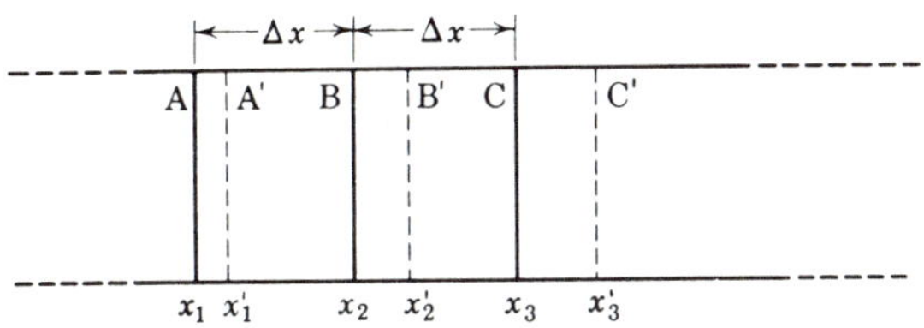

Fig.14.2

In Figure 14.2 the tube is represented as a right circular cylinder directed along the x axis. Let us concentrate attention on the element AB of length Δx, with ends having x coordinates x_1 and x_2, respectively. Thus

$$\Delta x = x_2 - x_1.$$

Now imagine a compression produces a displacement in the x direction. This moves A to A' at distance x_1' from the fixed chosen origin. We shall write $x_1' - x_1 = \xi$. To the first approximation

$$x_2' - x_2 = \xi + \frac{\partial \xi}{\partial x} \cdot \Delta x.$$

Therefore the separation of B' and A' is no longer that of B and A. Rather

$$x_2' - x_1' = \Delta x\left(1 + \frac{\partial \xi}{\partial x}\right). \tag{14.5}$$

[1] Actually the medium might equally well be a solid like rubber which is still more or less elastic for relatively large strains.

Carrying out the same process for BC (originally also equal to Δx) yields

$$x_3' - x_2' = \Delta x\left(1 + \frac{\partial \xi}{\partial x} + \frac{\partial^2 \xi}{\partial x^2} \cdot \Delta x\right). \tag{14.6}$$

In the process of expanding or contracting these volume elements there must be conservation of mass. Therefore we must write

$$\rho_0 S \Delta x = \rho S \Delta x\left(1 + \frac{\partial \xi}{\partial x}\right) = \text{constant}, \tag{14.7}$$

where ρ_0 is the mean equilibrium density of the medium in the undisturbed element AB, whereas ρ is the altered density in $A'B'$. We therefore obtain

$$\rho = \frac{\rho_0}{1 + \dfrac{\partial \xi}{\partial x}}, \tag{14.8}$$

and for the change in density associated with a change in x equal to Δx, if the latter is small enough, so that terms in $(\Delta x)^2$, etc., may be neglected,

$$\Delta \rho = \frac{-\rho_0 \dfrac{\partial^2 \xi}{\partial x^2} \Delta x}{\left(1 + \dfrac{\partial \xi}{\partial x}\right)^2} . \tag{14.9}$$

Now if we denote by p_e the excess compressive stress or pressure at B' (i.e., the difference between the actual pressure and the average equilibrium pressure) it follows from Hooke's law that

$$p_e = \frac{B}{\rho_0} \rho_e , \tag{14.10}$$

where B is the bulk modulus and ρ_e is the excess density at B'. Similarly the excess pressure at C' is

$$p_e' = \frac{B}{\rho_0} \rho_e' . \tag{14.11}$$

Hence the net force due to pressure on the volume element $B'C'$ in the positive x direction is

$$(p_e - p_e')S = \frac{B}{\rho_0}(\rho_e - \rho_e')S\,. \tag{14.12}$$

But

$$\rho_e - \rho_e' = -\Delta\rho\,, \tag{14.13}$$

and hence we may at once employ (14.9) to obtain for the equation of motion of the element in question

$$\frac{BS\dfrac{\partial^2\xi}{\partial x^2}\Delta x}{\left(1+\dfrac{\partial\xi}{\partial x}\right)^2} = \rho_0 S\Delta x\ddot{\xi}, \tag{14.14}$$

where $\ddot{\xi}$ is to a first approximation the second partial derivative of ξ with respect to t if the displacement is not too large.[1] We may consequently write the resultant equation of motion in the form

$$\ddot{\xi} = B/\rho_0 \cdot \frac{\partial^2\xi}{\partial x^2}\Bigg/\left(1+\frac{\partial\xi}{\partial x}\right)^2. \tag{14.15}$$

It will be noted that this is not precisely the wave equation in the form (14.2). The presence of the term $(1 + \partial\xi/\partial x)^2$ in the denominator on the right makes it a nonlinear partial differential equation, and we expect the solution to be more complicated in form than (14.3). In deriving this equation we made no assumption that $|\partial\xi/\partial x| \ll 1$. We merely threw away all terms in Δx of power higher than the first, i.e., concentrated on a *small* volume element. Consequently the equation (14.15) is a more precise description of what goes on in the disturbed medium than that obtained neglecting $\partial\xi/\partial x$ compared with unity. However, if we introduce the latter approximation, eq. (14.15) immediately reduces to

$$\ddot{\xi} = B/\rho_0 \cdot \frac{\partial^2\xi}{\partial x^2}, \tag{14.16}$$

which is in the standard form and corresponds to compressional waves traveling through the medium in the tube with velocity

$$c = \sqrt{B/\rho_0}\,. \tag{14.17}$$

[1]Cf. Sec. 14.8 for a more complete discussion of this point.

It is customary to refer to (14.15) as the equation for one-dimensional compressional waves of *finite* amplitude, whereas (14.16) refers to wave propagation for an infinitesimal disturbance. We shall not consider the solution for the finite amplitude case here.[1]

14.2. Harmonic Waves. Intensity. Although there is nothing essentially periodic about a wave in general, it can be readily shown that as soon as wave motion is set up in a *bounded* medium, periodicity in time and space will ensue. This suggests the study of periodic waves as of particular importance. The simplest type of periodic wave is the *harmonic,* in which the wave function is circular. Thus

$$\xi = A \cos k\,(z - ct) \tag{14.18}$$

denotes a harmonic progressive wave in the positive z direction. A is the amplitude of the wave or maximum displacement from equilibrium. The significance of k is readily seen from the diagram (Fig. 14.3) which represents a "snapshot" picture of the wave, where λ is the distance between successive points for which the phase differs by 2π, known as the *wave length.* Then

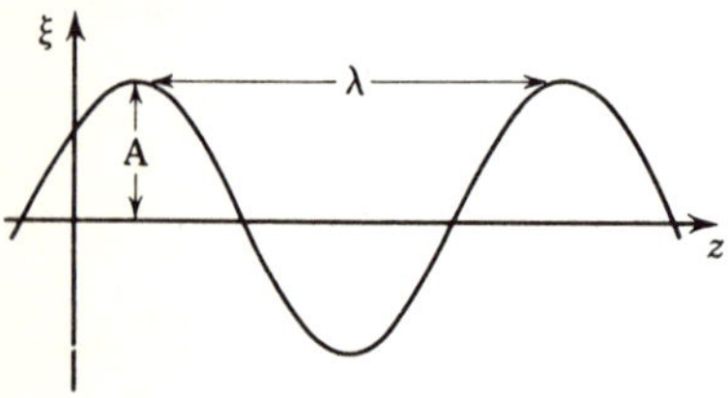

Fig. 14.3

$$k = 2\pi/\lambda. \tag{14.19}$$

The time it takes the disturbance at any point to recur to its original value is the period P of the wave and the number of such periods per second is the frequency ν. Then

$$\nu = \frac{1}{P} = \frac{kc}{2\pi} = \frac{c}{\lambda}. \tag{14.20}$$

This simple relation between frequency and wave length holds for harmonic waves of all kinds.

For many purposes it will be convenient to replace (14.18) by the complex notation

$$\xi = Ae^{i(\omega t - kz)}, \tag{14.21}$$

where $\omega = 2\pi\nu$, and is sometimes called the angular frequency. This is particularly useful when differentiation and integration of wave functions

[1] But cf. Prob. 15 at the end of the chapter. Cf. H. Lamb, *The Dynamical Theory of Sound* (New York, Dover Publications), pp. 177 ff.

are called for. Care must be taken, however, not to indulge in multiplication, squaring, etc., except with the real part.

Since every disturbance in a medium involves energy, wave motion implies a transmission of energy. Let us examine this for the string discussed in Sec. 14.1. The kinetic energy associated with the motion of the element dz of the string is clearly

$$K = \frac{\rho_l}{2} \dot{\xi}^2 \, dz. \tag{14.22}$$

The potential energy for the same element is the work done in the stretching of the element dz to its new length ds, where

$$ds = \sqrt{1 + \left(\frac{\partial \xi}{\partial z}\right)^2} \, dz. \tag{14.23}$$

Since the tension is assumed to remain unchanged at T during this process, the potential energy takes the form

$$\begin{aligned} V &= T \left[\sqrt{1 + \left(\frac{\partial \xi}{\partial z}\right)^2} \, dz - dz \right] \\ &= \frac{T}{2} \left(\frac{\partial \xi}{\partial z}\right)^2 dz, \end{aligned} \tag{14.24}$$

using in the expansion the same approximation employed in Sec. 14.1. The total energy per unit length, which is naturally termed the *energy density*, then becomes

$$w = \frac{\rho_l}{2} \dot{\xi}^2 + \frac{T}{2} \left(\frac{\partial \xi}{\partial z}\right)^2 . \tag{14.25}$$

This is a quantity which varies with both z and t. For a wave progressing along the string in the positive z direction, where

$$\xi = \xi(z - ct),$$

in which $c = \sqrt{T/\rho_l}$, w reduces to

$$w = \rho_l c^2 \xi'^2, \tag{14.26}$$

where

$$\xi' = \frac{d\xi}{d(z - ct)} .$$

As an illustration we shall specialize to a simple harmonic wave for which

$$\xi = A \cos(\omega t - kz),$$

Then

$$w = \rho_l c^2 A^2 k^2 \sin^2(\omega t - kz), \tag{14.27}$$

which fluctuates both in time and space. Physical significance can be given to the time average (over a time long compared with the period of the wave) or the space average (over a distance long compared with the wave length). In either case the result is

$$\overline{w} = \frac{1}{2}\rho_l \omega^2 A^2. \tag{14.28}$$

Since this is the *average* energy in unit length of the string due to the wave motion, the *average* rate of *flow* of energy along the string can be found by multiplication by the velocity c. This yields a quantity we shall call the *intensity* of the wave. Thus

$$I = \frac{1}{2}\rho_l c\omega^2 A^2. \tag{14.29}$$

The dependence on square of frequency and amplitude is characteristic of harmonic waves in general. There is another way of looking at the intensity which is instructive. We may view it as the average rate at which work is done by the disturbance in the string, or the average power contributed by the propagated disturbance. The latter is simply the average of the product of the velocity of the string at any point multiplied by the component of the force in the direction of the motion. Thus we ought to have

$$I = \overline{T\frac{\partial \xi}{\partial z} \cdot \dot{\xi}}. \tag{14.30}$$

If we substitute from the expression for ξ and average over a time τ long compared with the period we obtain again precisely (14.29).

The string is here treated as a medium without thickness. The question arises: What shall we do about the intensity for a medium like the fluid in the tube considered in Sec. 14.1? Here presumably we must talk about energy density as energy per unit *volume*. The kinetic energy density becomes

$$\frac{1}{2}\rho_0 \dot{\xi}^2, \tag{14.31}$$

and the potential energy density

$$\frac{1}{2}B\left(\frac{\partial \xi}{\partial x}\right)^2. \qquad (14.32)$$

The total average energy density for a harmonic wave in the x direction is then

$$\overline{w} = \frac{1}{2}\rho_0 \omega^2 A^2, \qquad (14.33)$$

corresponding closely to (14.34), save that the *actual* mean density ρ_0 appears in place of the line density. The intensity of the compressional wave in the tube then is

$$I = \frac{1}{2}\rho_0 c\omega^2 A^2, \qquad (14.34)$$

where now, of course, $c = \sqrt{\beta/\rho_0}$. It is of interest to observe that the intensity can be expressed in terms of the maximum excess pressure in the wave by the use of (14.10) coupled with (14.8). From the latter, if we assume $\left|\frac{\partial \xi}{\partial x}\right| \ll 1$, we get

$$\rho_e = -\rho_0 \frac{\partial \xi}{\partial x}, \qquad (14.35)$$

and hence from the former

$$p_e = -B\frac{\partial \xi}{\partial x} = -\rho_0 c^2 \frac{\partial \xi}{\partial x}. \qquad (14.36)$$

Using this for the harmonic wave for which (14.34) holds, we secure in its place the alternative expression for the intensity

$$I = \frac{1}{2}\frac{p_{e,\,max}^2}{\rho_0 c}, \qquad (14.37)$$

with $p_{e,max}$ denoting the amplitude of p_e. It must be emphasized that now the intensity is the average rate of flow of energy per unit area of the medium, where the unit area is perpendicular to the direction of wave propagation. We may also define intensity as the average rate at which power is transmitted by the medium per unit area. Thus

$$I = \overline{p_e \dot{\xi}}. \qquad (14.38)$$

Inspection shows that this gives the same result as (14.37). We shall recur to this in the study of acoustic waves in Sec. 14.8.

It is often of importance to represent the difference in intensity of two waves. A special notation has been invented for this which has proved of particular value in acoustics, although it is perfectly general. If two waves have intensities I_1 and I_2, respectively, they are said to differ in intensity level by δ *decibels,* where

$$\delta = 10 \log_{10} I_2/I_1. \tag{14.39}$$

Thus if $I_2 = 2I_1$, $\delta = 3.01$ decibels, usually abbreviated to *db*.

In our definition of intensity we probably did not put sufficient emphasis on the averaging process. Strictly speaking (14.30) holds only if $\tau \to \infty$. Actually the flow of energy in a harmonic wave fluctuates with the time and the fluctuation will be registered on a measuring instrument if the latter averages over too short a time. All measuring instruments take a certain time for the performance of the measurement, and if the quantity being measured changes during this time we expect the measurement to result in some kind of average of this quantity. This point is perhaps best brought out by an examination of the situation resulting from the superposition of two or more harmonic waves progressing in the same direction. If the waves are compressional waves in air (sound waves), it is well known that the phenomenon of beats may be observed if the two frequencies are close enough together. We shall find it simplest to limit ourselves to two harmonic waves of angular frequencies ω_1 and ω_2 progressing with the same velocity in the x direction. If their amplitudes are respectively A_1 and A_2, the resultant displacement takes the form

$$\xi = A_1 \cos \omega_1 (t - x/c) + A_2 \cos \omega_2 (t - x/c). \tag{14.40}$$

Let us now form $\overline{p_e \dot{\xi}}$ from

$$\dot{\xi} = -[A_1\omega_1 \sin \omega_1(t - x/c) + A_2\omega_2 \sin \omega_2 (t - x/c)], \tag{14.41}$$

$$p_e = -\rho_0 c^2 \frac{\partial \xi}{\partial x} = -\rho_0 c[A_1\omega_1 \sin \omega_1 (t - x/c) + A_2\omega_2 \sin \omega_2 (t - x/c)]. \tag{14.42}$$

The result for the intensity is

$$I = \frac{\rho_0 c}{2}(A_1{}^2\omega_1{}^2 + A_2{}^2\omega_2{}^2)$$

$$+ 2\rho_0 c A_1 A_2 \omega_1 \omega_2 \cdot \overline{\sin \omega_1 (t-x/c) \sin \omega_2 (t-x/c)}. \tag{14.43}$$

It is assumed that the time interval τ over which the averaging process is carried out is much greater than both $2\pi/\omega_1$ and $2\pi/\omega_2$, which are the periods of the two individual waves, respectively. If we carry out the time average indicated in the second term on the right in (14.43) the result is

$$\overline{\sin \omega_1(t - x/c) \cdot \sin \omega_2(t - x/c)}$$
$$= \frac{1}{\tau}\left[\frac{\sin (\omega_1 - \omega_2)(t - x/c)}{2(\omega_1 - \omega_2)} - \frac{\sin (\omega_1 + \omega_2)(t - x/c)}{2(\omega_1 + \omega_2)}\right]_0^\tau.$$

The second term in the bracket on the right is negligible compared with the first unless $\omega_1 >> \omega_2$ or $\omega_1 << \omega_2$. Let us rule out this case. Putting in the limits, we can write

$$\overline{\sin \omega_1(t - x/c) \cdot \sin \omega_2(t - x/c)}$$
$$= \frac{\sin (\omega_1 - \omega_2)(\tau - x/c) + \sin (\omega_1 - \omega_2)x/c}{2\tau(\omega_1 - \omega_2)}. \tag{14.44}$$

Now if

$$\tau >> 2\pi/(\omega_1 - \omega_2), \tag{14.45}$$

(14.44) shows that the second term on the right in (14.43) is negligibly small and hence in this case the intensity of the compound wave reduces to the *constant* value

$$I = \frac{\rho_0 c}{2}(A_1{}^2\omega_1{}^2 + A_2{}^2\omega_2{}^2), \tag{14.46}$$

which is just the sum of the individual intensities. On the other hand, it is perfectly possible that

$$\tau < 2\pi/(\omega_1 - \omega_2), \tag{14.47}$$

even when

$$\tau >> 2\pi/\omega_1, \; \tau >> 2\pi/\omega_2.$$

In this case the average defined by (14.38) becomes a fluctuating function of the time. Any measuring instrument which averages over a time interval satisfying (14.47) will clearly detect changes in intensity in the resultant wave It is these changes which the ear detects as beats in the acoustic case. If, however, the measuring instrument averages over a time interval satisfying (14.45) no fluctuations will be observed and a constant intensity will be measured.

14.3. Standing Waves. Fourier Analysis. The reference in the latter part of the preceding section to the superposition of waves traveling in the same direction prompts a consideration of what may happen when waves meet while traveling in opposite directions. We feel pretty sure this will happen, for example, when waves traveling in the positive x direction encounter an obstacle and undergo reflection. The simplest possible case is provided by a finite string of length l stretched with tension T between two fixed points at which no motion can ever take place.

The general harmonic solution for this set-up is, to use the convenient complex notation,

$$\xi = Ae^{i(\omega t - kz)} + Be^{i(\omega t + kz)}, \tag{14.48}$$

where the amplitudes A and B are in general also complex. Eq. (14.48) must satisfy the boundary conditions that $\xi = 0$ at $x = 0$ and l, i.e., at the ends of the string, at all t. Applying these we have

$$A + B = 0, \tag{14.49}$$

$$Ae^{-ikl} + Be^{ikl} = 0, \tag{14.50}$$

which lead to

$$\sin kl = 0. \tag{14.51}$$

The latter condition restricts the possible frequencies of harmonic wave propagation along the string to the discrete set

$$\nu_n = nc/2l, \tag{14.52}$$

where n takes on the positive integral values 1, 2, 3. . ., and c is given by (14.4). The lowest frequency corresponds to $n = 1$ and is called the *fundamental*; the others are *harmonics*, and the whole set form the characteristic frequencies or modes of vibration of the stretched string.

If we write $2A = a + ib$, the real part of ξ for the nth mode becomes

$$\xi_n = (a_n \sin 2\pi\nu_n t + b_n \cos 2\pi\nu_n t) \sin \frac{n\pi z}{l}. \tag{14.53}$$

It is clear that ξ_n is not a progressive wave function. Nevertheless it is still a function of time and distance along the string. We shall say it represents a *standing or stationary wave*. All points of the string still move harmonically in time with the exception of a discrete set for which

$$\sin n\pi z/l = 0. \tag{14.54}$$

These are called *nodes* or nodal points. The distance between successive ones is clearly $\lambda/2$.

Since the fundamental wave equation (14.2) is linear it follows that the sum of two or more particular solutions is also a solution. We are therefore tempted to assume that the complete harmonic solution of (14.2) subject to the assigned boundary conditions consists of the superposition of all the particular solutions of the form (14.53) as n runs from 1 to ∞. Thus we shall write

$$\xi = \sum_{n=1}^{\infty} (a_n \sin 2\pi\nu_n t + b_n \cos 2\pi\nu_n t) \sin \frac{n\pi z}{l}. \tag{14.55}$$

This assumption, called the *principle of superposition*, was first enunciated by Daniel Bernoulli in 1755. It raises some profound mathematical questions, among others the question of the existence of the summation on the right of (14.55). We shall not enter into these here. They are discussed in mathematical texts on Fourier's theorem and Fourier series.[1]

Clearly the values of the coefficients a_n and b_n in (14.55) depend on the initial conditions imposed on the string. Suppose that $\xi_0(z)$ gives the initial displacement of every point on the string and $\dot{\xi}_0(z)$ the initial velocity. We then have

$$\xi_0(z) = \sum_{n=1}^{\infty} b_n \sin \frac{n\pi z}{l}, \tag{14.56}$$

whereas

$$\dot{\xi}_0(z) = \sum_{n=1}^{\infty} 2\pi\nu_n a_n \sin \frac{n\pi z}{l}. \tag{14.57}$$

[1] Cf., e.g., P. Franklin, *Methods of Advanced Calculus* (New York, McGraw-Hill, 1944) pp. 401 ff.

The evaluation of b_n and a_n proceeds in the standard fashion of the Fourier analysis already used in Sec. 9.6. If we multiply (14.56) through by $\sin\frac{n'\pi z}{l}$, where n' is integral, and integrate from 0 to l we get

$$b_n = \frac{2}{l} \cdot \int_0^l \xi_0(z) \sin\frac{n\pi z}{l} dz. \tag{14.58}$$

Similarly from (14.57)

$$a_n = \frac{2}{\pi nc} \int_0^l \dot{\xi}_0(z) \sin\frac{n\pi z}{l} dz. \tag{14.59}$$

Thus if we know the initial displacement and velocity of every point on the string we can find the displacement and velocity at all subsequent times by substituting from (14.58) and (14.59) back into (14.55).

14.4. Wave Pulses and Their Representation by the Fourier Integral. In Sec. 14.3 we have utilized the Fourier series theorem to expand an assumed periodic function of the time, i.e., the transverse displacement at any point of a finite vibrating string, in an infinite series of harmonic components, i.e., components with frequencies given by eq. (14.52). This was assumed to provide a result valid over all time. We have also used the Fourier series technique to expand an *arbitrary* function of z in terms of harmonic components in *space*, as in eq. (14.56). This function did not need to be periodic in space; in fact as there discussed it represented the initial configuration of the string, which may be quite arbitrary as long as it is physically possible. However, we note that the function is defined over only a finite interval of space, namely between 0 and l. The implication is that an *arbitrary* function[1] of a particular variable may be expanded in a series of harmonic components only over a finite interval of the variable. On the other hand, a *periodic* function of a variable may be expanded in a series of harmonic components over an infinite range of variation of the variable. This is the somewhat loose physical way of expressing the content of the Fourier series theorem.

Now it is often desirable in physical vibration and wave problems to express an arbitrary function in terms of harmonic components with validity over an *infinite* range of variation of the independent variable. Thus we may wish to represent a single finite disturbance, called a *pulse,* in

[1]The function must still satisfy some further mathematical requirements, i.e., be piecewise regular. Cf. Franklin, *loc. cit.*, for details.

terms of harmonic components. Or we may wish to represent a finite wave train, i.e., a progressive wave disturbance which is confined to a limited region of space at any instant of time, in terms of harmonic components. For this purpose a different technique is required. This is provided by the so-called Fourier integral. We shall introduce it by considerations which are not mathematically rigorous but sufficiently plausible for our purposes and emphasize the physical meaning and value of the concept.

Let the function $f(x)$ be piecewise continuous (i.e., continuous save at a finite number of points) with its first derivative over the interval $-l \leqq x \leqq +l$. The Fourier series theorem guarantees that *in this interval* $f(x)$ can be expanded in the series

$$f(x) = \sum_{n=-\infty}^{\infty} a_n e^{in\pi x/l}, \tag{14.60}$$

where n is integral and we are using the complex exponential form for convenience. The coefficients a_n are given by

$$a_n = \frac{1}{2l} \int_{-l}^{+l} f(u) e^{-in\pi u/l} du, \tag{14.61}$$

in which u is a variable with the same range as x. If we substitute from (14.61) into (14.60), we get

$$f(x) = \sum_{n=-\infty}^{\infty} \frac{1}{2l} \cdot \int_{-l}^{+l} f(u) e^{-in\pi u/l} du \cdot e^{in\pi x/l} \tag{14.62}$$

Physically this means that $f(x)$ over the given interval is resolved into an infinite number of discrete harmonic components with periods given by $2l/n$. We now wish to extend the range of representation from $-\infty$ to $+\infty$ on the x axis.

Let $\pi/l = \Delta\alpha$, where α is a new parameter. Then (14.62) becomes

$$f(x) = \frac{1}{2\pi} \sum_{n=-\infty}^{\infty} \Delta\alpha \left[\int_{-l}^{+l} f(u) e^{-in\Delta\alpha \cdot u} du \right] e^{inx\Delta\alpha}, \tag{14.63}$$

which we may rewrite in the form

$$f(x) = \frac{1}{2\pi} \sum_{n=-\infty}^{\infty} \Delta\alpha \int_{-l}^{+l} f(u) e^{in\Delta\alpha(x-u)} du. \tag{14.64}$$

Now

$$\lim_{\Delta\alpha \to 0} \sum_{n=-\infty}^{\infty} \phi(n\Delta\alpha)\Delta\alpha = \int_{-\infty}^{+\infty} \phi(\alpha)d\alpha, \tag{14.65}$$

from the definition of the definite integral. Hence, since $l \to \infty$ as $\Delta\alpha \to 0$, we have

$$f(x) = \frac{1}{2\pi} \cdot \int_{-\infty}^{+\infty} d\alpha \int_{-\infty}^{+\infty} f(u)e^{i\alpha(x-u)}\,du. \tag{14.66}$$

This is the Fourier integral form for $f(x)$. The expression (14.66) hardly does justice to the physical significance of the expansion. It really amounts to a resolution of $f(x)$ into an infinite aggregate of continuously varying harmonic components. This is more clearly brought out by re-writing in the form

$$f(x) = \frac{1}{2\pi} \int_{-\infty}^{+\infty} e^{i\alpha x} d\alpha \int_{-\infty}^{+\infty} f(u)e^{-i\alpha u}\,du. \tag{14.67}$$

The coefficient or amplitude of the harmonic component with frequency $\alpha/2\pi$ is

$$\frac{1}{2\pi} \int_{-\infty}^{+\infty} f(u)e^{-i\alpha u}\,du. \tag{14.68}$$

It is customary to call

$$\overline{f(\alpha)} = \frac{1}{\sqrt{2\pi}} \int_{-\infty}^{+\infty} f(u)e^{-i\alpha u}\,du \tag{14.69}$$

the *Fourier transform* of $f(\alpha)$. It is thus possible to express $f(x)$ in terms of its Fourier transform as follows:

$$f(x) = \frac{1}{\sqrt{2\pi}} \int_{-\infty}^{+\infty} \overline{f(\alpha)}e^{i\alpha x}\,d\alpha. \tag{14.70}$$

To make the Fourier integral appear more concrete let us consider time as the independent variable in place of x and let α be replaced by ω. Then we shall be resolving an arbitrary function of t with suitable mathematical properties into harmonic components with continuously varying frequencies $\omega/2\pi$. Thus once more

$$f(t) = \frac{1}{2\pi} \int_{-\infty}^{+\infty} e^{i\omega t}\, d\omega \int_{-\infty}^{+\infty} f(u) e^{-i\omega u}\, du. \tag{14.71}$$

Suppose that $f(t)$ is both a real and an even function. Then (14.71) must be replaced by

$$f(t) = \frac{1}{2\pi} \int_{-\infty}^{+\infty} \cos \omega t\, d\omega \int_{-\infty}^{+\infty} f(u) \cos \omega u\, du. \tag{14.72}$$

Consulting Fig. 14.4 assume that the function $f(t)$ to be resolved into harmonic components is the finite pulse defined as follows:

$$f(t) = \cos \omega_0 t, \text{ for } |t| < T/2,$$
$$f(t) = 0, \text{ for } |t| > T/2. \tag{14.73}$$

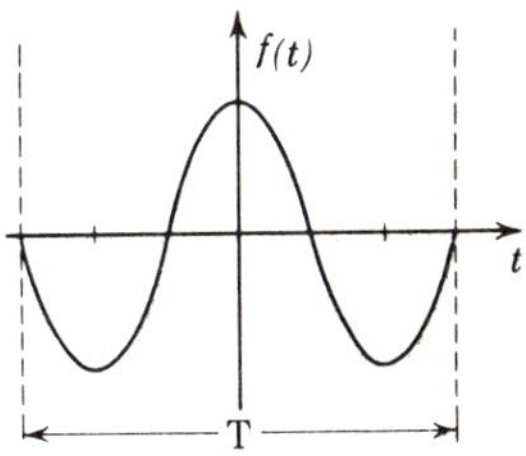

Fig. 14.4

Then the Fourier transform is

$$\overline{f(\omega)} = \frac{1}{\sqrt{2\pi}} \int_{-T/2}^{T/2} \cos \omega_0 u \cdot \cos \omega u\, du$$

$$= \frac{1}{\sqrt{2\pi}} \cdot \left[\frac{\sin \dfrac{(\omega_0 - \omega)T}{2}}{(\omega_0 - \omega)} + \frac{\sin \dfrac{(\omega_0 + \omega)T}{2}}{(\omega_0 + \omega)} \right]. \tag{14.74}$$

Hence

$$f(t) = \frac{1}{\pi} \int_0^{+\infty} \cos \omega t\, d\omega \left[\frac{\sin \dfrac{(\omega_0 - \omega)T}{2}}{(\omega_0 - \omega)} + \frac{\sin \dfrac{(\omega_0 + \omega)T}{2}}{(\omega_0 + \omega)} \right]. \tag{14.75}$$

where we let ω run from 0 to ∞ only since $\cos \omega t$ is an even function and only positive frequencies are meaningful. Now no matter what ω_0 is, the term containing

$$\frac{\sin \dfrac{(\omega_0-\omega)T}{2}}{(\omega_0-\omega)}$$

will predominate over the other for most of the range and we need consider it alone. Hence the amplitude of the harmonic component of frequency $\omega/2\pi$ is very closely

$$\frac{1}{\pi}\cdot\frac{\sin \dfrac{(\omega_0-\omega)T}{2}}{(\omega_0-\omega)} \tag{14.76}$$

If we plot (14.76) as a function of $\omega_0 - \omega$ we get the damped sine wave shown in Fig. 14.5. This function has zeros for

$$(\omega_0 - \omega)T = \pm 2n\pi, \tag{14.77}$$

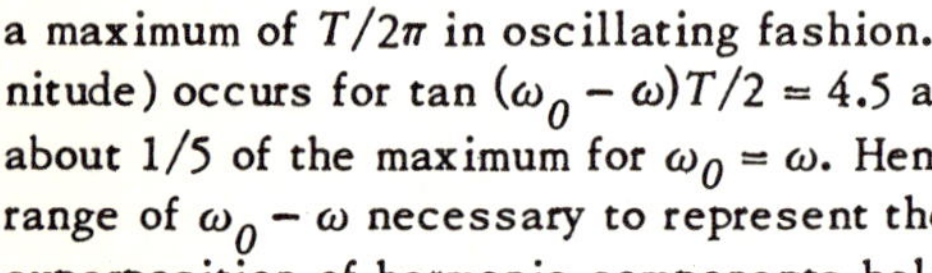

Fig. 14.5

where n is integral. As ω departs from ω_0 the amplitude of the corresponding harmonic component decreases in magnitude from a maximum of $T/2\pi$ in oscillating fashion. The next maximum (in magnitude) occurs for $\tan (\omega_0 - \omega)T/2 = 4.5$ and the amplitude there is about 1/5 of the maximum for $\omega_0 = \omega$. Hence if we denote by $\Delta\omega$ the range of $\omega_0 - \omega$ necessary to represent the pulse satisfactorily as a superposition of harmonic components holding for all t, we shall have

$$\Delta\omega \cdot \Delta t \geqq 2\pi, \tag{14.78}$$

if $T = \Delta t$ = duration time of the pulse. For most practical purposes it will be sufficient to go to the first zero in Fig. 14.5 and use the equal sign only in (14.78).

This has an interesting significance: If Δt is long, i.e., the pulse is of long duration, $\Delta\omega$ can be correspondingly small. Thus if $\Delta t \to \infty$, $\Delta\omega$ can become zero and the pulse can be represented throughout all time by the single harmonic component $\cos \omega_0 t$, as is indeed self-evident. On the other hand, if Δt is short, $\Delta\omega$ becomes large and the range of harmonic components in the representation becomes correspondingly great.

An important practical application of (14.78) occurs in the construction of an amplifier to amplify an electrical pulse. Depending on the sharpness of the pulse the amplifier must have a frequency response sufficiently wide to take care of the whole frequency band into which the pulse is resolved. Otherwise distortion will result.

Recollection of the quantum mechanical indeterminacy principle discussed in Sec. 11.8 suggests a connection with eq. (14.78). Let us rewrite the latter in the form

$$\Delta\nu \cdot \Delta t \geqq 1, \tag{14.79}$$

and attempt to apply it to the radiation of light from an atom. According to quantum mechanics light is emitted in the form of a photon during the transition from one energy state to a lower one of the atom in question and its frequency is equal to the difference between the energies divided by Planck's constant h. However, if the pulse analysis discussion applies to such a photon, we expect that its frequency will not be a precise value but will suffer a spread $\Delta\nu$ depending on the time of observation of the photon. One might suppose that this time of observation could be made as long as one desires and hence that $\Delta\nu$ could always be reduced to any desired small value. However, this is not the case: actually the time of observation of a given photon may not exceed the average lifetime of the state the transition from which gave rise to the photon. Such lifetimes can be predicted from quantum theory to be in general of the order of 10^{-8} sec. This means in general that $\Delta\nu$ will have the average value 10^8 sec^{-1}. The spectral line associated with the photon will therefore have a finite width, i.e., not be absolutely sharp. The figure predicted is indeed about the limit of sharpness of spectral lines which are actually observed and this serves as a confirmation of the view presented here. It also agrees with the general indeterminacy principle of quantum mechanics, since eq. (11.52b) bears a close resemblance to (14.79) and indeed tells the same story with proper interpretation of symbols.

So far we have applied the Fourier integral to the representation of a single vibration pulse, lasting for a finite interval of time, as a superposition of harmonic vibration components. An even more important problem is the harmonic analysis of a finite wave train, since in any actual wave problem we never deal with the infinite wave train represented by the usual sinusoidal solution of the wave equation.

Let the finite wave train be defined as follows:

$$\begin{aligned} f(x-ct) &= \cos k_0(x-ct), \quad \text{for } |x-ct| < L/2, \\ f(x-ct) &= 0, \qquad\qquad\qquad\quad \text{for } |x-ct| > L/2, \end{aligned} \tag{14.80}$$

where $k_0 = 2\pi/\lambda_0 = \omega_0/c$, as usual and ω_0 is the angular frequency in the wave train. The Fourier integral then becomes (the function being real and even)

$$f(x-ct) = \frac{1}{2\pi}\int_{-\infty}^{+\infty} \cos k(x-ct)\,dk \int_{-\infty}^{+\infty} f(u-c\tau)\cos k(u-c\tau)\,d(u-c\tau), \tag{14.81}$$

where u and τ are the new dummy parameters used in the definition of the amplitude of the harmonic component of wave parameter k. We can then write for this amplitude in the present case

$$A(k) = \frac{1}{2\pi}\int_{-L/2}^{L/2} \cos k_0(u-c\tau)\cos k(u-c\tau)\,d(u-c\tau). \tag{14.82}$$

The method used in (14.74) with the associated approximation yields

$$A(k) = \frac{\sin (k_0-k)L/2}{\pi(k_0-k)}. \tag{14.83}$$

The plot of $A(k)$ as a function of $k_0 - k$ is similar to Fig. 14.5. In particular the amplitude vanishes for

$$(k_0-k)L = \pm 2n\pi,$$

where n is any integer, or for

$$\Delta k = k_0 - k = 2\pi n/L. \tag{14.84}$$

The extent of the wave train is L, which we may denote by Δx. Then (14.84) becomes

$$\Delta x \cdot \Delta k \geqq 2\pi. \tag{14.85}$$

This is another indeterminacy relation which means that the longer the wave train the smaller is the range of wave parameters necessary for the resolution into harmonic waves extending throughout all space and over all time. On the other hand, a very short wave train needs a larger range of parameters. We can of course write (14.85) directly in terms of the wave length spread,

$$\Delta x \cdot \Delta(1/\lambda) \geqq 1, \tag{14.86}$$

or alternatively

$$|\Delta x| \cdot |\Delta\lambda| \geqq \lambda^2, \tag{14.87}$$

if the range $\Delta\lambda$ is not too large. The form of (14.86) tempts one to seek a relation with similar indeterminacy relations in quantum mechanics. As will be pointed out in Sec. 14.11 and developed in greater detail in Sec. 17.8, DeBroglie's wave mechanics associates with a particle of momentum mv a finite wave train with mean wave length λ given by

$$\lambda = h/mv. \tag{14.88}$$

On this view variation of wave length within the wave train corresponds to indeterminacy in the momentum of the associated particle, just as variation in the length of the train is associated with interminacy in the position of the particle. In fact, substitution from (14.88) into (14.86) yields

$$\Delta x \cdot \Delta(mv) \geqq h. \tag{14.89}$$

The similarity between this and the quantum mechanical inequality (11.52a) is obvious. Of course it must be emphasized that this interpretation of the indeterminacy principle is based on a pictorial view of quantum mechanics which is now of historical interest only, having been superseded by the more formal theory outlined in Chapter 11.

14.5. Wave Trains in a Dispersive Medium. Group Velocity. The Fourier integral representation of a finite wave train becomes considably more complicated when the wave velocity is a function of frequency, i.e., when the medium is *dispersive,* so that $\omega = \omega(k)$. When the wave train is resolved into the continuous series of harmonic components, there will be an average wave length λ_0 corresponding to an average wave parameter k_0. We now expand the wave phase $k(x - ct) = kx - \omega t$ in a Taylor's series about the mean phase $k_0 x - \omega_0 t$, obtaining

$$kx - \omega t = k_0 x - \omega_0 t + \left(\frac{dk}{dk}\right)_0 (k - k_0)x - \left(\frac{d\omega}{dk}\right)_0 (k - k_0)t + \ldots$$

$$= (k_0 x - \omega_0 t) + (k - k_0)\left[x - \left(\frac{d\omega}{dk}\right)_0 t + \ldots\right] \qquad (14.90)$$

We have here expressed the series in powers of $k - k_0$ and shall assume that the deviation of wave parameter k from the mean k_0 in the wave train is sufficiently small so that the higher powers of $k - k_0$ may be safely neglected. Note that $(d\omega/dk)_0$ means that the value of the derivative is to be taken for $k = k_0$. Now let us represent the finite wave train in perfectly general fashion by $f(x, t)$ and *assume* that it can be expressed in terms of the harmonic components $\cos (kx - \omega t)$ by means of the integral expansion

$$f(x, t) = \int_{-\infty}^{+\infty} A(k) \cos (kx - \omega t)\, dk. \qquad (14.91)$$

By substitution from (14.90) this becomes

$$f(x, t) = \cos (k_0 x - \omega_0 t) \int_{k_0 - \Delta k}^{k_0 + \Delta k} A(k) \cos (k - k_0)\left[x - \left(\frac{d\omega}{dk}\right)_0 t\right] dk$$

$$- \sin (k_0 x - \omega_0 t) \int_{k_0 - \Delta k}^{k_0 + \Delta k} A(k) \sin (k - k_0)\left[x - \left(\frac{d\omega}{dk}\right)_0 t\right] dk. \qquad (14.92)$$

In writing (14.92) we have further assumed that the expansion (14.91) is possible with

$$A(k) = 0, \text{ for } |k| \leqq k_0 + \Delta k; \qquad (14.93)$$

in other words, in the expansion it is not necessary for k to range more widely than finite Δk on either side of k_0. We now express (14.92) more simply in the obvious form

$$f(x, t) = F(x, t) \cos (k_0 x - \omega_0 t) - G(x, t) \sin (k_0 x - \omega_0 t). \qquad (14.94)$$

This can also be written as

$$f(x, t) = R(x, t) \cos [(k_0 x - \omega_0 t) + \alpha(x, t)], \tag{14.95}$$

where

$$R = \sqrt{F^2 + G^2}, \tag{14.96}$$

and

$$\tan \alpha = G/F. \tag{14.97}$$

The important thing is that the wave train $f(x, t)$ is here expressed as a single sinusoidal wave of frequency $\omega_0/2\pi$, but with amplitude $R(x, t)$ and phase $\alpha(x, t)$, which are functions of the time and space through the quantity $x - \left(\frac{d\omega}{dk}\right)_0 t$. If this latter quantity were to stay *constant*, we could think of the wave train as represented by the *single* sinusoidal disturbance with fixed amplitude and phase. Now this quantity will remain constant if we move along the x axis with velocity

$$U = \left(\frac{d\omega}{dk}\right)_0. \tag{14.98}$$

Hence to one moving along the x axis with this velocity the various component waves into which the train is being resolved will appear to superpose. There will then be a point where the superposition will result in maximum amplitude, the various waves heaping up, so to speak, at this point. The velocity with which this point of phase agreement and maximum amplitude moves through the medium, namely (14.98), is called the *group velocity*. The wave train has been resolved into a *group* of waves.

We have

$$\frac{1}{U} = \frac{d(\omega/c)}{d\omega} = \frac{1}{c} - \frac{\omega}{c^2} \cdot \frac{dc}{d\omega}, \tag{14.99}$$

whence

$$U = \frac{c}{1 - \frac{\omega}{c} \cdot \frac{dc}{d\omega}} = \frac{c}{1 - \frac{\nu}{c} \cdot \frac{dc}{d\nu}}. \tag{14.100}$$

Clearly if c is not a function of ν, $dc/d\nu = 0$ and $U = c$, i.e., the group velocity reduces to the phase velocity. In this case all waves in the group move with the same velocity. If, on the other hand, $dc/d\nu > 0$, $U > c$, where c here appears as the *mean* velocity for the waves of the group. If $dc/d\nu < 0$, $U < c$. Both cases arise in practice, as the following illustrations bring out.

Consider first a solid cylindrical rod which is subject to flexural deformation (cf. Fig. 14.16). We denote the lateral displacement by ξ. It can be shown that the equation of motion is[1]

Fig. 14.6

$$\rho\ddot{\xi} = -YR^2\frac{\partial^4\xi}{\partial x^4}, \tag{14.101}$$

if the rod lies along the x axis and ρ and Y have their ordinary significance in this chapter. R is the radius of gyration of the cross section of the rod with respect to a horizontal diameter. (14.101) is not the wave equation. Nevertheless we find it interesting to try a harmonic progressive wave solution in the form

$$\xi = A\cos(\omega t - kx). \tag{14.102}$$

This is found to satisfy the equation of motion if

$$c = \omega/k = (YR^2/\rho)^{1/4}\omega^{1/2}. \tag{14.103}$$

Evidently as far as lateral one-dimensional flexural waves are concerned, the rod acts as a dispersive medium. From (14.103) it follows at once that

$$\nu/c \cdot dc/d\nu = 1/2. \tag{14.104}$$

Hence the group velocity

$$U = 2c, \tag{14.105}$$

a physically somewhat remarkable result.

A more obvious example is provided by waves on the surface of water. According to Rayleigh[2] the velocity of such waves due to gravity and surface tension is given by

$$c = \sqrt{(g/k + Tk/\rho)\tanh kl}, \tag{14.106}$$

[1] Rayleigh, *Theory of Sound* (New York, Dover Publications, 1945), para. 163.
[2] Rayleigh, *loc. cit.* para. 353.

in which l is the depth of the water and T is the surface tension. The other symbols have the usual significance. For simplicity we consider first, waves in shallow water for which $kl \ll 1$ or $l \ll \lambda$. Then $\tanh kl \doteq kl$, and (14.106) becomes

$$c \doteq \sqrt{gl + k^2 Tl/\rho}. \tag{14.107}$$

Now for water T is of the order of 75 dynes/cm. Under the conditions first assumed, therefore the second term in the radical is negligible and

$$c \doteq \sqrt{gl} = \text{constant}. \tag{14.108}$$

In this case c becomes effectively constant and the group velocity is equal to the phase velocity, i.e., there is no dispersion.

If now, on the other hand, $l \gg \lambda$, so that $kl \gg 1$, $\tanh kl \doteq 1$, and

$$c \doteq \sqrt{g/k + Tk/\rho}. \tag{14.109}$$

For deep water long gravity waves, k is small and

$$c \doteq \sqrt{g/k}, \tag{14.110}$$

leading to

$$dc/d\omega = -g\omega^{-2},$$

and

$$U = c/2. \tag{14.111}$$

On the other hand, for ripples λ is small and the second term in the radical in (14.109) predominates, so that

$$c \doteq \sqrt{Tk/\rho}, \tag{14.112}$$

whence

$$U = 3c/2. \tag{14.113}$$

These different types of waves and values of group velocity have been studied experimentally and the above results verified in a general way.

Coming now to electromagnetic waves in general (cf. Chapter 16) and light waves in particular we find it useful to rewrite (14.100) in the form

$$U = c - \lambda \cdot dc/d\lambda, \tag{14.114}$$

which emphasizes the dependence of the phase velocity on the wave length. It will be recalled from elementary physics that in the so-called normal dispersion of light $dc/d\lambda > 0$. Therefore we expect for visible light in normal dispersion

$$U < c.$$

The effect of this on the experimental measurement of the velocity of light is interesting. When Michelson measured the velocity of light in CS_2 he found that whereas the index of refraction would imply a velocity 0.61 of that in air or vacuum, the actual fraction was 0.57. CS_2 is a dispersive medium in which $dc/d\lambda$ is a by no means negligible quantity. Hence Michelson concluded that his experiment actually measured the group velocity. This raises an interesting question about the relation between phase and group velocity, for it suggests that the phase velocity is involved in the geometrical bending of the wave represented by the index of refraction (cf. Sec. 16.7), whereas the group velocity is the measure of the rate of flow of energy in the beam. This demands further consideration which we are not in a position to supply here. Suffice it to say that the suggestion turns out to be substantially correct, although certain difficulties arise in the region of anomalous dispersion for which $dc/d\lambda < 0$ and hence from (14.114) $U > c$. According to the theory of relativity (Chapter 17) the maximum velocity at which energy can be transmitted in our physical universe is the velocity c of light in free space. Hence there would appear to be a definite inconsistency here. However we must remember that the region of anomalous dispersion is also an absorption band for the radiation. The question has been studied with some care by L. Brillouin.[1]

14.6. Nature and Motion of an Ideal Fluid. In this chapter so far we have concerned ourselves primarily with wave motion in elastic solids. It is now in order to introduce similar considerations for an ideal fluid, leading up to the discussion of wave motion in elastic liquids and gases and the related acoustical phenomena. An ideal fluid is one that does not support a shear and hence displays volume elasticity only. The only elastic disturbances propagated in such a medium are compressional in character.

Before proceeding to elastic waves in fluids we shall first treat briefly the general motion of an ideal fluid. This is the subject matter of hydrodynamics. We shall continue as far as possible to use the

[1] L. Brillouin, *Les Statistiques Quantiques* (Paris, 1930), Vol. II. pp. 370 ff. Also the same author, *Congres international d'electricite* (Paris, 1932), Vol. II, p. 739.

notation already developed in the description of elastic solids. Thus the displacement vector $\mathbf{\Delta}$ for a particle of fluid at x, y, z at time t is, as before

$$\mathbf{\Delta} = \mathbf{i}\xi + \mathbf{j}\eta + \mathbf{k}\zeta.$$

More important is the displacement velocity

$$v = \dot{\mathbf{\Delta}} \tag{14.115}$$

Operationally this is measured by the volume flow of fluid per unit time per unit area placed normal to the direction of flow, i.e., the direction of $\mathbf{v}$. If we concentrate our attention on a certain particle of fluid it will move from one place to another subject to the forces acting on it. To write the equation of motion it is necessary to develop an expression for the acceleration of the particle. This is evidently not just $\dot{\mathbf{v}}$, since the latter means the time rate of change of velocity at *one* point of the medium, but we must ascertain the rate of change of velocity as the particle moves on *through* the medium.

Assume that the velocity at point P with position vector $\mathbf{r}$ is $\mathbf{v}$ at time t, while the velocity at the neighboring point Q with position vector $\mathbf{r} + d\mathbf{r}$ is $\mathbf{v} + d\mathbf{v}$ at time $t + dt$. If we further suppose that the particle of fluid which was at P at time t has moved to Q at time $t + dt$, it follows that

$$d\mathbf{v} = \dot{\mathbf{v}}\, dt + \mathbf{v} \cdot \nabla\mathbf{v}\, dt, \tag{14.116}$$

in which $\dot{\mathbf{v}}\,dt$ is the contribution to the change in velocity due merely to the passage of time, whereas $\mathbf{v} \cdot \nabla\mathbf{v}\, dt$ is that due to the *motion* of the particle during the time interval dt. The resultant acceleration of the particle then becomes

$$\mathbf{a} = \dot{\mathbf{v}} + \mathbf{v} \cdot \nabla\mathbf{v}. \tag{14.117}$$

Suppose the particle occupies the volume element $d\tau$, so that its mass is $\rho d\tau$. If there exists an external body force of $\mathbf{F}$ per unit mass, the total external force on the particle is $\mathbf{F}\rho d\tau$. In addition to this external force the variation in pressure p across the boundaries of the element $d\tau$ produces a resultant force $-\nabla p d\tau$. This may be seen in the usual way by considering the element $d\tau$ as a rectangular parallelepiped and computing the resultant force due to pressure in the x, y, z directions. For the equation of motion we thus obtain

$$\dot{\mathbf{v}} + \mathbf{v} \cdot \nabla\mathbf{v} = \mathbf{F} - \frac{\nabla p}{\rho}. \tag{14.118}$$

We have already introduced previously (Sec. 12.1) the other fundamental equation governing the behavior of the moving fluid. This is the equation of continuity (cf. eq. 12.17)

$$\nabla \cdot (\rho \mathbf{v}) = -\dot{\rho}, \tag{14.119}$$

repeated here for emphasis and completeness. Every possible motion of an ideal fluid is subsumed in the two equations (14.118, 14.119).

It is possible here to consider only a few illustrations of the fluid equations. The most important application from the standpoint of the present chapter, i.e., that relating to wave motion, is discussed in Sec. 14.8. For the moment let us introduce some energy considerations for the special case of an incompressible fluid. Suppose in (14.118) that $\mathbf{F}$ is a conservative force and therefore

$$\mathbf{F} = -\nabla V, \tag{14.120}$$

where V is the corresponding potential and is assumed to be a function of x, y, z only and not an implicit function of the time. If we dot multiply both sides of (14.118) with $\mathbf{v}$ we get

$$\mathbf{v} \cdot \dot{\mathbf{v}} + \mathbf{v} \cdot (\mathbf{v} \cdot \nabla \mathbf{v}) = -\mathbf{v} \cdot \nabla V - \mathbf{v} \cdot \nabla p/\rho. \tag{14.121}$$

It is not difficult to show with a little vector analysis that the left-hand side can be so rewritten that the equation reads

$$\frac{\rho}{2} \cdot \frac{d}{dt}(\mathbf{v} \cdot \mathbf{v}) + \rho \frac{dV}{dt} = -\mathbf{v} \cdot \nabla p. \tag{14.122}$$

We have here utilized the fact that the time rate of change of any function of space and time, e.g., $\mathbf{v} \cdot \mathbf{v}$, as the fluid flows, is given by

$$\frac{d}{dt}(\mathbf{v} \cdot \mathbf{v}) = \frac{\partial}{\partial t}(\mathbf{v} \cdot \mathbf{v}) + \mathbf{v} \cdot \nabla(\mathbf{v} \cdot \mathbf{v}). \tag{14.123}$$

The presence of dV/dt in (14.122) is accounted for similarly. Let us integrate (14.122) over any portion of space occupied by the fluid and apply the divergence theorem to the right-hand side. The result is

$$\frac{d}{dt}\left[\int 1/2 \cdot \rho v^2 \, d\tau + \int \rho V \, d\tau\right] = -\int p\mathbf{v} \cdot \mathbf{n} \, dS, \tag{14.123}$$

where $\mathbf{n}$ as usual is the outward drawn unit normal to the surface. In writing the right-hand side in the preceding equation we have made use of the fact that the fluid is incompressible, i.e.,

$$\nabla \cdot \mathbf{v} = 0. \tag{14.124}$$

Now $\int 1/2 \cdot \rho v^2 \, d\tau$ is certainly the total kinetic energy of the fluid contained in the volume in question at any given instant and the time derivative represents the time rate of change of the kinetic energy as the fluid moves. It is natural then to interpret $\int \rho V \, d\tau$ as the total instantaneous potential energy of the fluid. The left-hand side consequently represents the time rate of change of the total energy of the fluid. If we examine the right-hand side, we find it is equal to the negative of the rate at which the pressure of the fluid under consideration does work on the fluid outside this volume. This is a kind of work-energy theorem for an incompressible fluid. If we like, we may think of the surface integral as indicating the rate of flow of energy across the boundary. Eq. (14.123) then says that the only way the energy of the fluid inside a given region can change is through the flow of energy across its surface. We shall meet a similar result of importance in the study of the electromagnetic field (cf. Sec. 15.4 on Poynting's theorem).

If the velocity of the fluid is constant in time we speak of the flow as *steady*. At each point in the fluid the direction and magnitude of $\mathbf{v}$ will then remain unchanged. If we construct curves whose tangents at each point are in the direction of $\mathbf{v}$ they will retain their identity in time. They are called *stream lines*. Steady flow can be studied by rewriting the fundamental equation of motion (14.118) in the form (assuming an external potential)

$$\dot{\mathbf{v}} + 1/2 \cdot \nabla(\mathbf{v} \cdot \mathbf{v}) + (\nabla \times \mathbf{v}) \times \mathbf{v} = -\nabla V - \frac{\nabla p}{\rho}, \tag{14.125}$$

which results from the vector identity for any two vectors **A** and **B**

$$\nabla(\mathbf{A} \cdot \mathbf{B}) = \mathbf{A} \cdot \nabla\mathbf{B} + \mathbf{B} \cdot \nabla\mathbf{A} - (\nabla \times \mathbf{B}) \times \mathbf{A} - (\nabla \times \mathbf{A}) \times \mathbf{B}. \tag{14.126}$$

This may be immediately verified from the definition of the curl of a vector given in Sec. 13.1 (eq. 13.20). For steady flow $\dot{\mathbf{v}} = 0$ and the behavior of the fluid hinges on the term $(\nabla \times \mathbf{v}) \times \mathbf{v}$. In Sec. 13.1 we saw that where the deformation of a medium is an infinitesimal rotation about some instantaneous axis the curl of the deformation is twice the angular rotation. From this it follows that the curl of the displacement velocity in the deformable medium is twice the instantaneous angular velocity.

We may then replace $\nabla \times \mathbf{v}$ by $2\,\boldsymbol{\omega}$ and further rewrite (14.125) for steady flow as

$$1/2 \cdot \nabla(\mathbf{v} \cdot \mathbf{v}) + 2\,\boldsymbol{\omega} \times \mathbf{v} = -\nabla V - \frac{\nabla p}{\rho}. \qquad (14.127)$$

Now it may be that the flow of the fluid is such that there is no rotational motion anywhere, i.e., $\boldsymbol{\omega}$ is identically zero. This is called *irrotational* flow and is a very important special case of fluid motion. For this (14.127) becomes, if ρ is constant,

$$\nabla(1/2 \cdot \rho v^2 + \rho V + p) = 0, \qquad (14.128)$$

which means that everywhere in the fluid

$$1/2 \cdot \rho v^2 + \rho V + p = \text{constant}. \qquad (14.129)$$

This is one form of the celebrated Bernoulli equation which states the conservation of energy per unit volume throughout an incompressible fluid in steady, irrotational flow. The applications of this equation are familiar from elementary physics and need no further comment here.

It is important to realize, however, that an equation like (14.129) holds even when the motion is not irrotational, as long as it is steady. Let us denote by $d\mathbf{s}$ the element of distance along a particular stream line at any point. Then clearly $d\mathbf{s} = \mathbf{v}\,dt$. Now dot multiply 14.127 with $d\mathbf{s}$. Since $\boldsymbol{\omega} \times \mathbf{v}$ is perpendicular to $\mathbf{v}$, the term $(\omega \times \mathbf{v}) \cdot d\mathbf{s} = 0$ identically and we get

$$\frac{1}{2}(\nabla v^2) \cdot d\mathbf{s} = -(\nabla V) \cdot d\mathbf{s} - \frac{(\nabla p) \cdot d\mathbf{s}}{\rho}. \qquad (14.130)$$

But from the definition of the gradient it follows that (14.130) may be written

$$d\left(\frac{1}{2}v^2\right) + dV + \frac{dp}{\rho} = 0. \qquad (14.131)$$

The integral of this for a homogeneous, incompressible fluid is obviously eq. (14.129), although to be sure the constant of integration is now associated with the particular stream line under consideration, whereas in (14.129) it holds for the whole fluid. We can of course apply (14.131) to a compressible fluid if we know how ρ depends on p and can integrate $\int dp/\rho$. This will then provide the application of Bernoulli's principle to compressible fluids.

14.7. Stokes' Theorem and Velocity Potential. It is clear that if $\mathbf{v} = \nabla\phi$, where ϕ is a kind of potential which depends on the coordinates, we have

$$\nabla \times \mathbf{v} = 0,$$

identically and the flow is irrotational. Incidentally, it may be shown that if the motion of an ideal fluid is irrotational initially it will remain so forever. We shall call ϕ a *velocity* potential.

It becomes of importance to demonstrate the converse of the above result, i.e., to show that if the curl of $\mathbf{v}$ vanishes identically, then $\mathbf{v}$ must be the gradient of a scalar function; in other words, a velocity potential must exist. To prove this we must introduce the theorem of Stokes, which expresses the line integral of a vector around a closed curve in terms of the surface integral of the normal component of the curl of the vector over any surface of which the closed curve is a boundary. In Fig. 14.7 we show a closed curve in space *ABCD* with a surface cap *S*. We let $d\mathbf{s}$ denote the vector element of arc along the curve. Our problem is to express the line integral of a vector $\mathbf{F}$ around the closed curve, viz.,

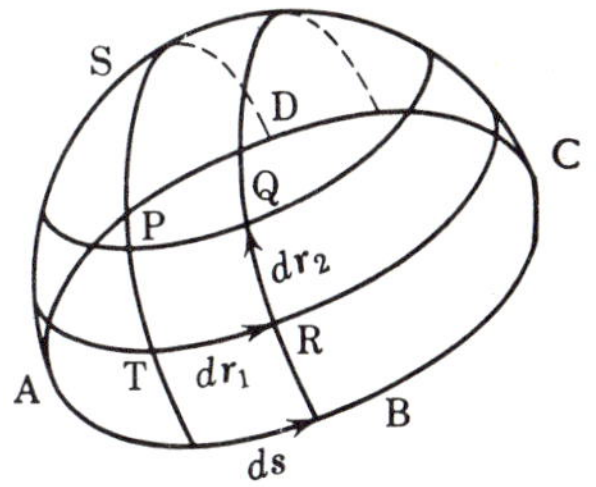

Fig. 14.7

$$\int \mathbf{F} \cdot d\mathbf{s}$$

in terms of an integral over the surface *S*. Let us divide up the surface *S* into elementary parallelograms. Then the line integral around the bounding curve is the sum of the corresponding line integrals around all the elementary parallelograms. For if the integration is conducted in a clockwise direction around each parallelogram (looking along the outward drawn normal) every side of every element is traversed once in one direction and once in the opposite direction, except for the sides which coincide with the boundary curve. If now we consider one of the parallelograms *PQRT* with sides $d\mathbf{r}_1$ and $d\mathbf{r}_2$ and let $(d\mathbf{F})_1$ be the change in $\mathbf{F}$ along $d\mathbf{r}_1$ and $(d\mathbf{F})_2$ that along $d\mathbf{r}_2$, we have

$$\begin{aligned}(d\mathbf{F})_1 &= (d\mathbf{r}_1 \cdot \nabla)\mathbf{F},\\ (d\mathbf{F})_2 &= (d\mathbf{r}_2 \cdot \nabla)\mathbf{F},\end{aligned} \tag{14.132}$$

as may be verified by direct expansion. We let $\mathbf{F}$ have value $\mathbf{F}$ at P, $\mathbf{F}+(d\mathbf{F})_1$ at Q, $\mathbf{F}+(d\mathbf{F})_2$ at T. The line integral around the parallelogram is then

$$\mathbf{F}\cdot d\mathbf{r}_1+[\mathbf{F}+(d\mathbf{F})_1]\cdot d\mathbf{r}_2-[\mathbf{F}+(d\mathbf{F})_2]\cdot d\mathbf{r}_1-\mathbf{F}\cdot d\mathbf{r}_2$$
$$=(d\mathbf{F})_1\cdot d\mathbf{r}_2-(d\mathbf{F})_2\cdot d\mathbf{r}_1. \qquad (14.133)$$

But from (14.132) this line integral becomes

$$\int_{PQRT}\mathbf{F}\cdot d\mathbf{r}=(d\mathbf{r}_1\cdot\nabla)\mathbf{F}\cdot d\mathbf{r}_2-(d\mathbf{r}_2\cdot\nabla)\mathbf{F}\cdot d\mathbf{r}_1. \qquad (14.134)$$

Now for any four vectors $\mathbf{A}$, $\mathbf{B}$, $\mathbf{C}$, $\mathbf{D}$ it can be shown that

$$(\mathbf{A}\times\mathbf{B})\cdot(\mathbf{C}\times\mathbf{D})=(\mathbf{A}\cdot\mathbf{C})(\mathbf{B}\cdot\mathbf{D})-(\mathbf{A}\cdot\mathbf{D})(\mathbf{B}\cdot\mathbf{C}), \qquad (14.135)$$

by the simple expedient of performing the allowable interchange of dot and cross. Hence from (14.134)

$$\int_{PQRT}\mathbf{F}\cdot d\mathbf{r}=(d\mathbf{r}_1\times d\mathbf{r}_2)\cdot(\nabla\times\mathbf{F}). \qquad (14.136)$$

But by definition $d\mathbf{r}_1\times d\mathbf{r}_2=\mathbf{n}\,dS$, where $\mathbf{n}$ is the outward drawn unit normal to the parallelogram element and dS is its area. Therefore

$$\int_{PQRT}\mathbf{F}\cdot d\mathbf{r}=(\nabla\times\mathbf{F})\cdot\mathbf{n}\,dS. \qquad (14.137)$$

The sum of the line integrals around all the elementary parallelograms will be the sum of all terms on the right of (14.137). Hence

$$\int_{ABCD}\mathbf{F}\cdot d\mathbf{s}=\int(\nabla\times\mathbf{F})\cdot\mathbf{n}\,dS. \qquad (14.138)$$

This is Stokes' theorem, whose fundamental significance lies in the fact that it relates a line integral to a surface integral.

We are now ready to note the consequence of the irrotationality of a vector. If $\nabla\times\mathbf{F}=0$ everywhere it follows from Stokes' theorem that the line integral of **F** about every closed curve vanishes. Hence if we consider two points A and B (cf. Fig. 14.8) the line integral from A to B is independent of the path, i.e., is the same for ACB as for ADB. It must therefore depend only on the initial and final points A and B. Since this is true for *all* points A and B, it follows that

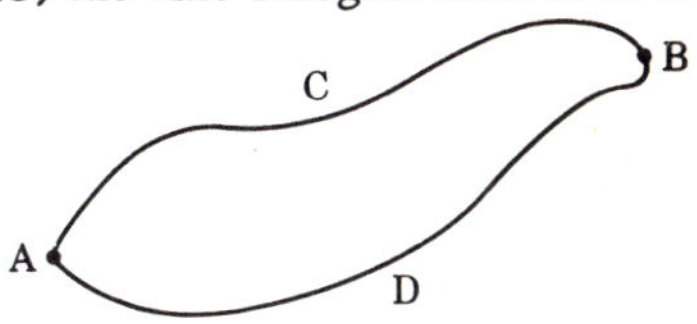

Fig. 14.8

$$\mathbf{F} \cdot d\mathbf{s} = d\phi, \tag{14.139}$$

where ϕ is a scalar point function. Hence we must have

$$\mathbf{F} = \nabla\phi. \tag{14.140}$$

Therefore in the irrotational motion of an ideal fluid there exists a potential for the velocity vector **v**.

14.8. Wave Motion in an Ideal Fluid. Acoustic Radiation. Leaving now the large scale motion of fluids, which to be sure has received and continues to receive much attention in hydraulics and aerodynamics, we revert to the situation which develops when a small displacement is introduced at any point in an ideal fluid, leading to a change in pressure or density there, small compared with the equilibrium values. We suppose this takes place irrotationally so that $\mathbf{v} = \nabla\phi$. What becomes of the equation of motion (14.118) with $\mathbf{F} = 0$, and the equation of continuity (14.119) under these conditions?

We introduce the condensation s, defined as

$$s = (\rho - \rho_0)/\rho_0 = \rho_e/\rho_0, \tag{14.141}$$

where ρ_0 is the mean equilibrium density and ρ_e is the excess density, a function of x, y, z, t. We shall be interested only in motions for which $|s| \ll 1$. The equation of continuity then takes the form

$$\nabla^2\phi + \dot{s} = \nabla^2\phi + \dot{\rho}_e/\rho_0 = 0. \tag{14.142}$$

In the equation of motion we shall assume that the magnitude of **v** and the rate of change of **v** with the space coordinates is so small that the term $\mathbf{v} \cdot \nabla\mathbf{v}$ may be neglected compared with $\dot{\mathbf{v}}$. We also introduce the excess pressure

$$p_e = p - p_0, \tag{14.143}$$

where p_0 is the mean equilibrium pressure. Eq. (14.118) then becomes (assuming possibility of interchange of differentiation with respect to space and time)

$$\nabla\dot{\phi} = -\frac{\nabla p_e}{\rho_0}. \tag{14.144}$$

It then follows that $-\dot{\phi}$ and p_e/ρ_0 can differ only by a quantity constant in space. Since we are interested only in those quantities which vary both in space and time, we can write

$$\dot{\phi} = -p_e/\rho_0 \tag{14.145}$$

for the equation of motion under the conditions just laid down. The equation of state of the fluid (cf. Chapter 10) may be written as a relation between excess pressure and excess density, viz.,

$$p_e/\rho_e = c^2, \tag{14.146}$$

where c^2 is a parameter which will in general depend on the temperature of the medium and may depend on pressure and density. The combination of (14.142), (14.145), and (14.146) leads finally to an equation for ϕ in the form

$$\nabla^2\phi - \ddot{\phi}/c^2 = 0. \tag{14.147}$$

This is the wave equation already encountered in (13.65) and in simpler form in (14.2). We are thus justified in concluding that the velocity potential ϕ of the ideal fluid is propagated as a wave with velocity c through the fluid. From (14.145) it follows that p_e and s satisfy the same differential equation and hence are also propagated as waves with the same velocity.

The velocity of the compressional wave in the fluid is given by (14.146). The reader should show from the definition of the bulk modulus B that for a compressional wave in a liquid

$$c = \sqrt{B/\rho}. \tag{14.148}$$

On the other hand, for a gas, since the isothermal bulk modulus is equal in this case to the pressure,

$$c = \sqrt{p/\rho}. \tag{14.149}$$

This is Newton's formula, which however does not agree with the observed velocity of sound in a gas. Laplace showed that it is more reasonable to assume that the compressional disturbance in a gas is adiabatic, leading to

$$c = \sqrt{\gamma p/\rho}, \tag{14.150}$$

where γ is the ratio of specific heat at constant pressure to that at constant volume. This agrees with the observed values of the velocity of sound in gases so well that the formula is often used for the computation of γ.

Although we have assumed (14.147) to be the general wave equation, we have not proved that the general solution of this three-dimensional equation corresponds to wave propagation. This is a somewhat elaborate task[1] into which we shall not enter. However, it is not difficult to show that certain special functions of x, y, z, t which represent wave propagation are solutions of (14.147). Thus consider the arbitrary wave function

$$\phi = f(\alpha x + \beta y + \gamma z - ct), \tag{14.151}$$

in which α, β, γ, and c are constants. Direct substitution shows that (14.151) is a solution of (14.147) provided

$$\alpha^2 + \beta^2 + \gamma^2 = 1. \tag{14.152}$$

In other words, α, β, γ, must behave like the direction cosines of a line. We shall call the argument of the arbitrary function in (14.151) the phase. At the instant t_0 the phase then has the value

$$\alpha x + \beta y + \gamma z - ct_0.$$

This is clearly constant over the surface

$$\alpha x + \beta y + \gamma z - ct_0 = K, \tag{14.153}$$

where K is some constant. This surface is a plane, the direction cosines of whose normal are α, β, γ. At a later instant $t_1 > t_0$ the same value of the phase will be found on a plane parallel to (14.153) and separated from it by the distance $c(t_1 - t_0)$. This justifies us indeed in saying that (14.151) represents wave propagation. Since at any instant the phase remains the same over a plane and this plane can be considered as moving through space in the direction of its normal, carrying its phase value with it, we may properly call the wave a *plane* wave. We shall name the equiphasal surface a *wave front.* These terms have been used in earlier sections but are here for the first time defined.

We have already seen in Sec. 14.3 that periodic waves are by far the more important variety. Hence in the discussion of compressional waves

[1]Cf., e.g., Jeans, *Electricity and Magnetism* (Cambridge Univ. Press, 4th ed., 1923), pp. 521 ff.

in fluid media, i.e., sound or acoustical waves as they are more commonly called, we shall restrict our attention in general to harmonic waves. From (14.151) it should be clear that the general expression for the velocity potential in a plane harmonic wave of frequency $\omega/2\pi = \nu$ progressing in the direction given by the direction cosines α, β, γ is

$$\phi = Ae^{i[\omega t - k(\alpha x + \beta y + \gamma z)]}, \tag{14.154}$$

in which A is the velocity potential amplitude, and $k = \omega/c = 2\pi/\lambda$, as usual. From the velocity potential it is not difficult to get the other wave propagated quantities, using eqs. (14.145) (for the excess pressure) and $\mathbf{v} = \nabla\phi$ (for the fluid velocity).

In particular we may note the important expression for the excess pressure

$$p_e = p_0 c^2 s = -\rho_0 \dot{\phi}. \tag{14.155}$$

Moreover from (14.142)

$$\dot{s} = -\nabla^2\phi = -\nabla \cdot \mathbf{v} = -\nabla \cdot \dot{\Delta}, \tag{14.156}$$

since we are here concerned with very small flow. Integrating with respect to the time and setting the constant of integration equal to zero we obtain

$$s = -\nabla \cdot \Delta, \tag{14.157}$$

which reminds us of eq. (13.60) in our general discussion of deformable media.

The other type of wave front of great interest in sound propagation is, of course, the *spherical.* Let us write eq. (14.147) in spherical coordinates and make the special assumption that ϕ is independent of latitude and longitude and depends on r alone. The ultimate result is

$$\frac{1}{r}\frac{\partial^2}{\partial r^2}(r\phi) = \ddot{\phi}/c^2. \tag{14.158}$$

Our study of the solution of eq. (14.2) shows us that the general solution of (14.158) is

$$\phi = \frac{1}{r} \cdot f_1(r-ct) + \frac{1}{r} \cdot f_2(r+ct). \tag{14.159}$$

Here it is of interest to note that $r\phi$ is a wave function while ϕ itself strictly speaking is not, since it depends on r outside of the combination $r \pm ct$. However, it is customary to say that ϕ is propagated as a spherical wave since, although its magnitude falls off inversely with the distance from the origin, there still exist equiphasal surfaces in the form of s spheres of radii $r = ct$.

The velocity potential for a spherical harmonic wave diverging from the origin is

$$\phi = \frac{A}{r} \cdot e^{i(\omega t - kr)}. \tag{14.160}$$

Here A is the amplitude at unit distance. The significance of ω and k remains as before. With a plus sign between ωt and kr we should be dealing with a harmonic spherical wave *converging* on the origin.

If we investigate the average rate of flow of energy across unit area or alternatively the product of average energy density and wave velocity, i.e., what we earlier denominated the intensity of the wave propagation, we reach a result similar to that already obtained in the simpler problem of a compressional wave traveling along a solid rod (cf. eq. 14.37). Thus the intensity of both plane and spherical progressive harmonic waves turns out to be

$$I = p^2_{e,max}/2\rho_0 c, \tag{14.161}$$

where $p_{e,\,max}$ is the maximum excess pressure at the point where the intensity is measured. In a plane wave in an ideal fluid this is constant, whereas in a spherical wave it of course falls off inversely as the distance from the wave origin. Eq. (14.161) is of great practical importance in acoustics since many intensity measuring instruments are based on it, as they actually measure the excess pressure.

It will be recalled that in the discussion of mechanical oscillations in Sec. 4.4 we introduced the concept of mechanical impedance in analogy with the corresponding electrical quantity. It turns out to be very useful to employ an analogous quantity in the flow of energy in an acoustic wave. If we treat excess pressure as analogous to electromotive force and acoustic volume current ($\dot{X} = Sv$, where S is the area of the wave front) as analogous to electric current, we naturally define

$$Z = p_e/\dot{X} \tag{14.162}$$

as the acoustic impedance. For the case of a plane progressive harmonic wave the reader should have no trouble in showing that

$$Z = \rho_0 c/S. \tag{14.163}$$

This is a real quantity, even though ϕ may be taken in its complex form (14.154). For many purposes it is convenient to use the impedance *for* unit area of wave front (not *per* unit area). This is referred to as the *specific* impedance

$$Z_s = \rho_0 c. \tag{14.164}$$

It turns out to be of particular significance in connection with the flow of acoustic radiation across boundaries separating media of different properties. The analogy with specific resistance in electric current flow should at once be clear.

As might be expected the acoustic impedance of a spherical wave is more complicated. Starting with (14.160) we derive

$$p_e = -\frac{i\omega A\rho_0}{r} \cdot e^{i(\omega t - kr)}, \tag{14.165}$$

and

$$\dot{X} = -\frac{SA}{r^2}(1 + ikr) \cdot e^{i(\omega t - kr)}. \tag{14.166}$$

Consequently from (14.162)

$$Z = \frac{\omega\rho_0 kr^2}{S(1 + k^2r^2)} + i\,\frac{\omega\rho_0 r}{S(1 + k^2r^2)}. \tag{14.167}$$

If the wave front is the whole sphere surrounding the point source $S = 4\pi r^2$. On the other hand, the more significant specific resistance is

$$Z_s = Z_{s1} + iZ_{s2}, \tag{14.168}$$

where

$$Z_{s1} = \frac{\omega\rho_0 kr^2}{1 + k^2r^2}; \quad Z_{s2} = \frac{\omega\rho_0 r}{1 + k^2r^2}. \tag{14.169}$$

This immediately brings to mind the complex electrical impedance, and by analogy we call Z_{s1} the specific acoustic resistance and Z_{s2} the specific acoustic reactance. It will be observed that as $r \to \infty$

$$Z_{s1} \to \rho_0 c, \quad Z_{s2} \to 0.$$

In words, at great distance from the spherical wave source the specific acoustic resistance approaches that of a plane wave and the specific acoustic reactance approaches zero. On the other hand, for r very small compared with the wave length λ, i.e., $kr \ll 1$, Z_{s2} outweighs Z_{s1} and both are small compared with $\rho_0 c$.

The point source is an ideal case. More important is the case of a source in the form of a solid pulsating sphere of radius a. We must now find a solution of the wave equation which satisfies the conditions prevailing at the boundary of the sphere. Let us assume that the spherical surface vibrates so that its velocity is

$$\dot{\xi}_a = \dot{\xi}_0 e^{i\omega t}. \tag{14.170}$$

By inspection it is seen that

$$\phi = \frac{A}{r} \cdot e^{i[\omega t - k(r-a)]} \tag{14.171}$$

satisfies (14.147), while at the same time it enables us to satisfy the condition (14.170) by proper choice of A. In fact we see that

$$A = -\dot{\xi}_0 a^2/(1 + ika) \tag{14.172}$$

will do it.

By its pulsation the vibrating sphere produces a force on the surrounding medium and so leads to its motion, and hence to radiation. But by the law of action and reaction it can do this only at the expense of a corresponding reaction force on itself. Let us compute this reaction force of the radiation on the sphere. From (14.171) with the value A in (14.172) we get for the excess pressure p_e at any value of $r > a$,

$$p_e = \frac{i\omega\rho_0 a^2 \dot{\xi}_0}{(1 + ika)r} \cdot e^{i[\omega t - k(r-a)]}. \tag{14.173}$$

For $r = a$, i.e., at the surface of the sphere, this becomes

$$p_{e,a} = \frac{i\omega\rho_0 a\dot{\xi}_0 e^{i\omega t}}{1 + ika}. \tag{14.174}$$

It is seen that the specific impedance at the surface of the sphere becomes

$$Z_s = \frac{\omega\rho_0 ka^2}{1 + k^2a^2} + \frac{i\omega\rho_0 a}{1 + k^2a^2}. \tag{14.175}$$

By rationalizing the denominator in (14.174) it develops that the total reaction force of the radiation on the sphere takes the form

$$F_a = 4\pi a^2 Z_{s1}\dot{\xi}_a + 4\pi a^2 Z_{s2}/\omega \cdot \ddot{\xi}_a, \tag{14.176}$$

where $\ddot{\xi}_a$ is the acceleration at the surface of the sphere. We note that the force is made up of two parts, the first of which is proportional to the velocity and hence may be looked upon as a *resisting* force, and the second proportional to the acceleration and may be considered an *inertia* force. It is natural to interpret the first as the contribution of the radiation to the mechanical resistance experienced by the sphere as a vibrator. The equation of motion of the latter may then be written

$$(m + 4\pi a^2 Z_{s2}/\omega)\ddot{\xi}_a + (R + 4\pi a^2 Z_{s1})\dot{\xi}_a + f\xi_a = F. \tag{14.177}$$

where m, R, and f are, respectively, the mass, mechanical damping or resistance coefficient, and stiffness of the vibrator (cf. Sec. 4.4). The term

$$4\pi a^2 Z_{s2}/\omega \tag{14.178}$$

then appears as the equivalent radiation mass. Because the vibrator is radiating into the medium its mass *appears* to be increased by this amount. The term

$$4\pi a^2 Z_{s1} \tag{14.179}$$

appears as the equivalent radiation resistance coefficient. The fact that the vibrator is radiating into the medium tends to damp its motion by the effect of this term.

Examination shows that for given a, at low frequencies the radiation mass tends to predominate over the radiation resistance. At high frequencies the reverse is true. Physically this means that at low frequencies the principal effect of the medium surrounding the vibrating sphere is to increase its effective inertia and hence increase its kinetic energy of vibration for given $\dot{\xi}_0$ without increasing the amount of energy being dissipated. On the other hand, at high frequencies the effect of the medium is to damp the motion of the sphere by permitting more radiation to escape into the medium. We may conclude that other things being equal the efficiency of a vibrating sphere as a radiator of acoustical radiation into a surrounding fluid medium increases with the frequency. It is easier, so to speak, to transmit high frequency radiation from a solid vibrator (usually termed in this connection a *transducer*) to a fluid medium than low frequency radiation. This has some very practical consequences, for example, in underwater sound. It may be shown that the same general principle holds good for a transducer of any shape, e.g., a vibrating flat circular plate, although the analysis is more complicated.

It also turns out that the same conclusions apply to any form of radiation, e.g., electromagnetic (cf. Chapter 16).

14.9. Acoustic Radiation in a Viscous Fluid. Acoustic Relaxation. In any actual fluid the propagation of compressional radiation is accompanied by irreversible dissipation or absorption of a kind not contemplated in the preceding section. We can understand this best by examining the special case of one-dimensional wave motion.

For the sake of review and emphasis let us develop the treatment of Sec. 14.8 for this special case. If we denote, as usual, the displacement from equilibrium in the fluid in the x direction by ξ, the excess pressure by p_e and the mean density by ρ_0, the hydrodynamic equation of motion (14.144) becomes

$$\rho_0 \ddot{\xi} = -\frac{\partial p_e}{\partial x}. \tag{14.180}$$

Writing Hooke's law in the form

$$p_e = -B\frac{\partial \xi}{\partial x} = B\rho_e/\rho_0, \tag{14.181}$$

where B is the adiabatic bulk modulus and ρ_e the excess density (cf. eqs. 14.141, and 14.157) we can obtain immediately the equation for ξ in its dependence on x and t

$$\ddot{\xi} = c^2 \frac{\partial^2 \xi}{\partial x^2}, \tag{14.182}$$

with

$$c = \sqrt{B/\rho_0} \tag{14.183}$$

as the wave velocity.

Now the observant reader will note a difficulty about (14.181) which was overlooked in the general discussion in Sec. 14.8. It is essentially a static equilibrium equation (in this respect like the equations of thermodynamics in general), which ignores the dynamical mechanism by which the density is changed when the excess pressure is altered by the external agency. This kind of neglect was indeed discussed in Sec. 4.6 where it was pointed out that the change in an external agency (e.g., pressure) does not instantaneously alter the density, but rather that time is necessary for the production of a finite change in the latter. This lag was referred to as a *relaxation* phenomenon. Taking account of this concept we ought to replace eq. (14.181) by the more correct relaxation equation

$$p_e = B\rho_e/\rho_0 + R\dot{\rho}_e, \tag{14.184}$$

in which the quantity R appears as the relaxation constant. For constant p_e this is an equation of the same type as (4.73) and the solution will obviously correspond to (4.74), with the *relaxation time,* or time taken by the density to reach the fraction $1 - 1/e$ of the equilibrium value corresponding to the given imposed excess pressure, given by

$$\tau = \rho_0 R/B. \tag{14.185}$$

An equation like (14.184) is encountered in elementary physics in the charging of a condenser. If the capacitance is C and the charging source of e.m.f. (E) is connected to the condenser by leads of resistance R, we have for the charge q at any instant

$$E = q/C + R\dot{q}, \tag{14.186}$$

in which E and q are the analogues of p_e and ρ_e respectively, in eq. (14.184). The electrical resistance is analogous to the relaxation parameter R and C is analogous to ρ_0/B. The relaxation time is RC, which is more commonly referred to in electrical theory as the time constant of the circuit. Although analogies of this kind must be treated with some care they can prove helpful in developing new points of view in physical problems.

Now in our present discussion p_e is not constant. We rewrite (14.184) in the form

$$p_e = -B\frac{\partial \xi}{\partial x} - R\rho_0\frac{\partial \dot{\xi}}{\partial x}, \tag{14.187}$$

whence elimination of p_e between (14.180) and (14.187) yields

$$\ddot{\xi} = B/\rho_0 \cdot \frac{\partial^2 \xi}{\partial x^2} + R\frac{\partial^2 \dot{\xi}}{\partial x^2}. \tag{14.188}$$

This differs from (14.182) by the correction term $R\frac{\partial^2 \dot{\xi}}{\partial x^2}$, which vanishes if the relaxation time is zero, viz., for vanishing R or infinite B.

Eq. (14.184) is more or less *ad hoc* and gives no value for R in terms of other properties of the fluid. Knowledge of the latter depends on the particular relaxation mechanism assumed. We shall derive R on the assumption that the mechanism is the *viscosity* of the fluid. It will be recalled that an ideal fluid does not support a shearing stress, although it may be squeezed. In an actual fluid in motion, however, adjacent layers tend to check each other's relative motion. This is a shearing effect. Experience indicates that the shearing stress with which such adjacent layers act on each other is proportional to the gradient of the velocity at right angles to the layers, and the constant ratio of stress to gradient is termed the viscosity or coefficient of viscosity of the fluid. We shall denote it here by μ' to distinguish it from the shear modulus for an elastic solid.

If we assume as a plausible postulate based on experience that the stress in a viscous fluid is a linear function of the *rate of change* of the strains (instead of the strains themselves, as in the elastic solid) we can immediately generalize the elastic solid stress and strain equations (13.51) to apply to the viscous fluid. Thus for the resultant dilatational stress components we may now write

$$\begin{aligned} X_{xx} &= -p + \lambda'\dot{\Theta} + 2\mu'\dot{\delta}_{xx}, \\ X_{yy} &= -p + \lambda'\dot{\Theta} + 2\mu'\dot{\delta}_{yy}, \\ X_{zz} &= -p + \lambda'\dot{\Theta} + 2\mu'\dot{\delta}_{zz}, \end{aligned} \tag{14.189}$$

where we have inserted in each case the total static pressure p in order to make the stress the resultant one and not merely that due to the flow. The postulational character of (14.189) must be emphasized. Note also that λ' merely appears as an unknown constant. If we add the three equations we have

$$\frac{X_{xx} + X_{yy} + X_{zz}}{3} = -p + \left(\lambda' + \frac{2}{3}\mu'\right)\dot{\Theta} \tag{14.190}$$

Now it seemed reasonable to Maxwell and Stokes to make the further assumption that

$$\frac{X_{xx} + X_{yy} + X_{zz}}{3} = -p, \tag{14.191}$$

whence

$$\lambda' = -\frac{2}{3}\mu'. \tag{14.192}$$

The assumption (14.191) means physically that the average dilatational stress in a moving fluid is still equal to the negative of the static pressure. Recent research has questioned this assumption,[1] but we shall adopt it here; it leads to the most simple result concerning the effect of viscosity on compressional wave propagation. Replacing p by $p_0 + p_e$ in the usual way, we may therefore write for X_{xx} (we shall confine our attention to the x direction)

$$X_{xx} = -(p_0 + p_e) + \frac{4}{3}\mu'\frac{\partial\dot{\xi}}{\partial x}. \tag{14.193}$$

The net dilatational stress in the x direction on the usual elementary fluid parallelepiped with dx as its length along the x axis is $\partial X_{xx}/\partial x \cdot dx$, and hence, employing (14.193), the equation of motion in the x direction becomes

$$\rho_0\ddot{\xi} = -\frac{\partial p_e}{\partial x} + \frac{4}{3}\mu'\frac{\partial^2\dot{\xi}}{\partial x^2}. \tag{14.194}$$

With $p_e = -\rho_0 c^2 \partial\xi/\partial x$, as usual, this becomes

$$\ddot{\xi} = c^2\frac{\partial^2\xi}{\partial x^2} + \frac{4}{3}\frac{\mu'}{\rho_0}\cdot\frac{\partial^2\dot{\xi}}{\partial x^2}. \tag{14.195}$$

This equation is in precisely the form (14.188), derived by the relaxation time method, and becomes identical with the latter if we make the identification

$$R = 4\mu'/3\rho_0. \tag{14.196}$$

The relaxation time for viscosity thus becomes

$$\tau = 4\mu'/3B, \tag{14.197}$$

or directly proportional to the viscosity.

It is now in order to investigate the effect of viscosity on the transmission of a compressional wave, in other words to solve eq. (14.195). We shall content ourselves with a harmonic solution in the form

$$\xi = \xi_0 e^{-\alpha x}\cos(\omega t - kx). \tag{14.198}$$

Substitution into (14.195) shows that this is a solution provided we have

$$2\alpha k = \frac{\omega^3\tau}{c^2(1+\omega^2\tau^2)}, \tag{14.199}$$

[1] Cf. L. Tisza, *Physical Review*, 61, 531, 1942.

$$k^2 - \alpha^2 = \frac{\omega^2}{c^2(1 + \omega^2\tau^2)}. \qquad (14.200)$$

Elimination of k between these two equations yields a biquadratic in α, which when solved, gives

$$\alpha = \frac{\omega^2 \tau V}{2c^2(1 + \omega^2\tau^2)}, \qquad (14.201)$$

where

$$V = \frac{\sqrt{2}c}{\omega\tau}[(1 + \omega^2\tau^2)(\sqrt{1 + \omega^2\tau^2} - 1)]^{1/2} \qquad (14.202)$$

is the revised phase velocity of the wave, i.e., $k = \omega/V$. It is seen that V depends on the frequency and hence viscosity produces dispersion. The term $e^{-\alpha x}$ leads to a progressive diminution of the amplitude of the wave in space, i.e., absorption. A viscous medium absorbs the energy of the wave. We call α the amplitude absorption coefficient. From the definition of wave intensity given in Sec. 14.8 it follows that 2α is the intensity absorption coefficient. It is the reciprocal of the distance in which the intensity is reduced to $1/e$ of its original value.

Now if $\omega\tau$ is sufficiently small to be neglected in comparison with unity, $V \approx c$. If the fluid under consideration is air, $\mu' = 1.8 \times 10^{-4}$ dyne sec/cm^2 at room temperature. The adiabatic bulk modulus for air is γp. Hence (14.197) indicates that for pressures above 1 dyne/cm^2, τ in seconds is a very small quantity compared with unity. For low frequencies (in fact up to rather high frequencies) $\omega\tau \ll 1$. In these circumstances, for air at room temperature, the amplitude absorption coefficient takes the form

$$\alpha = 2\mu'\omega^2/3\rho_0 c^3. \qquad (14.203)$$

To this approximation the absorption coefficient varies as the square of the frequency. The same general formula holds for a liquid like water.

In a general way experiment verifies the frequency dependence of α as predicted in (14.203) over the range of frequencies for which measurements are practicable. However, the observed values of α are practically always greater than the predicted ones, indicating that viscosity alone is unable to account for the whole of the observed absorption of sound in gases and liquids. There are of course other relaxation effects which may be expected to be operative. Among these are heat conduction and the rearrangement in molecular structure of the fluid as well as changes in the internal structure of the individual molecules. The last named effect leads to high selectivity in absorption with respect to frequency, i.e., the presence of definite absorption bands. This has actually been observed in several gases. The whole problem is rather too complicated for further investigation here.[1]

14.10. Boundary Problems in Wave Propagation. Selective Transmission and Filtration. So far in our treatment of wave motion we have supposed that the medium through which the propagation takes place is homogeneous. The question arises: What happens when the medium consists of parts possessing different densities and wave velocities? Intuition tells us that at the interface between two different media an elastic wave will suffer reflection. This will now be investigated.

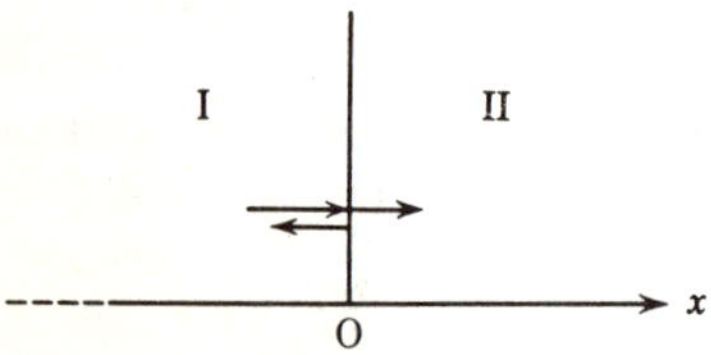

Fig. 14.9

Let us assume we have two media (denoted in Fig. 14.9 by I and II) separated by an interface perpendicular to the x axis at $x = 0$. Medium I has mean density ρ_{01} and wave velocity c_1, whereas the corresponding quantities for II are ρ_{02} and c_2. We imagine that a plane harmonic wave progresses from left to right in I and strikes the interface. For it the displacement will be denoted by

$$\xi_i = A_i e^{i(\omega t - k_1 x)}. \tag{14.204}$$

Because of the reflection there will be a plane wave of the same frequency traveling from right to left in medium I. We denote the displacement in this by

$$\xi_r = A_r e^{i(\omega t + k_1 x)}. \tag{14.205}$$

Since it is unlikely in general that *all* the original wave will be reflected, we must assume the existence of a transmitted wave in medium II proceeding from left to right with displacement denoted by

$$\xi_t = A_t e^{i(\omega t - k_2 x)}. \tag{14.206}$$

Note that

$$k_1 = \omega/c_1, \quad k_2 = \omega/c_2. \tag{14.207}$$

[1]Cf., e.g., L. Bergmann, *Ultrasonics* (New York, Wiley, 1938), pp. 104 ff. Also L. H. Hall, *Phys. Rev.*, 73, 775, 1948.

It is postulated that no change in the frequency of the wave takes place at the interface. This is an assumption fundamental to all harmonic wave motion.

Our task is to find the relations connecting the amplitudes A_i, A_r, A_t, due to the existence of boundary conditions which must be postulated, although they are intuitively pretty plausible. Clearly if the two media are to remain in contact at the interface, the resultant displacement must be the same in both media at the boundary. We refer to this as continuity in displacement. By the same token the interface can remain stable only if the excess pressure is the same in both media at the boundary. This is continuity in pressure. Since at the boundary $x = 0$, the displacement condition leads to

$$A_i + A_r = A_t. \tag{14.208}$$

Now since

$$p_e = \rho_0 c^2 s = -\rho_0 c^2 \frac{\partial \xi}{\partial x}, \tag{14.209}$$

for the excess pressure, we have

$$\begin{aligned} p_{ei} &= \ \ iA_i k_1 \rho_{01} c_1^2 e^{i(\omega t - k_1 x)}, \\ p_{er} &= -iA_r k_1 \rho_{01} c_1^2 e^{i(\omega t + k_1 x)}, \\ p_{et} &= \ \ iA_t k_2 \rho_{02} c_2^2 e^{i(\omega t - k_2 x)}. \end{aligned} \tag{14.210}$$

The pressure condition may therefore be written

$$(A_i - A_r)\rho_{01} c_1 = A_t \rho_{02} c_2 . \tag{14.211}$$

If we use the specific acoustic resistance notation of Sec. 14.8, we can rewrite (14.211) in the form

$$A_i - A_r = A_t Z_{12}, \tag{14.212}$$

where

$$Z_{12} = Z_{s2}/Z_{s1}. \tag{14.213}$$

The solution of (14.208) and (14.212) as simultaneous equations yields for A_r and A_t in terms of A_i the results

$$A_r = \left(\frac{1 - Z_{12}}{1 + Z_{12}}\right) A_i, \tag{14.214}$$

$$A_t = \frac{2A_i}{1 + Z_{12}}. \tag{14.215}$$

When the two media have the same specific acoustic resistances $A_r = 0$ and $A_t = A_i$, i.e., the wave is transmitted without reflection. The slightest difference in the Z values, however, leads to some reflection with a consequent loss in the transmitted amplitude.

The next question relates to the *intensity* of the transmitted wave in relation to that of the incident wave. This is the average rate of flow of energy normal to the wave front per unit area per unit time. In analogy with eq. (14.38) we now form

$$I = \overline{p_e \dot{\xi}}, \tag{14.216}$$

where we must use the *real parts* of p_e and $\dot{\xi}$ in carrying out the multiplication before averaging. In computing the incident intensity it will be instructive to consider the amplitude A_i as a complex quantity. This is in line with the use of the complex notation as in (14.204). Thus we write

$$A_i = a_i + ib_i. \tag{14.217}$$

Then

$$\dot{\xi}_{i,\,\text{real}} = -\omega[b_i \cos(\omega t - k_1 x) + a_i \sin(\omega t - k_1 x)], \tag{14.218}$$

while

$$p_{ei,\,\text{real}} = -\rho_{01} k_1 c_1^2 [b_i \cos(\omega t - k_1 x) + a_i \sin(\omega t - k_1 x)]. \tag{14.219}$$

Hence

$$I_i = \rho_{01} c_1 \omega^2 [b_i^2 \overline{\cos^2(\omega t - k_1 x)} + a_i^2 \overline{\sin^2(\omega t - k_1 x)}$$
$$+ 2a_i b_i \overline{\cos(\omega t - k_1 x) \sin(\omega t - k_1 x)}]. \tag{14.220}$$

From the averages already worked out previously, it follows that

$$I_i = \frac{1}{2}\rho_{01} c_1 \omega^2 (a_i^2 + b_i^2) = \frac{1}{2}\rho_{01} c_1 \omega^2 |A_i|^2. \tag{14.221}$$

In this notation the incident intensity is proportional to the square of the absolute value of the incident amplitude. The ratio of the transmitted intensity to the incident intensity is known as the *power transmission ratio* and often designated as P_r. We have then

$$P_r = \frac{Z_{s2}|A_t|^2}{Z_{s1}|A_i|^2}. \tag{14.222}$$

This is a general formula. In the present case, use of (14.215) leads at once to

$$P_r = \frac{4Z_{12}}{(1 + Z_{12})^2}. \tag{14.223}$$

The transmission is thus independent of frequency and dependent only on the relative specific acoustic resistance of the two media. The more nearly the resistances are matched, i.e., the more nearly equal they are, the closer P_r approaches to unity. If we plot P_r as a function of Z_{12}, the curve has a maximum value of unity at $Z_{12} = 1$ as Fig. 14.10 indicates. $1 - P_r$ denotes in each case the ratio of the reflected to incident intensity, as may be verified directly from (14.214). The reader will find it of interest to compare these results with the transmission of a plane wave through a medium confined to a rigid tube of area of cross section S_1 which encounters a sudden change in cross section to S_2 as in Fig. 14.11. If the medium does not change, the mathematical analysis proves closely analogous to that already worked out. However, it should be carefully noted that the boundary conditions are somewhat different. The pressure continuity condition remains the same, but the continuity in displacement must now be replaced by continuity in volume displacement, i.e., the product of displacement and area of cross section (compare the equation of continuity in steady hydrodynamic flow). If the power transmission ratio is redefined to mean the ratio of total transmitted power to total incident power, the value becomes the same as (14.223) with Z_{12} replaced by

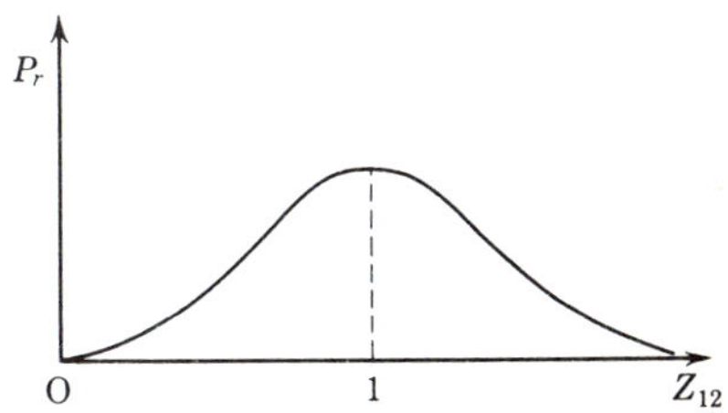

Fig. 14.10

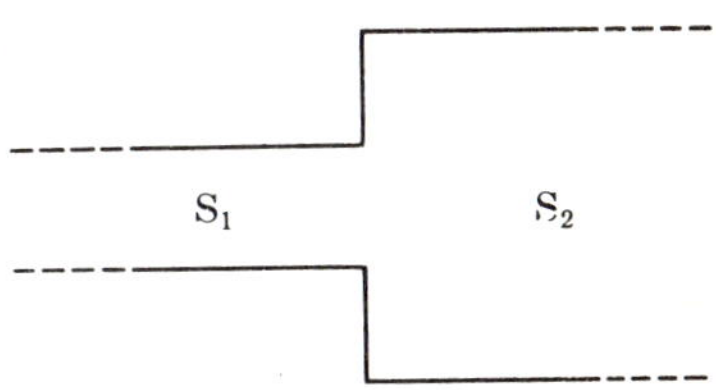

Fig. 14.11

S_2/S_1. The definition of P_r given here is the more generally accepted one.

It has been noted that the transmission in the cases just discussed is not selective with respect to frequency. But now let us consider a more general problem, namely the transmission of plane harmonic compressional waves through a plane parallel infinite stratified structure in which the layers alternate in their physical properties. The area of cross section remains the same for all the layers. In Fig. 14.12 we indicate schematically such a structure with alternating layers I, II, I, ... Medium I has specific acoustic resistance $\rho_{01}c_1$, and thickness $2l_1$. The corresponding quantities for II are indicated. To insure a systematic notation we characterize quantities referring to behavior at the midpoints of the successive strata by the subscripts 1, 2, 3, ... At interfaces the corresponding quantities are indicated with two subscripts 12, 21, etc. as shown. It is convenient to use the velocity potential as basic. If the waves in the structure travel in the positive and negative x directions (perpendicular to the interfaces), we have the general expression

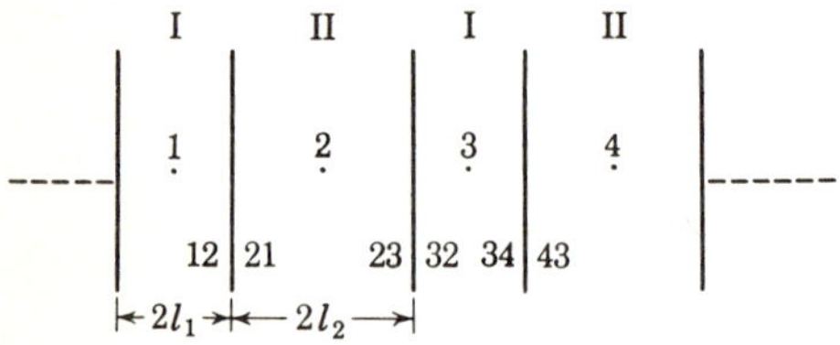

Fig. 14.12

$$\phi = Ae^{i(\omega t - kx)} + Be^{i(\omega t + kx)}. \tag{14.224}$$

It must be emphasized that the structure is assumed infinite in extent to both left and right. The choice of the first midpoint is purely arbitrary. To study the transmission through the structure it is necessary to determine the excess pressure p_3 in terms of the excess pressure p_1 and similarly for the volume currents $\dot{X}(= S\dot{\xi})$. From the fundamental formulas already developed (we here use p for p_e)

$$p = -i\omega\rho_0\left(Ae^{-ikx} + Be^{ikx}\right)e^{i\omega t}, \tag{14.225}$$

$$\dot{X} = ikS\left(-Ae^{-ikx} + Be^{ikx}\right)e^{i\omega t}. \tag{14.226}$$

Let the midpoint of the arbitrarily chosen first layer be taken as $x = 0$. Then

$$\begin{aligned} A &= \frac{i}{2}\left(p_1/\rho_{01}\omega + \dot{X}_1/k_1S,\right) \\ B &= \frac{i}{2}\left(p_1/\rho_{01}\omega - \dot{X}_1/k_1S.\right) \end{aligned} \tag{14.227}$$

wherein we have left out the common factor $e^{i\omega t}$ as superfluous for what follows. Substitution for A and B in (14.225) and (14.226) then yields (continuing to disregard the $e^{i\omega t}$)

$$\begin{aligned} p &= p_1 \cos k_1 x - iZ_1 \dot{X}_1 \sin k_1 x, \\ \dot{X} &= \dot{X}_1 \cos k_1 x - ip_1/Z_1 \cdot \sin k_1 x, \end{aligned} \tag{14.228}$$

where

$$Z_1 = \rho_{01} c_1 / S. \tag{14.229}$$

We are now ready to express the values of p and $\dot{X}$ at the first interface to the right of 1 in terms of p_1 and $\dot{X}_1$. We get from (14.228)

$$\begin{aligned} p_{12} &= p_1 \cos k_1 l_1 - iZ_1 \dot{X}_1 \sin k_1 l_1, \\ \dot{X}_{12} &= \dot{X}_1 \cos k_1 l_1 - ip_1/Z_1 \cdot \sin k_1 l_1. \end{aligned} \tag{14.230}$$

Since our choice of origin is arbitrary we may also write

$$\begin{aligned} p_{23} &= p_{21} \cos 2k_2 l_2 - iZ_2 \dot{X}_{21} \sin 2k_2 l_2, \\ \dot{X}_{23} &= \dot{X}_{21} \cos 2k_2 l_2 - ip_{21}/Z_2 \cdot \sin 2k_2 l_2. \end{aligned} \tag{14.231}$$

Finally in the same fashion

$$\begin{aligned} p_3 &= p_{32} \cos k_1 l_1 - iZ_1 \dot{X}_{32} \sin k_1 l_1, \\ \dot{X}_3 &= \dot{X}_{32} \cos k_1 l_1 - ip_{32}/Z_1 \sin k_1 l_1. \end{aligned} \tag{14.232}$$

It is now necessary to incorporate the boundary conditions. These are continuity of pressure and volume current. Thus

$$\begin{aligned} p_{12} &= p_{21}, \quad p_{23} = p_{32}, \\ \dot{X}_{12} &= \dot{X}_{21}, \quad \dot{X}_{23} = \dot{X}_{32}. \end{aligned} \tag{14.233}$$

Introduction of these into the above equations finally yields p_3 and $\dot{X}_3$ in terms of p_1 and $\dot{X}_1$. We get

$$\begin{aligned} p_3 &= Ap_1 - iBZ_1 \dot{X}_1, \\ \dot{X}_3 &= A\dot{X}_1 - iCp_1/Z_1. \end{aligned} \tag{14.234}$$

The coefficients A, B, C have the following form

$$A = \cos 2k_1l_1 \cos 2k_2l_2 - 1/2 \cdot (Z_1/Z_2 + Z_2/Z_1) \cdot \sin 2k_1l_1 \sin 2k_2l_2$$

$$B = \cos 2k_2l_2 \cdot \sin 2k_1l_1 + (Z_2/Z_1 \cdot \cos^2 k_1l_1 - Z_1/Z_2 \cdot \sin^2 k_1l_1) \cdot \sin 2k_2l_2$$

$$C = \cos 2k_2l_2 \cdot \sin 2k_1l_1 + (Z_1/Z_2 \cdot \cos^2 k_1l_1 - Z_2/Z_1 \sin^2 k_1l_1) \cdot \sin 2k_2l_2 \tag{14.235}$$

Actually eqs. (14.234) are more general than they appear. If the subscript j refers to any odd-numbered layer, we can write

$$p_{j+2} = Ap_j - iBZ_1\dot{X}_j, \quad \dot{X}_{j+2} = A\dot{X}_j - iC/Z_1 \cdot p_j, \tag{14.236}$$

where A, B, C are still given by (14.235). We wish to use these two equations to find the relation between p_{j+2} and p_j. Examination of (14.235) shows that

$$A^2 + BC = 1. \tag{14.237}$$

This suggests that we introduce an angle W (which may be real or complex) such that

$$A = \cos W. \tag{14.238}$$

Then

$$BC = \sin^2 W, \tag{14.239}$$

or

$$B = \sqrt{B/C} \cdot \sin W, \quad C = \sqrt{C/B} \cdot \sin W. \tag{14.240}$$

Substitution into (14.236) yields

$$p_{j+2} = p_j \cos W - iZ_1\sqrt{B/C}\,\dot{X}_j \sin W, \quad \dot{X}_{j+2} = \dot{X}_j \cos W - i\sqrt{C/B}/Z_1 \cdot p_j \sin W. \tag{14.241}$$

Since the structure is infinite in extent in both directions, if we take our stand at the *center* of any layer, we see the same arrangement in either direction. The mathematical expression for this is that $p_j/\dot{X}_j$, or the mid-layer impedance, is independent of j and the physical characteristics of the two types of layers. Let us call the constant ratio Z_0 and write

$$p_j/\dot{X}_j = Z_0. \tag{14.242}$$

Z_0 is usually referred to as the characteristic impedance of the structure. By dividing the first equation in (14.241) by the second we readily find that

$$Z_0 = Z_1\sqrt{B/C}. \tag{14.243}$$

This combined with (14.242) at once enables us to write

$$\begin{aligned} p_{j+2} &= p_j e^{-iW}, \\ \dot{X}_{j+2} &= \dot{X}_j e^{-iW}. \end{aligned} \tag{14.244}$$

which are the desired relations describing the transmission characteristics of the structure. If W is a real angle

$$|p_{j+2}| = |p_j|,$$

and p_{j+2} and p_j differ only in phase. In this case, clearly the structure transmits without attenuation. On the other hand, if W is a complex angle

$$|p_{j+2}| \neq |p_j|,$$

and the structure evidently attenuates, each pair of layers reducing the amplitude in the same proportion. Now W is real when $|\cos W| \leqq 1$, whereas W is complex when $|\cos W| > 1$. Note that $\cos W$ from (14.238) and (14.235) is always real. The transmission through the structure is thus best studied by plotting $\cos W$ as a function of frequency. If we do this we get a curve something like that shown in Fig. 14.13. In the unshaded regions $|\cos W| \leqq 1$ and hence we get unattenuated transmission, whereas in the shaded regions $|\cos W| > 1$ and attenuation results. The structure thus exhibits selective transmission with respect to frequency. It acts as a *filter* and refuses passage to certain frequency bands. Thus $0L_1$, L_2L_3 are *transmission* bands whereas L_1L_2 and L_3L_4 are *attenuation* bands. The width of the bands is controlled by the coefficient

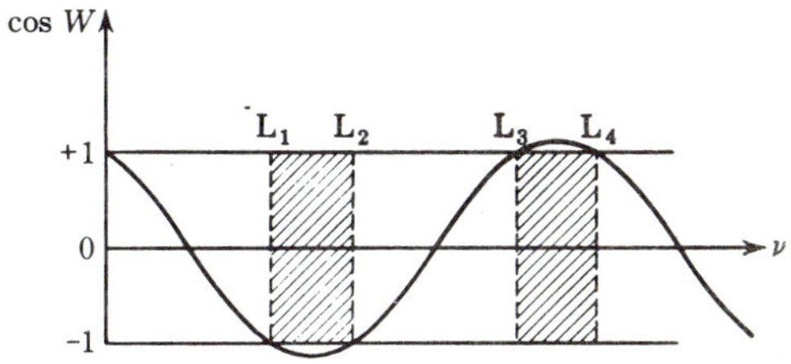

Fig. 14.13

$$1/2 \cdot (Z_1 Z_2 + Z_2/Z_1).$$

The method outlined above is applicable to many types of iterated structures. It can also be extended to apply to the more practical case of *finite* structures involving attenuation of propagation.[1] It is not confined to elastic waves but works for electromagnetic radiation as well (cf. Chapter 16).

14.11. Wave Fronts and Rays in Wave Motion. In Sec. 14.8 we introduced the concept of *wave front* in wave motion and specialized it to plane and spherical waves. Geometrically the whole study of wave motion in three-dimensional media reduces to the motion of the surfaces constituting the wave fronts. It is often possible to simplify the rather difficult analysis involved in this by the use of the curves which are normal to the wave fronts. These are called *rays.* Let us note the conditions under which their use is permissible.

The general equation for compressional waves in a fluid we found in the form (14.147), which is repeated here for reference

$$\nabla^2\phi - \ddot{\phi}/c^2 = 0.$$

We try the following harmonic solution of this equation

$$\phi = A(x, y, z) \cos 2\pi\nu[t - \psi(x, y, z)/c_0]. \qquad (14.245)$$

We are deliberately writing this in general form, with velocity potential amplitude $A(x, y, z)$ emphasizing the possible dependence of amplitude on position in the medium, and with the space part of the phase, i.e., $2\pi\nu\psi(x, y, z)/c_0$, also an arbitrary function of x, y, z. Note that ψ has the dimensions of distance and c_0 is a constant having the dimensions of velocity. The phase at the instant t has the constant value $2\pi\nu t_0$ over the surface with the equation

$$\psi(x, y, z) = c_0(t - t_0), \qquad (14.246)$$

and therefore this is the equation of the wave front at time t, where t_0 of course has an arbitrary value. Let us see what we can learn about ψ by substituting (14.245) into the wave equation. The substitution is tedious but straight-forward and yields, if for simplicity we replace the phase by the single symbol χ,

[1] For a survey of elastic wave filtration, cf. *Jour. App. Physics,* 9, 612, 1938; 10, 680, 1939.

$$\nabla^2 A \cos\chi + \frac{4\pi\nu}{c_0}\nabla\psi \cdot \nabla A \sin\chi - \frac{4\pi^2\nu^2}{c_0^2}A|\nabla\psi|^2 \cos\chi$$

$$+ \frac{2\pi\nu}{c_0}A\nabla^2\psi \sin\chi = -\frac{4\pi^2\nu^2}{c^2}A \cos\chi. \tag{14.247}$$

If this is to be identically satisfied for all t, we must have

$$\frac{\nabla^2 A}{A} - \frac{4\pi^2\nu^2}{c_0^2}|\nabla\psi|^2 + \frac{4\pi^2\nu^2}{c^2} = 0, \tag{14.248}$$

$$2\nabla\psi \cdot \nabla A + A\nabla^2\psi = 0. \tag{14.249}$$

Let us assume that the following approximations hold

$$\begin{aligned} \frac{\nabla^2 A}{A} &\ll \frac{4\pi^2\nu^2}{c_0^2}|\nabla\psi|^2, \\ \frac{\nabla^2 A}{A} &\ll \frac{4\pi^2\nu^2}{c^2}, \\ \frac{4\pi\nu}{c_0}\nabla\psi \cdot \nabla A &\ll \frac{4\pi^2\nu^2}{c_0^2}|\nabla\psi|^2 A, \\ \frac{2\pi\nu}{c_0}A|\nabla^2\psi| &\ll \frac{4\pi^2\nu^2}{c_0^2}|\nabla\psi|^2 A. \end{aligned} \tag{14.250}$$

Under these conditions eq. (14.248) reduces to

$$|\nabla\psi|^2 = c_0^2/c^2, \tag{14.251}$$

whereas (14.249) is made up of terms of much smaller order of magnitude than those left in (14.248) and hence may be neglected in further considerations. In eq. (14.251) we need not restrict c to be constant. In fact it may be a function of x, y, z. Thus we may expand and write the equation in the form

$$\left(\frac{\partial\psi}{\partial x}\right)^2 + \left(\frac{\partial\psi}{\partial y}\right)^2 + \left(\frac{\partial\psi}{\partial z}\right)^2 = c_0^2/[c(x, y, z)]^2. \tag{14.252}$$

This equation is usually known as the *eikonal* equation and ψ itself is known as the eikonal. The quantity c_0/c is usually called the *index of refraction of the medium* and denoted by n. If the velocity c is independent of space, n is a constant and (14.252) assumes a particularly simple form. The solution of this equation yields the wave fronts for the given medium in accordance with (14.246). Since, however, the equation is a partial differential equation, the solutions are diverse in character indicating the possibility of many different types of wave fronts. Only the imposed boundary conditions can specify the appropriate one in any given case.

Before investigating (14.252) further let us look into the significance of the approximations leading to it. It will first be necessary to examine further the geometrical significance of ψ and its connection with the velocity of propagation.

The value of the phase at time t at point $P(x, y, z)$

$$\Phi = 2\pi\nu(t - \psi/c_0).$$

The same value of the phase recurs at time $t + dt$ at the point $Q(x + dx, y + dy, z + dz)$ such that ψ becomes $\psi + d\psi$. Hence

$$dt = d\psi/c_0. \tag{14.253}$$

Now let the distance along the normal to surface ψ connecting the point P to its corresponding point on $\psi + d\psi$ be dl. The velocity with which the phase propagates itself along the normal is

$$V_p = dl/dt = c_0 \frac{dl}{d\psi}.$$

Hence

$$\frac{d\psi}{dl} = c_0/V_p. \tag{14.254}$$

But from the definition of the gradient

$$\frac{d\psi}{dl} = |\nabla\psi| = c_0/V_p. \tag{14.255}$$

Reverting to the approximation in the last line of (14.250) we see that this implies (since A, ν, and c_0 are arbitrary)

$$|\nabla^2\psi| \ll \frac{\nu}{c_0} |\nabla\psi|^2,$$

and hence

$$|\nabla^2\psi| << c_0\nu/V_p^{\,2}. \qquad (14.256)$$

Now

$$|\nabla^2\psi| = |\nabla \cdot \nabla\psi|,$$

and

$$\nabla\psi = \epsilon_n|\nabla\psi|,$$

where ϵ_n is the unit normal to the wave front. For simplicity take this along the x axis so that

$$\epsilon_n = \mathbf{i}.$$

Then from the fundamental equation connecting phase velocity, frequency, and wave length

$$|\nabla \cdot \nabla\psi| = \left|\frac{\partial}{\partial x}(c_0/V_p)\right| = c_0\left|\frac{\partial}{\partial x}\left(\frac{1}{\nu\lambda}\right)\right|.$$

Therefore (14.256) becomes

$$\left|\frac{\partial}{\partial x}\left(\frac{1}{\lambda}\right)\right| << 1/\lambda^2. \qquad (14.257)$$

But this means effectively that

$$\lambda\frac{\partial\lambda}{\partial x} << \lambda. \qquad (14.258)$$

or the change $d\lambda$ in λ over a distance of the order of magnitude of one wave length is negligible compared with the wave length itself. This is then a necessary condition in order that (14.252) may provide the solution of the wave equation. It is of course not the only one since (14.250) imposes a condition on the amplitude A. In fact inspection discloses that the change in A over a single wave length must be small compared with A.

It is interesting to observe that (14.252) can be derived by a geometrical argument based on the concept of the "ray" as the normal to the wave front. In Fig. 14.14 let P again be the point (x, y, z) on a particular wave front, and let PQ be the ray joining P to the point Q (i.e., normal to the wave front at P) on the new wave front at time $t + dt$. Let the direction cosines of PQ be α, β, γ. Then Q becomes the point

$$(x + \alpha V_p\, dt,\ y + \beta V_p\, dt,\ z + \gamma V_p\, dt)$$

and the equation of the wave front at $t + dt$ is

$$\psi(x + \alpha V_p\, dt,\ y + \beta V_p\, dt,\ z + \gamma V_p\, dt) = c_0(t - t_0 + dt). \quad (14.259)$$

If $V_p\, dt$ is sufficiently small we can expand the left-hand side in a Taylor's series and get

$$\psi(x, y, z) + \left(\alpha \frac{\partial \psi}{\partial x} + \beta \frac{\partial \psi}{\partial y} + \gamma \frac{\partial \psi}{\partial z}\right) V_p\, dt + \ldots = c_0(t - t_0) + c_0\, dt.$$

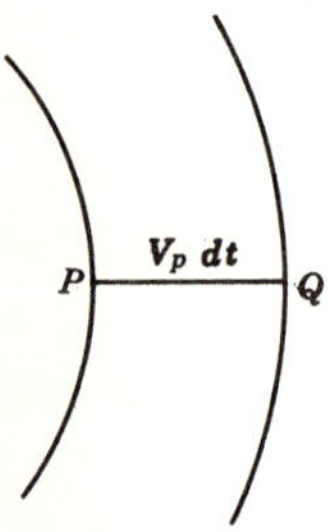

Fig. 14.14

If we confine ourselves to the first approximation, this leads to

$$\alpha \frac{\partial \psi}{\partial x} + \beta \frac{\partial \psi}{\partial y} + \gamma \frac{\partial \psi}{\partial z} = \frac{c_0}{V_p}. \quad (14.260)$$

But from geometry,

$$\alpha = \frac{\partial \psi}{\partial x}\Big/ |\nabla \psi|, \quad \beta = \frac{\partial \psi}{\partial y}\Big/ |\nabla \psi|, \quad \gamma = \frac{\partial \psi}{\partial z}\Big/ |\nabla \psi|. \quad (14.261)$$

Hence substitution into (14.260) gives

$$|\nabla \psi|^2 = c_0^{\,2}/V_p^{\,2}, \quad (14.262)$$

which is identical with (14.252) if we identify c with the phase propagation velocity V_p.

We can now proceed to find the differential equations which define the ray paths. Utilizing the index of refraction n, we may rewrite (14.261) in the form

$$\frac{\partial \psi}{\partial x} = \alpha n, \quad \frac{\partial \psi}{\partial y} = \beta n, \quad \frac{\partial \psi}{\partial z} = \gamma n. \quad (14.263)$$

It therefore follows that

$$\frac{\partial(\alpha n)}{\partial y} = \frac{\partial(\beta n)}{\partial x}, \quad \frac{\partial(\beta n)}{\partial z} = \frac{\partial(\gamma n)}{\partial y}, \quad \frac{\partial(\alpha n)}{\partial z} = \frac{\partial(\gamma n)}{\partial x}. \quad (14.264)$$

The rate of change of αn along the ray is

$$\frac{d(\alpha n)}{ds} = \frac{\partial(\alpha n)}{\partial x}\cdot\frac{dx}{ds} + \frac{\partial(\alpha n)}{\partial y}\cdot\frac{dy}{ds} + \frac{\partial(\alpha n)}{\partial z}\cdot\frac{dz}{ds}$$

$$= \alpha\frac{\partial(\alpha n)}{\partial x} + \beta\frac{\partial(\alpha n)}{\partial y} + \gamma\frac{\partial(\alpha n)}{\partial z}.$$

Carrying out the indicated differentiation on the right yields ultimately with the use of (14.264)

$$\frac{d(\alpha n)}{ds} = \frac{\partial n}{\partial x}, \qquad (14.265)$$

with the two similar equations following in like fashion,

$$\frac{d(\beta n)}{ds} = \frac{\partial n}{\partial y}, \qquad \frac{d(\gamma n)}{ds} = \frac{\partial n}{\partial z}. \qquad (14.266)$$

These three equations define the ray paths by giving α, β, γ, once n is known, as a function of x, y, z. In the special case in which n = constant, α, β, γ are also constants and the rays are straight lines. Other illustrations should be investigated by the reader.

The thoughtful reader will at once discern the analogy between the eikonal equation (14.252) and the Hamilton-Jacobi equation (9.35) for the special case of a single particle moving in a field of force characterized by the potential energy $V(x, y, z)$. In fact if as usual we denote the transformation function by S, the latter equation becomes (in rectangular coordinates)

$$\left(\frac{\partial S}{\partial x}\right)^2 + \left(\frac{\partial S}{\partial y}\right)^2 + \left(\frac{\partial S}{\partial z}\right)^2 = 2m\left[E - V(x, y, z)\right] \qquad (14.267)$$

This becomes mathematically identical with (14.252) if we identify S with ψ and write

$$[n(x, y, z)]^2 = 2m[E - V(x, y, z)] \qquad (14.268)$$

It would therefore appear that it should be possible to interpret the the solution of (14.252) as corresponding to the motion of a particle in a field of force of potential V such that (14.268) is satisfied. Closer study[1] shows that the motion of the particle takes place along the normals to the wave surfaces in the analogous wave propagation problem. These considerations are at the basis of the so-called wave mechanics of DeBroglie (recall Sec. 11.4). We have just seen that the problem of harmonic wave propagation yields to solution in terms of rays only if the wave-length variation is very small. Since the frequency remains constant, this in turn implies that the wave velocity c varies very little from place to place. This is the basic requirement for the use of the ray method or what is commonly called geometrical optics in the case of light propagation. When c varies rapidly from place to place so that the approximations leading to (14.252) are no longer justified, we must fall back on solutions of the wave equation in its usual form and the ray method ceases to have meaning. Effectively this means that every portion of the wave front at a given instant must be considered in deducing the form of the wave front at a subsequent instant (Huygens' principle, cf. Sec. 16.11). Analogously it is interesting to assume that one has a right to interpret the motion of a particle in a field of force characterized by the potential function $V(x, y, z)$ as given by the Hamilton-Jacobi equation (14.267) only if V does not vary too rapidly from place to place. Actually the last condition must be the subject of another hypothesis and essentially what DeBroglie assumed is that V must not vary appreciably over a distance of the order of magnitude of the dimensions of the moving particle. For particles of macroscopic size and fields ordinarily encountered this condition is satisfied and hence classical mechanics is justified. However for particles of atomic dimensions or smaller it seems altogether possible that the condition will not be satisfied and hence the Hamilton-Jacobi equation must be replaced by the corresponding wave equation. Proceeding on this general basis DeBroglie created a theory for the behavior of atomic particles which was later generalized into the quantum mechanics discussed in Chapter 11. Since the development of his theory was based on the theory of relativity we postpone consideration of the details to Sec. 17.8.

PROBLEMS

1. Set up the wave equation for compressional disturbances in a long solid rod and derive the expression for the wave velocity (eq. 13.73).

2. A plane harmonic wave $\xi = A \cos(\omega t - kx)$ progresses along a solid rod. What is the instantaneous tensile stress at any point? If the area of

[1] Cf. L. DeBroglie, *Wave Mechanics* (London, Macmillan, 1930).

cross section of the rod is S, at what average rate is energy transmitted by the wave?

3. When a plane harmonic wave as in Prob. 2 progresses along a solid rod, find the average kinetic energy density and the average potential energy density and the relation between them.

4. A string of length l is stretched with tension T between two fixed points. It is then pulled aside at its midpoint through distance b and let go. Find the amplitude coefficients a_n and b_n in the general expansion ξ (eq. 14.61) for $n = 1, 2, 3, 4$.

5. Write the fundamental equation of fluid motion (eq. 14.124) for the special case of steady flow, i.e., $\dot{\mathbf{v}} = 0$, and integrate, assuming that the external force is conservative. Obtain Bernoulli's theorem.

6. Prove Kelvin's circulation theorem, i.e., that in an ideal fluid if the initial motion is irrotational, it will remain so forever, assuming that the external force is conservative and the density is a function of pressure alone.

7. A plane harmonic progressive acoustic wave in the x direction in a fluid has the particle displacement in the form

$$\xi = A \cos (\omega t - kx),$$

where the various terms have the meaning assigned to them in the text. Find the expressions for the velocity potential, condensation, and excess pressure. If $\nu = 512$ cycles/sec and the wave is in air at 20°C, with excess pressure amplitude 10^{-1} dyne/cm^2, find A.

8. Using the data of the preceding problem, calculate the magnitude of the term $\mathbf{v} \cdot \nabla\mathbf{v}$ neglected in the use of the hydrodynamical equation of motion (eq. 14.124) in the derivation of the wave equation. Find the percentage error involved in the neglect.

9. Find the expression for the instantaneous potential energy density in an acoustic wave in a fluid in terms of the condensation. Show that for a plane wave the average kinetic energy density is equal to the average potential energy density. Prove that the intensity of a plane wave is given by the product of the average total energy density and the wave velocity.

10. A spherical harmonic wave of frequency 500 cycles/sec diverging from a point source in air has a velocity potential amplitude of 1 cm^2/sec

at 1 cm. from the source. Find the excess pressure amplitude 10 cm. from the source. What is the intensity of the wave 100 cm. from the source? Find the total power output of the source in watts.

11. Employing a solution of the form

$$\xi = \xi_0 e^{-\beta t} \cos(\omega t - kx)$$

for the equation for a compressional wave in a viscous medium (eq. 14.195), evaluate β and k in terms of ω. Discuss the nature of the dispersion by finding the wave velocity as a function of ω. Compare this expression with the corresponding one obtained by using the solution

$$\xi = \xi_0 e^{-\alpha x} \cos(\omega t - kx).$$

Also compare α and β and find a physical connection between them.

12. Prove that the intensity of a spherical compressional harmonic progressive wave is equal to $p_e^2/\rho_0 c$, where p_e is the excess pressure, ρ_0 the density of the medium, and c the compressional wave velocity.

13. A horizontal layer of oil (density 0.9 g/cm^3 and sound velocity 1150 meters/sec) of thickness l lies on the surface of some fresh water with its top surface exposed to air. Calculate the power transmission ratio for plane harmonic sound waves progressing from water to air through the oil layer at normal incidence. Find the frequencies for maximum and minimum transmission as a function of l.

14. Plane harmonic sound waves pass through a cylindrical tube containing an expansion as indicated in the accompanying figure. The areas of cross section are indicated in the figure. Find the power transmission through the expansion. It is assumed that the tubes to the left and right of the expansion extend effectively to infinity.

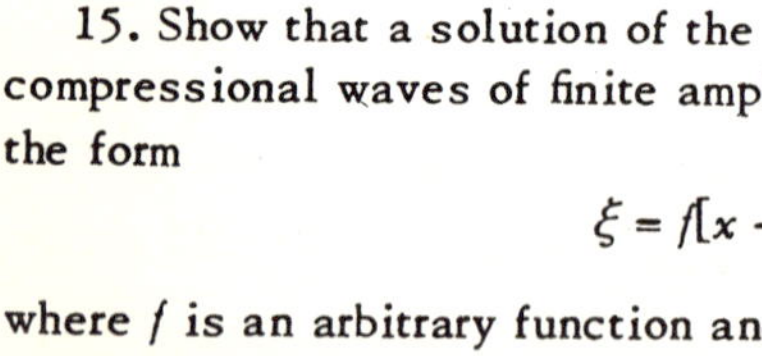

15. Show that a solution of the wave equation for one dimensional compressional waves of finite amplitude (eq. 14.15) can be written in the form

$$\xi = f[x - (c + \dot{\xi})t],$$

where f is an arbitrary function and c is the usual wave velocity for small disturbances. Comment on the physical significance of this type of solution.

16. Using analysis similar to that in Sec. 14.1 set up the wave equation for *spherical* compressional waves of finite amplitude.

17. Prove that the time rate of change of the total compressional wave energy in a given fixed volume may be expressed in the form

$$\rho_0 \int \dot{\phi}\mathbf{n} \cdot \nabla\phi \, dS,$$

where the integral is a surface integral taken over the whole closed surface surrounding the volume, and $\mathbf{n}$ is the unit normal to the surface. Compare with the general hydrodynamic equation (14.135).

18. The *circulation* in a fluid is defined as the integral around a closed path of the scalar product of the fluid velocity and the path element. Thus

$$C = \oint \mathbf{v} \cdot d\mathbf{l},$$

where $d\mathbf{l}$ is the path element. The notation indicates that the integration is taken around the complete path. Find the expression for C for the case of a simple vortex in which all particles of fluid move with the same angular speed ω about a common axis. Prove that in a frictionless fluid the circulation around any path remains constant in time.

19. The quantity

$$\psi_{AB} = \int_A^B \mathbf{n} \cdot \mathbf{v} \, d\mathbf{l},$$

where $\mathbf{n}$ is the unit *normal* to $d\mathbf{l}$, is called the *stream function,* for the fluid in which the velocity is $\mathbf{v}$, between the points A and B. If the path between A and B is a *stream* line, what is the value of ψ_{AB}?

CHAPTER XV

THE ELECTROMAGNETIC FIELD

15.1. Introductory Considerations, Historical and Methodological. It is usually admitted that field theory in physics reached its culmination during the latter part of the nineteenth century in its application by Maxwell and others to the phenomena of electricity and magnetism. These phenomena, although in their simplest aspects discovered in antiquity, stayed wrapped in mystery a much longer time than the fundamental facts of the motion of bodies, probably because it was only after the analytical development of mechanics that the quantitative study of electricity and magnetism could be profitably undertaken. Even today popular tradition still associates something mystifying with electricity, in what may be called its common sense aspects, which is not at all attributed to ordinary mechanics. One consequence of this is at once plain. We must not expect to find a single theory of electricity and magnetism as firmly established as the Newtonian or classical theory of mechanics. Moreover the theories that have been invented must be expected to involve concepts more abstract and less closely connected with immediate experience than the familiar concepts of mechanics.

As a matter of fact the evolution of the subject has witnessed a continued competitive struggle between the particle and field viewpoints. On the one hand, the simple phenomena of electrostatics and magnetostatics can be considered in terms of the mutual action of discrete electric charges and magnetic poles, a view naturally facilitated by the fact that the law of interaction in each case has the same mathematical form as the law of gravitation. Even in the dynamical phenomena of electric currents this view has received relatively satisfactory development in the electron theory. On the other hand, the propagation of electrical effects through space not occupied by charges or poles irresistibly suggests the relevancy of the field or continuum point of view.

It will of course be recalled that in the gravitational interaction of material particles in classical mechanics we have completely free choice of two alternative methods: (1) motion of particles subject to the resultant force of gravitation due to the presence of other particles and acting at a distance, and (2) motion of particles in the gravitational field due to a mass distribution and characterized by a field intensity or potential function (cf. Sec. 12.1). Although the points of view and mathematical analysis are different, the final answers are the same: definite orbits for the particles. It might be thought that essentially the same alternatives prevail in the case of electromagnetic phenomena. But the situation is actually not quite so simple in the latter case.

In the gravitational field there appears to be nothing analogous to the curious duality existing between electrical and magnetic effects, whereby magnetic actions ensue whenever electric charges move and the variation in magnetic flux leads to the flow of charge. Although this duality has been described more or less successfully in terms of action at a distance of moving charges, the field point of view has seemed a much less complicated and more natural mode of explanation. It is true that modern atomic theory has made a pure and unadulterated field theory untenable: it is now clear that the interaction of field and particles is the basic problem of the future of the subject. In particular the demands of the quantum theory must be met, and a complete quantum electrodynamics has yet to be evolved.

For most practical purposes the electromagnetic field equations to be discussed in this chapter have proved adequate, and where they break down, the reason can be plausibly assigned and the corresponding modifications can be introduced. We shall therefore look upon the field equations as effectively the logical postulates of the theory of electricity and magnetism. Together with the definitions of the concepts in them (including naturally their relation to measured quantities), they constitute the theory. Many deductions of the field equations from more "fundamental" postulates have been given. In fact this subject has long been the happy hunting ground of theorizers. We shall, however, forego the luxury of an assessment of such theories and content ourselves with an examination of the meaning of the equations in the light of the fundamental experiments of which they seem to be the simplest abstract description. We shall then follow the usual process of making mathematical deductions from the equations to derive new laws which can be or have been tested experimentally. Obviously no attempt will be made to provide a detailed treatise on electricity and magnetism: the stress as usual will be on concepts and methods with appropriately chosen illustrations.

15.2. The Field Equations. An electromagnetic field is a continuum characterized essentially by four vector functions of space and time, **E**, **H**, **D** and **B** which satisfy certain equations, called the field equations. These equations and their various solutions must describe adequately the known experimental facts of electricity and magnetism. Since the equations were originally suggested by experimental phenomena, we shall introduce them here by the historical method.

A point electric charge q or collection of such charges $q_1, q_2, \ldots q_n$ is said to produce in the empty space surrounding the charges an electric field whose intensity $\mathbf{E}$ at any point is defined as the force acting on a unit positive charge placed at that point. Since the law of force between charges is that of Coulomb (the inverse square law), Gauss' law of force flux (Sec. 12.3) immediately applies, and we have

$$\int \nabla \cdot \mathbf{E}\, d\tau = 4\pi \sum q_i, \tag{15.1}$$

where the integration on the left is taken over the volume containing the charges, which are measured in electrostatic units. Assuming a continuous distribution of charge with density ρ we can write (15.1) in the form

$$\nabla \cdot \mathbf{E} = 4\pi\rho\,. \tag{15.2}$$

At the same time, Coulomb's law also guarantees the existence of a potential function V such that

$$\mathbf{E} = -\nabla V, \tag{15.3}$$

whence (15.2) becomes Poisson's equation or

$$\nabla^2 V = -4\pi\rho\,, \tag{15.4}$$

which we have already encountered in Secs. 12.3 and 12.4. The whole study of electric fields in free space which do not change with the time (i.e., electrostatic fields) is contained in the preceding four equations and hence is virtually identical with the theory of the gravitational field developed in Chapter 12.

Since $\nabla \times \nabla V = 0$ identically, it follows from (15.3) that $\mathbf{E}$ is an irrotational vector in the sense of Sec. 13.5 (cf. Sec. 14.7).

From elementary physics we recall that the region around a magnet or collection of magnets is called a magnetic field. This is characterized by an intensity vector $\mathbf{H}$, which at any point is the force exerted on a unit north pole placed at that point. Since the law of Coulomb also holds for the interaction of magnetic poles, $\mathbf{H}$ in free space, as long as there are no changes with time, obeys precisely the same relations as **E**. A difference enters, however, when we consider that magnetic charges always occur attached to magnets, i.e., in pairs of north and south magnetic poles; one never encounters free and isolated magnetic poles.

Consider space full of magnetic material and concentrate on the rectangular parallelepiped in Fig. 15.1. Since this contains magnetic dipoles we shall define the magnetic moment as a vector whose magnitude is the product of the pole strength and separation of the poles in the dipole. If we add vectorially the magnetic moments for all the dipoles in unit volume, the result will be called the intensity of *magnetization* (total magnetic moment per unit volume) and denoted by $\mathcal{I}$. This is a function of x, y, z in magnetostatics, and in general a function of t also. Then $\mathcal{I}_x\, dy\, dz$ denotes the net magnetic charge passing through the face $OABC$ in the positive x direction. Correspondingly

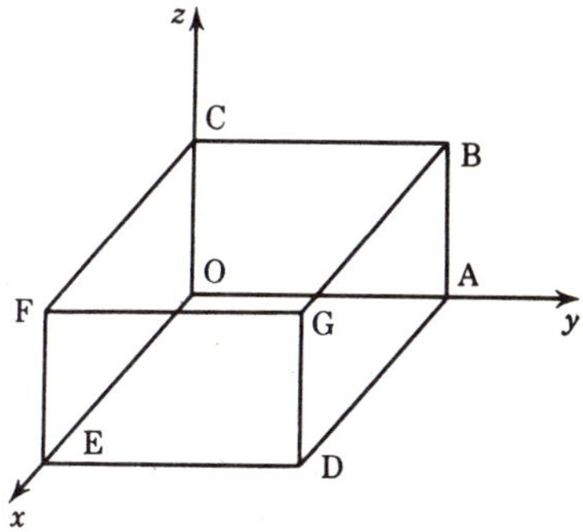

Fig. 15.1

that passing *out* of the parallel face $EFGD$ is $\left(\mathcal{I}_x + \dfrac{\partial \mathcal{I}_x}{\partial x}\, dx\right) \cdot dy\, dz.$

The net positive charge passing into the volume element in the x direction is then $-\dfrac{\partial \mathcal{I}_x}{\partial x}\, dx\, dy\, dz$ and for all three coordinate directions this becomes $-\nabla \cdot \mathcal{I}\, d\tau$. Hence the total magnetic charge due to the magnets placed in a given volume is

$$-\int \nabla \cdot \mathcal{I}\, d\tau, \tag{15.5}$$

where the integration is taken over the volume. If now we apply Gauss' force flux law to the surface surrounding this volume, we get

$$\int \nabla \cdot \mathbf{H}\, d\tau = -4\pi \int \nabla \cdot \mathcal{I}\, d\tau, \tag{15.6}$$

or

$$\int \nabla \cdot (\mathbf{H} + 4\pi \mathcal{I})\, dr = 0. \tag{15.7}$$

The temptation is strong to introduce a new vector

$$\mathbf{B} = \mathbf{H} + 4\pi \mathcal{I}, \tag{15.8}$$

which is called the magnetic induction. Evidently the divergence theorem over any closed surface whatever yields

$$\int \mathbf{B} \cdot \mathbf{n}\, dS = 0, \tag{15.9}$$

where $\mathbf{n}$ is the outward drawn unit normal. Consequently if we represent $\mathbf{B}$ by equivalent lines of flow (as in a fluid) we are forced to consider the lines as closed lines: they never originate or terminate in any finite region. Another way of expressing this is

$$\nabla \cdot \mathbf{B} = 0, \tag{15.10}$$

or $\mathbf{B}$ is a *solenoidal* vector. (Cf. 13.66)

A word about units is in order here. In the case of electrical charges, ρ is given in e.s.u. of charge per cm^3 and $\mathbf{E}$ in dynes per e.s.u. The potential V is given in ergs per e.s.u. In the magnetic case, $\mathbf{H}$ is given in dynes per pole and $\mathcal{I}$ in poles per cm^2. Actually from Coulomb's law these are the same unit. It is common to introduce the line per cm^2 as a measure of $\mathbf{H}$ and $\mathbf{B}$. This is equivalent to the dyne per pole. In treatises on electromagnetic theory much attention is quite properly paid to units. While fully recognizing the importance of this matter in the practical application of the theory and in the comparison of theory and experiment, we do not propose to enter into a lengthy discussion here, as it is our intention to focus emphasis on the fundamental physical principles which are quite independent of the units in which the various quantities are expressed. Where important parameters enter because of the characteristic units employed (the most important of these will be the number c, representing the ratio of the electromagnetic to the electrostatic unit of charge) we shall devote the appropriate emphasis.

From elementary physics the reader will recall the experiment of Oersted indicating the presence of a magnetic field in the neighborhood of a conductor carrying an electric current. We shall define current as rate of flow of charge. Thus

$$I = \frac{dq}{dt}. \tag{15.11}$$

Ampère showed that in order to satisfy Oersted's results the intensity of the field must be that which would be due to a magnetic shell with a boundary coinciding with the circuit through which the current flows and with strength equal to the current. He was then led to the result that the work required to convey a unit pole once around a current-carrying wire in any closed path is equal to $4\pi I$, where I is expressed in so-called electromagnetic units. Thus

$$\int \mathbf{H} \cdot d\mathbf{s} = 4\pi I, \tag{15.12}$$

where the integral on the left is a *line integral.* It will be desirable to review this demonstration briefly here. We shall not strive for rigor but try to make clear the fundamental idea. It is first necessary to obtain the potential of the magnetic field due to a dipole. In Fig. 15.2 the dipole is represented by the positive and negative magnetic poles $+m$ and $-m$ separated by distance l. The distance from the midpoint O on the line of the poles to some point P in the field is r and θ is the angle OP makes with the line of the poles. From Sec. 12.3 we can at once write for the potential at P

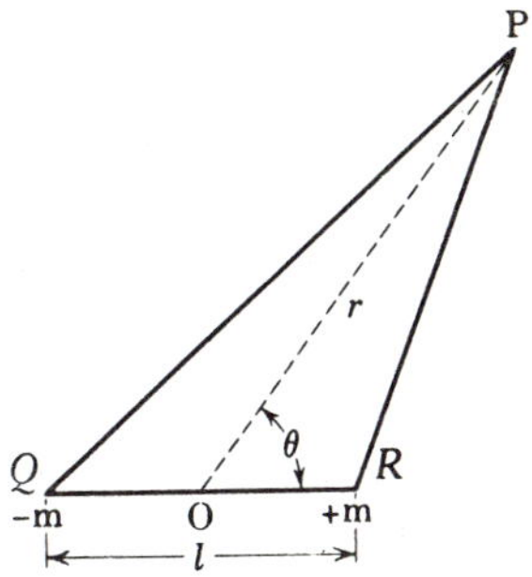

Fig. 15.2

$$V_P = \frac{m}{\sqrt{r^2 + l^2/4 - rl\cos\theta}} - \frac{m}{\sqrt{r^2 + l^2/4 + rl\cos\theta}}. \tag{15.13}$$

We are interested primarily in the case in which $r >> l$. Hence we can rewrite (15.13) in the approximate form

$$V_P = ml\cos\theta/r^2 = M\cos\theta/r^2 = \mathbf{M} \cdot \mathbf{r}/r^3. \tag{15.14}$$

The quantity $\mathbf{M}$ occurring here is the *magnetic moment* of the dipole. Strictly speaking it is a vector directed along the axis of the dipole from $-m$ to $+m$, and this accounts for the vector form in (15.14).

We may think of a magnetic shell as a thin sheet in which a large number of small magnetic dipoles are embedded with their axes normal to the surface at every point. The word "small" here means that l is small compared with the surface dimensions of the sheet. The shell will then have a magnetic moment per unit area and the magnitude of this is called its *strength*, represented by Φ. It follows that the magnetic potential due to element of area dS of a magnetic shell at point P distant r from dS is

$$dV_p = \Phi\, dS\, \mathbf{n} \cdot \mathbf{r}/r^3, \tag{15.15}$$

where $\mathbf{n}$ is the unit vector normal to dS. Now θ is the angle between the normal to the surface of the shell and $\mathbf{r}$ and hence $dS \cos\theta$ is the projection of dS normal to $\mathbf{r}$. Hence we can rewrite (15.15)

$$dV_p = \Phi\, d\omega, \tag{15.16}$$

in which $d\omega = dS \cos\theta/r^2$ is the solid angle subtended by dS at P. The total magnetic potential at P is therefore

$$V_p = \int \Phi\, d\omega \tag{15.17}$$

taken over the whole shell. If the shell has constant strength (15.17) yields

$$V_p = \Phi \int d\omega = \Phi\Omega, \tag{15.18}$$

where Ω is the total solid angle subtended at P by the whole shell. Now suppose we take a unit pole around a path which begins at point A on one side of a magnetic shell and ends at B on the other side just opposite A (cf. Fig. 15.3). Since work done is the product of magnetic pole strength by difference in magnetic potential, it follows that the work done in carrying the unit pole from A to B is

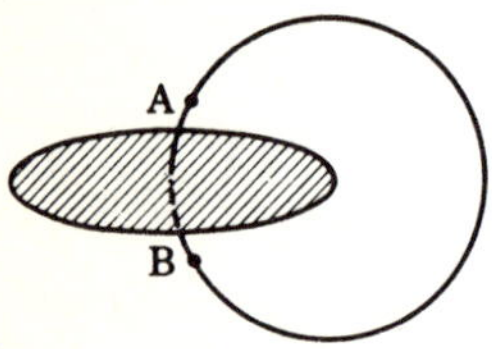

Fig. 15.3

$$\Phi(\Omega_B - \Omega_A). \tag{15.19}$$

But as A approaches B, the quantity in the parenthesis approaches 4π. Hence the work required to carry a unit pole once around a magnetic shell of constant strength Φ is given by

$$W = 4\pi\Phi. \tag{15.20}$$

If with Ampère we identify Φ with the current I in so-called electromagnetic units, eq. (15.12) follows from Ampère's fundamental assumption. It may profitably be rewritten in terms of the current density $\mathbf{j}$ (a vector whose direction is that of the instantaneous charge-velocity and whose magnitude is equal to the current per unit area). Thus (15.12) becomes

$$\int \mathbf{H} \cdot d\mathbf{s} = 4\pi \int \mathbf{j} \cdot \mathbf{n}\, dS, \tag{15.21}$$

where dS is now the element of surface through which the flow takes place and which is surrounded by the path of the pole. By the use of Stokes' theorem (cf. Sec. 14.7) the line integral in (15.21) can be replaced by a surface integral over the same surface which appears on the right-hand side. Thus

$$\int \nabla \times \mathbf{H} \cdot \mathbf{n}\, dS = 4\pi \int \mathbf{j} \cdot \mathbf{n}\, dS. \tag{15.22}$$

Since this must hold for surfaces of arbitrary size, it follows that the law of Ampère can be put into the differential form

$$\nabla \times \mathbf{H} = 4\pi \mathbf{j}. \tag{15.23}$$

Clearly, in general, $\mathbf{H}$ is not an irrotational vector.

Now if electric current is thought of as flow of charge we can write

$$\mathbf{j} = \rho \mathbf{v}, \tag{15.24}$$

where ρ is the charge density and $\mathbf{v}$ the actual velocity. If this is substituted into (15.23) and the divergence of both sides is taken, the result is

$$\nabla \cdot \nabla \times \mathbf{H} = 4\pi \nabla \cdot (\rho \mathbf{v}) = 0, \tag{15.25}$$

since the divergence of the curl of any vector is identically zero. This indicates (cf. Sec. 12.3) that the motion of electric charge in the electric current is analogous to the motion of an incompressible fluid. It would not allow us to consider currents in which the charge density varies with the time as in the simple discharge of a condenser and many other electrical current phenomena. Maxwell was therefore led to assume that (15.23) must be modified to satisfy the complete equation of continuity, viz. (cf. Chapter 12)

$$\nabla \cdot (\rho \mathbf{v}) = -\dot{\rho}. \tag{15.26}$$

But from eq. (15.2) we have

$$\dot{\rho} = \frac{1}{4\pi}\nabla \cdot \dot{\mathbf{E}}. \tag{15.27}$$

Maxwell suggested then that (15.23) should be modified to read

$$\nabla \times \mathbf{H} = 4\pi\rho\mathbf{v} + \dot{\mathbf{E}}, \tag{15.28}$$

since if the divergence of both sides is taken the result is identically zero through the use of (15.26). Of course, eq. (15.28) will remain correct if *any* solenoidal vector is added to the right-hand side. In (15.28) the units must be electromagnetic throughout. If $\rho\mathbf{v}$ and $\mathbf{E}$ on the right are to be measured in electrostatic units, we must modify (15.28) by dividing the right-hand side by c, the ratio between the electromagnetic unit of charge and the electrostatic unit of charge; this number is approximately 3×10^{10}. We then obtain

$$\nabla \times \mathbf{H} = 4\pi\frac{\rho\mathbf{v}}{c} + \frac{\dot{\mathbf{E}}}{c}. \tag{15.29}$$

As a matter of fact this equation is *not* adequate in general to describe the behavior of moving charges in material media, and we have now to investigate its further modification.

Material media, as far as electric phenomena are concerned, may be divided into two classes: (1) conductors in which electric charges (specifically negative electrons and negative and positive ions) are free to move, and (2) insulators or dielectrics in which the presence of an electric field does not cause the actual large-scale motion of electric charges but does displace negative with respect to positive charge and hence introduces *polarization* into the medium. The electric moment per unit volume is called the polarization and is usually denoted by **P.** It has the same properties as the intensity of magnetization in magnetism, and in fact the method followed in eqs. (15.5) - (15.8) inclusive enables us to rewrite Gauss' law for a dielectric subject to an electric field of intensity **E** in the form

$$\int \nabla \cdot (\mathbf{E} + 4\pi\mathbf{P})\, d\tau = 4\pi \sum q_i, \tag{15.30}$$

where $\sum q_i$ denotes the sum of the real or true charges present inside the surface enclosing the volume over which the integral is taken. Similarly

$$-\int \nabla \cdot \mathbf{P}\, d\tau$$

represents the polarization charge. This suggests the introduction of the new vector **D** where

$$\mathbf{D} = \mathbf{E} + 4\pi\mathbf{P}. \tag{15.31}$$

This is called the *electric displacement.* Clearly for a dielectric medium eq. (15.2) must be replaced by

$$\nabla \cdot \mathbf{D} = 4\pi\rho. \tag{15.32}$$

Consequently eq. (15.29) for a material medium must be rewritten as

$$\nabla \times \mathbf{H} = \frac{4\pi\rho\mathbf{v}}{c} + \frac{\dot{\mathbf{D}}}{c}, \tag{15.33}$$

and this is the general form of Ampère's equation.

We next consider the results of Faraday's experiments on the production of electric currents by the use of magnetism. These demonstrated that whenever the flux of induction changes through a closed circuit, an electromotive force is induced which is instantaneously equal in magnitude to the time rate of change of flux and in such a direction as to oppose by its action the change of flux which is producing it.

The electromotive force in a closed circuit is defined to be the work done in the motion of a unit positive charge once around the circuit. It is therefore the line integral of the electric intensity. We may therefore express the content of Faraday's law of electromagnetic induction in the form

$$\int \mathbf{E} \cdot d\mathbf{s} = -\frac{\partial}{\partial t}\int \mathbf{B} \cdot \mathbf{n}\, dS. \tag{15.34}$$

By the use of Stokes' theorem and the usual adjustment of units we can write this in the form

$$\nabla \times \mathbf{E} = -\frac{\dot{\mathbf{B}}}{c}. \tag{15.35}$$

The equations of the electromagnetic field may then be summarized as follows:

$$\begin{aligned} \nabla \cdot \mathbf{D} &= 4\pi\rho, & \nabla \cdot \mathbf{B} &= 0, \\ \nabla \times \mathbf{H} &= \frac{1}{c}(4\pi\rho\mathbf{v} + \dot{\mathbf{D}}), & \nabla \times \mathbf{E} &= -\frac{\dot{\mathbf{B}}}{c}. \end{aligned} \tag{15.36}$$

It is customary to supplement this list with the expression for the total force on a unit positive charge in an electromagnetic field, namely

$$\mathbf{F} = \mathbf{E} + \frac{1}{c}\mathbf{v} \times \mathbf{B}. \tag{15.37}$$

Finally we may introduce the *dielectric constant* κ and the permeability μ defined for homogeneous, isotropic media in such a way that

$$\mathbf{D} = \kappa\mathbf{E}, \quad \mathbf{B} = \mu\mathbf{H}. \tag{15.38}$$

The whole of electromagnetism for homogeneous, isotropic media is contained in the physical interpretation of eqs. (15.36), (15.37), and (15.38).

It may be noted parenthetically that by the use of the so-called Heaviside-Lorentz units the factor 4π may be made to disappear from eqs. (15.36).

15.3. Physical Meaning of the Field Quantities. Before proceeding to deduce some consequences of the field equations and to discuss their solution, it is perhaps in order to consider the meaning to be attached to the quantities **E**, **H**, **D**, and **B** which enter into them. We have indeed mentioned the conventional definitions of **E** and **H** as forces acting on unit electric charge and unit magnetic pole, respectively. Field quantities defined in this way are, however, subject to an obvious difficulty. If we follow the definition literally and place a unit charge, for example, at a point in an electric field in order to measure the force on it, we are not really measuring the intensity of the field as it was before the charge was placed there, since the charge itself, no matter how small, will disturb the original field. In practice we neglect such considerations by assuming that the test charge can be chosen so small as to have negligible disturbing effect. Theoretically we can indeed get around the difficulty by treating **E** as the limit of the force divided by the charge as the charge goes to zero. In a strict field theory in which charges, masses etc., are indefinitely divisible into smaller parts this view is cogent. Unfortunately the success of the atomic point of view forces us to attribute discreteness to every continuum, even if the granularity is too small in many cases to affect large-scale phenomena. There will still remain the logical difficulty of defining **E** and **H** for a point on an electron and this will become serious as soon as we try to write the field equations for a medium containing charged discrete particles.

A possible logical way out of the difficulty is to treat **E**, **H**, **B**, and **D** as defined by the equations (15.36, 37, 38) which they satisfy. In other words, we assume the existence of a set of field quantities which satisfy certain partial differential equations and we then endeavor to extract as much physical significance as possible by assigning operational meaning to deductions from the equations. It then turns out that operationally we never have to refer to forces on unit charges or unit poles. The equations tell us that corresponding to a given distribution of ρ and $\rho\mathbf{v}$ as functions of space and time in media characterized by μ and κ there will be solutions **E**, **H**, **B**, and **D**, whose arbitrary character is mitigated by the boundary conditions of the problem. From these solutions we ought to be able to get valuable information about the charges and currents, e.g., the force interaction of currents and the shape of the paths of charged particles.

The reader interested in the logical and historical development of the theory of the electromagnetic field will find worthwhile discussions in the writings of J. C. Maxwell, Ritz, Lorentz, and others.[1] The basic theoretical ideas in this subject are still only about a century and a quarter old. It is therefore not surprising that there still exists considerable diversity of opinion about the logical structure of the theory of electromagnetism. (Recall the comments in Sec. 15.1). Just as in mechanics, there are many different ways of presenting the subject. However they all ultimately lead to the field equations, which we shall therefore take as basic. We must indeed be careful not to conclude that *every* electromagnetic phenomenon can be directly derived from these equations without further recourse to experience. For example, the reader will look in vain for Ohm's law among the field equations. This does not mean that there is an incompatibility between Ohm's law and the field equations. Actually when charges flow through certain media, a special relation exists between $\rho\mathbf{v}$ and **E** which serves as a kind of boundary condition on the field equations. We shall discuss the interesting result of this in a subsequent section.

[1]Cf. the paper by W. F. G. Swann, "Relativity and Electrodynamics," *Rev. Mod. Phys.*, 2, 243, 1930.

15.4. Some Physical Consequences of the Field Equations. Wave Propagation and the Poynting Flux. Let us first confine ourselves to media in which there are no charges or currents. The equations (15.36) then become

$$\nabla \cdot \kappa \mathbf{E} = \nabla \cdot \mu \mathbf{H} = 0,$$

$$\nabla \times \mathbf{H} = \frac{\kappa}{c}\dot{\mathbf{E}}, \quad \nabla \times \mathbf{E} = -\frac{\mu}{c}\dot{\mathbf{H}}. \tag{15.39}$$

If we take the curl of the third equation and assume that μ and κ are constant in the space considered, we obtain ultimately

$$\nabla^2 \mathbf{H} = \frac{\mu\kappa}{c^2}\ddot{\mathbf{H}}. \tag{15.40}$$

Similar treatment of the fourth equation yields

$$\nabla^2 \mathbf{E} = \frac{\mu\kappa}{c^2}\ddot{\mathbf{E}}. \tag{15.41}$$

From Sec. 13.5 and the further discussion in Chapter 14 we thus conclude that under the conditions specified both **E** and **H** are propagated as waves through space with the wave velocity

$$v_c = \frac{c}{\sqrt{\mu\kappa}}. \tag{15.42}$$

In free space, i.e., space devoid of matter, experience indicates that μ and κ are numerically equal to unity. The number c representing the ratio between the electromagnetic unit of charge and the electrostatic unit of charge is very close to 3×10^{10}. This number should therefore be the velocity in cm/sec of electromagnetic radiation in free space. Actually this is the velocity of light in free space, leading to the suggestion that visible light is a form of electromagnetic radiation. This is the famous prediction of Maxwell.

Reverting to eqs. (15.39), note that we may also write

$$\mathbf{H} \cdot (\nabla \times \mathbf{E}) = -\frac{\mu}{c}\mathbf{H} \cdot \dot{\mathbf{H}}, \tag{15.43}$$

$$\mathbf{E} \cdot (\nabla \times \mathbf{H}) = \frac{\kappa}{c}\mathbf{E} \cdot \dot{\mathbf{E}}. \tag{15.44}$$

If we subtract (15.44) from (15.43) the result is

$$\nabla \cdot (\mathbf{E} \times \mathbf{H}) = -\frac{1}{c}(\mu \mathbf{H} \cdot \dot{\mathbf{H}} + \kappa \mathbf{E} \cdot \dot{\mathbf{E}}), \tag{15.45}$$

or

$$c\nabla \cdot (\mathbf{E} \times \mathbf{H}) = -\frac{\partial}{\partial t}\left[\frac{1}{2}(\mu H^2 + \kappa E^2)\right]. \tag{15.46}$$

Integrating over a finite volume of space and employing the divergence theorem leads to

$$c\int (\mathbf{E} \times \mathbf{H}) \cdot \mathbf{n}\, dS + \frac{\partial}{\partial t}\int \frac{1}{2}(\mu H^2 + \kappa E^2)\, d\tau = 0, \tag{15.47}$$

where the surface integral is taken over the surface bounding the given volume. Let us first examine this by supposing that **E** and **H** are zero on the surface. In this special case therefore

$$\frac{1}{2}\int (\mu H^2 + \kappa E^2)\, d\tau = \text{const.} \tag{15.48}$$

This suggests that $1/2 \cdot (\mu H^2 + \kappa E^2)$ may be interpreted as proportional to the *energy* per unit volume associated with the presence of the electric and magnetic fields. This presumption is made plausible by the following consideration.

The mutual potential energy of a continuous distribution of charge of density ρ, i.e., the work necessary to form the distribution against the mutual repulsion of the parts is

$$W = \frac{1}{2}\int \rho\phi\, d\tau, \tag{15.49}$$

where ϕ is the potential produced by the distribution at the point occupied by charge element $\rho\, d\tau$. The integration is taken over the volume occupied by the whole distribution. Now since

$$\nabla \cdot \mathbf{D} = 4\pi\rho, \tag{15.50}$$

we get

$$W = \frac{1}{8\pi} \int \phi \nabla \cdot \mathbf{D}\, d\tau. \qquad (15.51)$$

From Green's theorem (Sec. 12.6)

$$\int \phi \nabla \cdot \mathbf{D}\, d\tau = \int \phi \mathbf{D} \cdot \mathbf{n}\, dS - \int \nabla \phi \cdot \mathbf{D}\, d\tau, \qquad (15.52)$$

where the surface integral is taken over the surface bounding the volume occupied by the charge. Actually nothing prevents us from taking this integral over the surface of a sphere of very large radius and letting the radius go to infinity. In this case the surface integral vanishes because of the fact that ϕ and $\mathbf{D}$ go to zero faster with r than the surface element increases with r. This gives

$$W = \frac{1}{8\pi} \int \mathbf{E} \cdot \mathbf{D}\, d\tau = \frac{1}{8\pi} \int \kappa E^2\, d\tau, \qquad (15.53)$$

which agrees with our presumption with respect to the energy density in an isolated static electric field, at any rate.

The situation with respect to the volume integral of μH^2 in (15.48) is more complicated even in the static case since magnetic particles occur only in the form of dipoles. Moreover we ought not to have to restrict our treatment to the static case anyway. As a matter of fact we can give a simple illustrative case by considering the magnetic field due to the current in a solenoid. The inductance of such a coil of length l and cross-sectional area S with n turns from elementary physics is

$$L = 4\pi n^2 S\mu/l,$$

and hence the energy associated with the development of current I in the solenoid is

$$\frac{1}{2} L I^2 = 2\pi n^2 I^2 S\mu/l.$$

Since the magnetic field intensity $\mathbf{H}$ at the center of the coil has magnitude

$$H = 4\pi nI/l$$

it is clear that the energy is

$$\frac{\mu}{8\pi} SlH^2,$$

whence the energy density is $\frac{\mu}{8\pi}H^2$, agreeing with our presumption in (15.48). This is by no means a proof but it is a good plausibility consideration.

It will be worth while to give a demonstration of the expression for the energy of a collection of magnetic dipoles. We first need the expression for the potential energy of a magnet of moment **M** in a magnetic field of potential V. Reverting to Fig. 15.2, let $-m$ be placed at Q and $+m$ at R. The work required to place $+m$ at R in the field is given by definition by mV_R, where V_R is the value of V at R, and the work to put $-m$ at Q is given similarly by $-mV_Q$. Hence the resultant work to establish the dipole in the field is

$$m(V_R - V_Q),$$

and if the poles are sufficiently close together this can be written

$$ml(V_R - V_Q)/l = M\frac{dV}{dl}, \tag{15.54}$$

if we denote the dipole separation by the vector **l**. But we may also write

$$\frac{dV}{dl} = \frac{d\mathbf{l}}{dl}\cdot\nabla V, \tag{15.55}$$

where $d\mathbf{l}/dl$ appears as the unit vector along **l**. Now the magnetic moment

$$\mathbf{M} = (ml)\, d\mathbf{l}/dl. \tag{15.56}$$

Therefore we can write for the mutual potential energy of the dipole

$$w = \mathbf{M}\cdot\nabla V. \tag{15.57}$$

If we bring together dipoles with moments $\mathbf{M}_1, \mathbf{M}_2, \ldots \mathbf{M}_n$, the total mutual potential energy is

$$W = \frac{1}{2} \sum \mathbf{M}_j \cdot (\nabla V)_j, \tag{15.58}$$

in which $(\nabla V)_j$ means the gradient of the magnetic potential in the neighborhood of the jth dipole. The factor 1/2 is inserted to avoid counting the energy twice over in the summation. If now we think of the dipoles as continuously distributed we may write $\mathbf{M} = \mathcal{I}_p \, d\tau$, where $\mathcal{I}_p$, as usual is the magnetic moment per unit volume and $d\tau$ the volume element. Then (15.58) takes the form

$$W = \frac{1}{2} \int \mathcal{I}_p \cdot \nabla V \, d\tau. \tag{15.59}$$

The total intensity of magnetization in the medium containing the dipoles is not $\mathcal{I}_p$ but rather $\mathcal{I}$, where

$$\mathcal{I} = \mathcal{I}_p + \chi \mathbf{H}, \tag{15.60}$$

and χ is the magnetic susceptibility. (Note that the field $\mathbf{H}$ induces polarization.) Hence

$$\mathcal{I}_p = \mathcal{I} - \chi \mathbf{H} = \frac{\mathbf{B} - \mu \mathbf{H}}{4\pi}, \tag{15.61}$$

since

$$1 + 4\pi\chi = \mu. \tag{15.62}$$

We therefore have for the energy

$$W = \frac{1}{8\pi} \int (\mathbf{B} - \mu \mathbf{H}) \cdot \nabla V \, d\tau. \tag{15.63}$$

Consider

$$\int \nabla \cdot (V\mathbf{B}) \, d\tau = \int \mathbf{B} \cdot \nabla V \, d\tau + \int V \nabla \cdot \mathbf{B} \, d\tau,$$

and reflect that $\nabla \cdot \mathbf{B} = 0$. With this and the use of the divergence theorem it follows that

$$\int \mathbf{B} \cdot \nabla V \, d\tau = \int V\mathbf{B} \cdot \mathbf{n} \, dS, \tag{15.64}$$

in which the surface integral is taken over a surface enclosing the whole volume containing the dipoles. We can push this surface to infinity if we choose and since, at infinity, V varies as $1/r^2$ and $\mathbf{B}$ vanishes, the surface integral vanishes and the left-hand side also. The result is that

$$W = -\frac{1}{8\pi} \int \mu\mathbf{H} \cdot \nabla V \, d\tau.$$

But

$$\mathbf{H} = -\nabla V$$

by definition of the magnetic potential, and hence finally

$$W = \frac{1}{8\pi} \cdot \int \mu H^2 \, d\tau, \tag{15.65}$$

in agreement with our presumption in (15.48). Once more we should temper our enthusiasm over this demonstration. It applies to static fields, whereas in (15.47) and (15.48) we are dealing with fields which vary with the time.

Let us go back to eq. (15.47) and *assume* that the second term on the left *is* the time rate of change of the total energy in the volume of the medium under consideration due to the presence of the electric and magnetic fields $\mathbf{E}$ and $\mathbf{H}$. The principle of the conservation of energy demands that a change of energy in the volume must be compensated by a corresponding change elsewhere, and we therefore look to the first term on the right to provide this compensation. If the energy in the volume enclosed by the surface is changing and there are no sources or sinks of energy in the volume, the only way to conserve energy is by means of a flow across the bounding surface. We are therefore led to interpret

$$\frac{c}{4\pi} \int (\mathbf{E} \times \mathbf{H}) \cdot \mathbf{n} \, dS \tag{15.66}$$

as the rate of electromagnetic energy flow across the bounding surface. Hence

$$\frac{c}{4\pi}\mathbf{E} \times \mathbf{H} \tag{15.67}$$

is a vector whose magnitude represents the rate of flow of energy per unit area in the electromagnetic field. The presumption is strong that the actual flow shall be considered as directed along this vector, which is normal to both **E** and **H**. This vector is called the Poynting vector, and the associated surface integral (15.66) is called the Poynting flux. If we think of the propagation of an electromagnetic wave as taking place along the Poynting vector, it follows that such a wave must be transverse. We shall give a more direct demonstration of this for plane waves in Sec. 16.2.

It is interesting to note that the theory of the electromagnetic field can be developed by assuming (15.47) as a fundamental principle and then applying conservation of energy. The field equations can then be derived without difficulty. This is the method of Planck.[1] Since this provides an alternative view of the field equations it may be worth while to survey it briefly here.

Starting with (15.47) as the fundamental principle, we transform the surface integral into a volume integral and carry out the indicated time differentiation. If the result is to hold for every volume element, we finally arrive at the *identity*

$$c(\mathbf{H} \cdot \nabla \times \mathbf{E} - \mathbf{E} \cdot \nabla \times \mathbf{H}) + \kappa\mathbf{E} \cdot \dot{\mathbf{E}} + \mu\mathbf{H} \cdot \dot{\mathbf{H}} = 0. \tag{15.68}$$

Since this is now an identity we must equate the coefficients of **E** and **H** to zero separately and get at once

$$\begin{aligned} \nabla \times \mathbf{E} &= -\mu\dot{\mathbf{H}}/c, \\ \nabla \times \mathbf{H} &= \kappa\dot{\mathbf{E}}/c. \end{aligned} \tag{15.69}$$

These are the third and fourth field equations (15.39). Of course, these hold for the special case of media containing no charges or currents. We can, however, derive the more general form (15.36) by generalizing (15.47) to include energy dissipation inside the volume. For this purpose we assume that Joule's law holds and that the rate of electric power transformation per unit volume can be written in the form

$$\mathbf{E} \cdot \rho\mathbf{v}$$

[1] Cf. Max Planck, *Introduction to the Theory of Electricity and Magnetism* (Vol. 3 of *Introduction to Theoretical Physics*, London, Macmillan, 1932).

The energy conservation equation now takes the general form

$$c(\mathbf{H}\cdot\nabla\times\mathbf{E}-\mathbf{E}\cdot\nabla\times\mathbf{H})+\kappa\mathbf{E}\cdot\dot{\mathbf{E}}+\mu\mathbf{H}\cdot\dot{\mathbf{H}}+4\pi\mathbf{E}\cdot\rho\mathbf{v}=0. \quad (15.70)$$

Instead of (15.69) we now get

$$\begin{aligned}\nabla\times\mathbf{E}&=-\dot{\mathbf{B}}/c,\\ \nabla\times\mathbf{H}&=\dot{\mathbf{D}}/c+4\pi\rho\mathbf{v}/c,\end{aligned} \quad (15.71)$$

corresponding to the third and fourth equations in (15.36). Planck then goes on to deduce from (15.71) what he calls the principle of conservation of electric charge and in so doing effectively derives the first and second equations in (15.36), i.e., the divergence equations which we have seen are the differential representation of Coulomb's law applied to electric charges and magnetic poles.

15.5. Energy of a Collection of Currents. Since we have been discussing energy in the electromagnetic field this is an appropriate place to introduce the energy of the field associated with an aggregate of electric currents. In Sec. 8.7 this energy was introduced in the form

$$W_e=\frac{1}{2}\sum_{j=1}^{n}N_j i_j, \quad (15.72)$$

where i_j is the current in the jth circuit, and N_j the flux of magnetic induction through this circuit. We shall show that this expression follows from (15.65) for the energy involved in a magnetic field $\dot{\mathbf{H}}$. The assumption is now made that the field is due entirely to currents. Consider the jth current and recall that the lines of magnetic induction which thread this current are *closed* lines. We may therefore take for our volume element $d\tau$ in (15.65) a tube of cross section A and length $d\mathbf{s}$. HA will be the flux N through the tube. We then carry out the volume integration in (15.65) by summing over all the tubes that thread i_j, thus getting for the contribution of i_j to the energy

$$W_j=\frac{N_j}{8\pi}\int\mathbf{H}\cdot d\mathbf{s}. \quad (15.73)$$

Now $\int \mathbf{H} \cdot d\mathbf{s}$ is taken around a path which threads the current. By Ampere's law (15.12) this is $4\pi i_j$. Hence $W_j = \frac{1}{2} N_j i_j$. Therefore the total energy will be obtained by summing this over all the currents, and we get

$$W = \frac{1}{2} \sum N_j i_j, \tag{15.74}$$

as in (15.72). This method is due essentially to Jeans.

15.6. General Solution of the Field Equations. Vector and Scalar Retarded Potentials. It is now in order to discuss the general solution of the field equations in their form (15.36). By a solution we shall mean the following: suppose in any medium we have a distribution of charge defined by the density ρ and a distribution of currents characterized by $\rho\mathbf{v}$. The problem will be to find the resulting expressions for $\mathbf{E}$ and $\mathbf{H}$ as functions of ρ and $\rho\mathbf{v}$. The method used involves the introduction of the so-called scalar and vector potentials to be denoted by ϕ and $\mathbf{A}$, respectively.

Since $\mathbf{B}$ is a solenoidal vector it can be represented as the curl of vector, thus

$$\mathbf{B} = \nabla \times \mathbf{A}, \tag{15.75}$$

since identically $\nabla \cdot \nabla \times \mathbf{A} = 0$. We shall call $\mathbf{A}$ the *vector potential.* It is not uniquely defined by (15.75). In fact the addition of the gradient of any scalar point function would still allow it to satisfy the fundamental condition. Now from the fourth field equation in (15.36) there results

$$\nabla \times (\mathbf{E} + \dot{\mathbf{A}}/c) = 0, \tag{15.76}$$

which means that $\mathbf{E} + \frac{1}{c}\dot{\mathbf{A}}$ can be written as the negative gradient of a scalar function which we shall denote by ϕ and call the *scalar potential.* Therefore

$$\mathbf{E} + \dot{\mathbf{A}}/c = -\nabla\phi. \tag{15.77}$$

It also is not uniquely defined because of the lack of uniqueness in $\mathbf{A}$.

If now we assume that in the particular medium being considered μ and κ are constants in space and time we have

$$\mathbf{H} = \frac{1}{\mu}\nabla \times \mathbf{A}, \quad \mathbf{D} = -\kappa(\nabla\phi + \dot{\mathbf{A}}/c), \tag{15.78}$$

and if we substitute into the equations (15.36) the result is

$$\nabla \times \nabla \times \mathbf{A} = \frac{\mu}{c}(4\pi\rho\mathbf{v}) - \frac{\mu\kappa}{c}(\nabla\dot{\phi} + \ddot{\mathbf{A}}/c), \tag{15.79}$$

and

$$\nabla \cdot \mathbf{D} = 4\pi\rho = -\kappa(\nabla^2\phi + \nabla \cdot \dot{\mathbf{A}}/c). \tag{15.80}$$

Since **A** and ϕ are not originally uniquely defined, in order to simplify the above equations Lorentz introduced the following relation connecting them.

$$\nabla \cdot \mathbf{A} = -\mu\kappa\dot{\phi}/c. \tag{15.81}$$

This reduces (15.80) to

$$\nabla^2\phi - \mu\kappa\ddot{\phi}/c^2 = -4\pi\rho/\kappa, \tag{15.82}$$

while (15.79) becomes

$$\nabla^2\mathbf{A} - \frac{\mu\kappa}{c^2}\ddot{\mathbf{A}} = -\frac{\mu}{c}4\pi\rho\mathbf{v}. \tag{15.83}$$

These equations remind us of the wave equation but are not homogeneous. When they are solved for **A** and ϕ as functions of space and time, substitution into (15.77) and (15.78) will yield **E** and **H**. The solution of (15.82) and (15.83) is best carried out by the use of the Fourier integral, which has already been introduced in Sec. (14.4).

For simplicity we confine ourselves to the scalar equation (15.82), though precisely the same method is applicable to (15.83). In eq. (15.82) ϕ is of course a function of x, y, z, t. The Fourier integral enables us to express the functional dependence on t in terms of harmonic components. Thus (cf. eq. 14.71),

$$\phi(x, y, z, t) = \frac{1}{2\pi}\int_{-\infty}^{\infty} e^{i\omega t}\,d\omega \int_{-\infty}^{\infty} \phi(x, y, z, u)e^{-i\omega u}\,du. \tag{15.84}$$

In similar fashion

$$\nabla^2\phi = \frac{1}{2\pi}\int_{-\infty}^{\infty} e^{i\omega t}\, d\omega \int_{-\infty}^{\infty} \nabla^2\phi(x, y, z, u)e^{-i\omega u}\, du. \tag{15.85}$$

$$\ddot{\phi} = -\frac{1}{2\pi}\int_{-\infty}^{\infty} e^{i\omega t}\, d\omega \int_{-\infty}^{\infty} \omega^2\phi(x, y, z, u)e^{-i\omega u}\, du, \tag{15.86}$$

$$\rho = \frac{1}{2\pi}\int_{-\infty}^{\infty} e^{i\omega t}\, d\omega \int_{-\infty}^{\infty} \rho(x, y, z, u)e^{-i\omega u}\, du. \tag{15.87}$$

The original equation for ϕ then becomes

$$\frac{1}{2\pi}\int_{-\infty}^{\infty} e^{i\omega t}\, d\omega \int_{-\infty}^{\infty} e^{-i\omega u}\Big[\nabla^2\phi(u)$$

$$+\frac{\mu\kappa}{c^2}\omega^2\phi(u) + \frac{4\pi}{\kappa}\rho(u)\Big]\, du = 0, \tag{15.88}$$

where we have written $\phi(u)$ for $\phi(x, y, z, u)$ etc., Eq. (15.88) will be satisfied if we have identically

$$\nabla^2\phi(u) + \frac{\mu\kappa}{c^2}\omega^2\phi(u) + \frac{4\pi\rho(u)}{\kappa} = 0, \tag{15.89}$$

in which space coordinates alone enter. To obtain the solution of (15.89) we use the method of Green and consider the equation

$$\nabla^2 V + k^2 V = 0, \tag{15.90}$$

in which V is a function of space coordinates only and k is a real parameter. We fix an arbitrary point P and let r be the distance of any other point in space from P. Then a particular solution of (15.90) is

$$V = \frac{e^{\pm ikr}}{r}, \tag{15.91}$$

as is clear by direct inspection. Now let us introduce Green's theorem in the form (12.56), and let V be given by (15.91) while U is for the moment arbitrary. Let us draw about point P a sphere of radius a and another sphere of arbitrarily larger radius a'. Apply the volume integration in Green's relation to the volume included between the two spheres. We have then

$$\int \frac{e^{\pm ikr}}{r}(\nabla^2 U + k^2 U)\, d\tau$$

$$= \int_{\text{outer sphere}} \left[\frac{e^{\pm ikr}}{r}\nabla U - U\nabla\left(\frac{e^{\pm ikr}}{r}\right)\right] \cdot \mathbf{n}\, dS + \int_{\text{inner sphere}} \left[U\frac{\partial}{\partial a}\left(\frac{e^{\pm ika}}{a}\right) - \frac{e^{\pm ika}}{a}\frac{\partial U}{\partial a}\right] a^2\, d\omega. \qquad (15.92)$$

We have written $\mathbf{n} \cdot \nabla U = \dfrac{\partial U}{\partial a}$ at the surface of the sphere of radius a and replaced dS there by $a^2\, d\omega$, where $d\omega$ is the element of solid angle. We allow the sphere about P to shrink to zero and the outer sphere to go to infinity; if U goes to zero as $r \to \infty$ as strongly as $1/r^2$, the first integral on the right will vanish. Actually we may not wish to restrict U to vanish as strongly as this. We expect potential functions to vanish at ∞ as $1/r$. In this case the first integral on the right will certainly converge to a finite constant even if its value is not zero, and this is sufficient for the use we wish to make of the result. The second integral then reduces to $-4\pi U_p$, where U_p is the value of U at the point P. Hence

$$\int \frac{e^{\pm ikr}}{r} \cdot (\nabla^2 U + k^2 U)\, d\tau = -4\pi U_p, \qquad (15.93)$$

with the possible addition of a constant from the first integral on the right-hand side of eq. (15.92). Now U was originally arbitrary. If we choose $U = \phi$ in (15.89) we obtain

$$\phi_P = \frac{1}{4\pi}\int \frac{4\pi\rho(u)}{\kappa r}\, e^{\pm i\sqrt{\mu\kappa}\,\omega r/c}\, d\tau. \qquad (15.94)$$

This enables us to write (from eq. 15.84)

$$\phi_p(t) = \frac{1}{2\pi}\int\int_{-\infty}^{\infty} d\tau\, d\omega \int_{-\infty}^{\infty} \frac{\rho(u)}{\kappa r} e^{i\omega\left[t-u\pm\frac{r}{c/\sqrt{\kappa\mu}}\right]} du. \qquad (15.95)$$

Comparing this with (15.87), viz.,

$$\rho(t) = \frac{1}{2\pi}\int_{-\infty}^{\infty} d\omega \int_{-\infty}^{\infty} \rho(u) e^{i\omega(t-u)}\, du,$$

we see that (15.95) can be put into the form

$$\phi_P(t) = \frac{1}{\kappa}\int \frac{1}{r}\cdot\rho\left(t\pm\frac{r}{v_c}\right) d\tau. \qquad (15.96)$$

where $v_c = c/\sqrt{\mu\kappa}$. Let us see what this solution means, taking first the minus sign in the parentheses. If we wish to evaluate the scalar potential ϕ at point P at time t, we must take the charge density ρ at the volume element $d\tau$ distant r from P *at the time* $t - r/v_c$ and *not* the time t. The same must be done for all charge elements contributing to the potential. From (15.42) v_c is the velocity of the electromagnetic radiation in the medium in question. The resulting potential is termed a *retarded potential.* If the plus sign were used in (15.96) the result would be an *advanced* potential, of doubtful physical significance.[1] We shall therefore restrict ourselves to the use of the minus sign.

The same method applied to the vector potential equation (15.83) yields

$$\mathbf{A}_P(t) = \frac{\mu}{c}\int \frac{1}{r}\cdot\left[\rho\mathbf{v}\left(t\pm\frac{r}{v_c}\right)\right] d\tau, \qquad (15.97)$$

where the bracket notation in the integrand means that the whole product $\rho\mathbf{v}$ is to be taken at the time $t \pm \dfrac{r}{v_c}$. Indeed it is conventional to write in the usual case of the retarded potential

[1] P. A. M. Dirac has recently suggested that the combined use of retarded and advanced potentials may help to clear up some difficulties in the theory of electromagnetic radiation from charged particles. Cf. *Proc. Roy. Soc. London,* A 167, 148, 1938.

$$[\rho(t - r/v_c)] = [\rho], \tag{15.98}$$

$$[\rho\mathbf{v}(t - r/v_c)] = [\rho\mathbf{v}]. \tag{15.99}$$

The physical significance of the solutions (15.96) and (15.97) will be brought out most clearly by the consideration of some examples. This will be undertaken in the next chapter.

PROBLEMS

1. Find the equation of the lines of force in the xy plane of the field due to two equal stationary charges $+q$ and $-q$ at $(-a, 0)$ and $(+a, 0)$, respectively, on the x axis. Find the expression for the potential at any point P in the xy plane and specialize for the case where $2a \ll r$, where r = distance from the origin to P.

2. A uniformly magnetized body is one for which the intensity of magnetization $\mathcal{I}$ is the same at every point. If the earth is taken to a first approximation as a uniformly magnetized sphere, find the magnetic potential and field intensity at any point on the surface. How will the dip or magnetic inclination vary with the latitude?

3. A homogeneous right circular cylindrical conductor of radius a cm and infinitely long carries a steady current of I e.m.u. of uniform current density. Find **H** for any point inside and outside the conductor.

4. Apply the field equations to show that the normal component of the electric displacement is continuous at the interface of two dielectric media of different dielectric constants. Find the corresponding boundary conditions governing **E**, **H**, and **B**.

5. Calculate the Poynting flux for the case of a long straight homogeneous wire of radius a cm and carrying steady current I in e.m.u. Compare with the rate at which heat is being produced in the wire.

6. In a parallel plate condenser in which the plates are circular and of radius R and separated by distance l, the charge alternates according to the relation $q = q_0 \cos \omega t$. If the medium has dielectric constant κ, permeability $= \mu =$ unity, find **D** and **H** at every point between the plates. (N.B. neglect end effects and assume symmetry about the axis of the condenser.)

7. Discuss the experimental verification of eq. (15.42) for hydrogen and water.

8. A magnetic field is directed along the x axis. It is independent of position in space but its magnitude is directly proportional to the time. Find the vector potential. If the scalar potential is zero, find the induced electric field.

9. Electrons are emitted from the negative plate of a plane parallel plate condenser with negligible initial velocities under the action of ultraviolet light. Let the potential difference between the plates be V. A uniform magnetic field $\mathbf{H}$ is applied parallel to the condenser plates.

(a) If the separation of the plates is l, find the magnitude of the magnetic field intensity $\mathbf{H}$ which will just keep the electrons from reaching the positive plate.

(b) Find the path of the electrons.

10. Use the Fourier integral to resolve the step function

$$f(t) = 1 \text{ for } |t| < l$$
$$f(t) = 1/2 \text{ for } |t| = l$$
$$f(t) = 0 \text{ for } |t| > l$$

into a set of harmonic components. Find the amplitude of the component with frequency $\omega/2\pi$. Plot this as a function of ω and comment on the dependence of the plot on l (i.e., discuss the extreme condition of l very small and l very large).

11. Show that the electromagnetic field equations may be solved in terms of a function $\mathbf{\Pi}(x, y, z, t)$ by means of the relations

$$\mathbf{E} = \nabla\nabla \cdot \mathbf{\Pi} - \ddot{\mathbf{\Pi}}, \quad \mathbf{H} = \nabla \times \dot{\mathbf{\Pi}}$$

where $\mathbf{\Pi}$ is any solution of the equation

$$\nabla \times \nabla \times \mathbf{\Pi} - \nabla\nabla \cdot \mathbf{\Pi} + \ddot{\mathbf{\Pi}} = \mathbf{P}$$

and $\mathbf{P}$ is the vector such that

$$\rho\mathbf{v} = \dot{\mathbf{P}} \quad \text{and} \quad \rho = -\nabla \cdot \mathbf{P}.$$

What is the relation between $\mathbf{\Pi}$ and $\mathbf{A}$? (Assume free space)

12. Show that the equation of motion of a charged particle in an electromagnetic field can be expressed in Lagrangian form if one chooses as the Lagrangian function

$$L = mv^2/2 - e(\phi - \mathbf{A} \cdot \mathbf{v}/c)$$

where **A** and ϕ are the vector and scalar potentials, respectively.

13. A sphere of radius a is filled with a volume distribution of static charge with uniform density ρ. Find the potential and intensity of the electric field at a point distant r from the center of the sphere. Consider both $r < a$ and $r > a$.

14. The static charge on a conductor resides wholly on its surface. Prove that the electric field intensity inside a statically charged conductor is equal to zero. Find the expressions for **E** and **D** in a dielectric at a point outside but immediately adjacent to a conductor charged with surface density σ.

15. Show that the potential of the electric field due to a volume distribution of charge of density ρ and a distribution of dipoles of resultant polarization **P** is given by

$$V = \int \frac{(\rho - \nabla \cdot \mathbf{P})}{r} \, d\tau,$$

where r is the distance from the volume element $d\tau$ (at which the density is ρ and the polarization **P**) to the point at which the potential is calculated. What does the expression become if one uses mks units instead of the cgs system? In the former, charge is in coulombs and potential expressed in joules/coulomb or volts.

16. A point charge $+q$ is placed at the point A distant u from the center, O, of an uncharged and grounded conducting sphere of radius a, where $u > a$. Show that if a charge $q' = -aq/u$ is placed at a point on the line OA inside the sphere distant $b = a^2/u$ from O, the potential at all points on the surface of the sphere will be zero. The charge q' is called the electrical *image* in the sphere of the charge q. The effect of this charge and the charge q is to produce the same distribution of potential as the charge q and the *induced* charge on the conductor.

17. A charge $+q$ is placed at a perpendicular distance l from a very large (ideally infinite) grounded conducting plane with no initial charge. Where is the image charge located and what is its magnitude? Show that the image is equivalent to an *induced* surface charge of density

$$\sigma = -ql/2\pi r^3,$$

where r is the distance from the point charge q.

18. Calculate the intensity of the magnetic field due to a dipole of moment **M** at a point whose distance from the center of the dipole is large compared with the dipole separation. Express it as the sum of two vectors, one directed *along* the dipole moment and the other along the line joining the center of the dipole to the point in question. How will the resulting expression differ for an electric dipole of moment **p**?

19. Find the electrostatic energy of a charged conducting sphere by the use of eq. (15.53). Compare with the elementary physics expression for the same quantity, employing the capacitance of the sphere.

20. Show that for a distribution of steady currents, i.e., $\rho\mathbf{v}$ independent of time, the vector potential takes the form

$$\mathbf{A} = \frac{\mu}{c}\int \frac{\rho\mathbf{v}\, d\tau}{r},$$

where r is the distance from $d\tau$ to the point at which $\mathbf{A}$ is to be given. Prove that $\mathbf{A}$ is a solenoidal vector, recalling that for steady currents $\nabla \cdot (\rho\mathbf{v}) = 0$.

21. Find the vector potential for a steady current of magnitude I flowing in a straight wire of length l.

22. Use eq. (15.37) and the vector potential from Prob. 21 to calculate the force per unit length between two parallel wires of length l ($\gg$ their separation) carrying currents I_1 and I_2, respectively.

23. Show that the energy of a steady current distribution of density **i**, may be written in the form

$$W = \frac{1}{2}\int \mathbf{A} \cdot \mathbf{i}\, d\tau,$$

where **A** is the vector potential associated with the currents.

24. Starting with eq. (15.47) and employing the vector and scalar potentials, show that

$$c\int (\mathbf{E} \times \mathbf{H}) \cdot \mathbf{n}\, dS = 4\pi\int \left(\phi\dot{\rho} + \rho\mathbf{v} \cdot \frac{\partial \mathbf{A}}{\partial t}\right) d\tau.$$

Comment on the physical significance of the right side.

25. The *momentum* **M** of an electromagnetic field is defined in such a way that its time rate of change is equal to the negative of the volume integral of the total force on the matter containing charges and carrying currents. Thus (conservation of momentum)

$$\frac{d\mathbf{M}}{dt} = -\int \rho \mathbf{F}\, d\tau,$$

where **F** is given by eq. (15.37). Show that

$$\frac{d\mathbf{M}}{dt} = \frac{d}{dt}\int \frac{\mu\kappa}{c^2}\mathbf{S}\, d\tau - \frac{\kappa}{4\pi}\int [\mathbf{E}\nabla\cdot\mathbf{E} - \mathbf{E}\times(\nabla\times\mathbf{E})]\, d\tau$$

$$-\frac{\mu}{4\pi}\int [\mathbf{H}\nabla\cdot\mathbf{H} - \mathbf{H}\times(\nabla\times\mathbf{H})]\, d\tau,$$

where **S** is the Poynting vector (eq. 15.67) $c/4\pi \cdot \mathbf{E}\times\mathbf{H}$. Show further that if the tensors T^m_{ij} and T^e_{ij} are introduced as follows

$$T^m_{ij} = H_i H_j - \frac{1}{2}H^2\delta_{ij},$$

$$T^e_{ij} = E_i E_j - \frac{1}{2}E^2\delta_{ij},$$

(with $\delta_{ij} = 1$ for $1 = j$ and zero for $i \neq j$),

$$\operatorname{div} T^m = \mathbf{H}\nabla\cdot\mathbf{H} - \mathbf{H}\times(\nabla\times\mathbf{H}),$$

$$\operatorname{div} T^e = \mathbf{E}\nabla\cdot\mathbf{E} - \mathbf{E}\times(\nabla\times\mathbf{E}).$$

(Cf. Prob. 16, Chap. 13.) Hence prove that $\frac{d\mathbf{M}}{dt}$ can be expressed in the form

$$\frac{d\mathbf{M}}{dt} = \frac{d}{dt}\int \frac{\mu\kappa}{c^2}\mathbf{S}\, d\tau - \frac{1}{4\pi}\int \sum_{i=1}^{3} (\kappa T^e + \mu T^m)_{ij}\, dS_j.$$

The tensor $\frac{1}{4\pi}(\kappa T^e + \mu T^m)$ is the Maxwell *stress tensor*. Give a physical interpretation of the above result.

CHAPTER XVI

ELECTROMAGNETIC RADIATION

16.1. Radiation from a Linear Oscillating Current Element. As an illustration of the solution of the electromagnetic field equations by the method of Sec. 15.6 consider a linear current element QOL of length l along the z axis with its center at the origin (cf. Fig. 16.1). Let alternating current

$$I = I_0 \sin \omega t \tag{16.1}$$

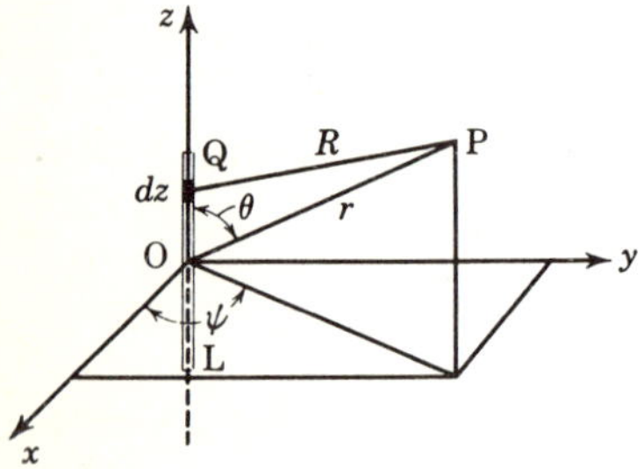

Fig. 16.1

flow through the element. We wish to find the resulting field intensities in free space at point P distant r from the origin. Since here $\rho = 0$ the problem will be solved by the evaluation of the vector potential $\mathbf{A}$ by means of eq. (15.97). Since the current density $\rho \mathbf{v}$ is directed along the z axis, $\mathbf{A}$ reduces to A_z and we have for point P at time t

$$A_z = \frac{I_0}{cS} \int_{-l/2}^{l/2} \frac{\sin \omega(t - R/c)}{R} S \, dz, \tag{16.2}$$

where R is the distance from P to element of length dz of the conductor. Since free space is involved $\mu = \kappa = 1$ and $v_c = c$. The quantity S is the area of cross section of the conductor, which cancels out. The integral in (16.2) is most simply evaluated by considering the (complex) form (with $k = \omega/c$)

$$\int_{-l/2}^{l/2} \frac{e^{i(\omega t - kR)}}{R} dz, \tag{16.3}$$

to the imaginary part of which it is proportional. Taking out the time dependent part we have to evaluate

$$\int_{-l/2}^{l/2} \frac{e^{-ikR}}{R}\, dz, \tag{16.4}$$

where $R^2 = r^2 + z^2 - 2rz \cos \theta$. We shall now assume that $r >> l/2$, and hence that (16.4) may be approximated by

$$\int_{-l/2}^{l/2} \frac{e^{-ikR}}{R}\, dz = \frac{e^{-ikr}}{r} \int_{-l/2}^{l/2} \frac{e^{ikz \cos \theta}}{1 - \dfrac{z \cos \theta}{r}}\, dz$$

$$\doteq \frac{e^{-ikr}}{r} \int_{-l/2}^{l/2} e^{ikz \cos \theta}\left(1 + \frac{z}{r} \cos \theta\right) dz. \tag{16.5}$$

If in the integration we further assume that $kl << 1$, which implies that the wave length of the radiation from the oscillating current is very large compared with the length l, we finally get

$$\int_{-l/2}^{l/2} \frac{e^{-ikR}}{R}\, dz = \frac{le^{-ikr}}{r}, \tag{16.6}$$

as might indeed have been suspected directly from an inspection of (16.4), under the assumed approximations. Consequently under the indicated approximation the vector potential at P has the components

$$A_x = A_y = 0; \quad A_z = \frac{I_0 l}{cr} \sin \ \omega(t - r/c). \tag{16.7}$$

It will be more useful to obtain the components of $\mathbf{A}$ in spherical coordinates, i.e., A_r, A_θ, A_ψ. Examination of Fig. 16.1 discloses that

$$A_r = A_z \cos \theta, \quad A_\theta = -A_z \sin \theta, \quad A_\psi = 0. \tag{16.8}$$

Moreover

$$\nabla \times \mathbf{A} = \mathbf{r}_1 \left[\frac{1}{r^2 \sin \theta}\left\{\frac{\partial}{\partial \theta}(r \sin \theta A_\psi) - \frac{\partial}{\partial \psi}(rA_\theta)\right\}\right] +$$

$$+ \boldsymbol{\theta}_1 \left[\frac{1}{r \sin\theta} \left\{ \frac{\partial A_r}{\partial \psi} - \frac{\partial}{\partial r} (r \sin\theta A_\psi) \right\} \right]$$

$$+ \boldsymbol{\psi}_1 \left[\frac{1}{r} \left\{ \frac{\partial}{\partial r} (r A_\theta) - \frac{\partial A_r}{\partial \theta} \right\} \right], \tag{16.9}$$

where $\mathbf{r}_1$, $\boldsymbol{\theta}_1$, and $\boldsymbol{\psi}_1$ are the usual unit vectors (cf. Sec. 2.6). From this we readily show that the components of the magnetic field intensity $\mathbf{H}$ are

$$H_r = (\nabla \times \mathbf{A})_r = 0, \tag{16.10}$$

$$H_\theta = (\nabla \times \mathbf{A})_\theta = 0, \tag{16.11}$$

$$H_\psi = (\nabla \times \mathbf{A})_\psi = \frac{I_0 l}{cr} \sin\theta \left[k \cos(\omega t - kr) + \frac{1}{r} \sin(\omega t - kr) \right]. \tag{16.12}$$

To obtain the components of $\mathbf{E}$, we must use eq. (15.77) and hence first find the scalar potential ϕ, which, however, is given by the Lorentz relation, (15.81), viz.,

$$\dot{\phi} = -c \nabla \cdot \mathbf{A}.$$

We have therefore

$$\dot{\phi} = -c \left[\frac{1}{r^2} \cdot \frac{\partial}{\partial r} (r^2 A_r) + \frac{1}{r \sin\theta} \cdot \frac{\partial}{\partial \theta} (A_\theta \sin\theta) + \frac{1}{r \sin\theta} \cdot \frac{\partial A_\psi}{\partial \psi} \right] \tag{16.13}$$

By substitution from (16.7) and (16.8) and time integration there results

$$\frac{\phi}{c} = -\frac{I_0 l \cos\theta}{\omega c r^2} \cos(\omega t - kr) + \frac{I_0 l k \cos\theta}{\omega c r} \sin(\omega t - kr). \tag{16.14}$$

Since

$$E_r = -(\nabla\phi)_r - \dot{A}_r/c,$$

we obtain finally for the r component of $\mathbf{E}$

$$E_r = \frac{2 I_0 l \cos\theta}{\omega r} \left[\frac{k}{r} \sin(\omega t - kr) - \frac{1}{r^2} \cos(\omega t - kr) \right] \tag{16.15}$$

The same method yields

$$E_\theta = \frac{I_0 l \sin\theta}{\omega r}\left[k^2 \cos(\omega t - kr) + \frac{k}{r}\sin(\omega t - kr) - \frac{1}{r^2}\cos(\omega t - kr)\right] \tag{16.16}$$

and

$$E_\psi = 0. \tag{16.17}$$

It is of interest to observe the form taken by the magnetic and electric field components at distances $r \gg \lambda$, where $\lambda = 2\pi/k$. These become

$$H_\theta = 0, \quad H_r = 0, \quad H_\psi \to \frac{I_0 lk}{cr}\sin\theta \cdot \cos(\omega t - kr), \tag{16.18}$$

$$E_\theta \to \frac{I_0 lk^2 \sin\theta}{\omega r}\cos(\omega t - kr), \quad E_r \to 0, \quad E_\psi = 0. \tag{16.19}$$

The statement $E_r \to 0$ means that it falls off as $1/r^2$ and has no non-vanishing terms in $1/r$. From Sec. (14.8) we conclude that the field at great distances from the current element is propagated as a spherical electromagnetic wave in which **E** and **H** are in the same phase and at right angles to each other. Moreover their amplitudes are equal. The field closer to the current element is obviously much more complicated.

16.2. Properties of Electromagnetic Radiation in a Medium Containing No Charge or Currents. We now return to the discussion in Sec. 15.4 to examine further the properties of electromagnetic waves. Let us first confine our attention to plane waves in which E and H are functions of x and t only. Then eqs. (15.40) and (15.41) become respectively

$$\frac{\partial^2 \mathbf{H}}{\partial x^2} = \frac{\ddot{\mathbf{H}}}{v_c^2}, \quad \frac{\partial^2 \mathbf{E}}{\partial x^2} = \frac{\ddot{\mathbf{E}}}{v_c^2} \tag{16.20}$$

where v_c is given by eq. (15.42). The solutions of (16.20) are in the form (for waves in the positive x direction)

$$\begin{aligned} E_x &= E_x(x - v_c t), & H_x &= H_x(x - v_c t), \\ E_y &= E_y(x - v_c t), & H_y &= H_y(x - v_c t), \\ E_z &= E_z(x - v_c t), & H_z &= H_z(x - v_c t). \end{aligned} \tag{16.21}$$

Since $\nabla \cdot \mathbf{E} = \nabla \cdot \mathbf{H} = 0$ it follows that

$$\frac{\partial E_x}{\partial x} = \frac{\partial H_x}{\partial x} = 0, \tag{16.22}$$

and hence E_x and H_x are constant and not propagated. We shall therefore put them equal to zero. A plane electromagnetic wave is thus a *transverse* wave.

If we go back to the fourth equation in (15.39) we conclude that

$$\frac{\partial E_z}{\partial x} = E_z{}' = -\frac{\mu v_c}{c} H_y{}' = -\sqrt{\frac{\mu}{\kappa}} H_y{}', \tag{16.23}$$

where the prime indicates differentiation with respect to the argument $x - v_c t$. Similarly we find

$$E_y{}' = \sqrt{\frac{\mu}{\kappa}} H_z{}'. \tag{16.24}$$

By integration we obtain

$$E_z = -\sqrt{\frac{\mu}{\kappa}} H_y; \quad E_y = \sqrt{\frac{\mu}{\kappa}} H_z. \tag{16.25}$$

If then we form the product $\mathbf{E} \cdot \mathbf{H}$, the result is

$$\mathbf{E} \cdot \mathbf{H} = E_y H_y + E_z H_z = 0. \tag{16.26}$$

Consequently $\mathbf{E}$ is perpendicular to $\mathbf{H}$. If we calculate the Poynting flux we find that

$$\frac{c}{4\pi} \mathbf{E} \times \mathbf{H} = \mathbf{i} \frac{c}{4\pi} \sqrt{\frac{\mu}{\kappa}} (H_y{}^2 + H_z{}^2) = \mathbf{i} \frac{c}{4\pi} \sqrt{\frac{\kappa}{\mu}} (E_y{}^2 + E_z{}^2). \tag{16.27}$$

Thus the energy propagation in the wave is along the positive x direction, as it should be. We have already seen that the total electromagnetic energy is given by

$$\frac{1}{8\pi} \int \mu (H_x{}^2 + H_y{}^2 + H_z{}^2)\, d\tau + \frac{1}{8\pi} \int \kappa (E_x{}^2 + E_y{}^2 + E_z{}^2)\, d\tau.$$

In the present case this reduces to

$$\frac{1}{8\pi}\left[\int \mu(H_y^{\,2} + H_z^{\,2})\,d\tau + \int \kappa(E_y^{\,2} + E_z^{\,2})\,d\tau\right].$$

But from (16.27) this reduces to

$$\frac{1}{4\pi}\int \mu(H_y^{\,2} + H_z^{\,2})\,d\tau = \frac{1}{4\pi}\int \kappa(E_y^{\,2} + E_z^{\,2})\,d\tau. \tag{16.28}$$

Hence (16.27) says that in the propagation of a plane electromagnetic wave the rate of flow of energy per unit area per second as measured by the Poynting vector is equal to the phase velocity of the wave multiplied by the energy density. We have already noted a similar result for elastic waves in Sec. 14.2.

Spherical electromagnetic waves are fully exemplified by the limiting case of the previous section. If we reexamine eqs. (16.18) and (16.19) we note that the wave is propagated along the r direction while both E_r and H_r vanish to the approximation which neglects $1/r^2$ compared with $1/r$. This indicates the approximately transverse character of the wave. Moreover we can readily show that at great distances from the current element the Poynting vector $\frac{c}{4\pi}\mathbf{E}\times\mathbf{H}$ is directed along $\mathbf{r}$ and is equal in magnitude to the product of the wave velocity and the energy density. We have indeed

$$\begin{aligned}\mathbf{E} &= \mathbf{r}_1 E_r + \boldsymbol{\theta}_1 E_\theta + \boldsymbol{\psi}_1 E_\psi,\\ \mathbf{H} &= \mathbf{r}_1 H_r + \boldsymbol{\theta}_1 H_\theta + \boldsymbol{\psi}_1 H_\psi,\end{aligned} \tag{16.29}$$

where the unit vectors $\mathbf{r}_1$, $\boldsymbol{\theta}_1$, and $\boldsymbol{\psi}_1$ have already been defined in Sec. 2.6 and have the same product properties as $\mathbf{i}$, $\mathbf{j}$, $\mathbf{k}$, e.g., $\mathbf{r}_1 \times \boldsymbol{\theta}_1 = \boldsymbol{\psi}_1$, etc. Consequently

$$\mathbf{E}\times\mathbf{H} = \begin{vmatrix} \mathbf{r}_1 & \boldsymbol{\theta}_1 & \boldsymbol{\psi}_1 \\ E_r & E_\theta & E_\psi \\ H_r & H_\theta & H_\psi \end{vmatrix} \tag{16.30}$$

If we use (16.18) and (16.19) we get at once

$$\frac{c}{4\pi}\mathbf{E}\times\mathbf{H} = \frac{\mathbf{r}_1}{4\pi}\frac{I_0^2 l^2 k^2}{cr^2}\sin^2\theta\cos^2(\omega t - kr). \tag{16.31}$$

But this is precisely equal to c times the energy density, viz.,

$$\frac{c}{8\pi}(E_\theta^2 + H_\psi^2). \tag{16.32}$$

16.3. Electromagnetic Waves in Conducting Media. Skin Effect. So far in this chapter we have confined ourselves to the special case of electromagnetic waves in free space or at any rate in media in which there exist no static or moving charges. Let us go back to the field equations in general form (15.36), and assume that the medium to which they apply is a conductor of electricity in which Ohm's law is obeyed and in which there is no free charge, i.e., $\nabla\cdot\mathbf{E} = 0$. In this case we can write

$$\rho\mathbf{v} = \sigma\mathbf{E}, \tag{16.33}$$

where σ is here the electrical conductivity. The third and fourth equations in (15.36) now become

$$\nabla\times\mathbf{H} = (4\pi\sigma\mathbf{E} + \kappa\dot{\mathbf{E}})/c, \tag{16.34}$$

$$\nabla\times\mathbf{E} = -\mu\dot{\mathbf{H}}/c. \tag{16.35}$$

The usual method of taking the curl of the first and using the second now yields

$$\nabla^2\mathbf{H} = \frac{4\pi\mu\sigma}{c^2}\dot{\mathbf{H}} + \frac{\kappa\mu}{c^2}\ddot{\mathbf{H}}, \tag{16.36}$$

with a similar equation for **E**. If we restrict our attention to plane waves by assuming that the components of **E** and **H** are functions of x alone, E_x and H_x turn out to be constants which may appropriately be put equal to zero. Then we have to deal with an equation of the form

$$\frac{\partial^2 E_y}{\partial x^2} = \frac{\ddot{E}_y}{v_c^2} + \frac{4\pi\sigma}{\kappa v_c^2}\dot{E}_y \tag{16.37}$$

We are tempted to consider harmonic time dependence and a solution of the form

$$E_y = f(x)e^{i\omega t} \tag{16.38}$$

Substitution yields

$$f(x) = Ae^{-i\sqrt{\omega^2 - 4\pi i\sigma\omega/\kappa} \cdot x/v_c} + Be^{+i\sqrt{\omega^2 - 4\pi i\sigma\omega/\kappa} \cdot x/v_c}, \quad (16.39)$$

where A and B are arbitrary constants. Let us write

$$\sqrt{\kappa\mu\omega^2 - 4\pi i\mu\sigma\omega} = \omega(n - ik), \quad (16.40)$$

whence

$$\begin{aligned} n &= \sqrt{\frac{\mu}{2}\left(\sqrt{\kappa^2 + 4\sigma^2 T^2} + \kappa\right)}, \\ k &= \sqrt{\frac{\mu}{2}\left(\sqrt{\kappa^2 + 4\sigma^2 T^2} - \kappa\right)}, \end{aligned} \quad (16.41)$$

in which $T = 1/\nu = 2\pi/\omega$. This allows us to write, with appropriate attention to signs, for the part of E_y corresponding to wave propagation in the positive x direction

$$E_y = Ae^{-\frac{\omega kx}{c}} e^{i\left(\omega t - \frac{\omega nx}{c}\right)} \quad (16.42)$$

which is a wave progressing with phase velocity

$$V = \frac{c}{n} . \quad (16.43)$$

Here n appears as the effective index of refraction of the conducting medium. The wave is damped, and the decrease of amplitude with distance is measured by $e^{\frac{-\omega kx}{c}}$, so that the effective absorption coefficient is

$$\propto = \frac{\omega k}{c} \quad (16.44)$$

These equations have some interesting physical consequences. In the first place they make it clear that an electrical conductor absorbs electromagnetic radiation with a damping coefficient depending on the frequency. Moreover the velocity itself will in general depend on the frequency and therefore the conductor will appear as a dispersive medium. Metals are good conductors of electricity and hence we may expect to find them absorbers of electromagnetic waves. This is certainly true both for the relatively long radio waves as well as the short waves of visible light, since metals are opaque to such radiation unless the thickness of the material is extremely small. On the other hand, it may be objected that water (particularly salt water) is a good conductor of electricity, yet it is transparent to visible light. This would appear as a definite contradiction to the result of this section, were it not for the fact that even relatively thin layers of water absorb radio waves and fairly thick layers absorb visible light to a considerable extent, as the "color" of sea water amply indicates. The dependence of absorption on frequency will then take care of this situation, although we must be careful not to conclude too hastily that (16.41), (16.43), and (16.44) will apply to all such cases. As a matter of fact they are not very good approximations for water, and we have here an illustration of the fact that the direct application of electromagnetic theory to *continuous* material media may not lead to correct predictions: it may prove necessary to replace the continuous character of the medium by a discrete structure and modify the theory by considering the interaction between the radiation and the particles of this structure. We shall meet this situation again further on in this section and shall discuss it more in detail in Chapter 17.

In the case of most metals σ in e.s.u. is of the order of 10^{18} and, although we do not know what κ is, we may safely assume it can be neglected in (16.41). Under these conditions, to a good approximation

$$n \doteq k = \sqrt{2\pi\mu\sigma/\omega}, \tag{16.45}$$

and

$$\propto = \sqrt{2\pi\mu\sigma\omega}/c. \tag{16.46}$$

Now imagine a plane electromagnetic wave propagated in the z direction in a metal with conductivity σ. The **E** and **H** vectors must be in the xy plane. Take **E** in the x direction and **H** in the y direction. Using real parts we then have

$$H_y = H_0 e^{-\propto z} \cos(\omega t - \propto z). \tag{16.47}$$

Since κ is being neglected we can calculate the current density $\rho\mathbf{v}$ from (16.34) as

$$(\rho\mathbf{v}) = \frac{c}{4\pi} \cdot \nabla \times \mathbf{H}. \tag{16.48}$$

Now $H_x = H_z = 0$ and hence $\nabla \times \mathbf{H}$ reduces to $-\mathbf{i}\, \partial H_y/\partial z$, so that the current flow is wholly in the x direction and $(\rho \mathbf{v})_y = (\rho \mathbf{v})_z = 0$. Then

$$(\rho \mathbf{v})_x = \frac{c\alpha}{4\pi} H_0 e^{-\alpha z} [\cos(\omega t - \alpha z) - \sin(\omega t - \alpha z)]\,. \qquad (16.49)$$

The current density is independent of x but falls off with z exponentially. This means that the flow of charge is confined to a narrow sheet of metal parallel to the xy plane, a situation which has been labeled the *skin effect.* The reader may show, for example, that in copper for a wave length of 1 cm in air or vacuum the value of $1/\alpha$ or the depth of penetration for which the current density amplitude drops to $1/e$th of its maximum value is approximately 4×10^{-5} cm. For $\lambda = 10^3$ cm, this distance is increased to 1.2×10^{-3} cm.

16.4. Electromagnetic Waves Along Transmission Lines. The Hollow Pipe. Electromagnetic Filtration. Electromagnetic radiation normally spreads out in all directions from the source. It can be guided, however, from one place to another by the use of conductors of suitable shape. We shall here examine the transmission of electrical energy through the space enclosed by a hollow, perfectly conducting pipe in the form of a circular cylinder of radius R whose axis is taken as the z axis.

Since there will be no free charge or current, and since we shall assume $\kappa = \mu = 1$, the field equations appear in the form

$$\begin{aligned} \nabla \cdot \mathbf{E} &= 0, & \nabla \cdot \mathbf{H} &= 0, \\ \nabla \times \mathbf{E} &= -\dot{\mathbf{H}}/c, & \nabla \times \mathbf{H} &= \dot{\mathbf{E}}/c. \end{aligned} \qquad (16.50)$$

We shall be interested only in field quantities which are harmonic functions of the time with frequency $\omega/2\pi$. Hence writing $k = \omega/c$, as usual, we have from (16.50)

$$\begin{aligned} \nabla \cdot \mathbf{E} &= 0, & \nabla \cdot \mathbf{H} &= 0, \\ \nabla \times \mathbf{E} &= -ik\mathbf{H}, & \nabla \times \mathbf{H} &= ik\mathbf{E}. \end{aligned} \qquad (16.51)$$

We are looking for solutions of the above which correspond to harmonic waves traveling in both directions along the tube. These will be classified in three types depending on the behavior of the components E_z and H_z. Thus

$$\text{Type 1:} \quad H_z = 0, \quad E_z \neq 0.$$

Type 2: $E_z = 0, \quad H_z \neq 0.$

Type 3: $E_z = H_z = 0.$

We begin with Type 1. Taking the curl of $\nabla \times \mathbf{E}$ in (16.51) as usual, we wind up with

$$\nabla^2\mathbf{E} + k^2\mathbf{E} = 0. \tag{16.52}$$

Now since we are interested in harmonic propagation along the length of the tube, we must assume that E_x, E_y, and E_z depend on z through a term of the form $e^{-ik_3 z}$ (where we must indicate that the coefficient of z in the exponent is not necessarily equal to k, since the wave velocity in the z direction is not necessarily c). Hence the z component equation takes the form

$$\frac{\partial^2 E_z}{\partial x^2} + \frac{\partial^2 E_z}{\partial y^2} + (k^2 - k_3{}^2)E_z = 0. \tag{16.53}$$

To obtain the equations giving the variation of E_x and E_y with z we must write the two curl equations in component form, getting

$$\begin{aligned} -\frac{\partial H_y}{\partial z} &= ikE_x, \\ \frac{\partial H_x}{\partial z} &= ikE_y, \\ \frac{\partial H_y}{\partial x} - \frac{\partial H_x}{\partial y} &= ikE_z, \end{aligned} \tag{16.54}$$

and

$$\begin{aligned} -ikH_x &= \frac{\partial E_z}{\partial y} - \frac{\partial E_y}{\partial z}, \\ -ikH_y &= \frac{\partial E_x}{\partial z} - \frac{\partial E_z}{\partial x}, \\ -ikH_z &= \frac{\partial E_y}{\partial x} - \frac{\partial E_x}{\partial y} = 0. \end{aligned} \tag{16.55}$$

Differentiating the second equation of (16.55) through by z and utilizing the first equation of (16.54) we arrive at

$$\frac{\partial^2 E_x}{\partial z^2} + k^2 E_x = \frac{\partial}{\partial x}\left(\frac{\partial E_z}{\partial z}\right). \tag{16.56}$$

A similar process performed on the first equation of (16.55) with the help of the second equation of (16.54) yields

$$\frac{\partial^2 E_y}{\partial z^2} + k^2 E_y = \frac{\partial}{\partial y}\left(\frac{\partial E_z}{\partial z}\right). \tag{16.57}$$

Since $\partial^2 E_x/\partial z^2 = -k_3{}^2 E_x$, etc., we immediately have from (16.56) and (16.57)

$$E_x = \left(\frac{1}{k^2 - k_3{}^2}\right) \frac{\partial}{\partial x}\left(\frac{\partial E_z}{\partial z}\right),$$
$$E_y = \left(\frac{1}{k^2 - k_3{}^2}\right) \frac{\partial}{\partial y}\left(\frac{\partial E_z}{\partial z}\right). \tag{16.58}$$

Having determined the vector **E** we may proceed to express **H** in terms of it. The combination of the first equation in (16.55) with the second equation of (16.54) (utilizing the fact that $\partial^2 H_x/\partial z^2 = -k_3{}^2 H_x$) gives

$$H_x = \left(\frac{ik}{k^2 - k_3{}^2}\right) \frac{\partial E_z}{\partial y} \tag{16.59}$$

Similarly we can show that

$$H_y = \left(\frac{-ik}{k^2 - k_3{}^2}\right) \frac{\partial E_z}{\partial x} \tag{16.60}$$

We must now pay some attention to boundary conditions. Since the tangential component of **E** must be continuous at a boundary surface,[1] it follows that it must be zero at the surface of a conductor, since **E** vanishes inside the conductor. Therefore in the hollow tube here considered **E** must be *normal* to all surfaces. It follows that **H** must be tangential. This limits the possible solutions of the partial differential equations above. In particular E_x and E_y must vanish on the end surfaces of the line (assumed perpendicular to the axis of the tube). Therefore when we go over to real parts E_z must depend on z through $\cos k_3 z$

[1] This is assumed here but demonstrated in Sec. 16.7.

and not sin $k_3 z$, so that, as (16.58) indicates, E_x and E_y can vary as sin $k_3 z$ and be able to vanish at the origin.

We are now prepared to write the final solution in formal fashion

$$E_z = g(x, y) \cos k_3 z, \tag{16.61}$$

where $g(x, y)$ satisfies the equation

$$\frac{\partial^2 g}{\partial x^2} + \frac{\partial^2 g}{\partial y^2} + k_g^2 g = 0, \tag{16.62}$$

under the boundary condition $g(x, y) = 0$ at the surrounding surfaces. The characteristic or eigenvalues of k_g are determined by the solution of (16.62) subject to the given boundary condition. Evidently we must also have

$$k^2 = k_3^2 + k_g^2 \tag{16.63}$$

From (16.58) there results

$$\begin{aligned} E_x &= -\frac{k_3}{k_g^2} \frac{\partial g}{\partial x} \sin k_3 z, \\ E_y &= -\frac{k_3}{k_g^2} \frac{\partial g}{\partial y} \sin k_3 z, \end{aligned} \tag{16.64}$$

while from (16.59) and (16.60) we get

$$\begin{aligned} H_x &= \frac{ik}{k_g^2} \frac{\partial g}{\partial y} \cos k_3 z, \\ H_y &= -\frac{ik}{k_g^2} \frac{\partial g}{\partial x} \cos k_3 z. \end{aligned} \tag{16.65}$$

We can now examine the Type 2 situation in similar fashion. We need not go through the details but merely present the results as follows

$$H_z = h(x, y) \sin k_3 z, \tag{16.66}$$

where $h(x, y)$ satisfies the equation

$$\frac{\partial^2 h}{\partial x^2} + \frac{\partial^2 h}{\partial y^2} + k_h^2 h = 0, \tag{16.67}$$

under the boundary condition $\partial b(x, y)/\partial n = 0$ at the surrounding surfaces. The reason for the change in boundary condition is inherent in the existence of the relations (analogous to 16.56 and 16.57)

$$\frac{\partial^2 H_x}{\partial z^2} + k^2 H_x = (k^2 - k_3^2) H_x = \frac{\partial}{\partial x}\left(\frac{\partial H_z}{\partial z}\right),$$
$$\frac{\partial^2 H_y}{\partial z^2} + k^2 H_y = (k^2 - k_3^2) H_y = \frac{\partial}{\partial y}\left(\frac{\partial H_z}{\partial z}\right). \tag{16.68}$$

Now since **H** is tangential to the boundary (cf. Sec. 16.7) it follows that H_x and H_y must vanish on the cylindrical surface, and from (16.68) the gradient of H_z normal to this surface must be zero. This is the basis for $\partial b/\partial n = 0$. Continuing, we obtain

$$k^2 = k_h^2 + k_3^2, \tag{16.69}$$

$$\left.\begin{aligned} E_x &= \frac{-ik}{k_h^2} \frac{\partial b}{\partial y} \sin k_3 z, \\ E_y &= \frac{ik}{k_h^2} \frac{\partial b}{\partial x} \sin k_3 z, \end{aligned}\right\} \tag{16.70}$$

$$\left.\begin{aligned} H_x &= \frac{k_3}{k_h^2} \frac{\partial b}{\partial x} \cos k_3 z, \\ H_y &= \frac{k_3}{k_h^2} \frac{\partial b}{\partial y} \cos k_3 z. \end{aligned}\right\} \tag{16.71}$$

If the cylinder is of finite length, extending from $z = 0$ to $z = l$, the condition of vanishing E_x and E_y at the ends leads to

$$\sin k_3 l = 0, \quad k_3 = n\pi/l, \tag{16.72}$$

where n is an integer. This relation gives the allowed frequencies for electromagnetic radiation in the cylindrical cavity. It is of course reminiscent of the normal modes of a stretched string.

So far the results have actually been independent of the precise shape of the cylinder. However to obtain specific results it is, of course, necessary to solve eqs. (16.62) and (16.67) and hence to specify the nature of the cylinder. Let us assume it is circular with radius R. It will be convenient to write eq. (16.62) in cylindrical coordinates r, ϕ, z (cf. Sec. 2.2). Eq. (16.62) then takes the form

$$\frac{1}{r}\frac{\partial}{\partial r}\left(r\frac{\partial g}{\partial r}\right) + \frac{1}{r^2}\frac{\partial^2 g}{\partial \phi^2} + k_g{}^2 g = 0. \tag{16.73}$$

We try for a solution in the form

$$g = G(r)\Phi(\phi), \tag{16.74}$$

whence we obtain the two equations for $G(r)$ and $\Phi(\phi)$

$$\frac{d^2\Phi}{d\phi^2} + p^2\Phi = 0, \tag{16.75}$$

$$r^2\frac{d^2G}{dr^2} + r\frac{dG}{dr} + (k_g{}^2 r^2 - p^2)G = 0. \tag{16.76}$$

Here p^2 appears as a constant parameter. Since Φ is necessarily single-valued, p must be an integer, and we write for the solution of (16.75)

$$\Phi = \Phi_0 e^{\pm ip\phi}. \tag{16.77}$$

Eq. (16.76) has long been known as Bessel's equation,[1] and its general solution has the form

$$G = AJ_p(k_g r) + BK_p(k_g r), \tag{16.78}$$

where $J_p(k_g r)$ is called Bessel's function of the first kind of order p and K_p Bessel's function of the second kind of order p. These can be expressed in series form as follows

$$J_p(u) = \sum_{k=0}^{\infty} \frac{(-1)^k}{k!(p+k)!}\left(\frac{u}{2}\right)^{p+2k} \tag{16.79}$$

[1] Cf. R. S. Burington and C. C. Torrance, *Higher Mathematics* (New York, McGraw-Hill, 1939), pp. 430 ff.

$$K_p(u) = J_p(u) \log u - 1/2 \cdot \sum_{j=0}^{p-1} \frac{(p-j-1)!}{j!} \left(\frac{u}{2}\right)^{-p+2j}$$

$$-1/2 \cdot \sum_{k-0}^{\infty} \frac{(-1)^k}{k!(n+k)!} \left[(1 + 1/2 + \ldots\right.$$

$$\left. + 1/k) + \left(1 + 1/2 + \ldots + \frac{1}{p+k}\right)\right] \left(\frac{u}{2}\right)^{p+2k} \tag{16.80}$$

As far as our present problem is concerned $K_p(u)$ goes to infinity for $x = 0$. Hence it cannot play a role in our solution. Therefore the solution of (16.73) becomes

$$g(r, \phi) = J_p(k_g r) e^{\pm ip\phi}. \tag{16.81}$$

For the type 1 modes ($E_z \neq 0$), the boundary condition on g makes it vanish on the surface of the cylinder. Hence the k_g are determined by the equation

$$J_p(k_g R) = 0. \tag{16.82}$$

The type 1 modes are usually referred to as E modes. If the roots of $J_p(u)$ are denoted by u_{jp}, where $j = 1$ corresponds to the smallest root, etc., we can write (employing 16.63) for the allowed values of k

$$k^2{}_{Enjp} = u^2{}_{jp}/R^2 + n^2\pi^2/l^2. \tag{16.83}$$

Tables may be consulted[1] for the values of u_{jp}. We may note merely that the E mode of lowest frequency corresponds to $n = 0$, $p = 0$, $j = 1$, and gives

$$k_{E010} = 2.405/R.$$

The corresponding wave length is

$$\lambda_{E010} = 2.61R.$$

[1]Cf. Jahnke and Emde, *Tables of Functions with Formulae and Curves* (American Ed., New York, Dover Publications, 1943) p. 168.

It will be evident that the frequencies do not form a harmonic series.

Type 2 modes ($H_z \neq 0$) are usually known as H modes. Eq. (16.67) is of the same form as (16.62) and the solution (16.81) still applies. However the boundary condition on $h(r, \phi)$ is different. Thus we now have at the cylindrical surface

$$\frac{dh}{dr} = 0.$$

Hence the modes are determined by the relations

$$\frac{dJ_p(k_h r)}{dr} = 0 \tag{16.84}$$

at $r = R$. The roots of the first derivative of $J_p(u)$ are not generally tabulated in books.[1] However, good approximation formulas are available in Jahnke and Emde (*loc. cit.*, p. 143). Let us denote the roots of $dJ_p(v)/dv$ by v_{jp}. Then the allowed values of k are given by

$$k^2_{Hnjp} = v^2_{jp}/R^2 + n^2\pi^2/l^2. \tag{16.85}$$

As an illustration, for $p = 0$, $j = 1$, $n = 1$ we get

$$k^2_{H110} = 14.6/R^2 + \pi^2/l^2.$$

Note that for an H mode, $n > 0$.

Let us revert for a moment to the relation (16.63) which we may write conveniently in terms of wave length (since $k = 2\pi/\lambda$)

$$\frac{1}{\lambda^2} = \frac{1}{\lambda_3^{\,2}} + \frac{1}{\lambda_g^{\,2}}. \tag{16.86}$$

Here λ is the wave length of the radiation of frequency $\omega/2\pi$ in free space. λ_3 is the effective wave length in the hollow pipe, i.e., the wave length with which propagation takes place *along* the pipe. To attach a meaning to λ_g, observe that if $\lambda < \lambda_g$, $1/\lambda_3^{\,2}$ is positive and λ_3 is real, i.e., we get actual wave transmission down the pipe. On the other hand, if $\lambda > \lambda_g$, $\lambda_3^{\,2}$ is negative and λ_3 is imaginary. But an imaginary wave length (i.e., imaginary k value) means that there is no propagation, but

[1] But cf. D. B. Smith, L. M. Rodgers, and E. H. Traub, *Jour. Franklin Inst.*, 237, 301, 1944.

rather attenuation in the z direction. Hence certain wave lengths in free space are transmitted along the pipe while certain others are attenuated: the pipe acts as a *filter* for electromagnetic radiation. Short wave lengths (or high frequencies) are favored over long wave lengths (or low frequencies). In line with our earlier study of elastic wave transmission, we naturally term the hollow pipe a high pass electromagnetic wave filter. The cut-off wave length is then λ_g. It is the longest wave length for which transmission without attenuation takes place along the pipe. Since we have already seen that the longest wave length for an E mode is $2.61R$, it follows that for pipes of ordinary feasible dimensions, only radiation of rather short wave length is capable of transmission. Such pipes are useful in the propagation of what have come to be called microwaves.

We have assumed the walls of the cylinder to be perfectly conducting. If they are not, it is clear that $\mathbf{E}$ will no longer have to be normal to the walls, and some energy flow will take place into the walls with subsequent attenuation. Even in the transmission band of the filter we will not then have perfect transmission.

16.5. The Coaxial Cable as a Transmission Line. Another illustration of a medium for the confinement of electromagnetic radiation is the space between two right concentric circular cylindrical conductors. As a matter of fact we do not need to restrict our discussion to circular cylinders as many of the results to be obtained apply to pairs of cylinders in general. However, probably the most interesting case is the coaxial cable consisting of a circular cylindrical conductor of radius R_1 surrounded by a hollow conducting concentric circular cylindrical tube of radius R_2, the intervening space being air or some other dielectric.

Here we shall find harmonic modes corresponding to vanishing values of both E_z and H_z in distinction to the results of the previous section for the hollow pipe. Let us go back to (16.54) and (16.55) and assume that E_z is zero as well as H_z. Then we obtain

$$-\frac{\partial H_y}{\partial z} = ikE_x, \quad \frac{\partial H_x}{\partial z} = ikE_y, \quad \frac{\partial H_y}{\partial x} - \frac{\partial H_x}{\partial y} = 0, \tag{16.87}$$

as well as

$$ikH_x = \frac{\partial E_y}{\partial z}, \quad -ikH_y = \frac{\partial E_x}{\partial z}, \quad \frac{\partial E_y}{\partial x} - \frac{\partial E_x}{\partial y} = 0. \tag{16.88}$$

From the third equation in (16.88) we conclude that we can introduce a function $U(x, y, z)$ such that

$$E_x = -\frac{\partial U}{\partial x}, \quad E_y = -\frac{\partial U}{\partial y}. \tag{16.89}$$

Then the first and second equations of (16.88) demand that

$$ikH_x = -\frac{\partial}{\partial y}\left(\frac{\partial U}{\partial z}\right), \quad ikH_y = \frac{\partial}{\partial x}\left(\frac{\partial U}{\partial z}\right). \tag{16.90}$$

Combination of these with (16.87) yields

$$\begin{aligned} \frac{\partial^2}{\partial z^2}\left(\frac{\partial U}{\partial x}\right) + k^2\left(\frac{\partial U}{\partial x}\right) &= 0, \\ \frac{\partial^2}{\partial z^2}\left(\frac{\partial U}{\partial y}\right) + k^2\left(\frac{\partial U}{\partial y}\right) &= 0, \\ \frac{\partial^2}{\partial x^2}\left(\frac{\partial U}{\partial z}\right) + \frac{\partial^2}{\partial y^2}\left(\frac{\partial U}{\partial z}\right) &= 0. \end{aligned} \tag{16.91}$$

The requirement that E_x and E_y vanish at the ends of the structure, i.e., at $z = 0$ and $z = l$, means that U must contain the factor $\sin n\pi z/l$, where n is integral. Let us therefore set

$$U = u(x, y) \sin n\pi z/l. \tag{16.92}$$

From the third equation in (16.91) it follows that

$$\frac{\partial^2 u}{\partial x^2} + \frac{\partial^2 u}{\partial y^2} = 0, \tag{16.93}$$

i.e., u satisfies Laplace's equation. Now the boundary conditions demand that E_x and E_y shall vanish at the surface of the cylinders. Hence u must have constant values there, independent of x and y, as is clear from (16.89). We here see at once the great difference between the hollow pipe and the coaxial line. In the former there is only one surface. But if u has to satisfy (16.93) inside the pipe and reduce to a constant value on the surface of the pipe, it must have the same constant value everywhere inside. We have showed this for the special case of a sphere in Sec. 12.4. Potential theory and the general properties of harmonic functions (i.e., solutions of Laplace's equation) guarantee it for a surface of arbitrary form.[1] Hence all components of electric and magnetic field vanish inside the pipe and there exist no modes in this case for which E_z and H_z are both zero.

On the other hand, if the hollow pipe is replaced by the coaxial cable there are two surfaces bounding the region in which (16.93) holds, and it is perfectly possible for u to have the constant value $u = u_1$ on the inside surface and the different constant value $u = u_2$ on the outside surface, with a nonconstant solution $u(x, y)$ in the space between the surfaces. Hence in this case E_z and H_z can both vanish without leading to zero values for E_x, E_y, H_x, H_y. Moreover the boundary value problem given by eq. (16.93) and the values of u on the two surfaces for any section normal to the z axis does not involve the parameter k. Hence the permissible values of the latter are determined entirely by the length of the line and are given indeed by $n\pi/l$. The corresponding wavelengths are $\lambda_n = l/2n$. Although we are here dealing with coaxial cylinders, the same result holds for cylinders of arbitrary form. This is an important result, since it means we can have transmission for the mode corresponding to $n = 0$ or $\lambda = \infty$. This however is essentially the λ_g or cut-off wave length of Sec. 16.4. Hence the cut-off wave length in this mode (called a principal mode) is infinite and in this mode all wave lengths can be transmitted. This is one of the great practical advantages of the coaxial cable. For $n = 0$, $U = 0$, and hence $E_x = E_y = 0$. But as (16.90) indicates we can still have in this case

$$H_x = -\frac{\partial u}{\partial y}, \quad H_y = \frac{\partial u}{\partial x}. \tag{16.94}$$

This means the existence of a magnetic field independent of z, i.e., a steady field, which must be due to steady current moving in one direction on the inner cylinder and in the opposite direction on the outer cylinder.

Let us return to the more general modes in which n need not be zero. The solution of (16.93) subject to the cylindrical boundary conditions is

$$u = K_1 \log r + K_2, \tag{16.95}$$

where K_1 and K_2 are constants chosen so that

$$K_1 \log R_1 + R_2 = u_1, \quad K_1 \log R_2 + K_2 = u_2. \tag{16.96}$$

[1]Cf., e.g., H. B. Phillips, *Vector Analysis* (New York, John Wiley, 1933), Chap. VII, particularly p. 161.

This solution may be obtained by inspection. The general uniqueness theorem (cf. Sec. 12.4) guarantees it is the only one meeting the imposed conditions. Hence

$$U = (K_1 \log r + K_2) \sin n\pi z/l, \tag{16.97}$$

and from (16.89) we have

$$\begin{aligned} E_x &= -\frac{K_1 x}{r^2} \cdot \sin n\pi z/l, \\ E_y &= -\frac{K_1 y}{r^2} \cdot \sin n\pi z/l. \end{aligned} \tag{16.98}$$

But this means that **E** is directed radially in toward the axis and varies in magnitude inversely with r. Moreover from (16.90) we have

$$\begin{aligned} H_x &= -\frac{K_1 y}{r^2} \cos n\pi z/l, \\ H_y &= \frac{K_1 x}{r^2} \cos n\pi z/l. \end{aligned} \tag{16.99}$$

This means that **H** is directed circumferentially and also varies in magnitude inversely with r. In fact,

$$H = \sqrt{H_x^{\,2} + H_y^{\,2}} = \frac{K_1}{r} \cos n\pi z/l. \tag{16.100}$$

Now if we take the line integral of the magnitude of **H** around the periphery of the inner conductor of the cable we must get $4\pi I$, where I is the magnitude of the current through the conductor (cf. eq. 15.12). Hence, using amplitudes only,

$$2\pi R_1 H = \frac{2\pi R_1 K_1}{R_1} = 4\pi I,$$

or

$$K_1 = 2I, \tag{16.101}$$

whence (16.100) gives for the amplitude of the magnetic field

$$H = \frac{2I}{r}. \tag{16.102}$$

Similarly the amplitude of the electric field is

$$E = \frac{2I}{r}. \tag{16.103}$$

We can calculate the amplitude of the potential difference between the two cylinders by forming the line integral of the amplitude of electric field intensity along a radius from the inner to the outer cylinder. Thus

$$V = \int_{R_1}^{R_2} E\,dr = 2\int_{R_1}^{R_2} \frac{I}{r}\,dr. \tag{16.104}$$

We thus obtain

$$V = 2I \log (R_2/R_1). \tag{16.105}$$

The ratio of V to I is called the *characteristic impedance* of the coaxial line. We write

$$Z_0 = V/I = 2 \log (R_2/R_1). \tag{16.106}$$

If this is to be expressed in ohms the factor 2 must be replaced by 60.

16.6. Transmission Lines from the Standpoint of Circuit Theory. The discussion of the preceding two sections has provided an illustration of the application of electromagnetic field theory to the transmission of electric energy along conductors. In many practical cases of transmission it has proved convenient to use circuit analysis, i.e., effectively Ohm's law generalized to variable currents in circuits with resistance, inductance, and capacitance. It is of interest to note the relation between this point of view and that of field theory.

An electrical network is a combination of elements consisting of resistances, inductances, and capacitances forming a set of complete circuits called loops or meshes. If harmonic electromotive forces are impressed on certain elements of the network, harmonic currents flow through the loops and the task of network theory is to calculate the steady state value of these currents. This is done by the application of Kirchhoff's laws:

(1) The total electromotive force around any closed loop or mesh is zero.

(2) The algebraic sum of the currents entering any branch point of the network is zero.

Actually the network problem is mathematically identical with that of the forced vibrations of coupled mechanical systems discussed in Sec. 6.5, albeit generalized to include an arbitrary number of separate systems.[1] A transmission line may be considered as a network consisting of an iterated sequence of similar loops. A standard illustration is provided schematically in Fig. 16.2 which shows a series of current elements with inductance L and resistance R shunted by elements with capacitance C and conductance G. At low frequencies it proves possible to discuss the flow of current and transmission of power along the iterated network by treating the circuit elements as concentrated or "lumped," i.e., by employing essentially forced oscillation equations. However at higher frequencies the elements of each network can no longer be considered as lumped but rather must be treated as continuously distributed, and the ordinary differential equations of oscillation theory must be replaced by the partial differential equations of medium or field theory. It turns out to be possible to extend the lumped impedance theory to the continuous case by rather simple means. We introduce the concepts of resistance, inductance, conductance, and capacitance *per unit length* of the line and in what follows interpret the symbols mentioned above in this sense. If we consider an infinitesimal length dz of the line as shown in Fig. 16.3, the change in e.m.f. between beginning and end of the portion dz is given by the drop in potential difference in the line. Thus

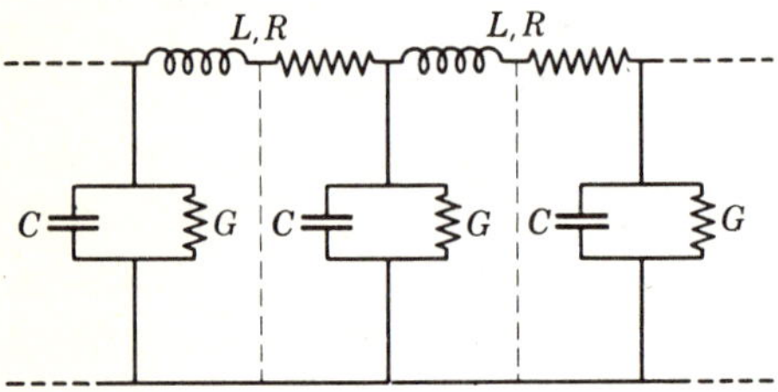

Fig. 16.2

$$\frac{\partial V}{\partial z}\, dz = -RI\, dz - L\, dz \cdot \frac{\partial I}{\partial t}. \tag{16.107}$$

[1] A good discussion of network theory will be found in W. P. Mason, *Electromechanical Transducers and Wave Filters* (New York, Van Nostrand Co., 1942), Chap. II. The application to transmission lines is also discussed in R. I. Sarbacher and W. A. Edson, *Hyper and Ultra High Frequency Engineering* (New York, John Wiley, 1943), Chap. 9.

Similarly the change in current between the ends of the portion dz is given by the current through the shunt. Thus

$$\frac{\partial I}{\partial z}\, dz = -G\, dz \cdot V - C\, dz \cdot \frac{\partial V}{\partial t}. \tag{16.108}$$

We look for a solution of (16.107) and (16.108) in which the e.m.f. and current have the form

$$V = V_0 e^{i(\omega t - kz)}, \quad I = I_0 e^{i(\omega t - kz)}, \tag{16.109}$$

i.e., correspond to wave propagation along the line with e.m.f. and current amplitudes V_0 and I_0, respectively. The equations then become

$$ikV_0 = (R + i\omega L)I_0,$$

$$ikI_0 = (G + i\omega C)V_0. \tag{16.110}$$

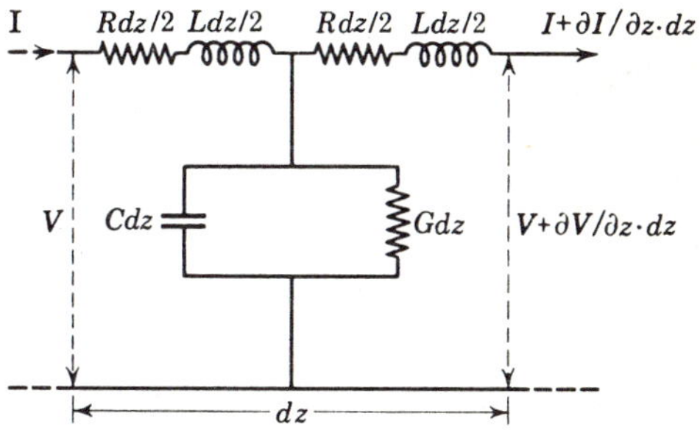

Fig. 16.3

The solution of these equations is possible only for

$$k^2 = -(R + i\omega L)(G + i\omega C). \tag{16.111}$$

Thus we get in general a complex propagation parameter, which means transmission with attenuation. However, if R and G both vanish (no resistance in the main line and infinite resistance in the shunt), there results

$$k = \omega\sqrt{LC}. \tag{16.112}$$

This corresponds to wave propagation without attenuation and with phase velocity

$$c = 1/\sqrt{LC}.$$

It is of some importance to investigate the connection between the preceding discussion and electromagnetic field theory. We revert to the coaxial cable of Sec. 16.5. To avoid ambiguity $V(z, t)$ as potential difference will be taken to mean the line integral of $\mathbf{E}$ on a path from one cylinder to the other in a plane normal to the axis (i.e., of constant z). We have already seen that the components E_x and E_y are derivable from a potential (16.89), and hence the line integral in question is independent of the path as long as z is constant. We have already used this definition of V in (16.104). We interpret current in terms of the line integral of $\mathbf{H}$ around a closed path enclosing the inner cylinder in a plane of constant z. This was used in the preceding section in getting (16.102). The only question that might arise refers to the displacement current $\dot{\mathbf{D}}$. But this makes no contribution to the line integral because it is normal to the conductor at every point.

We next have to interpret the inductance per unit length, which we have denoted above as L. If current I flows, the magnetic field energy in the volume enclosed between the planes z and $z + dz$ can be written in the form $(L\ dz)I^2/2$. But the same quantity can also be written as the volume integral

$$\frac{1}{8\pi}\int \mathbf{H}\cdot\mathbf{H}\ d\tau,$$

where the integration is taken over the volume between the planes. The equating of the two equivalent expressions leads to

$$L = \int \mathbf{H}\cdot\mathbf{H}\ d\tau/4\pi I^2\ dz, \tag{16.114}$$

which serves as the field theory definition of L. The capacitance C per unit length can be obtained in similar fashion by equating two equivalent expressions for the electric field energy in the volume enclosed by the planes z and $z + dz$, viz., $(C\ dz)\ V^2/2$ and

$$\frac{1}{8\pi}\int \mathbf{E}\cdot\mathbf{E}\ d\tau.$$

This gives

$$C = \int \mathbf{E}\cdot\mathbf{E}\ d\tau/4\pi V^2\ dz. \tag{16.115}$$

This is the field theory definition of C.

16.7. Transmission of Electromagnetic Waves Across Boundaries. Fresnel Formulas. To discuss the transmission of an electromagnetic wave across the interface separating media of different properties obviously demands the use of boundary conditions. We have already had occasion to employ these in Sec. 16.4. Now we wish to establish their validity before proceeding to the problem of this section. In Fig. 16.4 let S represent a section of the surface separating medium I characterized

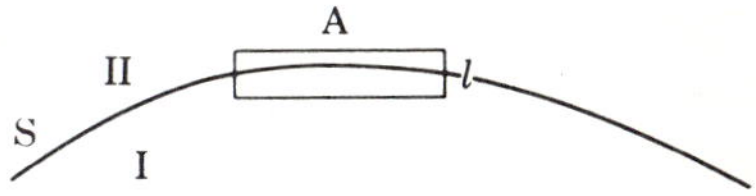

Fig. 16.4

by dielectric constant κ_1 and permeability μ_1 and medium II characterized by corresponding quantities κ_2 and μ_2. Consider the pillbox volume with circular area A at top and bottom and thickness l so constructed that the top is in medium II and the bottom in medium I. Apply the first of the electromagnetic equations (15.36) to this pillbox by integrating over its volume. Thus

$$\int \nabla \cdot \mathbf{D}\, d\tau = 4\pi q, \tag{16.116}$$

where q is the total charge enclosed by the box. We shall now imagine that the box has its volume shrunk to zero by letting $l \to 0$ while A stays constant. We also suppose that the charge density on the interface is zero. Replacing the left side of (16.116) by the associated surface integral by means of the divergence theorem, we have

$$\int \mathbf{D} \cdot \mathbf{n}\, dS = 0, \tag{16.117}$$

where $\mathbf{n}$ is the unit outward drawn normal. As the box is shrunk to zero the only nonvanishing contribution to the surface integral comes from the flat surfaces. The result is

$$\left(D_{n_1} - D_{n_2}\right) A = 0, \tag{16.118}$$

where D_{n_1} is the component of $\mathbf{D}$ normal to the interface in medium I, etc. Therefore since $A \neq 0$,

$$D_{n_1} = D_{n_2}, \tag{16.119}$$

and we reach the conclusion that at an uncharged surface separating two dielectric media, the normal component of the electric displacement is continuous. The use of the equation $\nabla \cdot \mathbf{B} = 0$ shows that the same result is true for the normal component of **B**.

To study the behavior of **E** we proceed differently. We construct a rectangular contour in a plane which cuts the interface normally. The long sides of the contour are in the two media respectively parallel to the interface. The short sides are perpendicular to the interface. Using the fourth equation in (15.36) we form the surface integral of both sides over the area of the rectangle, viz.,

$$\int \nabla \times \mathbf{E} \cdot \mathbf{n}\, dS = -\frac{1}{c} \int \dot{\mathbf{B}} \cdot \mathbf{n}\, dS,$$

or employing Stokes' theorem

$$\int \mathbf{E} \cdot d\mathbf{s} = -\frac{1}{c} \int \dot{\mathbf{B}} \cdot \mathbf{n}\, dS. \tag{16.120}$$

If now the area of the rectangle is shrunk to zero by letting the short sides go to zero, keeping the long sides equal to L, the right-hand side of (16.120) will surely vanish (since $\dot{\mathbf{B}}$ is finite) and we shall have

$$\left(E_{t_1} - E_{t_2}\right) L = 0,$$

whence, since $L \neq 0$,

$$E_{t_1} = E_{t_2}, \tag{16.121}$$

or the tangential component of **E** is continuous at the boundary. The same can be proved of the tangential component of **H** by the use of the third equation in (15.36), provided there are no currents in the interface.

Let us now apply these boundary conditions to the transmission of a plane electromagnetic wave from a dielectric medium in which the wave velocity is V_1 to another in which the wave velocity is V_2. We shall assume that the boundary is plane and for the sake of generality discuss the case of oblique incidence.

Let us take the boundary plane as the xy plane and for simplicity suppose that the normal to the incident wave front (the incident ray; cf. Sec. 14.11) lies in the yz plane. This is called the plane of incidence.

If the normal makes the angle ϕ_i with the z axis, its direction cosines are $\sin\phi_i$, $\cos\phi_i$. It is now necessary to assume that at the boundary there will be in general some reflection as well as some transmission. Now the reflected wave normal and the z axis will determine a plane whose angle with the yz plane we shall call θ'. At the same time the plane determined by the transmitted wave normal and the z axis will be supposed to make the angle θ'' with the yz plane. The situation is schematically depicted in Fig. 16.5, where the reflected ray is shown making the angle ϕ_r with the z axis and the transmitted ray the angle ϕ_t with the z axis. It follows from eq. (14.154) that the components of the incident electric field intensity all have the phase part

$$e^{i[\omega t - k_i(y \sin\phi_i + z\cos\phi_i)]}.$$

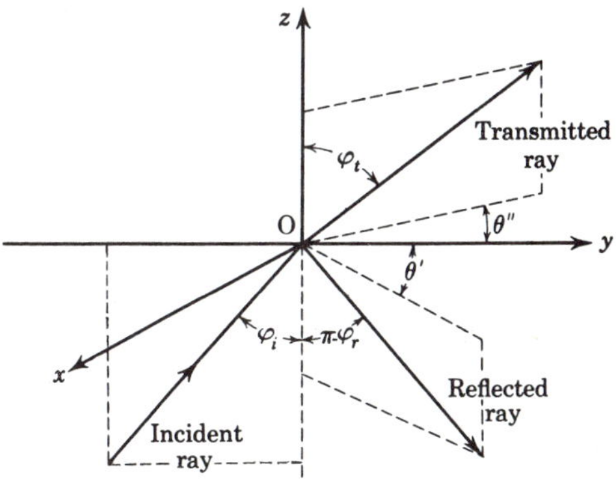

Fig. 16.5

Consequently we may write for the incident electric field intensity

$$\mathbf{E}_i = \mathbf{A}_i e^{i[\omega t - k_i(y\sin\phi_i + z\cos\phi_i)]}, \tag{16.122}$$

where $k_i = \omega/V_1$, and $\mathbf{A}_i$ is the incident electric vector amplitude. Similarly the reflected electric field intensity may be written in the form

$$\mathbf{E}_r = \mathbf{A}_r e^{i[\omega t - k_r(y\sin\phi_r\cos\theta' + x\sin\phi_r\sin\theta' + z\cos\phi_r) + \delta_r]}, \tag{16.123}$$

and the transmitted intensity in the form

$$\mathbf{E}_t = \mathbf{A}_t e^{i[\omega t - k_t(y\sin\phi_t\cos\theta'' - x\sin\phi_t\sin\theta'' + z\cos\phi_t) + \delta_t]}. \tag{16.124}$$

We have introduced the presumptive phase changes on reflection and transmission, δ_r and δ_t, respectively. We have, however, assumed that the reflected and transmitted frequencies are the same as the incident frequency. Actually this follows from the boundary conditions. For the continuity at the boundary (i.e., $z = 0$) of the tangential components of E leads to

$$E_{ix} + E_{rx} = E_{tx}, \qquad E_{iy} + E_{ry} = E_{ty}. \tag{16.125}$$

These must hold for all values of t, x, and y. But this cannot possibly be unless the value of ω is the same in (16.122, 123, and 124). The same argument leads to

$$\delta_r \text{ and } \delta_t = 0 \text{ or } \pi. \tag{16.126}$$

Moreover we cannot satisfy the boundary conditions at $z = 0$ unless we have

$$k_i \sin\phi_i = k_r \sin\phi_r \cos\theta' = k_t \sin\phi_t \cos\theta'', \tag{16.127}$$

$$k_r \sin\phi_r \sin\theta' = -k_t \sin\phi_t \sin\theta'' = 0. \tag{16.128}$$

It follows that θ' and $\theta'' = 0$ or π, (16.129)

$$k_i \sin\phi_i = k_t \sin\phi_t = k_r \sin\phi_r. \tag{16.130}$$

In (16.129) the value π is excluded since it is inconsistent with (16.127) Recalling that $k_r = k_i$, we are led to the result that

$$\phi_i = \pi - \phi_r, \tag{16.131}$$

$$\frac{\sin\phi_i}{\sin\phi_t} = \frac{V_1}{V_2} = \sqrt{\kappa_2/\kappa_1} = n, \tag{16.132}$$

where κ_1 and κ_2 are the dielectric constants of the two media, respectively, and n is the index of refraction of medium II with respect to medium I. We take $\mu_1 = \mu_2 = 1$. The content of the above may be summarized as follows: The reflected, transmitted, and incident wave normals or rays all lie in the same plane. The angle of incidence ϕ_i is equal to the angle of reflection $\pi - \phi_r$. Finally the ratio of the sine of the angle of incidence to that of the angle of transmission (or refraction as it is usually called) is a constant and equal to the ratio of the wave velocity in the incident medium to that in the transmitted region. The equations (16.129, 131, 132) contain the so-called laws of reflection and refraction of a plane electromagnetic wave at a plane boundary. It is important to recognize that they result from the fact that the boundary conditions must hold over the whole boundary and are independent of the precise form of the conditions. This deduction of Snell's law is an important element in the success of the electromagnetic theory of light.

We proceed to examine the more detailed consequences of the boundary conditions. Confining our attention to the yz plane we draw the incident, reflected and transmitted rays as in Fig. 16.6. For convenience we resolve the amplitude of the incident electric vector $\mathbf{E}_i$ into components $\mathbf{E}_{i1}$ perpendicular to the direction of propagation and in the plane of incidence, and $\mathbf{E}_{i2}$ perpendicular to the direction of propagation and perpendicular to the plane of incidence. The same is done with the reflected and transmitted rays. We may then write for the x, y, and z components of the incident vector

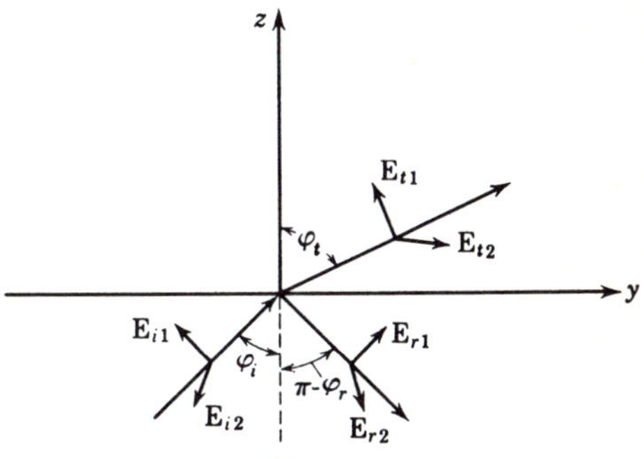

Fig. 16.6

$$E_{iy} = -E_{i1} \cos \phi_i e^{i\Psi_i},$$

$$E_{ix} = E_{i2} e^{i\Psi_i}, \tag{16.133}$$

$$E_{iz} = E_{i1} \sin \phi_i e^{i\Psi_i},$$

where

$$\Psi_i = \omega t - k_i(y \sin \phi_i + z \cos \phi_i). \tag{16.134}$$

To get the corresponding components of the incident magnetic vector $\mathbf{H}_i$ we must use the field equations. Thus

$$-\frac{1}{c}\dot{H}_{iy} = \frac{\partial E_{ix}}{\partial z} - \frac{\partial E_{iz}}{\partial x} = -ik_i \cos \phi_i E_{i2} e^{i\Psi_i}. \tag{16.135}$$

or

$$H_{iy} = \frac{c}{V_1} \cos \phi_i E_{i2} e^{i\Psi_i}. \tag{16.136}$$

Similarly we obtain

$$H_{ix} = \frac{c}{V_1} E_{i1} e^{i\Psi_i}, \tag{16.137}$$

$$H_{iz} = -\frac{c}{V_1} E_{i2} \sin \phi_i e^{i\Psi_i}. \tag{16.138}$$

The corresponding expressions for the reflected wave turn out to be

$$\begin{aligned} E_{ry} &= E_{r1} \cos \phi_i e^{i\Psi_r}, \\ E_{rx} &= E_{r2} e^{i\Psi_r}, \\ E_{rz} &= E_{r1} \sin \phi_i e^{i\Psi_r}, \end{aligned} \tag{16.139}$$

$$\begin{aligned} H_{ry} &= \frac{-c}{V_1} E_{r2} \cos \phi_i e^{i\Psi_r}, \\ H_{rx} &= \frac{c}{V_1} E_{r1} e^{i\Psi_r}, \\ H_{rz} &= \frac{-c}{V_1} E_{r2} \sin \phi_i e^{i\Psi_r}, \end{aligned} \tag{16.140}$$

where

$$\Psi r = \omega t - k_i(y \sin \phi_i - z \cos \phi_i). \tag{16.141}$$

Finally for the transmitted wave we have

$$\begin{aligned} E_{ty} &= -E_{t1} \cos \phi_t e^{i\Psi_t}, \\ E_{tx} &= E_{t2} e^{i\Psi_t}, \\ E_{tz} &= E_{t1} \sin \phi_t e^{i\Psi_t}, \end{aligned} \tag{16.142}$$

$$H_{ty} = \frac{c}{V_2} E_{t2} \cos \phi_t e^{i\Psi_t},$$

$$H_{tx} = \frac{c}{V_2} E_{t1} e^{i\Psi_t}, \tag{16.143}$$

$$H_{tz} = \frac{-c}{V_2} E_{t2} \sin \phi_t e^{i\Psi_t},$$

where

$$\Psi_t = \omega t - k_t(y \sin \phi_t + z \cos \phi_t). \tag{16.144}$$

We are now ready to apply the boundary conditions. The continuity of the tangential components of **E** and **H** yields

$$\begin{aligned} E_{ix} + E_{rx} &= E_{tx}, \\ E_{iy} + E_{ry} &= E_{ty}, \\ H_{ix} + H_{rx} &= H_{tx}, \\ H_{iy} + H_{ry} &= H_{ty}. \end{aligned} \tag{16.145}$$

The continuity of the normal components of **D** and **B** yields

$$\begin{aligned} \kappa_1(E_{iz} + E_{rz}) &= \kappa_2 E_{tz}, \\ H_{iz} + H_{rz} &= H_{tz}. \end{aligned} \tag{16.146}$$

Of the above six relations only four are independent. When we use (16.133) to (16.144) these four reduce to

$$(E_{i1} - E_{r1}) \cos \phi_i = E_{t1} \cos \phi_t, \tag{16.147}$$

$$E_{i2} + E_{r2} = E_{t2}, \tag{16.148}$$

$$\sqrt{\kappa_1}(E_{i1} + E_{r1}) = \sqrt{\kappa_2} E_{t1}, \tag{16.149}$$

$$\sqrt{\kappa_1} \cos \phi_i (E_{i2} - E_{r2}) = \sqrt{\kappa_2} \cos \phi_t E_{t2}, \tag{16.150}$$

where we have also used the law of refraction and have set $\sqrt{\kappa_1} = c/V_1$, etc. We may solve the above four equations for E_{r1} and E_{t1} in terms of E_{i1}, as well as E_{r2} and E_{t2} in terms of E_{i2}. After some reduction involving trigonometric transformations the final results are

$$E_{r1} = E_{i1} \cdot \frac{\tan(\phi_i - \phi_t)}{\tan(\phi_i + \phi_t)}, \tag{16.151}$$

$$E_{t1} = E_{i1} \cdot \frac{2\cos\phi_i \sin\phi_t}{\cos(\phi_i - \phi_t)\sin(\phi_i + \phi_t)}, \tag{16.152}$$

$$E_{r2} = E_{i2} \cdot \frac{\sin(\phi_t - \phi_i)}{\sin(\phi_t + \phi_i)}, \tag{16.153}$$

$$E_{t2} = E_{i2} \cdot \frac{2\cos\phi_i \sin\phi_t}{\sin(\phi_t + \phi_i)}. \tag{16.154}$$

Equations (16.151) – (16.154) inclusive are known as Fresnel's formulas after their original discoverer. He deduced them for light waves from his elastic solid theory and hence gave the amplitude coefficients a different interpretation.

We can immediately draw certain consequences of interest. If $\phi_i + \phi_t = \pi/2$, $E_{r1} = 0$ and the electric vector in the reflected wave is perpendicular to the plane of incidence. The reflected wave is said to be *plane polarized* in the plane of incidence. The condition for this is evidently that the reflected and transmitted rays shall be perpendicular. For this it follows that the angle of incidence must satisfy the relation

$$\tan\phi_i = n, \tag{16.155}$$

where n, as usual, is the index of refraction for the wave going from medium I to medium II. The relation (16.155) is known as Brewster's law for polarization by reflection. The angle arc tan n is called the polarizing angle. Under these conditions

$$E_{r2} = E_{i2}\cos 2\phi_i = E_{i2}\left(\frac{1 - n^2}{1 + n^2}\right), \tag{16.156}$$

$$E_{t2} = \frac{2E_{i2}}{1+n^2}, \tag{16.157}$$

$$E_{t1} = \frac{E_{i1}}{n}. \tag{16.158}$$

Let us next examine *normal* incidence, with $\phi_i = \phi_t = 0$. Then

$$E_{r1} = E_{i1}\left(\frac{n-1}{n+1}\right); \quad E_{t1} = \frac{2E_{i1}}{n+1}; \tag{16.159}$$

$$E_{r2} = -E_{i2}\left(\frac{n-1}{n+1}\right); \quad E_{t2} = \frac{2E_{i2}}{n+1}. \tag{16.160}$$

The average rate of flow of energy per unit area in the incident wave, i.e., the intensity, is given by the average incident Poynting flux or

$$\overline{\frac{c}{4\pi}|(\mathbf{E}_i \times \mathbf{H}_i)|} = \frac{c\sqrt{\kappa_1}}{8\pi}(E_{i1}{}^2 + E_{i2}{}^2) = I_i, \tag{16.161}$$

Similarly the average rate of flow of energy per unit area in the reflected wave is

$$\frac{c\sqrt{\kappa_1}}{8\pi}(E_{r1}{}^2 + E_{r2}{}^2) = I_r, \tag{16.162}$$

and in the transmitted wave is

$$\frac{c\sqrt{\kappa_2}}{8\pi}(E_{t1}{}^2 + E_{t2}{}^2) = I_t. \tag{16.163}$$

The ratio of reflected to incident intensity then becomes

$$I_r/I_i = \left(\frac{n-1}{n+1}\right)^2, \tag{16.164}$$

whereas the ratio of transmitted to incident intensity is

$$I_t/I_i = 4n/(n+1)^2, \tag{16.165}$$

Note that the sum equals unity, as is required by the conservation principle.

The reader is advised to investigate the case of grazing incidence ($\phi_i = \pi/2$) and to show that here the reflection is complete but with a change in phase of π.

16.8. Total Internal Reflection. Elliptic Polarization. The possibility of total internal reflection arises when $\sqrt{\kappa_2} < \sqrt{\kappa_1}$, so that the index of refraction n is less than unity. Then for angles of incidence such that

$$\sin \phi_i > n \tag{16.166}$$

the angle of refraction is imaginary, or we may write

$$\cos \phi_t = \pm i\sqrt{\sin^2 \phi_i/n^2 - 1}. \tag{16.167}$$

An imaginary value for $\cos \phi_t$ presumably means that all the radiation is reflected and none transmitted. This is borne out by an examination of the ratio of E_{r1} to E_{i1}. Thus now we have

$$E_{r1}/E_{i1} = \frac{\cos \phi_i \sin \phi_i \mp i \sin \phi_t \sqrt{\sin^2 \phi_i/n^2 - 1}}{\cos \phi_i \sin \phi_i \pm i \sin \phi_t \sqrt{\sin^2 \phi_i/n^2 - 1}}, \tag{16.168}$$

which means that E_{r1} and E_{i1} have the same magnitude and differ only in phase. Investigation shows that the expression for the incident wave intensity, eq. (16.161) is perfectly general, i.e., holds for *any* angle of incidence and also holds even if E_{i1} and E_{i2} are complex. In the latter case indeed, we have to treat $E_{i1}{}^2$ as $|E_{i1}|^2$, etc. Consequently since from (16.168) we have

$$|E_{r1}|^2 = |E_{i1}|^2, \tag{16.169}$$

and since (16.153) likewise yields under these conditions

$$|E_{r2}|^2 = |E_{i2}|^2, \tag{16.170}$$

it follows that the intensity of the incident wave is equal to the intensity of the reflected wave. This means that when (16.166) holds, the intensity of the transmitted wave is zero. This does not mean, however, that there is *no* transmitted electric or magnetic vector. Before investigating this let us note the significance of the phase difference between the reflected and incident field intensities.

If we write

$$\frac{a - ib}{a + ib} = Re^{-i\theta}, \tag{16.171}$$

it develops that $R = 1$ and $\theta = \arctan \dfrac{2ab}{a^2 - b^2} = 2 \arctan b/a$. Hence we may set

$$E_{r1} = E_{i1} e^{-2i \arctan\left(\sqrt{\sin^2 \phi_i / n^2 - 1} / n \cos \phi_i\right)}. \tag{16.172}$$

Similarly

$$E_{r2} = E_{i2} e^{-2i \arctan\left(n\sqrt{\sin^2 \phi_i / n^2 - 1} / \cos \phi_i\right)}. \tag{16.173}$$

Hence if E_{i1} and E_{i2} are in phase, leading to *plane* polarized incident radiation in a plane through the incident ray making an angle of 45° with the plane of incidence, E_{r1} and E_{r2} will be out of phase by the amount

$$\delta = 2\left[\arctan\left(n\sqrt{\sin^2 \phi_i/n^2 - 1}/\cos \phi_i\right) - \arctan\left(\sqrt{\sin^2 \phi_i/n^2 - 1}/n \cos \phi_i\right)\right]. \tag{16.174}$$

This means that in general the reflected radiation is *elliptically polarized.*

If we now go back to (16.142), we can write for E_{ty} in the case of total reflection

$$E_{ty} = iE_{t1} \sqrt{\sin^2 \phi_i/n^2 - 1} \cdot e^{i\omega\left[t - \frac{\sqrt{\kappa_2}}{c} y \sin \phi_t\right]} \cdot e^{-\omega z \sqrt{\kappa_2}/c \cdot \sqrt{\sin^2 \phi_i/n^2 - 1}}. \tag{16.175}$$

Note that to make physical sense we must use the minus sign before the radical in (16.167) in substituting $\cos \phi_t$. The result is a wave progressing in the y direction with velocity $c/\sqrt{\kappa_2} \sin \phi_t$, but with amplitude damped out more or less rapidly. The damping coefficient being a linear function of z assures that the wave will not penetrate far into the second medium as long as $\sin \phi_i$ is not too close to n. The same considerations apply to E_{ty} and E_{tz} and we see also that the wave in the second medium in nontransverse since there is no phase change with z. By direct calculation of the Poynting flux normal to the surface it can be shown that there is no *average* transfer of energy normally across the surface. Nevertheless it appears that energy is transferred in the nontransverse wave *along* the surface. The existence of radiation in the second medium has actually been exhibited experimentally. In an experiment due to Hall (cf. Fig. 16.7) the curved surface of a lens of large radius of curvature was pressed against the

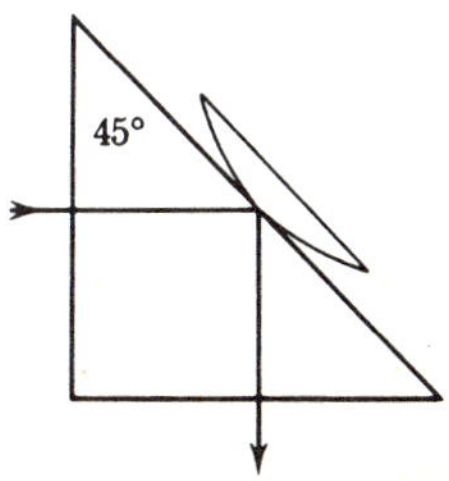

Fig. 16.7

hypotenuse face of a 45° glass prism. When light is incident at 45° on this face so that total reflection ensues in accordance with (16.166), one nevertheless detects a Newton ring pattern through the lens: this necessarily implies some radiation penetrating into the thin air layer between lens and prism.[1]

16.9. Transmission of Electromagnetic Waves Through Stratified Media. Filtration. The transmission of electromagnetic radiation through a succession of media with different properties is a problem analogous to that of the transmission of elastic (i.e., acoustic) radiation (cf. Sec. 14.10). We can best understand this by examining the transmission at normal incidence through a plane parallel slab of dielectric of thickness l in which the wave velocity is V_2 compared with V_1 in the surrounding medium. The situation is depicted in Fig. 16.8 which shows the slab of medium II embedded in medium I. If a plane harmonic electromagnetic wave is incident normally on the face at $x = 0$, we can denote the incident electric and magnetic field intensities as

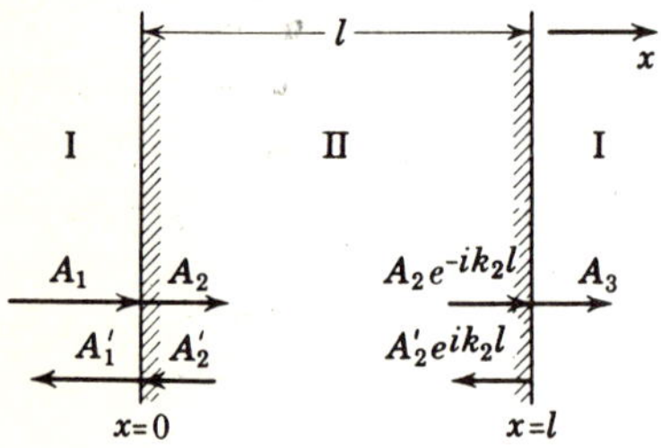

Fig. 16.8

$$E_i = A_1 e^{i(\omega t - k_1 x)},$$
$$H_i = B_1 e^{i(\omega t - k_1 x)}. \tag{16.176}$$

But B_1 and A_1 are not independent. In fact from the field equations (cf. 16.133 – 16.138) we see that

$$B_1 = n_1 A_1, \tag{16.177}$$

where $n_1 = c/V_1$ = index of refraction of the first medium (with respect to vacuum). Similarly for the wave reflected at the boundary surface at $x = 0$ we have

$$E_r = A_1' e^{i(\omega t + k_1 x)},$$
$$H_r = -n_1 A_1' e^{i(\omega t + k_1 x)}. \tag{16.178}$$

[1] The whole problem is somewhat complicated and deserves more careful treatment than we can give it here. Cf. J. Picht, *Ann. der Physik*, **395**, 432, 1929. Also Noether, *Ann. der Physik*, **11**, 141, 1931.

All field intensity vectors lie of course in planes parallel to the boundary surface. The figure indicates the amplitudes assumed inside the slab at both $x = 0$ and $x = l$. We denote by A_3 the amplitude of the electric vector in the transmitted radiation to the right of $x = l$. We are than able to write the boundary conditions in the following form:
At $x = 0$

$$\begin{aligned} A_1 + A_1' &= A_2 + A_2', \\ n_1(A_1 - A_1') &= n_2(A_2 - A_2'). \end{aligned} \tag{16.179}$$

At $x = l$,

$$\begin{aligned} A_2 e^{-ik_2 l} + A_2' e^{ik_2 l} &= A_3, \\ n_2\left(A_2 e^{-ik_2 l} - A_2' e^{ik_2 l}\right) &= n_1 A_3. \end{aligned} \tag{16.180}$$

The mathematical problem is the elimination of A_1', A_2, and A_2' to obtain A_3 in terms of A_1 and hence the transmitted electric field intensity in terms of the incident. We omit the algebra and go at once to the result which is

$$A_3/A_1 = \frac{1}{\cos k_2 l + i \sin k_2 l \cdot (n_{12}^2 + 1)/2n_{12}}, \tag{16.181}$$

where $k_2 = \omega/V_2$ and $n_{12} = n_2/n_1$ = index of refraction of medium II relative to medium I. The power transmission coefficient or ratio of transmitted energy per unit area per second to the incident is

$$P = |A_3|^2 / |A_1|^2 = \frac{1}{1 + \sin^2 k_2 l \left[\left(\frac{n_{12}^2 + 1}{2n_{12}}\right)^2 - 1\right]}. \tag{16.182}$$

The reader should work out the numerical details for a special case. We see at any rate that the transmission is selective with respect to frequency. In particular for frequencies such that $\sin k_2 l = 0$, we get $P = 1$ or complete transmission. On the other hand, when $\sin^2 k_2 l = 1$, we get minimum transmission.

The suggestion is strong that if we were to imagine a structure composed of alternating layers of two different dielectric media, it should show actual filtration properties like the elastic wave filter discussed in Sec. 14.10. In fact the mathematical treatment is precisely the same in the two cases. All we need to do is to rewrite the equations in Sec. 14.10 replacing p by E, X by H, and $\rho_0 c/S$ by $1/\sqrt{\kappa}$. This effectively makes the electromagnetic analogue of the acoustic impedance E/H in this case. The transmission of the harmonic radiation through the equivalent infinite structure is then governed, in analogy with eqs. (14.235) and (14.238) by

$$\cos W = \cos 2k_1 l_1 \cos 2k_2 l_2$$
$$- 1/2 \cdot (\sqrt{\kappa_1/\kappa_2} + \sqrt{\kappa_2/\kappa_1}) \sin 2k_1 l_1 \cdot \sin 2k_2 l_2. \qquad (16.183)$$

Note that we can also write $\sqrt{\kappa_1/\kappa_2} + \sqrt{\kappa_2/\kappa_1}$ as $n_1/n_2 + n_2/n_1$, where n_1 and n_2 are the indices of refraction of the two layer media for electromagnetic radiation respectively relative to vacuum. The stratified structure obviously acts like a filter with alternate transmission bands (for which $|\cos W| \leqq 1$) and attenuation bands (for which $|\cos W| > 1$). If the electromagnetic waves are visible light waves with frequency of the order of 10^{15} cycles, the k values are very large compared with unity. Hence if the layer thicknesses $2l_1$, $2l_2$ are of macroscopic dimensions (i.e., of order 1 cm) the transmission and attenuation bands will be very narrow. To obtain a filter with wide bands in which the filtration properties might be of practical use in the case of visible light it is clearly necessary to use layer thicknesses very small compared with 1 cm. Such filters have been constructed.[1]

It may be of interest to point out that in the case of visible light the problem just discussed corresponds to the interference of harmonic light waves superposed with appropriate phase relations. In most physical optics books the interference problem is handled in geometrical fashion rather than by the analytical method suggested in this section. Reference may be made in this connection to the Fabry-Perot interferometer, for example.

16.10. Metallic Reflection. The special case of the reflection of electromagnetic radiation at a *metallic* surface is of considerable interest. To study this we must revert to the analysis of Sec. 16.3 and utilize a complex index of refraction (eqs. 16.40 and 16.42). Thus if we denote the latter by n' we now have

$$n' = n - ik, \qquad (16.184)$$

[1]Cf. R. B. Lindsay, *Phys. Rev.*, 60, 169, 1941; and C. H. Cartwright and A. F. Turner, *Phys. Rev.*, 55, 595, 1128, 1939.

where n and k are given by eq. (16.41). The dielectric constant κ is unknown for metals, but the conductivity σ is so large that we may safely assume that at not too high frequencies

$$4\sigma^2/\nu^2 \gg \kappa^2, \tag{16.185}$$

so that

$$n \doteq k \doteq \sqrt{\sigma/\nu}. \tag{16.186}$$

This approximation enables us to study the intensity of the reflected radiation from a metal. Confining our attention to normal incidence, in (16.159) we replace n by n' and after some reduction obtain for the ratio of reflected to incident intensity

$$I_r/I_i = \frac{1 - 1/n + 1/2n^2}{1 + 1/n + 1/2n^2} = \frac{1 - \sqrt{\nu/\sigma} + \nu/2\sigma}{1 + \sqrt{\nu/\sigma} + \nu/2\sigma}. \tag{16.187}$$

In the special case of infrared radiation for which $\nu/\sigma \ll 1$, (16.187) yields the approximate result

$$I_r/I_i = 1 - 2\sqrt{\nu/\sigma}. \tag{16.188}$$

This has been experimentally verified for radiation for which $\nu \leqq 10^{13}$ cycles/sec. It indicates that the reflecting power decreases with increasing frequency, i.e., the reflection is selective with respect to frequency.

It should be emphasized that (16.187) breaks down completely for visible light. This is not so surprising in view of the remarks in Sec. (16.3) where it was pointed out that a medium which transmits electromagnetic radiation in accordance with the field equations so far developed at relatively low frequencies may not obey these equations at very high frequencies and small wave lengths. In other words, the granularity or discreteness of the medium may have to be taken into account at the higher frequencies. We meet this situation particularly in connection with the dispersion of light, treated in Chapter 17.

16.11. Diffraction of Electromagnetic Radiation. It is well known that waves of all kinds in traveling through space tend to bend around obstacles or are *diffracted.* We now give a brief summary of the theory of diffraction. The discussion is, of course, not restricted to electromagnetic radiation, but in view of its importance in the case of light it seems appropriate to take it up at this point.

We shall first go back to the general wave equation

$$\nabla^2\psi = \ddot{\psi}/V^2, \tag{16.189}$$

where ψ is here treated as a scalar function of x, y, z, t. It might be a component of the electric or magnetic vector. The task we wish to set ourselves is the evaluation of the function ψ at any point P in space in terms of the values of ψ and its space and time derivatives over any given closed surface. It will be convenient to assume that ψ is a harmonic function of the time, i.e., can be expressed in the form

$$\psi(x, y, z, t) = Q(x, y, z)e^{i\omega t}, \tag{16.190}$$

where the frequency as usual is $\nu = \omega/2\pi$. Our discussion of the Fourier integral in Sec. 14.4 shows that the above assumption is not so restrictive as might at first appear. Eq. (16.190) then becomes

$$\nabla^2 Q + k^2 Q = 0, \tag{16.191}$$

where $k = \omega/V$. We have already met this equation in Sec. 15.6 (eq. 15.90) in connection with the general solution of the electromagnetic field equations. There we indicated a method of handling an equation of this form by the use of Green's theorem (eq. 12.56). We proceed to rewrite the latter. If U_1 and U_2 are functions of x, y, z which have no singularities inside a closed region S, then

$$\int \left(U_1\nabla^2 U_2 - U_2\nabla^2 U_1\right) d\tau = \int \left(U_1\frac{\partial U_2}{\partial n} - U_2\frac{\partial U_1}{\partial n}\right) dS,$$

where the integral on the left is a volume integral taken over the space enclosed by the surface and the integral on the right is taken over the surface. $\partial U/\partial n$ is the derivative of U normal to the surface (the normal, as usual, is outward drawn). Let us put

$$U_1 = e^{-ikr}/r, \tag{16.192}$$

where r is the distance to $d\tau$ from the point P, and suppose $U_2 = Q$. As in Sec. 15.6 we take as the volume over which the integration is to be conducted that between S and a small sphere S' about P. The situation is shown in Fig. 16.9. Now if we form $\nabla^2 U_1$ we find at once that

$$\nabla^2 U_1 + k^2 U_1 = 0, \tag{16.193}$$

Hence the volume integral in question becomes $\int U_1(\nabla^2 U_2 + k^2 U_2)\, d\tau$ and this vanishes by virtue of (16.191). Hence the surface integral must likewise vanish. It is made up of the sum of two parts, one over S and the other over S'. If we consider the latter first, it develops that as the radius of S' goes to zero, the only contribution left to the surface integral over S' is $-4\pi Q_p$, where Q_p is the value of Q at the point P. Consequently the vanishing of the whole surface integral can be represented in the form

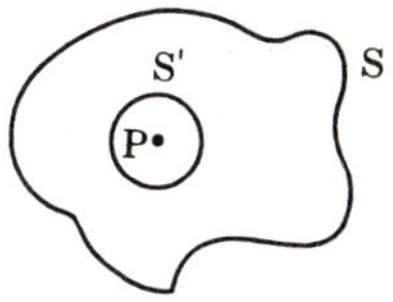

Fig. 16.9

$$4\pi Q_p = \int \left[\frac{e^{-ikr}}{r} \cdot \frac{\partial Q}{\partial n} - Q \frac{\partial}{\partial n}\left(\frac{e^{-ikr}}{r}\right) \right] dS. \tag{16.194}$$

This can be rewritten as

$$4\pi Q_p = \int \left[\frac{e^{-ikr}}{r} \frac{\partial Q}{\partial n} - Qe^{-ikr} \frac{\partial}{\partial n}\left(\frac{1}{r}\right) + ikQ \cdot \frac{e^{-ikr}}{r} \cdot \frac{\partial r}{\partial n} \right] dS, \tag{16.195}$$

by expansion of the integrand. Finally from (16.190) we have for ψ_p the result

$$\psi_p = \frac{1}{4\pi} \int \left[\frac{e^{i\omega(t-r/V)}}{r} \cdot \frac{\partial Q}{\partial n} - Qe^{i(\omega t - r/V)} \frac{\partial}{\partial n}\left(\frac{1}{r}\right) + ikQ \frac{e^{i(\omega t - r/V)}}{r} \cdot \frac{\partial r}{\partial n} \right] dS. \tag{16.196}$$

Now $Qe^{i\omega(t-r/V)}$ is the same as ψ taken at time $t - r/V$ instead of t. We may in fact to indicate this write

$$Qe^{i\omega(t-r/V)} = [\psi]. \tag{16.197}$$

The use of the bracket to represent a *retarded* quantity has already been employed in Sec. 15.6. Hence we may rewrite (16.196) in the form

$$\psi_p = \frac{1}{4\pi}\int\left\{\frac{1}{r}\left[\frac{\partial\psi}{\partial n}\right] - \frac{\partial}{\partial n}\left(\frac{1}{r}\right)[\psi] + \frac{1}{rV}\frac{\partial r}{\partial n}\left[\frac{\partial\psi}{\partial t}\right]\right\} dS. \quad (16.198)$$

Physically (16.198) emphasizes that ψ_p is made up of the sum of contributions from the various surface elements dS of the surface S. The reader will recognize this as analogous to Huygens' principle in wave propagation in elementary physics. In fact it appears as the generalized form of this principle. Moreover, (16.198) stresses that to get the value of the disturbance at time t at point P we must consider the corresponding values of ψ, $\partial\psi/\partial n$, and $\partial\psi/\partial t$ on the surface S at time $t - r/V$, where r/V is the time necessary for the disturbance to travel from dS to P, a distance r.

For the sake of illustration suppose O in Fig. 16.10 is a point source of spherical harmonic waves and we wish to calculate the resultant disturbance at P enclosed by the closed surface S. Draw OR from O to R, a point on S at which the surface element dS is constructed. Call $PR = r$ and $RO = r'$. Let the angle between the outward drawn normal at dS and PR be θ and that between the same normal and RO be θ'. Now the disturbance at R from the point source at O is given by

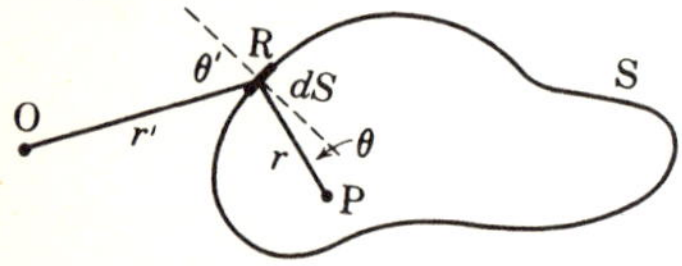

Fig. 16.10

$$\psi = \frac{A}{r'}\cos(\omega t - kr'), \quad (16.199)$$

and

$$\frac{\partial\psi}{\partial n} = -\frac{\partial\psi}{\partial r'}\cos\theta' = A\cos\theta'\left[\frac{\cos(\omega t - kr')}{r'^2} - \frac{kr'}{r'^2}\sin(\omega t - kr')\right]. \quad (16.200)$$

If the wave length $\lambda(= 2\pi/k)$ is small compared with r', which will almost certainly be true for visible light, the second term in the bracket will far outweigh the first and we shall have, approximately

$$\frac{\partial\psi}{\partial n} \doteq -\frac{Ak\cos\theta'}{r'}\cdot\sin(\omega t - kr'). \quad (16.201)$$

Also

$$\frac{\partial\left(\frac{1}{r}\right)}{\partial n} = -\frac{\cos\theta}{r^2}, \quad (16.202)$$

and

$$\frac{\partial\psi}{\partial t} = -\frac{A\omega}{r'} \sin(\omega t - kr'). \qquad (16.203)$$

To use (16.198) we need the retarded values. These are easily obtained and we finally write ψ_P in the form

$$\psi_P = \frac{1}{4\pi}\iint\left\{\frac{-Ak\cos\theta'}{rr'}\sin\left[\omega t - k(r+r')\right]\right.$$

$$+\frac{A\cos\theta}{r'r^2}\cos\left[\omega t - k(r+r')\right]$$

$$\left.-\frac{Ak\cos\theta}{r'r}\sin\left[\omega t - k(r+r')\right]\right\}dS. \qquad (16.204)$$

Once again we use the approximation that $\lambda \ll r'$, so that the second term of the integrand is negligible compared with the other two. Then to this approximation

$$\psi_P = -\frac{A}{2\lambda}\int\frac{1}{r'r}(\cos\theta + \cos\theta')\sin\left[\omega t - k(r+r')\right]dS. \qquad (16.205)$$

This means that we can interpret the resultant disturbance at P as the superposition of spherical wavelets from each element dS, each with amplitude $\frac{A}{\lambda r'r}\left(\frac{\cos\theta + \cos\theta'}{2}\right)$. There is to be sure a phase difference between the incident wave at R and the effective wavelet from R to P represented in the difference between $\cos(\omega t - kr')$ and $\sin[\omega t - k(r+r')]$. This is the form of Huygens' principle applicable to the present problem. Note, however, that its mathematical form removes a difficulty present in the simplified version of Huygens' principle discussed in elementary physics. If the angles θ and θ' are small, as is the case in most applications in optics, $(\cos\theta + \cos\theta')/2$ is very close to unity and the contributions in the various directions are not seriously dependent on direction. But in the direction directly *behind* dS, i.e., in the direction from which the original wave from O is advancing, $\cos\theta' = -\cos\theta$ and the above sum is very close to zero. This explains why in applying Huygens' principle we do not have the right to construct the wavelets whose envelope is to form the new wave front by drawing them *behind* the advancing wave front.

All problems in wave diffraction are inherent in eq. (16.205) or equations like it.

PROBLEMS

1. Derive the standard electrodynamical expression for the rate of radiation from an accelerated electron, viz., $e^2f^2/6\pi c^3$ where f = acceleration of the electron. Find the average rate of radiation from a line linear harmonic oscillator with frequency ν and maximum electric moment eA.

2. In the case of a cavity with perfectly reflecting walls containing electromagnetic radiation propagated equally in all directions, show that the radiation pressure on the walls is one-third the average energy density of the radiation.

3. Use Poynting's theorem to calculate the rate of energy radiation at great distances from the oscillating current element treated in Sec. 16.1

4. Apply Fresnel's formulas (eqs. 16.151 – 16.154) to the case of grazing incidence and calculate the change of phase on reflection.

5. A plane wave of yellow light ($\lambda = 5890A^\circ$) is incident on a plane crown glass surface at an angle of 45°. Find the ratio between the reflected and incident intensities.

6. Light passes at normal incidence through a plane parallel glass plate of thickness l. Find the expression for the transmitted intensity in terms of the incident intensity.

7. A plane harmonic electromagnetic wave in the positive x direction may be expressed in terms of the electric vector as follows

$$E_y = E_{y,\max} e^{i(\omega t - kx)},$$

where $\omega = 2\pi\nu$ and $k = 2\pi/\lambda = \omega/c$. Find the associated magnetic vector. Derive the expression for the intensity of the wave. If the amplitude $E_{y,\max}$ is 0.5 volt/cm, how much energy is transported by the wave front across 1 cm^2 per second if $\nu = 10^{15}$ cycles/sec?

8. Light falls at normal incidence on a metal surface. If n and k are, respectively, the index of refraction and absorption coefficient, show that the ratio of reflected to incident intensity is

$$R = \frac{n^2 + k^2 - 2n + 1}{n^2 + k^2 + 2n + 1}.$$

Show that for a metal approximately $n = k = \sqrt{\sigma/\nu}$, where σ is the conductivity and ν the frequency. Calculate the value of R for copper for light of $\lambda = 12 \times 10^{-4}$ cm. Compare with the experimental value.

9. Prove that an oscillating electric dipole whose moment p is a sinusoidal function of the time is equivalent to a sinusoidal current $I\,dl$ whose amplitude is equal to the angular frequency multiplied by the maximum dipole moment and which is $90°$ out of phase with the oscillating dipole. Relate this result to the contents of Sec. 16.1.

10. A point source of light radiates with a power output of 500 watts. Find the root-mean-square average values of **E** and **H** in the electromagnetic field at 10 meters from the source in a dielectric medium with $\kappa = 2$. What is the intensity of the wave at this point? What is the average radiation density? What is the volume density of the electromagnetic field momentum (cf. Prob. 25, Chapter 15)? Calculate the divergence of the Maxwell electromagnetic stress tensor.

11. An electromagnetic wave of frequency 1 megacycle (10^6 cycles/sec) is propagated through a block of copper. How far will the wave go before its amplitude is reduced in the ratio $1/e$? What is the phase velocity of the wave?

12. What is the penetration depth in the skin effect for aluminum at 100 megacycles?

13. In the transmission line discussed in Sec. 16.6, derive the expression for the propagation velocity if the conductance G is neglected but the resistance R is taken into account.

14. A transmission line consists of a pair of parallel cylindrical copper wires each of radius a and with their axes spaced a distance l apart. The distributed constants are

$$R = \frac{83.2 \times 10^{-9}\sqrt{\nu}}{a} \text{ ohms/cm}$$

$$L = 4 \log l/a \times 10^{-9} \text{ henries/cm}$$

$$C = \frac{(10/9) \times 10^{-12}}{4 \log l/a} \text{ farads/cm}$$

Calculate the propagation velocity in cm/sec for frequency ν. What interesting fact about this emerges if R is neglected? Taking account of R, calculate the transmission loss per cm as a function of ν and the dimensions.

15. A transmission line consists of a copper coaxial cable with radii R_1 and R_2 respectively. The distributed constants are

$$R = 41.6 \times 10^{-9}\sqrt{\nu}\,(1/R_1 + 1/R_2) \text{ ohms/cm}$$

$$L = 2 \log R_2/R_1 \times 10^{-9} \text{ henries/cm}$$

$$C = \frac{(10/9) \times 10^{-12}}{2 \log R_2/R_1} \text{ farads/cm.}$$

Calculate the propagation velocity in cm/sec for frequency ν. Taking account of R, calculate the transmission loss per cm as a function of ν and the dimensions.

16. A hollow conducting circular cylindrical pipe is 1 meter long and has a radius of 1 cm. Calculate the fundamental frequencies for the **E** and **H** modes. Also calculate the next highest frequency for each type. What is the cut-off wave length?

17. Obtain and plot the propagation function cos W for the case of an effectively infinite optical filter consisting of alternate layers of air of thickness 10^{-4} cm and glass of thickness 10^{-3} cm and average refractive index 1.5. Obtain the transmission and attenuation band widths.

CHAPTER XVII

ELECTRON THEORY AND RELATIVITY

17.1. Historical Survey of the Discovery of the Electron. In the previous two chapters we have discussed electrical and in particular electromagnetic phenomena as illustrations of continuum or field physics. Historically considered the success of this viewpoint has been great, especially in connection with the propagation of electrical effects. Nevertheless in the treatment of electromagnetic propagation in conducting media (i.e., absorption and dispersion) as well as that of the reflection of electromagnetic waves by metals we noticed a lack of agreement with experiment save at very long wave lengths: it has seemed natural to attribute these discrepancies to an absence of strict continuity and homogeneity of the medium, i.e., to the presence of discrete structure in the medium, whose interaction with the radiation was being neglected and which cannot be safely ignored when the wave length of the radiation is of the same order of magnitude as or smaller than the granularity. Now it is hard to see how the discreteness of the medium, i.e., an atomic structure, can have any noteworthy effect on electromagnetic radiation unless the particles of the structure are themselves electrically charged. This naturally suggests an investigation of the nature of the charge associated with the particles, i.e., its sign, magnitude, etc.

It will be recalled from elementary physics that Faraday's laws of electrolysis strongly suggest not only an atomic hypothesis for the constitution of matter but also a discrete character for the charges associated with the atoms or ions. One of the simplest interpretations of Faraday's experiments is that all atoms moving through an electrolytic cell carry positive or negative charges which are multiples of a fundamental smallest charge. This charge was named the "electron" in 1891 by C. J. Stoney, who also gave an estimate of its magnitude as 0.3×10^{-10} e.s.u., based on the value of the faraday and Avogadro's number.

For the isolation of the electronic charge, i.e., the discovery and study of a particle which acts as if it were the specific carrier of the charge, we have to look to the phenomena connected with the conduction of electricity through gases. It will be recalled, that when a sufficiently high voltage is placed across terminals separated by a gas at a pressure of the order of 10^{-3} mm of mercury, the discharge takes the form of invisible radiation which makes itself evident by the fluorescence it produces in the walls of the surrounding glass vessel. Since this radiation appears always to emanate from the cathode of the discharge tube, the rays were called *cathode rays* by Crookes, who studied them extensively around 1880. Their properties appeared to be consistent with the assumption that they consist of a stream of negatively charged particles whose average velocity depends on the applied voltage. From the deflections produced in such streams by magnetostatic and electrostatic fields, J. J. Thomson in 1897 was first able to assign a value to the ratio of the charge to the mass. It turned out to be about 1.77×10^7 e.m.u./gram or 5.3×10^{17} e.s.u./gram. If one assumes that the charge on each cathode ray particle is the same as the fundamental electrolytic charge mentioned in the previous paragraph it follows that the mass of the particle is a very small fraction, i.e., of the order of 10^{-3} to 10^{-4} of the mass of a hydrogen atom. On the other hand, if the mass of the particle were assumed to be the same as the mass of the hydrogen atom, its charge would have to be some 10^3 to 10^4 times the charge of the hydrogen atom in electrolysis. The importance of an independent measurement of the charge on the cathode ray particle was thus obvious. A method for this measurement was suggested by J. J. Thomson and the details carried out during the first decade of the twentieth century by a number of investigators, notably by R. A. Millikan. These led to the conclusion that all cathode ray particles, since called electrons, have the same negative electrical charge, viz., 4.8×10^{-10} e.s.u.

During the last few years of the nineteenth century the discovery of radioactivity by Becquerel and the theory proposed by Lorentz to explain the Zeeman effect of the splitting of spectral lines by a magnetic field helped materially to strengthen confidence in the existence of the electron as a fundamental charged particle present in all matter.

17.2. The Lorentz Electron. Dynamical Equation and Equation of Motion. Mass and Momentum. The question now arises as to what form the equations of electromagnetic theory will take if electric charges are concentrated on particles like electrons and ions. Since the matter exhibiting electromagnetic phenomena is now no longer a continuum we cannot expect to use with success a pure field theory. Nevertheless the fact that the classical field equations (eqs. 15.36–38 in Sec. 15.1) describe successfully such a large domain of macroscopic behavior suggests that their application to the electron theory of matter be attempted with the least possible modification that will work. We expect, for example, that at points not occupied by atomic particles the electron theory of electromagnetism will reduce to the classical theory for free space. It is therefore plausible to assume that the equations (15.36–38) are of universal validity. In them ρ still means charge per unit volume, only now we must think of such charges as located on particles. But the distribution on each particle must be thought of as continuous. A great variety of assumptions are possible here. We shall discuss only that of Lorentz in accordance with which the electron is assumed to be a spherical shell when observed in a reference system in which it is at rest. The shell has a continuous distribution of charge. The force on the element of charge $\rho\, d\tau$ placed in an electromagnetic field characterized by **E** and **H** is *defined* to be $\rho \mathbf{E}\, d\tau$, measured in the reference system in which at the moment considered the charge element is at rest. Measured in the reference system with respect to which the charge element has the velocity **v**, the corresponding force is *defined* to be $\rho\, d\tau(\mathbf{E} + 1/c \cdot (\mathbf{v} \times \mathbf{H}))$. The connection between this and eq. (15.37) is clear.

What we now wish to do is to set up an equation of motion of the electron which will serve the same purpose as the equation of motion in the classical mechanics of particles. But we are stopped at once because we do not know what mass to attribute to an electron. We have no assurance that mass as defined for the macroscopic particles of classical mechanics will apply to an elementary charged particle like an electron. From the observed experimental variation of e/m for an electron with velocity, there is good reason indeed for supposing that the definition of mass given in Chapter 3 will no longer apply. The dynamics of the electron needs to be approached from a different point of view. We might indeed expect that the situation will be different from that prevailing with ordinary macroscopic uncharged particles, i.e., the particles of classical mechanics. An ordinary mass particle produces a gravitational field (cf. Sec. 12.3) which extends throughout all space. At the same time the particle moves under the influence of the forces produced by other fields. The motion is given, of course, by the celebrated equation $\mathbf{F} = m\mathbf{a}$, one methodological interpretation of which has already been given in Chapter 3. But we can give another interpretation. We can say that the mass particle's own force field produces a reaction force on it equal to the negative of the product of the mass of the particle and its acceleration in the primary inertial system. We may indeed refer to this force as the *kinetic reaction*. It arises because of the fact that the mass particle is accelerated and would not exist if the particle were moving with constant velocity or were at rest. From this point of view the motion of the particle is described by equating the sum of the external force on the particle and its kinetic reaction to *zero*. We might go so far as to call this sum the *total* force on the particle, where **F** is the external force and $-m\mathbf{a}$, the kinetic reaction, is the internal force or force on the particle due to its own field.

Now the classical theory of gravitation merely postulates the nature of the gravitational force field and does not specify how this field is propagated through space. Hence we are unable to calculate the force on a mass particle due to its own field when it is accelerating in the inertial system. If we could we should presumably be able to calculate the mass of the particle in terms of its size and shape, the velocity of propagation of the gravitational field, and the constant of gravitation. As it is, in order to assign a mass to a particle we are compelled in classical mechanics to study its interaction with other particles. Of course it was in this way that we introduced the concept of mass in mechanics.

The situation in electromagnetism is different, since we assume that electrically charged particles produce fields whose changes are governed by the Maxwell field equations. These lead to propagation of the forces due to the field with definite finite velocity. Consequently for an assumed configuration of charge constituting a charged particle like an electron we can calculate the equivalent internal force or the reaction of the electron's field on itself. Then if we follow the classical physics pattern and equate this to the negative of the *external* force on the electron we shall have an equation governing the motion of the electron. In particular if the internal force turns out to be proportional to the acceleration, the coefficient can be interpreted as the mass of the electron.

The actual calculation of the force on a moving electron due to its own field is a rather elaborate business and will not be undertaken here.[1] If the electron is the Lorentz model described in the first part of this section, the result for the internal force $\mathbf{F}_i$ is an infinite series of terms of which three are given here.

$$\mathbf{F}_i = -\frac{e^2}{6\pi a c^2}\mathbf{f} + \frac{e^2}{6\pi c^3}\dot{\mathbf{f}} - \frac{e^2 a}{9\pi c^4}\ddot{\mathbf{f}} + \ldots \quad (17.1)$$

[1] Cf., e.g., L. Page and N. I. Adams, *Electrodynamics* (New York, Dover Publications), pp. 176 ff. For a more general treatment reference may be made to W. Heitler, *Quantum Theory of Radiation* (Oxford Univ. Press, 2d Ed., 1944), Chap. I.

In this expression **f** is the acceleration of the electron (neglecting its variation from point to point over the surface of the particle) and a is the radius. The dynamical equation of motion of the electron then takes the form

$$e\mathbf{E}_0 = \frac{e^2}{6\pi ac^2}\mathbf{f} - \frac{e^2}{6\pi c^3}\dot{\mathbf{f}} + \frac{e^2 a}{9\pi c^4}\ddot{\mathbf{f}} + \dots, \tag{17.2}$$

where $\mathbf{E}_0$ is the external field intensity at the surface of the electron and assumed the same over the whole surface. It must be emphasized that (17.2) is valid only in the reference system in which the electron is instantaneously at rest.

Before we discuss the consequences of the above equation we should note once more that it comes from a rather specialized model of the electron and that if we were to assume some other model we should expect a different set of coefficients on the right-hand side. Moreover we also note that as an equation of motion (17.2) leaves something to be desired because of the presence of the derivatives of the acceleration **f**. This indeed would make the introduction of a mass of the electron illusory were it not for the fact that for many motions of interest, e.g., harmonic motions of not too high frequency, the higher order terms are negligible. This is due to the large value of c and the presumably small value of the radius a. To this approximation we can consider

$$m_0 = e^2/6\pi ac^2 \tag{17.3}$$

as the *mass* of the electron in the system of axes with respect to which it is instantaneously at rest. Before it can be evaluated we must know the radius a. All we can very well say about this on *a priori* grounds is that it must be much smaller than atomic dimensions. From a rather wide variety of sources we know that the radii of molecules and atoms is of the order of magnitude of 10^{-8} cm. Hence we must conclude that the radius of the electron is less than 10^{-8} cm. Therefore (17.3) indicates that

$$m_0 > 10^{-33} \text{ gram.}$$

Now as a matter of fact the experimentally determined ratio of charge to mass as already noted in Sec. 17.1 is 5.3×10^{17} e.s.u./gram. If we assume that the charge is 4.8×10^{-10} e.s.u. it follows that m_0 should be 9×10^{-28} grams. On this basis eq. (17.3) fixes the radius a of the Lorentz electron. The result is

$$a = 1.88 \times 10^{-13}\,\text{cm},$$

which certainly fulfills the elementary requirement of a particle able to fit into an atom.

But now it is necessary to face the question of what happens to the mass when the motion of the electron is referred to a system of axes with respect to which it is in motion with instantaneous velocity $\mathbf{v}$. This involves a transformation to a moving reference system. We discussed this in Sec. 3.2 in connection with the fundamentals of classical mechanics. It turns out, however, that the transformation of Newtonian relativity introduced there will not do for electromagnetic phenomena. Rather we need the transformation equations of the special theory of relativity of Einstein. These will be developed in Sec. 17.4. For the moment we shall merely note that their use leads to the following more general equation to replace (17.2) (neglecting $\mathbf{f}$ and all higher derivatives),

$$e\left[\mathbf{E}_0 + \frac{1}{c}(\mathbf{v} \times \mathbf{H}_0)\right] = \frac{d}{dt}\left(\frac{m_0\mathbf{v}}{\sqrt{1-v^2/c^2}}\right), \tag{17.4}$$

where v is, as usual, the magnitude of vector velocity $\mathbf{v}$. We may then look upon

$$m_0\mathbf{v}\Big/\sqrt{1-v^2/c^2}$$

as the momentum of the electron. The general expression for the mass on this basis then becomes

$$m = m_0\Big/\sqrt{1-v^2/c^2},$$

which indicates that $m = m_0$ only when $v = 0$, i.e., in the system in which the electron is momentarily at rest. For $v \neq 0$, $m > m_0$, or the mass increases with velocity. This result has been amply verified by e/m measurements by Bucherer and others. It has some interesting consequences. In the first place it implies that $v \leqq c$, or the velocity of light in free space must be the upper limit to the velocity of electrons. Moreover it suggests that it is no longer possible to express the force on the electron as the product of mass times acceleration in unique fashion. Indeed if we carry out the indicated time differentiation in (17.4) we get

$$\frac{d}{dt}\left(\mathbf{v}\Big/\sqrt{1-v^2/c^2}\right) = \dot{\mathbf{v}}\Big/\sqrt{1-v^2/c^2} + \mathbf{v}(\mathbf{v}\cdot\dot{\mathbf{v}})\Big/c^2(1-v^2/c^2)^{3/2}. \tag{17.5}$$

Let us write

$$\mathbf{v} = v\mathbf{v}_1,$$

where $\mathbf{v}_1$ is the unit vector in the direction of the velocity. Then

$$\dot{\mathbf{v}} = \dot{v}\mathbf{v}_1 + v\dot{\mathbf{v}}_1 . \tag{17.6}$$

If we substitute this into (17.5) the following force equation results after some reduction

$$e\left[\mathbf{E}_0 + \frac{1}{c}(\mathbf{v} \times \mathbf{H}_0)\right] = \frac{m_0 v \dot{\mathbf{v}}_1}{\sqrt{1-v^2/c^2}} + \frac{m_0 \dot{v} \mathbf{v}_1}{(1-v^2/c^2)^{3/2}} . \tag{17.7}$$

Both terms on the right are acceleration terms but $\dot{v}\mathbf{v}_1$ is the acceleration component in the direction of the velocity, whereas $v\dot{\mathbf{v}}_1$ is the acceleration component at right angles to the velocity. On this picture the electron therefore has two masses, one associated with the acceleration in the direction of motion and other associated with the acceleration normal thereto. The first has been called the longitudinal mass m_l and the second the transverse mass m_t. We therefore write

$$m_l = m_0/(1-v^2/c^2)^{3/2}, \quad m_t = m_0/\sqrt{1-v^2/c^2} . \tag{17.8}$$

Of these, experiments appear to point to m_t as the fundamental mass of importance. Its meaning in the theory of relativity will be discussed in a subsequent section.

17.3. The Field Equations of Electron Theory. Invariance and the Lorentz-Einstein Transformation Equations. In the previous section we decided to retain the fundamental electromagnetic field equations even for media containing charged particles, e.g., electrons. We saw that this makes necessary some modification in the fundamental ideas of motion for such particles. Before we proceed to apply these to special cases it will be desirable to have another look at the equations themselves and in particular to consider the problem of the medium in which they hold. In nineteenth century physics under the influence of materialistic notions this medium was denominated the ether of space and was supposed to permeate all material substances. Its chief business was supposed to be to propagate electromagnetic effects in much the same way in which ordinary material media propagate elastic effects. At the same time it was only natural to attempt to ascertain other properties of the ether. In particular the question arose: is the ether a reference system equivalent to the fixed stars? If it is, then one cannot hope to detect one's uniform motion with respect to it by any mechanical effect (Newtonian relativity), but on the other hand one might suppose it possible to detect such motion by wave propagation, just as one can detect one's motion with respect to a source of sound by means of the Doppler effect.

Maxwell suggested that if the earth moves with respect to a stationary ether, this motion would be detected by sending a light signal a certain distance in the direction of the motion and comparing the time taken for its return after reflection by a mirror with that taken by the travel of a signal through the same distance at right angles to the direction of motion. This is the basis of the celebrated Michelson, Morley and Miller experiment. Maxwell pointed out that the two times might be expected to differ by an extremely small amount and there was some doubt whether a sufficiently sensitive time measuring device existed. However, Michelson applied his interferometer to the test of the suggestion. The schematic diagram of the interferometer is shown in Fig. 17.1, in which P_1 and P_2 are two plane parallel plates of glass, with P_1 half-silvered on the side facing P_2. Light from the source S is incident at 45° on P_1 at A and part is reflected from the half-silvered surface to the mirror M_1, while the rest is transmitted through P_2 to the mirror M_2. It is arranged so that AM_1 $(=l_1)$ is perpendicular to AM_2 $(=l_2)$. The beams of light striking M_1 and M_2 are reflected there and ultimately rejoin at A whence they are further reflected into an eyepiece at O. As l_1 changes with respect to l_2, interference fringes are observed at O. If the interferometer is stationary in the ether and $l_1 = l_2$, the times taken by the two split parts of the light beam are the same.

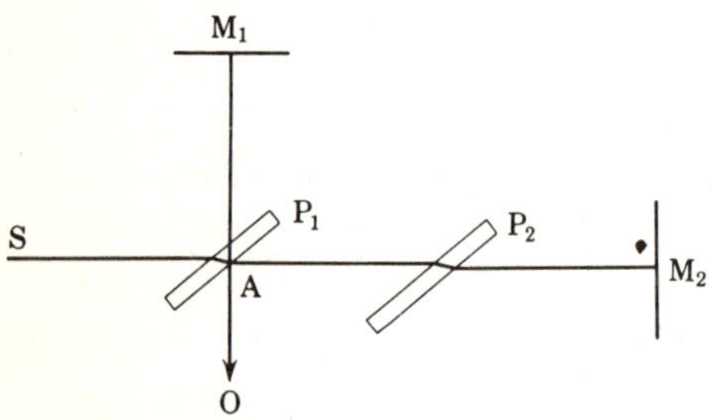

Fig. 17.1

However, when the apparatus moves with constant velocity v in the direction AM_2, the total time for a light signal to travel from A to M_2 and back is

$$\frac{l_2}{c-v} + \frac{l_2}{c+v} = \frac{2cl_2}{c^2-v^2}$$

The time taken for a signal to travel from A to M_1 and back, taking into account the motion, is

$$\frac{2cl_1\sqrt{1-v^2/c^2}}{c^2-v^2}.$$

The difference between these two times is

$$(\Delta t)_1 = \frac{2}{c(1-v^2/c^2)}\left[l_2 - \sqrt{1-v^2/c^2}\cdot l_1\right]. \tag{17.9}$$

Now we have no idea what the direction of motion of the apparatus really is. Hence there is no way in which to test this equation. However, if we rotate the arm AM_2 through 90°, AM_1 becomes the direction of the assumed motion and the time difference between the two paths becomes

$$(\Delta t)_2 = \frac{2}{c(1-v^2/c^2)}\left[l_2\sqrt{1-v^2/c^2} - l_1\right]. \tag{17.10}$$

The difference between $(\Delta t)_1$ and $(\Delta t)_2$ may be denoted by $\Delta^2 t$. It has the value

$$\Delta^2 t = \frac{2(l_1+l_2)}{c(1-v^2/c^2)}\left[1 - \sqrt{1-v^2/c^2}\right]. \tag{17.11}$$

For $l_1 = l_2 = 1$ meter and v of the order of 30 km/sec (the velocity of the earth in its orbit around the sun, relative to axes fixed in the sun) $\Delta^2 t$ comes out of the order of 10^{-16} sec. The interferometer was sensitive enough to detect a time difference of this order as the apparatus was rotated in the fashion described. When the experiment was first tried by Michelson in 1881, the values obtained were only from 1/10 to 1/4 of those expected on the basis of the above assumed velocity of the apparatus. It was concluded that the results were due to experimental error and that effectively $\Delta^2 t$ is really zero, so that it is impossible by the use of light signals to detect the motion of the earth through the ether. The experiment has been repeated many times since, first by Michelson and Morley in 1887 and later by D. C. Miller,[1] who spent many years on it. Although the results were always less than predicted, in no case was a strictly zero value found for $\Delta^2 t$ and Miller in particular insisted on the significance of the small positive result. However, most experimenters and commentators on the experiment have concluded that the results show the impossibility of detecting by electromagnetic means the motion of the earth through the ether. As a matter of fact other experiments designed during the early 1900's to detect the motion

[1] Cf. D. C. Miller, "The Ether Drift Experiment and the Determination of the Absolute Motion of the Earth," *Rev. Mod. Phys.*, 5, 203, 1933.

of the earth through the ether[1] failed to yield positive results of anything like the expected magnitude, and this served to strengthen the general conclusion that by no electromagnetic means can one detect motion with respect to the ether.

A little reflection will indicate that the above results imply the indifference of the equations of the electromagnetic field with respect to different reference systems moving relative to each other with constant velocity. This means that the equations must remain invariant in form under the transformation from one such reference system to another. Now the only transformation of this kind so far discussed is the Newtonian transformation of Sec. 3.2. If this were the proper transformation in going from one inertial system to another moving with respect to the first with constant velocity, then after the transformation the fundamental field equations (15.36) must retain the same form. Specifically we must have

$$\nabla' \cdot \mathbf{D}' = 4\pi\rho', \quad \nabla' \cdot \mathbf{B}' = 0,$$
$$\nabla' \times \mathbf{H}' = \frac{1}{c}(4\pi\rho'\mathbf{v}' + \dot{\mathbf{D}}'), \quad \nabla' \times \mathbf{E}' = -\frac{\dot{\mathbf{B}}'}{c}, \tag{17.12}$$

where the primes refer to transformed quantities, e.g., $\nabla' \times \mathbf{E}'$ has the components $(\partial E_z'/\partial y' - \partial E_y'/\partial z')$, etc. Actually the Newtonian transformation does not lead to (17.12). Lorentz,[2] who examined this problem found that the simplest transformation which gives (17.12) or complete invariance to the field equations is the following:

$$x' = \frac{x - ut}{\sqrt{1-u^2/c^2}}, \quad y' = y, \quad z = z', \quad t' = \frac{t - ux/c^2}{\sqrt{1-u^2/c^2}}, \tag{17.13}$$

in which u is the relative velocity of the two systems, assumed for simplicity to lie along the common x axis. To these must be added the further transformation equations

$$E_x' = E_x, \quad E_y' = \frac{E_y - uH_z/c}{\sqrt{1-u^2/c^2}}, \quad E_z' = \frac{E_z + uH_y/c}{\sqrt{1-u^2/c^2}},$$
$$H_x' = H_x, \quad H_y' = \frac{H_y + uE_z/c}{\sqrt{1-u^2/c^2}}, \quad H_z' = \frac{H_z - uE_y/c}{\sqrt{1-u^2/c^2}}, \tag{17.14}$$

[1] Cf. E. T. Whittaker, *A History of the Theories of the Ether and Electricity* (London, Longmans, Green, 1910), pp. 437 ff.

[2] H. A. Lorentz, *Proc. Amsterdam Acad.*, 6, 809, 1904.

$$v_x{}' = \frac{v_x - u}{1 - v_x u/c^2}, \quad v_y{}' = \frac{v_y}{\sqrt{1-u^2/c^2}(1 - v_x u/c^2)},$$
$$v_z{}' = \frac{v_z}{\sqrt{1-u^2/c^2}(1 - v_x u/c^2)}, \quad \rho' = \frac{\rho(1 - v_x u/c^2)}{\sqrt{1-u^2/c^2}}. \tag{17.15}$$

The further fundamental assumption was made by Lorentz that the charge associated with any volume element remains invariant under the transformation. It is this indeed which leads to the expression for ρ' in terms of ρ in (17.15) along with the velocity transformation equations. The reader is advised to show by direct calculation that the use of the above transformation equations actually yields (17.12) and the required invariance.[1]

Eqs. (17.13) are now called the Lorentz-Einstein transformation equations. It is well to pause a moment to comment on their significance. They must hold in order to assure that the electromagnetic field equations shall remain invariant under a transformation to a moving system. This invariance in turn is necessary if we are to account for our apparent inability to detect the uniform motion of a reference system by electromagnetic means. We note the difference between the equations and those of Newton for the same type of problem. In particular we see that the natural Newtonian assumption $t' = t$ can no longer hold. Time is not measured in the same way in two inertial systems. However, we also note that if $u/c \ll 1$, i.e., as the relative velocity of the systems becomes very small compared with the velocity of light, the equations (17.13) do reduce to the Newtonian form. Similar remarks apply to the velocity equations (17.15) which indeed are derivable from (17.13) as will presently be shown.

Eqs. (17.14) bring out a situation which is ordinarily overlooked in elementary electromagnetism: the field quantities **E** and **H** which we ordinarily think of as distinct entities are really inextricably linked, since what appears as a purely electric field intensity in one reference system appears as a combination of electric and magnetic field intensities in the other.

The considerations of this section serve largely to discredit the concept of the ether as a medium for the propagation of electromagnetic effects. They also suggest that the transformation equations (17.13) may have an even more fundamental significance transcending the equations of electromagnetism. This was actually pointed out by Einstein in 1905 and is the subject of the next section.

[1] Or cf. O. W. Richardson, *The Electron Theory of Matter* (Cambridge University Press, 2d Ed., 1916), pp. 307 ff.

17.4. The Special Theory of Relativity of Einstein. We have already stressed in Sec. 3.2 the invariance of the equations of classical mechanics with respect to any transformation of the frame of reference to another moving with respect to the first with constant velocity. Physically this means that it is impossible to detect motion at constant velocity relative to any frame of reference by purely mechanical means. We are all aware that no mechanical experiment performed on a train moving at constant velocity can indicate whether or not the train is actually moving. It is only when the train speeds up or slows down that mechanical effects take place which enable us to decide within the train whether or not it is moving.

Now the Michelson-Morley-Miller experiment effectively attempted to employ electromagnetic means to detect motion at constant velocity with respect to a frame of reference. If the results of this experiment and others of similar character are interpreted as negative in character, it follows that our ability to detect such motion is certainly severely restricted if not actually nonexistent, since between them mechanics and electromagnetism pretty thoroughly subsume our whole physical experience. It therefore does not appear too daring to introduce the general hypothesis that physical laws of *every* kind have the same form in all inertial systems, i.e., in all reference systems differing only in the fact that they are moving at constant velocity with respect to each other. This is the fundamental postulate that Einstein[1] introduced to generalize the Newtonian theory of relativity. It is the basic hypothesis of what has come to be called Einstein's restricted or special theory of relativity. Supplementing it, Einstein also introduced the further postulate that the velocity of light in free space has the same value in all inertial systems (assuming the use of the same units of space and time). This is really a consequence of the basic postulate with the help of the observed physical fact that the velocity of light in free space is constant in some *one* inertial system.

It is now necessary to discuss further the meaning of these hypotheses. The measurement of velocity implies the measurement of time and this in turn implies the use of a clock which accompanies the experimenter. The measurement of the time associated with an event taking place at the location of the observer would appear to provide no great difficulty, although of course the running of the clock is an arbitrary affair. However, the time associated with an event at another point in space than the one occupied by the observer is another matter and there

[1] A. Einstein, *Annalen der Physik*, 17, 891, 1905.

exists no uniquely given way of determining it. Hence Einstein makes the following postulates:

(1) A clock is placed at every point of space.

(2) All these clocks are identical in structure and run the same way when they are examined by an observer at one place.

(3) Before being used for measuring time, the clocks must be put in synchronism with each other.

The process of synchronization is again arbitrary but Einstein adopts the following scheme as sufficiently plausible. Consider two points P_1 and P_2 with their respective clocks attached rigidly to them. Let a light signal be emitted from P_1 at time t_1 as indicated on P_1's clock. Suppose this signal reaches P_2 at time t_2 as indicated on P_2's clock. By means of a mirror it is then reflected back to P_1, and arrives there at time t_1' as indicated on P_1's clock. The two clocks are said to be synchronized if

$$t_2 = \frac{1}{2}(t_1 + t_1'), \tag{17.16}$$

i.e., this is the time which the clock at P_2 must be adjusted to read to satisfy the condition of synchronism. It is further assumed that if the clock at P_2 is in synchronism with that at P_1, then the clock at P_1 is in synchronism with that at P_2, i.e., the relation is a reciprocal one. Moreover if clocks at P_2 and P_3 are synchronized with one at P_1, they are synchronized with each other.

The time associated with an event taking place at any point in space is then defined as that given by the clock located at that point, assumed to have been put in synchronism with all other clocks.

Our task now is to use the fundamental hypotheses of relativity to derive the transformation equations connecting two inertial systems moving with respect to each other at constant velocity. For this purpose it should be sufficient to utilize the assumed invariance under transformation of any physical law whatever.

We begin by assuming that the transformation from x, y, z, t in one system to the corresponding x', y', z', t' in the other is *linear*, in order to insure that to each event in the one system there corresponds one and only one event in the other. It is most simple and involves no real loss in generality to assume that the two coordinate systems S and S' move along a common x axis indicated in Fig. 17.2 with constant relative velocity u. Two of the transformation equations are then immediately given by $y = y'$, $z = z'$. It remains to find the equations connecting

x and t with x' and t'. We suppose that the origins O and O' coincide when $t = t' = 0$. The clock attached to O is assumed to record the value zero when it coincides with O' and the clock fixed at O' records the same value at that instant. The clocks at all points in S are synchronized and the same is true of all clocks in S'. The origin O' of S' has an x coordinate (in the system S) at time t measured by the clock located at x given by $x = ut$. Hence the kinematic equation for the motion of O' as measured in system S is $x - ut = 0$. But measured in S' the motion of O' is simply $x' = 0$. These two equations are equivalent and must hold

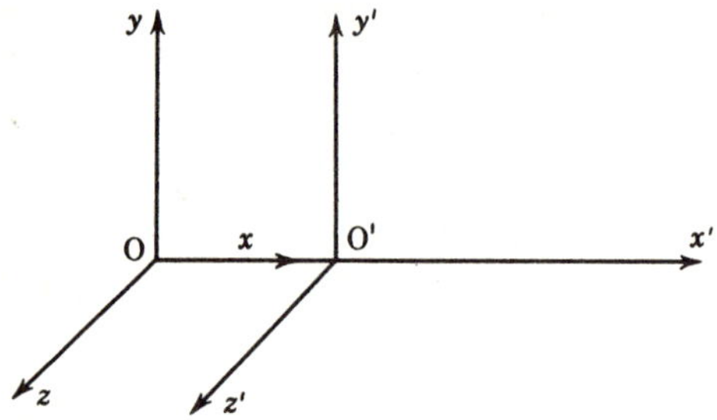

Fig. 17.2

at the same time. Hence we must be able to write

$$x' = \gamma(x - ut), \tag{17.17}$$

where γ is a constant which we must evaluate. If now we apply a precisely similar argument to the motion of O, we obtain

$$x = \gamma' (x' + ut'). \tag{17.18}$$

But the fundamental postulate of relativity demands that

$$\gamma = \gamma', \tag{17.19}$$

since if this equality did not hold, the physical law describing the relative motion of the two reference systems would have a different form depending on the system in which it is expressed. As a matter of fact on Newtonian relativity $\gamma = \gamma' = 1$. This also implies $t = t'$, but we can no longer assume this. Let us find t' in terms of x and t from (17.17) and (17.18). The result is

$$t' = \gamma t + (1-\gamma^2)x/u\gamma. \tag{17.20}$$

Taking differentials yields

$$dt' = \gamma\, dt + (1-\gamma^2)\, dx/u\gamma, \tag{17.21}$$

while from (17.17) itself we have

$$dx' = \gamma(dx - u dt). \tag{17.22}$$

Now dx'/dt' is the instantaneous velocity of a particle with reference to system S' while dx/dt gives the corresponding quantity in S. Hence we have by division

$$\frac{dx'}{dt'} = \frac{dx/dt - u}{1 + dx/dt \cdot (1-\gamma^2)/u\gamma^2}. \tag{17.23}$$

This is a perfectly general expression for any kind of moving particle. Apply it to the special case of a light particle or photon, whose velocity by the principle of relativity is the constant c in both systems. Eq. (17.23) therefore leads to

$$c = \frac{c - u}{1 + c(1-\gamma^2)/u\gamma^2}, \tag{17.24}$$

or

$$\gamma = 1\Big/\sqrt{1 - u^2/c^2}. \tag{17.25}$$

The Lorentz-Einstein transformation equations are immediately obtainable by substitution back into (17.17) and (17.20), viz.,

$$x' = \frac{x - ut}{\sqrt{1-\beta^2}}, \quad y' = y, \quad z' = z, \quad t' = \frac{t - \beta x/c}{\sqrt{1-\beta^2}}, \tag{17.26}$$

where we have introduced β for u/c. Solving for the unprimed coordinates in terms of the primed, we obtain the reciprocal transformation

$$x = \frac{x' + ut'}{\sqrt{1-\beta^2}}, \quad y = y', \quad z = z', \quad t = \frac{t' + \beta x'/c}{\sqrt{1-\beta^2}}. \tag{17.27}$$

We are now ready to consider some of the kinematic consequences of these equations. First take the measurement of length. We measure the length of a rod AB which is assumed to be at rest in system S and lies along the x axis, by noting the x coordinates, x_A and x_B, of the two ends. Then

$$l = x_B - x_A$$

is the required length. Since the rod is at rest in S, it makes no difference at what time (as indicated in the clocks of S) the measurements are performed. However, the situation is different in S'. Thus from (17.27) we have

$$x_B - x_A = [(x'_B - x'_A) + u(t'_B - t'_A)]\Big/\sqrt{1-\beta^2}.$$

If then we denote $x'_B - x'_A$ by l', or the length of the rod as measured in S' we get

$$l' = l\sqrt{1-\beta^2} - u(t'_B - t'_A).$$

It then appears that the length l' depends on the times t'_B and t'_A at which x'_B and x'_A are measured respectively. Hence it is not precisely defined. However, it is natural to take $t'_B = t'_A$. Then

$$l' = l\sqrt{1-\beta^2}, \tag{17.28}$$

from which it appears that the length as measured in the system S' with respect to which the rod is moving with speed u is shorter than measured in the system in which it is at rest in the ratio $\sqrt{1-\beta^2} : 1$. It is interesting to observe that in 1892, Fitzgerald put forth the hypothesis that a moving object actually shortens in the direction of its motion, in order to account for the presumed null result of the Michelson-Morley experiment. The contraction in question has precisely the magnitude given by (17.28).

An interesting correlative of the contraction in length is the change in time measurement associated with motion. We consider a clock C' fixed at O', and in addition a sequence of synchronized clocks fixed in S at the points $x_1, x_2, x_3 \ldots$. Let the time when C' passes x_1 be t_1 on the clock fixed at x_1 and t'_1 on C'. Similarly let the time when C' passes x_2 be t_2 on the clock fixed at x_2 and t'_2 on C'. Then from (17.26)

$$t'_2 - t'_1 = [(t_2 - t_1) - \beta/c \cdot (x_2 - x_1)]\Big/\sqrt{1-\beta^2}. \tag{17.29}$$

But

$$x_2 - x_1 = u(t_2 - t_1),$$

whence (17.29) becomes

$$t'_2 - t'_1 = \sqrt{1-\beta^2} \cdot (t_2 - t_1). \tag{17.30}$$

This means that the time interval on the moving clock C' is shorter than the time interval between the same two events as measured by the clocks fixed in S. Now we can turn the thing around and have a clock C fixed at

O pass in turn a sequence of synchronized clocks fixed in S' at points $x_1', x_2', x_3', \ldots$. Now t_1 denotes the time on C when it passes x_1', while t_1' is the corresponding time on the clock fixed at x_1' for the same event. Then (17.29) is replaced by

$$t_2 - t_1 = [(t_2' - t_1') + \beta/c \cdot (x_2' - x_1')]/\sqrt{1-\beta^2}. \tag{17.31}$$

But now

$$x_2' - x_1' = -u(t_2' - t_1'),$$

and therefore in this case

$$t_2 - t_1 = \sqrt{1-\beta^2} \cdot (t_2' - t_1'). \tag{17.32}$$

Once more the time interval on the clock considered to be moving is shorter than that on the clock considered to be stationary. These results are commonly summarized in the statement that a moving clock always runs "slow." This statement by itself is misleading and the precise results (17.29) and (17.32) are much to be preferred.

For a discussion of the conclusions of relativity on the length of rods and time intervals in moving systems *Foundations of Physics* (pp. 34 ff.) may be consulted.

Next we shall discuss the composition of velocities according to relativity. In ordinary Newtonian relativity if a particle has velocity v relative to system S, its velocity v' relative to system S' is simply $v' = v - u$. However from (17.23) we see by substituting the value of γ given by (17.25) that

$$v' = \frac{v - u}{1 - \beta v/c}. \tag{17.33}$$

Hence the law of composition of velocities differs from the conventional one of classical mechanics, unless $\beta \ll 1$. The connection with (17.15) is now obvious. It is an interesting result of (17.33) that if $v = c$, $v' = c$ also. Thus the composition of the velocity of light with any velocity of a material particle yields again the velocity of light. Hence no velocity of a material particle may exceed the velocity of light. This indeed may be expressed in more general fashion if we are willing to restrict all actual physical happenings to taking place in *positive* time intervals. Then the maximum speed of propagation of any physical event is c. The demonstration is left to the reader.

17.5. Relativity Dynamics. We shall now try to develop the equations of motion of a charged particle so as to satisfy the requirements of relativity. Let us assume that the equations of motion of the electron in an electric field with components E_x, E_y, E_z with reference to a system of axes S' in which the electron is instantaneously at rest are

$$m_0 \frac{d^2x'}{dt'^2} = eE_x', \qquad m_0 \frac{d^2y'}{dt'^2} = eE_y', \qquad m_0 \frac{d^2z'}{dt'^2} = eE_z', \tag{17.34}$$

where m_0 is the rest mass of the electron. We are here essentially following (17.2). But now we wish to find the corresponding equations of motion for the electron referred to the system S relative to which S' is moving with constant velocity u in the x direction. We proceed to use the relativity transformation equations. Thus from (17.33)

$$\frac{dx'}{dt'} = \frac{\frac{dx}{dt} - u}{1 - u\frac{dx}{dt}\Big/c^2}.$$

We differentiate with respect to t' and get

$$\frac{d^2x'}{dt'^2} = \frac{d}{dt'}\left(\frac{dx/dt - u}{1 - u\frac{dx}{dt}\Big/c^2}\right) = \frac{d}{dt}\left(\frac{\frac{dx}{dt} - u}{1 - u\frac{dx}{dt}\Big/c^2}\right)\cdot\frac{dt}{dt'}$$

$$= \frac{\sqrt{1-\beta^2}}{1-\beta\frac{dx}{dt}\Big/c}\cdot\frac{d}{dt}\left(\frac{\frac{dx}{dt} - u}{1-\beta\frac{dx}{dt}\Big/c}\right).$$

Carrying out the indicated differentiation, we have

$$\frac{d^2x'}{dt'^2} = \frac{\sqrt{1-\beta^2}}{\left(1-\beta\frac{dx}{dt}\Big/c\right)^3}\left[\left(1-\beta\frac{dx}{dt}\Big/c\right)\frac{d^2x}{dt^2} + \beta/c\cdot\left(\frac{dx}{dt} - u\right)\frac{d^2x}{dt^2}\right],$$

which on further reduction yields

$$\frac{d^2x'}{dt'^2} = \frac{(1-\beta^2)^{1/2}\,(1-u\beta/c)}{\left(1-\beta\frac{dx}{dt}\Big/c\right)^3}\cdot\frac{d^2x}{dt^2}. \tag{17.35}$$

Next we get

$$\frac{dy'}{dt'} = \frac{\frac{dy}{dt}\sqrt{1-\beta^2}}{1-\beta\frac{dx}{dt}/c},$$

and finally

$$\frac{d^2y'}{dt'^2} = \frac{(1-\beta^2)}{(1-\beta\frac{dx}{dt}/c)^3}\left[(1-\beta\frac{dx}{dt}/c)\frac{d^2y}{dt^2} + \frac{dy}{dt}\cdot\frac{\beta}{c}\cdot\frac{d^2x}{dt^2}\right]. \quad (17.36)$$

Also

$$\frac{d^2z'}{dt'^2} = \frac{(1-\beta^2)}{(1-\beta\frac{dx}{dt}/c)^3}\left[(1-\beta\frac{dx}{dt}/c)\frac{d^2z}{dt^2} + \frac{dz}{dt}\cdot\frac{\beta}{c}\cdot\frac{d^2x}{dt^2}\right]. \quad (17.37)$$

Since the electron is instantaneously at rest in S' and S' is moving at constant velocity u with respect to S, this velocity is also the velocity of the electron in the system S and hence in the above equations $dx/dt = u = \beta c$, while $dy/dt = dz/dt = 0$. We can therefore rewrite the equations as follows

$$\frac{d^2x'}{dt'^2} = \frac{1}{(1-\beta^2)^{3/2}}\frac{d^2x}{dt^2},$$

$$\frac{d^2y'}{dt'^2} = \frac{1}{(1-\beta^2)}\frac{d^2y}{dt^2}, \quad (17.38)$$

$$\frac{d^2z'}{dt'^2} = \frac{1}{(1-\beta^2)}\frac{d^2z}{dt^2}.$$

We are now ready to write the equations of motion in the system S. We recall from (17.14) that Lorentz, to assure the invariance of the field equations under the transformation indicated, i.e., in order to assure that the relativity condition be satisfied, has to assume

$$E'_x = E_x, \quad E'_y = \frac{E_y - uH_z/c}{\sqrt{1-\beta^2}}, \quad E'_z = \frac{E_z + uH_y/c}{\sqrt{1-\beta^2}}.$$

We therefore write (17.34) in the form

$$\frac{m_0}{(1-\beta^2)^{3/2}} \cdot \frac{d^2x}{dt^2} = eE_x,$$

$$\frac{m_0}{(1-\beta^2)^{1/2}} \cdot \frac{d^2y}{dt^2} = e(E_y - uH_z/c), \tag{17.39}$$

$$\frac{m_0}{(1-\beta^2)^{1/2}} \cdot \frac{d^2z}{dt^2} = e(E_z + uH_y/c).$$

It appears at once that the mass associated with acceleration in the direction of the velocity is

$$m_0/(1-\beta^2)^{3/2}, \tag{17.40}$$

whereas the mass associated with acceleration in the direction at right angles to the velocity is

$$m_0/(1-\beta^2)^{1/2}. \tag{17.41}$$

These are the longitudinal and transverse masses m_l and m_t, respectively, already mentioned in (17.8). We now wish to consolidate (17.39) into a single vector equation.

Employing **f** as the vector acceleration, i.e.,

$$\mathbf{f} = \mathbf{i}\ddot{x} + \mathbf{j}\ddot{y} + \mathbf{k}\ddot{z},$$

the reader may show that (17.39) can be written as the single vector relation

$$\frac{m_0}{(1-\beta^2)^{3/2}}\left[\mathbf{f} + \frac{1}{c^2}(\mathbf{f}\times\mathbf{u})\times\mathbf{u}\right] = e\left[\mathbf{E} + \frac{1}{c}(\mathbf{u}\times\mathbf{H})\right]. \tag{17.42}$$

But the left-hand side can be further simplified. Expand the triple vector product and compare with (17.5). The result is that (17.42) becomes

$$\frac{d}{dt}\left(\frac{m_0\mathbf{u}}{\sqrt{1-\beta^2}}\right) = e\left[\mathbf{E} + \frac{1}{c}(\mathbf{u}\times\mathbf{H})\right]. \tag{17.43}$$

This is the equation already introduced in (17.4). We see that we can throw the relativistic force equation into the equivalent Newtonian form (time rate of change of momentum equals resultant force) if we define the mass as

$$m_0/\sqrt{1-\beta^2}.$$

The relativistic mass therefore appears as the equivalent of the transverse mass of electrodynamics.

The preceding discussion of relativity dynamics has been restricted to the electron and has leaned heavily on the Lorentz theory. It is customary to extend the analysis to mass particles in general. This has been done, for example, by H. Minkowski.[1] By what amounts to a redefinition of force the fundamental equation of dynamics can still be put in the form (we go back to the use of **v** for velocity)

$$\frac{d}{dt}\left(\frac{m_0\mathbf{v}}{\sqrt{1-\beta^2}}\right) = \mathbf{F}. \tag{17.44}$$

It is now of interest to investigate the definition of energy in relativistic dynamics. We proceed as in classical mechanics, i.e., we find the work done on the particle by the resultant force during its motion. From (17.44) it follows that

$$\int_{\mathbf{r}_0}^{\mathbf{r}_1} \mathbf{F}\cdot d\mathbf{r} = m_0 \int_{\mathbf{r}_0}^{\mathbf{r}_1} \frac{d}{dt}\left(\mathbf{v}\Big/\sqrt{1-v^2/c^2}\right)\cdot d\mathbf{r}, \tag{17.45}$$

where **v** is, of course, the velocity of the particle in the primary inertial system. We shall define the right-hand side as the change in the kinetic energy during the motion. By some algebraic manipulation based on $d\mathbf{r}/dt = \mathbf{v}$, it develops that

$$\frac{d}{dt}\left(\mathbf{v}\Big/\sqrt{1-v^2/c^2}\right)\cdot d\mathbf{r} = d\left(c^2\Big/\sqrt{1-\beta^2}\right). \tag{17.46}$$

Therefore the kinetic energy in relativistic dynamics appears in the form

$$K = m_0c^2\Big/\sqrt{1-\beta^2} + \text{constant}. \tag{17.47}$$

It is therefore indeterminate to an arbitrary constant. However, it is natural to assume that when $v = 0$, $K = 0$. This fixes the constant as $-m_0c^2$ and we finally write

$$K = m_0c^2\left(1\Big/\sqrt{1-\beta^2}-1\right). \tag{17.48}$$

[1]Cf. *Foundations of Physics*, p. 347.

If we denote the variable mass of the particle by $m = m_0/\sqrt{1-\beta^2}$, we can express the kinetic energy still more simply as

$$K = (m - m_0)c^2 . \tag{17.49}$$

There is an interesting interpretation of this equation. Let us rewrite it in the form

$$m = m_0 + K/c^2 . \tag{17.50}$$

It then appears that the actual mass of the moving particle is the sum of the *rest* mass and a contribution made up of its kinetic energy divided by c^2. As the kinetic energy changes, the mass changes. This of course suggests that loss or gain of kinetic energy by any means may be interpreted as loss or gain of mass. Einstein has generalized this by the hypothesis that with every mass m there is associated a kinetic energy mc^2. This makes the total energy W of a free particle (expanding the radical in eq. 17.48)

$$W = mc^2 = m_0c^2 + \frac{1}{2}m_0v^2 + \ldots , \tag{17.51}$$

where m_0c^2 is called the *intrinsic* energy of the particle, existing as it does even when the particle is at rest. Einstein has attributed even further significance to the relation (17.51) by showing by an elementary argument[1] that to a given amount of energy of electromagnetic radiation W in a cavity, there must be associated inertia whose mass measure is W/c^2.

The relation

$$W = mc^2 \tag{17.52}$$

has taken a key place in modern theories of the structure of nuclei of atoms.

17.6. Advanced Mechanics of Charged Particles in Electromagnetic Fields. We shall now find it of some interest to study the motion of a charged particle, e.g., an electron, in an electromagnetic field from the standpoint of the Lagrangian and Hamiltonian methods discussed in Chapters 8 and 9. It turns out to be necessary to invoke again the scalar potential ϕ and the vector potential $\mathbf{A}$ defined in Sec. 15.6, so that, to repeat for convenience, the field vectors in free space are given by

[1] Cf. *Foundations of Physics*, p. 351.

$$\mathbf{H} = \nabla \times \mathbf{A},$$

$$\mathbf{E} = -\nabla\phi - \dot{\mathbf{A}}/c.$$

We now look for a Lagrangian function which is of such a form that we can write eq. (17.43) in Lagrangian form. This is largely a matter of inspection. We try

$$L = -m_0 c^2 \sqrt{1-v^2/c^2} - e\phi + \frac{e}{c}\mathbf{A}\cdot\mathbf{v}, \tag{17.53}$$

where

$$\mathbf{v} = \mathbf{i}\dot{x} + \mathbf{j}\dot{y} + \mathbf{k}\dot{z}; \qquad v^2 = \dot{x}^2 + \dot{y}^2 + \dot{z}^2. \tag{17.54}$$

Let us confine our attention to the x component equation. We have

$$\frac{\partial L}{\partial \dot{x}} = m_0\dot{x}/\sqrt{1-v^2/c^2} + \frac{e}{c}A_x,$$

and

$$\frac{d}{dt}\left(\frac{\partial L}{\partial \dot{x}}\right) = \frac{d}{dt}\left(m_0\dot{x}/\sqrt{1-v^2/c^2}\right) + \frac{e}{c}\left(\frac{\partial A_x}{\partial t} + \dot{x}\frac{\partial A_x}{\partial x} + \dot{y}\frac{\partial A_x}{\partial y} + \dot{z}\frac{\partial A_x}{\partial z}\right).$$

Also

$$\frac{\partial L}{\partial x} = -e\frac{\partial \phi}{\partial x} + \frac{e}{c}\left(\dot{x}\frac{\partial A_x}{\partial x} + \dot{y}\frac{\partial A_y}{\partial x} + \dot{z}\frac{\partial A_z}{\partial x}\right).$$

Consequently

$$\frac{d}{dt}\left(\frac{\partial L}{\partial \dot{x}}\right) - \frac{\partial L}{\partial x} = \frac{d}{dt}\left(m_0\dot{x}/\sqrt{1-v^2/c^2}\right) + e\left(\frac{\partial \phi}{\partial x} + \frac{1}{c}\frac{\partial A_x}{\partial t}\right)$$

$$-\frac{e}{c}\dot{y}\left(\frac{\partial A_y}{\partial x} - \frac{\partial A_x}{\partial y}\right) + \frac{e}{c}\dot{z}\left(\frac{\partial A_x}{\partial z} - \frac{\partial A_z}{\partial x}\right). \tag{17.55}$$

But from the definition of the vector potential, this can be rewritten

$$\frac{d}{dt}\left(\frac{\partial L}{\partial \dot{x}}\right) - \frac{\partial L}{\partial x} = \frac{d}{dt}\left(m_0\dot{x}/\sqrt{1-v^2/c^2}\right) - eE_x - \frac{e}{c}(\mathbf{v}\times\mathbf{H})_x. \tag{17.56}$$

Hence the Lagrangian equation in the form

$$\frac{d}{dt}\left(\frac{\partial L}{\partial \dot{x}}\right) - \frac{\partial L}{\partial x} = 0,$$

becomes simply

$$\frac{d}{dt}\left(m_0\dot{x}\Big/\sqrt{1-v^2/c^2}\right) = e\left[E_x + \frac{1}{c}(\mathbf{v}\times\mathbf{H})_x\right], \tag{17.57}$$

which is precisely the x component of eq. (17.43). This shows that we have chosen an appropriate Lagrangian function.

The next step is to find a Hamiltonian function for the particle. The generalized momentum p_x is at once given by

$$p_x = \frac{\partial L}{\partial \dot{x}} = m_0\dot{x}\Big/\sqrt{1-v^2/c^2} + \frac{e}{c}A_x.$$

Hence the Hamiltonian H is (cf. Sec. 9.1)

$$H = \dot{x}p_x + \dot{y}p_y + \dot{z}p_z - L$$

$$= m_0c^2\sqrt{1 + \frac{1}{m_0^2c^2}\left[(p_x - \frac{e}{c}A_x)^2 + (p_y - \frac{e}{c}A_y)^2 + (p_z - \frac{e}{c}A_z)^2\right]} + e\phi, \tag{17.58}$$

after some algebraic manipulation. The equation representing the constancy of H or the conservation of energy then becomes

$$(p_x - \frac{e}{c}A_x)^2 + (p_y - \frac{e}{c}A_y)^2 + (p_z - \frac{e}{c}A_z)^2 + m_0^2c^2 = (W - e\phi)^2/c^2. \tag{17.59}$$

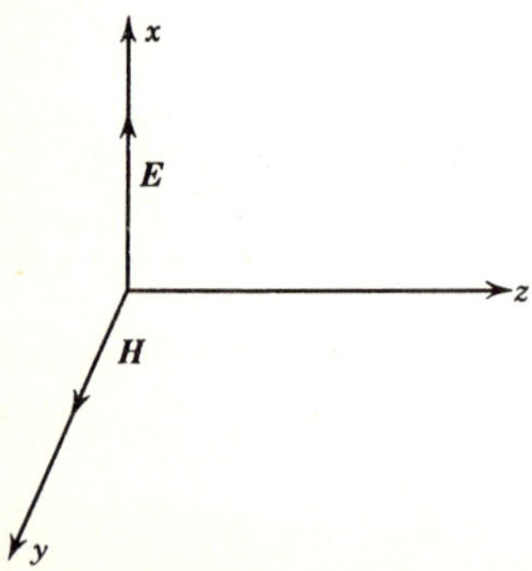

Fig. 17.3

We are now ready to apply the Hamilton-Jacobi method discussed in Sec. 9.3 to the determination of the motion of charged particles. It will be simplest to limit the discussion to a special case The most common one is that of motion through uniform electric and magnetic fields **E** and **H** constant in time and at right angles to each other. We assume that **E** is directed along the x axis and **H** along the y axis, as indicated in Fig. 17.3. For the scalar potential we have simply $\phi = -Ex$, where we have put the additive constant of integration equal to zero. The vector potential is of course not so specifically defined and we may use several different but equally applicable functions. We choose as most appropriate for the mathematical treatment $\mathbf{A} = -\mathbf{k}Hx$. Eq. (17.59) then takes the form

$$p_x^2 + p_y^2 + (p_z + \frac{e}{c} Hx)^2 + m_0^2c^2 = (W + eEx)^2/c^2. \quad (17.60)$$

Replacing p_x by $\frac{\partial S}{\partial x}$, etc., we have for the Hamilton-Jacobi equation

$$\left(\frac{\partial S}{\partial x}\right)^2 + \left(\frac{\partial S}{\partial y}\right)^2 + \left(\frac{\partial S}{\partial z} + \frac{e}{c}Hx\right)^2 + m_0^2c^2 = (W + eEx)^2/c^2. \quad (17.61)$$

Now the Hamiltonian for this motion does not contain y or z. Hence p_y and p_z are constants. Therefore

$$\frac{\partial S}{\partial y} = \frac{m_0\dot{y}}{\sqrt{1-\beta^2}} = \text{constant} = a_1/c,$$

$$\frac{\partial S}{\partial z} = \frac{m_0\dot{z}}{\sqrt{1-\beta^2}} - \frac{e}{c}Hx = \text{constant} = a_2/c. \quad (17.62)$$

If we substitute these into (17.61) the result is

$$\frac{\partial S}{\partial x} = \frac{1}{c} \sqrt{(W + eEx)^2 - m_0^2c^4 - a_1^2 - (a_2 + eHx)^2}. \quad (17.63)$$

The variables are thus separated and we may solve to find

$$S = a_1y/c + a_2z/c$$

$$+ \frac{1}{c}\int \sqrt{W^2 - a_1^2 - a_1^2 - m_0^2c^4 + 2ex(EW - a_2H) + e^2x^2(E^2 - H^2)}\, dx. \quad (17.64)$$

The orbit of the particle is then found in the usual way (cf. Sec. 9.3) from the equations

$$\frac{\partial S}{\partial W} = t, \quad \frac{\partial S}{\partial a_1} = C_1, \quad \frac{\partial S}{\partial a_2} = C_2. \quad (17.65)$$

However, for simplicity we shall confine ourselves to the determination of the projection of the orbit on the xz plane. The equation of this is clearly $\partial S/\partial a_2$ = constant or

$$z - \int \frac{(a_2 + eHx)\,dx}{\sqrt{(W + eEx)^2 - m_0{}^2c^4 - a_1{}^2 - (a_2 + eHx)^2}} = C_2c = K. \quad (17.66)$$

The integration becomes simple, if we let $x + \dfrac{a_2}{eH} = u$ so that

$$z - K = \int \frac{u\,du}{\sqrt{Au^2 + Bu + C}}, \quad (17.67)$$

with

$$A = E^2/H^2 - 1, \quad B = \frac{2EW}{eH^2} - \frac{2E^2a_2}{eH^3},$$

$$C = \frac{W^2 + E^2/A^2 \cdot a_2{}^2 - a_1{}^2 - m_0{}^2c^4}{e^2H^2}.$$

We shall assume that $E/H < 1$ so that $A < 0$. From Peirce's Tables 161 and 169, eq. (17.67) becomes

$$z - K = \left[\frac{\sqrt{Au^2 + Bu + C}}{A} + \frac{B}{2A\sqrt{-A}} \text{ arc sin } \left(\frac{2Au + B}{\sqrt{B^2 - 4AC}}\right)\right]. \quad (17.68)$$

Introduce the parameter θ so that

$$\sin\theta = \frac{2Au + B}{\sqrt{B^2 - 4AC}}, \quad \cos\theta = \frac{2\sqrt{-A}\,\sqrt{Au^2 + Bu + C}}{\sqrt{B^2 - 4AC}}. \quad (17.69)$$

Then the orbit will be characterized by the following parametric equations

$$\sqrt{-A}\,(z - K) = \frac{\sqrt{B^2 - 4AC}}{2A}\cos\theta + \frac{B}{2A}\theta, \quad (17.70)$$

$$u = \frac{\sqrt{B^2 - 4AC}\sin\theta - B}{2A}. \quad (17.71)$$

Eqs. (17.70) and (17.71) considered as equations defining $\sqrt{-A}\,(z - K)$ and u are the parametric equations of a *cycloid.* The details are left for the reader to fill in. In particular the reader should show that for $E = 0$, so that $A = -1$ and $B = 0$, the path is the familiar circle corresponding to motion through a magnetic field alone.

The above method is sufficiently general to handle a wide variety of motions through electric and magnetic fields. Since relativistic mechanics has been employed throughout, there is no limitation with respect to the velocity save that imposed by relativity itself.

17.7. Electron Theory of Dispersion of Electromagnetic Waves. We have already called attention in Sec. 16.3 to the inadequacy of the field theory of electromagnetism in accounting for the details of the transmission of electromagnetic waves through material conducting media. In 16.10 a similar difficulty arose in connection with metallic reflection. We might indeed have emphasized (Prob. 7 of Chapter 15) that the fundamental equation $V = c/\sqrt{\mu\kappa}$ for the velocity of electromagnetic radiation in a material medium breaks down rather badly for short waves. In every case in fact the predictions of the classical field theory agree with experiment better for *very long* wave length radiation than for short waves. This in itself would suggest that whereas for long waves a material medium may act more or less like a continuum for the propagation of radiation, for short wave lengths it is necessary to postulate a *discrete* structure whose interaction with the radiation will account for the peculiarities of the transmission. As a matter of fact we have seen in the present chapter that this structure of material media has long been postulated on other grounds. The problem of the present section is to see how the electron theory of the structure of matter can account for the observed dispersion and absorption of electromagnetic radiation and in particular light.

We shall assume that the homogeneous isotropic medium under consideration consists of atoms made up of positively charged nuclei and negative electrons. We shall further assume that no permanent dipoles are present. When an electric field $\mathbf{E}$ is applied polarization $\mathbf{P}$ is produced, measured by the induced electric moment per unit volume. The resulting electric displacement $\mathbf{D}$ is given by eq. (15.31), viz.,

$$\mathbf{D} = \mathbf{E} + 4\pi\mathbf{P} = \kappa\mathbf{E}\,.$$

In this case $\mathbf{P}$ and $\mathbf{E}$ have the same direction and we can write for the dielectric constant

$$\kappa = 1 + 4\pi P/E\,. \tag{17.72}$$

It is customary to introduce the so-called *polarizability* α, defined as the ratio between the electric moment produced in a single atom to the inducing field intensity. If there are N atoms per unit volume, we then have

$$P = N\alpha E, \tag{17.73}$$

and (17.72) becomes

$$\kappa = 1 + 4\pi N\alpha. \tag{17.74}$$

Now let us suppose that a plane harmonic electromagnetic wave traverses the medium in the positive x direction so that the electric vector has the form $E_0 e^{(\omega t - kx)}$. Under the influence of this field the electron will suffer oscillatory motion. It will be postulated that the associated magnetic vector produces only a negligible effect since its action depends on the velocity of the electron. Moreover if the amplitude of motion of the electron is of the order of magnitude of the diameter of an atom, for long wave length radiation (even for visible light in fact) the term e^{-ikx} in the wave function will be of negligible significance. Consequently the action of the wave effectively reduces to the time oscillating part $E_0 e^{i\omega t}$. We have already solved the problem of the forced oscillations of a particle under the action of a harmonic force (Sec. 4.4). If the mass of the electron is m, its effective "stiffness" when pulled aside from its equilibrium position f, and the damping coefficient R, the equation of motion is (the charge is e)

$$m\ddot{x} + R\dot{x} + fx = eE_0 e^{i\omega t}. \tag{17.75}$$

The steady state solution for the displacement of the electron from equilibrium is

$$x = \frac{eE_0 e^{i\omega t}}{i\omega R + (f - m\omega^2)}, \tag{17.76}$$

leading to a dipole moment ex or Nex per unit volume. The use of (17.72) then leads to

$$\kappa = 1 + \frac{4\pi N e^2}{i\omega R + (f - m\omega^2)}. \tag{17.77}$$

The result is a complex dielectric constant, which indicates both dispersion and absorption. If now we return to (16.40) and set $\sigma = 0$ and $\mu = 1$ we get

$$(n - ik)^2 = 1 + \frac{4\pi Ne^2}{i\omega R + (f - m\omega^2)}. \tag{17.78}$$

Let us call $f/m = \omega_0^2$, where ω_0 is 2π times the "natural" frequency of vibration of the electron as an oscillator. Under the assumption that $R\omega/m$ is very small we obtain the following approximate result

$$n \doteq 1 + \frac{2\pi Ne^2}{m(\omega_0^2 - \omega^2)}, \tag{17.79}$$

where we have also used the fact that the second term on the right in (17.79) is very much smaller than 1 for the assumed values of N, e, and m and for reasonable values of ω and ω_0. Of course ω must not be too close to ω_0 in this case. Using $\nu = V/\lambda$, where V is the velocity and λ the wave length, we have

$$n = 1 + \frac{Ne^2}{2\pi mV^2(1/\lambda_0^2 - 1/\lambda^2)}. \tag{17.80}$$

If $\lambda >> \lambda_0$ this can be written in the form

$$n = 1 + M(1 + \lambda_0^2/\lambda^2 + \lambda_0^4/\lambda^4 + \ldots), \tag{17.81}$$

where $M = Ne^2\lambda_0^2/2\pi mV^2$. Eq. (17.81) yields the normal dispersion curve for visible light. For $\lambda_0 = \lambda$ we encounter the phenomenon of absorption and the associated anomalous dispersion. The reader will find it instructive to plot (17.80) as a function of λ, noting the behavior in the vicinity of $\lambda = \lambda_0$. Then n should also be plotted for the case in which R is not neglected and the difference noted.

The theory sketched above applies pretty well to the dispersion and absorption of electromagnetic radiation in gases in which the force on the electron due to the wave does not need to be supplemented by the interaction forces produced by other charges in its neighborhood. However, the latter forces must be considered in the case of liquids and solids. We shall not work out the details,[1] but merely point out that these forces may most simply be accounted for by replacing

$$P/E = \frac{e^2N}{i\omega R + (f - m\omega^2)}$$

[1] Cf. Slater and Frank, *Introduction to Theoretical Physics* (New York, McGraw-Hill, 1933), pp. 278 f.

by

$$P/(E + \frac{4}{3}\pi P) = \frac{e^2 N}{i\omega R + (f - m\omega^2)} .$$

This leads to the revised expression for κ

$$\frac{\kappa - 1}{\kappa + 2} = \frac{4\pi}{3} \cdot \frac{e^2 N}{i\omega R + (f - m\omega^2)} = \frac{(n - ik)^2 - 1}{(n - ik)^2 + 2}. \tag{17.82}$$

If k is negligible we can then say that for a given frequency

$$\frac{1}{\rho}\left(\frac{n^2 - 1}{n^2 + 1}\right) = \text{constant}, \tag{17.83}$$

where ρ is the density. This is the Lorenz-Lorentz law which is observed to hold for many substances. Eq. (17.78) must now be modified to read

$$(n - ik)^2 = 1 + \frac{4\pi N e^2}{i\omega R + (f - m\omega^2) - \frac{4\pi N e^2}{3}}, \tag{17.84}$$

which amounts to a decrease in the effective natural frequency of the electron.

We have for simplicity restricted our discussion to a single electron with a definite natural frequency. It would be more accurate to introduce groups of electrons with a characteristic natural frequency associated with each group. Then eq. (17.84) takes the form

$$(n - ik)^2 = 1 + 4\pi \cdot \sum_j \frac{N_j e^2}{i\omega R_j + (f_j - m\omega^2) - \frac{4\pi}{3} N_j e^2}, \tag{17.85}$$

where there are N_j electrons of type j characterized by a natural frequency $\omega_j/2\pi = \sqrt{f_j/m}$, and damping constant R_j.

We have assumed that the electrons whose interaction with the radiation produces absorption and dispersion of the waves act like linear oscillators. Actually they are presumably bound to atoms and move with respect to the atomic nuclei and other electrons in accordance with the principles of the theory of atomic structure. However, it is not in principle difficult to extend the above theory to the electrons in the Bohr atomic model and the fundamental results remain as before. The characteristic frequencies would then appear as the frequencies of orbital motion of the electrons. Unfortunately this turns out not to agree with experiment, which appears to demand that the characteristic frequencies be those given by the Bohr frequency condition discussed in Sec. 9.5 (viz., $(E_1 - E_2)/h = \nu_{12}$ is the frequency associated with an atomic transition from energy state E_1 to energy state E_2). Modern quantum mechanics (cf. Chapter 11) has provided a theory of dispersion which rectifies this difficulty. It is, however, too involved for discussion here.[1]

17.8. Relativity and Wave Mechanics. One of the interesting features of relativity is that it owes its origin to electromagnetic theory, which is essentially a field or medium theory. Nevertheless, as we have seen in the preceding sections of this chapter, relativity has had an important part to play in the study of the motion of particles, particularly when their velocity approaches that of light in free space. In a sense therefore we may look upon relativity as a kind of bridge connecting field theory with particle theory. This aspect is strengthened by the fact that it was the use of relativity which led DeBroglie to the invention of his wave mechanics, which we have already mentioned (Sec. 14.11) as a forerunner of quantum mechanics and an important link between particle and field physics. Although the theory has been absorbed into quantum mechanics, it will be worthwhile to sketch the elementary development as an illustration of the special theory of relativity.

DeBroglie's fundamental assumption is that any physical particle which is stationary in an inertial system must be represented by a periodic function of space and time extending throughout all space and possessing the same phase throughout. Thus if we denote the position coordinates and time of the particle with reference to the system in which it is at rest (its *proper* system, so called) by x_0, y_0, z_0, t_0, respectively, the analytic representation of the particle becomes

$$w(x_0, y_0, z_0, t_0) = f(x_0, y_0, z_0)e^{2\pi i\nu_0 t_0} . \tag{17.86}$$

This is essentially a simple harmonic vibration with frequency ν_0, assumed to be characteristic of the particle, and amplitude $f(x_0, y_0, z_0)$, a function of position in space. Nothing is said about the physical meaning to be attached to the vibration. The hypothesis is highly abstract. It acquires more significance as soon as we refer the particle to an inertial system with respect to which it is moving with constant velocity. For simplicity let the speed be v along the x axis. The new representation of the particle will then be found by the use of the relativity transformation equations (17.13) or (17.26), in which the primed coordinates

[1] Cf. *Foundations of Physics*, p. 474.

are now to be treated as x_0, y_0, z_0, t_0, respectively. In carrying out the transformation on (17.86) we must not forget that the frequency ν_0 depends on the time and hence must be transformed in accordance with the relation

$$dt_0 = dt/\sqrt{1-\beta^2}\,.$$

Therefore in the new inertial system the frequency ν_0 must be replaced by ν, where

$$\nu = \nu_0/\sqrt{1-\beta^2}\,. \tag{17.87}$$

When the transformation has been carried out we get for the analytical representation of the particle in the new system

$$w'(x, y, z, t) = f\left(\frac{x - vt}{\sqrt{1-\beta^2}}, y, z\right) e^{2\pi i\nu(t - vx/c^2)}\,, \tag{17.88}$$

which will be recognized as a plane progressive harmonic wave of frequency ν moving in the positive x direction with velocity V given by

$$V = c^2/v\,. \tag{17.89}$$

Obviously $V \geqq c$ since $v \leqq c$. This might seem to contradict the fundamental result of relativity about the relation between velocities of particles and c. DeBroglie gets around this by assuming that the wave merely pilots the particle and does not itself carry energy with velocity V. The wave length of the wave is of greatest interest. This is obtained by the use of two fundamental relations: (1) the quantum theory assumption connecting the energy W with the frequency ν, viz.,

$$W = h\nu\,, \tag{17.90}$$

and (2) the Einstein mass energy relation

$$W = mc^2\,.$$

The combination of these two with the harmonic wave relation $\lambda = V/\nu$ and (17.89) yields

$$\lambda = h/mv\,. \tag{17.91}$$

This relation has been amply verified by experiments on the diffraction of electrons and other charged particles at the surfaces of single metal crystals. As a matter of fact it finds a natural place in the theory of quantum mechanics and does not depend on DeBroglie's original conception for its theoretical derivation. If the reader turns back to Chapter 11 and consults eq. (11.19) he will find the ψ function for a particle constrained to move on a finite straight line but otherwise free to be given by a function of x which behaves like a harmonic wave as far as its spatial dependence is concerned. If indeed we multiplied ψ by $e^{i\omega t}$, the parameter k in (11.19) and (11.20) would appear as

$$k = 2\pi/\lambda,$$

where λ is the equivalent wave length of the wave. But if we set $E = p^2/2m$ in (11.20) where p is the momentum of the particle, we obtain at once

$$\lambda = h/p,$$

which is precisely the DeBroglie result.

With this intimation of the possible relation between the particle and field concepts in physics we shall close the survey of the fundamentals of theoretical physics presented in this book. Both points of view are of great utility and importance. The task of the theoretical physics of the future is to weld them into a single unified theory.

PROBLEMS

1. Show that the relativistic energy of a free electron can be expressed in the following two ways:

$$E^2 = p^2c^2 + m_0{}^2c^4$$

$$E = \mathbf{v} \cdot \mathbf{p} + m_0c^2\sqrt{1-\beta^2}.$$

Here p is the momentum of the electron, $\mathbf{v}$ its velocity, m_0 its rest mass, c the velocity of light in free space, and $\beta = v/c$.

2. Go through the analysis showing that eqs. (17.12) hold for the Lorentz-Einstein transformation equations (17.13)–(17.15).

3. Show that in the case of an electron moving along the x axis the longitudinal and transverse mass are given respectively by the relations

$$m_l = \frac{1}{c}\frac{\partial G}{\partial \beta}; \qquad m_t = \frac{1}{c}\frac{G}{\beta},$$

where G is the momentum of the electron treated as wholly electromagnetic in origin.

4. Prove that on the basis of the special theory of relativity the maximum speed of propagation of any physical effect is equal to the velocity of light in free space.

5. Show that if one replaces t by the pure imaginary $ict = l$, the Lorentz–Einstein transformation equations of special relativity reduce (so far as the x and l equations are concerned) to the equations for the rotation of axes in the x, l plane through the imaginary angle $\psi = \text{arc tan}\ i\beta$.

6. Discuss the Einstein demonstration that to the radiation energy E there must be assigned the equivalent mass E/c^2.

7. Let the frequency of a light wave in reference system S be ν. Show that the frequency in reference system S' is

$$\nu' = \frac{\nu(1 - v \cos \alpha/c)}{\sqrt{1 - v^2/c^2}},$$

where v is the relative velocity of S and S' and α is the angle between the direction of propagation of the light wave and the line of relative motion of S and S'. Obtain the corresponding formula on classical, non-relativistic theory (Doppler effect).

8. Derive the relativistic form of the expression for the Compton effect wave length shift, i.e.,

$$\Delta\lambda = 2\lambda_0 \sin^2 \phi/2,$$

where $\lambda_0 = h/m_0c$ and ϕ is the angle of scattering of the incident quantum.

9. Use the result of Prob. 1, Chapter 16, for the rate of radiation from an accelerated electron to calculate the rate at which energy in a beam of electromagnetic radiation incident on an electron is scattered by the electron. Thus derive the formula for the intensity of the scattered radiation

$$I = \frac{8\pi}{3} I_0(e^2/mc^2)^2,$$

where I_0 is the intensity of the incident radiation. Use this to determine an effective "radius" of the electron.

10. Two electrons each with initial speed v approach each other along the same straight line. Using relativistic dynamics calculate their distance of closest approach. Compare this with the effective "radius" given by eq. 17.3 and that resulting from Prob. 9.

11. An electron is projected with arbitrary velocity in a uniform magnetic field of intensity **H** directed along the z axis. Find its trajectory.

12. Use the analysis in Sec. 17.7 to calculate the absorption coefficient for electromagnetic radiation passing through a gas.

13. An electron moves in a quantized elliptical orbit about a positive nucleus as in the Bohr theory of the hydrogen atom. Find the expression for the fraction of the whole period of revolution which the electron spends on the average in a spherical shell of thickness dr at distance r from the nucleus. (Hint: Use the expression for the eccentric anomaly, Sec. 9.6. Hence show that if the electron's motion is assumed to be equivalent to a spherical distribution of negative charge about the nucleus the average fraction of the total charge e in the spherical shell dr at distance r is

$$de = \frac{er\,dr}{\pi a^2 \sqrt{\epsilon^2 - 1 + \dfrac{2r}{a} - \dfrac{r^2}{a^2}}}\,.$$

Plot the distribution function. Integrate over r to get the total charge; this of course should be e.)

14. Use the results of Prob. 13 to derive the expression for the potential produced at any point distant r from the nucleus by the spherical distribution to which the electron's motion has been reduced. Using appropriate numerical values (cf. Sec. 9.5) plot the potential as a function of r.

15. In the DeBroglie wave mechanics (Sec. 17.8) show that the DeBroglie "matter" waves suffer dispersion and find the dependence of the phase velocity V on the frequency ν. In one form of the theory it was assumed that the motion of the particle is represented by a group of harmonic waves with slightly different frequencies clustered about the average frequency ν. Prove that in this case the group velocity is equal to the velocity of the particle (cf. Sec. 14.5).

16. Prove that the DeBroglie wave length for an electron with kinetic energy K electron volts and moving with velocity small compared with the velocity of light in free space is given by

$$\lambda = \frac{12.24}{\sqrt{K}} \times 10^{-8} \text{ cm.}$$

How must this be corrected for the case in which the velocity v is not small compared with c?

17. Find the group velocity corresponding to the Cauchy law of dispersion (eq. 17.81).

INDEX

SOME DOVER SCIENCE BOOKS

CHANCE, LUCK AND STATISTICS: THE SCIENCE OF CHANCE,
Horace C. Levinson
Theory of probability and science of statistics in simple, non-technical language. Part I deals with theory of probability, covering odd superstitions in regard to "luck," the meaning of betting odds, the law of mathematical expectation, gambling, and applications in poker, roulette, lotteries, dice, bridge, and other games of chance. Part II discusses the misuse of statistics, the concept of statistical probabilities, normal and skew frequency distributions, and statistics applied to various fields—birth rates, stock speculation, insurance rates, advertising, etc. "Presented in an easy humorous style which I consider the best kind of expository writing," Prof. A. C. Cohen, Industry Quality Control. Enlarged revised edition. Formerly titled *The Science of Chance.* Preface and two new appendices by the author. Index. xiv + 365pp. 5⅜ x 8. Paperbound $2.00

BASIC ELECTRONICS,
prepared by the U.S. Navy Training Publications Center
A thorough and comprehensive manual on the fundamentals of electronics. Written clearly, it is equally useful for self-study or course work for those with a knowledge of the principles of basic electricity. Partial contents: Operating Principles of the Electron Tube; Introduction to Transistors; Power Supplies for Electronic Equipment; Tuned Circuits; Electron-Tube Amplifiers; Audio Power Amplifiers; Oscillators; Transmitters; Transmission Lines; Antennas and Propagation; Introduction to Computers; and related topics. Appendix. Index. Hundreds of illustrations and diagrams. vi + 471pp. 6½ x 9¼.
Paperbound $2.75

BASIC THEORY AND APPLICATION OF TRANSISTORS,
prepared by the U.S. Department of the Army
An introductory manual prepared for an army training program. One of the finest available surveys of theory and application of transistor design and operation. Minimal knowledge of physics and theory of electron tubes required. Suitable for textbook use, course supplement, or home study. Chapters: Introduction; fundamental theory of transistors; transistor amplifier fundamentals; parameters, equivalent circuits, and characteristic curves; bias stabilization; transistor analysis and comparison using characteristic curves and charts; audio amplifiers; tuned amplifiers; wide-band amplifiers; oscillators; pulse and switching circuits; modulation, mixing, and demodulation; and additional semiconductor devices. Unabridged, corrected edition. 240 schematic drawings, photographs, wiring diagrams, etc. 2 Appendices. Glossary. Index. 263pp. 6½ x 9¼. Paperbound $1.25

GUIDE TO THE LITERATURE OF MATHEMATICS AND PHYSICS,
N. G. Parke III
Over 5000 entries included under approximately 120 major subject headings of selected most important books, monographs, periodicals, articles in English, plus important works in German, French, Italian, Spanish, Russian (many recently available works). Covers every branch of physics, math, related engineering. Includes author, title, edition, publisher, place, date, number of volumes, number of pages. A 40-page introduction on the basic problems of research and study provides useful information on the organization and use of libraries, the psychology of learning, etc. This reference work will save you hours of time. 2nd revised edition. Indices of authors, subjects, 464pp. 5⅜ x 8.
Paperbound $2.75

THE RISE OF THE NEW PHYSICS (formerly THE DECLINE OF MECHANISM), *A. d'Abro*
This authoritative and comprehensive 2-volume exposition is unique in scientific publishing. Written for intelligent readers not familiar with higher mathematics, it is the only thorough explanation in non-technical language of modern mathematical-physical theory. Combining both history and exposition, it ranges from classical Newtonian concepts up through the electronic theories of Dirac and Heisenberg, the statistical mechanics of Fermi, and Einstein's relativity theories. "A must for anyone doing serious study in the physical sciences," *J. of Franklin Inst.* 97 illustrations. 991pp. 2 volumes.
T3, T4 Two volume set, paperbound $5.50

THE STRANGE STORY OF THE QUANTUM, AN ACCOUNT FOR THE GENERAL READER OF THE GROWTH OF IDEAS UNDERLYING OUR PRESENT ATOMIC KNOWLEDGE, *B. Hoffmann*
Presents lucidly and expertly, with barest amount of mathematics, the problems and theories which led to modern quantum physics. Dr. Hoffmann begins with the closing years of the 19th century, when certain trifling discrepancies were noticed, and with illuminating analogies and examples takes you through the brilliant concepts of Planck, Einstein, Pauli, de Broglie, Bohr, Schroedinger, Heisenberg, Dirac, Sommerfeld, Feynman, etc. This edition includes a new, long postscript carrying the story through 1958. "Of the books attempting an account of the history and contents of our modern atomic physics which have come to my attention, this is the best," H. Margenau, Yale University, in *American Journal of Physics.* 32 tables and line illustrations. Index. 275pp. 5⅜ x 8.
T518 Paperbound $2.00

GREAT IDEAS AND THEORIES OF MODERN COSMOLOGY, *Jagjit Singh*
The theories of Jeans, Eddington, Milne, Kant, Bondi, Gold, Newton, Einstein, Gamow, Hoyle, Dirac, Kuiper, Hubble, Weizsäcker and many others on such cosmological questions as the origin of the universe, space and time, planet formation, "continuous creation," the birth, life, and death of the stars, the origin of the galaxies, etc. By the author of the popular *Great Ideas of Modern Mathematics.* A gifted popularizer of science, he makes the most difficult abstractions crystal-clear even to the most non-mathematical reader. Index. xii + 276pp. 5⅜ x 8½ T925 Paperbound $2.00

GREAT IDEAS OF MODERN MATHEMATICS: THEIR NATURE AND USE, *Jagjit Singh*
Reader with only high school math will understand main mathematical ideas of modern physics, astronomy, genetics, psychology, evolution, etc., better than many who use them as tools, but comprehend little of their basic structure. Author uses his wide knowledge of non-mathematical fields in brilliant exposition of differential equations, matrices, group theory, logic, statistics, problems of mathematical foundations, imaginary numbers, vectors, etc. Original publications, appendices. indexes. 65 illustr. 322pp. 5⅜ x 8. T587 Paperbound $2.00

THE MATHEMATICS OF GREAT AMATEURS, *Julian L. Coolidge*
Great discoveries made by poets, theologians, philosophers, artists and other non-mathematicians: Omar Khayyam, Leonardo da Vinci, Albrecht Dürer, John Napier, Pascal, Diderot, Bolzano, etc. Surprising accounts of what can result from a non-professional preoccupation with the oldest of sciences. 56 figures. viii + 211pp. 5⅜ x 8½. S1009 Paperbound $2.00

FIVE VOLUME "THEORY OF FUNCTIONS" SET BY KONRAD KNOPP

This five-volume set, prepared by Konrad Knopp, provides a complete and readily followed account of theory of functions. Proofs are given concisely, yet without sacrifice of completeness or rigor. These volumes are used as texts by such universities as M.I.T., University of Chicago, N. Y. City College, and many others. "Excellent introduction . . . remarkably readable, concise, clear, rigorous," *Journal of the American Statistical Association.*

ELEMENTS OF THE THEORY OF FUNCTIONS,
Konrad Knopp
This book provides the student with background for further volumes in this set, or texts on a similar level. Partial contents: foundations, system of complex numbers and the Gaussian plane of numbers, Riemann sphere of numbers, mapping by linear functions, normal forms, the logarithm, the cyclometric functions and binomial series. "Not only for the young student, but also for the student who knows all about what is in it," *Mathematical Journal.* Bibliography. Index. 140pp. 5⅜ x 8. Paperbound $1.50

THEORY OF FUNCTIONS, PART I,
Konrad Knopp
With volume II, this book provides coverage of basic concepts and theorems. Partial contents: numbers and points, functions of a complex variable, integral of a continuous function, Cauchy's integral theorem, Cauchy's integral formulae, series with variable terms, expansion of analytic functions in power series, analytic continuation and complete definition of analytic functions, entire transcendental functions, Laurent expansion, types of singularities. Bibliography. Index. vii + 146pp. 5⅜ x 8. Paperbound $1.35

THEORY OF FUNCTIONS, PART II,
Konrad Knopp
Application and further development of general theory, special topics. Single valued functions. Entire, Weierstrass, Meromorphic functions. Riemann surfaces. Algebraic functions. Analytical configuration, Riemann surface. Bibliography. Index. x + 150pp. 5⅜ x 8. Paperbound $1.35

PROBLEM BOOK IN THE THEORY OF FUNCTIONS, VOLUME 1.
Konrad Knopp
Problems in elementary theory, for use with Knopp's *Theory of Functions,* or any other text, arranged according to increasing difficulty. Fundamental concepts, sequences of numbers and infinite series, complex variable, integral theorems, development in series, conformal mapping. 182 problems. Answers. viii + 126pp. 5⅜ x 8. Paperbound $1.35

PROBLEM BOOK IN THE THEORY OF FUNCTIONS, VOLUME 2,
Konrad Knopp
Advanced theory of functions, to be used either with Knopp's *Theory of Functions,* or any other comparable text. Singularities, entire & meromorphic functions, periodic, analytic, continuation, multiple-valued functions, Riemann surfaces, conformal mapping. Includes a section of additional elementary problems. "The difficult task of selecting from the immense material of the modern theory of functions the problems just within the reach of the beginner is here masterfully accomplished," *Am. Math. Soc.* Answers. 138pp. 5⅜ x 8. Paperbound $1.50

AN INTRODUCTION TO THE GEOMETRY OF N DIMENSIONS,
D. H. Y. Sommerville
An introduction presupposing no prior knowledge of the field, the only book in English devoted exclusively to higher dimensional geometry. Discusses fundamental ideas of incidence, parallelism, perpendicularity, angles between linear space; enumerative geometry; analytical geometry from projective and metric points of view; polytopes; elementary ideas in analysis situs; content of hyper-spacial figures. Bibliography. Index. 60 diagrams. 196pp. 5⅜ x 8.
Paperbound $1.50

ELEMENTARY CONCEPTS OF TOPOLOGY, *P. Alexandroff*
First English translation of the famous brief introduction to topology for the beginner or for the mathematician not undertaking extensive study. This unusually useful intuitive approach deals primarily with the concepts of complex, cycle, and homology, and is wholly consistent with current investigations. Ranges from basic concepts of set-theoretic topology to the concept of Betti groups. "Glowing example of harmony between intuition and thought," David Hilbert. Translated by A. E. Farley. Introduction by D. Hilbert. Index. 25 figures. 73pp. 5⅜ x 8. Paperbound $1.00

ELEMENTS OF NON-EUCLIDEAN GEOMETRY,
D. M. Y. Sommerville
Unique in proceeding step-by-step, in the manner of traditional geometry. Enables the student with only a good knowledge of high school algebra and geometry to grasp elementary hyperbolic, elliptic, analytic non-Euclidean geometries; space curvature and its philosophical implications; theory of radical axes; homothetic centres and systems of circles; parataxy and parallelism; absolute measure; Gauss' proof of the defect area theorem; geodesic representation; much more, all with exceptional clarity. 126 problems at chapter endings provide progressive practice and familiarity. 133 figures. Index. xvi + 274pp. 5⅜ x 8. Paperbound $2.00

INTRODUCTION TO THE THEORY OF NUMBERS, *L. E. Dickson*
Thorough, comprehensive approach with adequate coverage of classical literature, an introductory volume beginners can follow. Chapters on divisibility, congruences, quadratic residues & reciprocity. Diophantine equations, etc. Full treatment of binary quadratic forms without usual restriction to integral coefficients. Covers infinitude of primes, least residues. Fermat's theorem. Euler's phi function, Legendre's symbol, Gauss's lemma, automorphs, reduced forms, recent theorems of Thue & Siegel, many more. Much material not readily available elsewhere. 239 problems. Index. I figure. viii + 183pp. 5⅜ x 8.
Paperbound $1.75

MATHEMATICAL TABLES AND FORMULAS,
compiled by Robert D. Carmichael and Edwin R. Smith
Valuable collection for students, etc. Contains all tables necessary in college algebra and trigonometry, such as five-place common logarithms, logarithmic sines and tangents of small angles, logarithmic trigonometric functions, natural trigonometric functions, four-place antilogarithms, tables for changing from sexagesimal to circular and from circular to sexagesimal measure of angles, etc. Also many tables and formulas not ordinarily accessible, including powers, roots, and reciprocals, exponential and hyperbolic functions, ten-place logarithms of prime numbers, and formulas and theorems from analytical and elementary geometry and from calculus. Explanatory introduction. viii + 269pp. 5⅜ x 8½. Paperbound $1.25

CELESTIAL OBJECTS FOR COMMON TELESCOPES,
Rev. T. W. Webb
Classic handbook for the use and pleasure of the amateur astronomer. Of inestimable aid in locating and identifying thousands of celestial objects. Vol I, The Solar System: discussions of the principle and operation of the telescope, procedures of observations and telescope-photography, spectroscopy, etc., precise location information of sun, moon, planets, meteors. Vol. II, The Stars: alphabetical listing of constellations, information on double stars, clusters, stars with unusual spectra, variables, and nebulae, etc. Nearly 4,000 objects noted. Edited and extensively revised by Margaret W. Mayall, director of the American Assn. of Variable Star Observers. New Index by Mrs. Mayall giving the location of all objects mentioned in the text for Epoch 2000. New Precession Table added. New appendices on the planetary satellites, constellation names and abbreviations, and solar system data. Total of 46 illustrations. Total of xxxix + 606pp. 5⅜ x 8. Vol. 1 Paperbound $2.25, Vol. 2 Paperbound $2.25
The set $4.50

PLANETARY THEORY,
E. W. Brown and C. A. Shook
Provides a clear presentation of basic methods for calculating planetary orbits for today's astronomer. Begins with a careful exposition of specialized mathematical topics essential for handling perturbation theory and then goes on to indicate how most of the previous methods reduce ultimately to two general calculation methods: obtaining expressions either for the coordinates of planetary positions or for the elements which determine the perturbed paths. An example of each is given and worked in detail. Corrected edition. Preface. Appendix. Index. xii + 302pp. 5⅜ x 8½. Paperbound $2.25

STAR NAMES AND THEIR MEANINGS,
Richard Hinckley Allen
An unusual book documenting the various attributions of names to the individual stars over the centuries. Here is a treasure-house of information on a topic not normally delved into even by professional astronomers; provides a fascinating background to the stars in folk-lore, literary references, ancient writings, star catalogs and maps over the centuries. Constellation-by-constellation analysis covers hundreds of stars and other asterisms, including the Pleiades, Hyades, Andromedan Nebula, etc. Introduction. Indices. List of authors and authorities. xx + 563pp. 5⅜ x 8½. Paperbound $2.50

A SHORT HISTORY OF ASTRONOMY, *A. Berry*
Popular standard work for over 50 years, this thorough and accurate volume covers the science from primitive times to the end of the 19th century. After the Greeks and the Middle Ages, individual chapters analyze Copernicus, Brahe, Galileo, Kepler, and Newton, and the mixed reception of their discoveries. Post-Newtonian achievements are then discussed in unusual detail: Halley, Bradley, Lagrange, Laplace, Herschel, Bessel, etc. 2 Indexes. 104 illustrations, 9 portraits. xxxi + 440pp. 5⅜ x 8. Paperbound $2.75

SOME THEORY OF SAMPLING, *W. E. Deming*
The purpose of this book is to make sampling techniques understandable to and useable by social scientists, industrial managers, and natural scientists who are finding statistics increasingly part of their work. Over 200 exercises, plus dozens of actual applications. 61 tables. 90 figs. xix + 602pp. 5⅜ x 8½.
Paperbound $3.50

PRINCIPLES OF STRATIGRAPHY,
A. W. Grabau
Classic of 20th century geology, unmatched in scope and comprehensiveness. Nearly 600 pages cover the structure and origins of every kind of sedimentary, hydrogenic, oceanic, pyroclastic, atmoclastic, hydroclastic, marine hydroclastic, and bioclastic rock; metamorphism; erosion; etc. Includes also the constitution of the atmosphere; morphology of oceans, rivers, glaciers; volcanic activities; faults and earthquakes; and fundamental principles of paleontology (nearly 200 pages). New introduction by Prof. M. Kay, Columbia U. 1277 bibliographical entries. 264 diagrams. Tables, maps, etc. Two volume set. Total of xxxii + 1185pp. 5⅜ x 8. Vol. 1 Paperbound $2.50, Vol. 2 Paperbound $2.50,
The set $5.00

SNOW CRYSTALS, *W. A. Bentley and W. J. Humphreys*
Over 200 pages of Bentley's famous microphotographs of snow flakes—the product of painstaking, methodical work at his Jericho, Vermont studio. The pictures, which also include plates of frost, glaze and dew on vegetation, spider webs, windowpanes; sleet; graupel or soft hail, were chosen both for their scientific interest and their aesthetic qualities. The wonder of nature's diversity is exhibited in the intricate, beautiful patterns of the snow flakes. Introductory text by W. J. Humphreys. Selected bibliography. 2,453 illustrations. 224pp. 8 x 10¼. Paperbound $3.25

THE BIRTH AND DEVELOPMENT OF THE GEOLOGICAL SCIENCES,
F. D. Adams
Most thorough history of the earth sciences ever written. Geological thought from earliest times to the end of the 19th century, covering over 300 early thinkers & systems: fossils & their explanation, vulcanists vs. neptunists, figured stones & paleontology, generation of stones, dozens of similar topics. 91 illustrations, including medieval, renaissance woodcuts, etc. Index. 632 footnotes, mostly bibliographical. 511pp. 5⅜ x 8. Paperbound $2.75

ORGANIC CHEMISTRY, *F. C. Whitmore*
The entire subject of organic chemistry for the practicing chemist and the advanced student. Storehouse of facts, theories, processes found elsewhere only in specialized journals. Covers aliphatic compounds (500 pages on the properties and synthetic preparation of hydrocarbons, halides, proteins, ketones, etc.), alicyclic compounds, aromatic compounds, heterocyclic compounds, organophosphorus and organometallic compounds. Methods of synthetic preparation analyzed critically throughout. Includes much of biochemical interest. "The scope of this volume is astonishing," *Industrial and Engineering Chemistry*. 12,000-reference index. 2387-item bibliography. Total of x + 1005pp. 5⅜ x 8. Two volume set, paperbound $4.50

THE PHASE RULE AND ITS APPLICATION,
Alexander Findlay
Covering chemical phenomena of 1, 2, 3, 4, and multiple component systems, this "standard work on the subject" (*Nature*, London), has been completely revised and brought up to date by A. N. Campbell and N. O. Smith. Brand new material has been added on such matters as binary, tertiary liquid equilibria, solid solutions in ternary systems, quinary systems of salts and water. Completely revised to triangular coordinates in ternary systems, clarified graphic representation, solid models, etc. 9th revised edition. Author, subject indexes. 236 figures. 505 footnotes, mostly bibliographic. xii + 494pp. 5⅜ x 8.
Paperbound $2.75